Flora of North America

FLORA OF NORTH AMERICA EDITORIAL COMMITTEE

Flora of North America

North of Mexico

Edited by FLORA OF NORTH AMERICA EDITORIAL COMMITTEE

VOLUME 22

Magnoliophyta: Alismatidae, Arecidae, Commelinidae (in part), and Zingiberidae

BUTOMALES   (Flowering-rush order)

HYDROCHARITALES   (Canadian-waterweed order)

NAJADALES   (Arrow-grass order)

ARECALES   (Palm order)

ARALES   (Aroid order)

COMMELINALES   (Spiderwort order)

ERIOCAULALES   (Eriocaul order)

JUNCALES   (Rush order)

TYPHALES   (Cat-tail order)

BROMELIALES   (Bromelia order)

ZINGIBERALES   (Ginger order)

NEW YORK   OXFORD   •   OXFORD UNIVERSITY PRESS   •   2000

Oxford University Press

Oxford   New York

Athens   Auckland   Bangkok   Bogotá   Buenos Aires   Calcutta
Cape Town   Chennai   Dar es Salaam   Delhi   Florence   Hong Kong   Istanbul
Karachi   Kuala Lumpur   Madrid   Melbourne   Mexico City   Mumbai
Nairobi   Paris   São Paulo   Singapore   Taipei   Tokyo   Toronto   Warsaw

and associated companies in

Berlin   Ibadan

Published by Oxford University Press, Inc.,
198 Madison Avenue, New York, New York 10016
www.oup.com

Library of Congress Cataloging-in-Publication Data
(Revised for volume 22)
Flora of North America north of Mexico
edited by Flora of North America Editorial Committee.
Includes bibliographical references and indexes.
Contents: v. 1. Introduction—v. 2. Pteridophytes and gymnosperms—
v. 3. Magnoliophyta: Magnoliidae and Hamamelidae—
v. 22. Magnoliophyta: Alismatidae, Arecidae, Commelinidae (in part), and Zingiberidae

ISBN 0-19-513729-9 (v. 22)
1. Botany—North America.
2. Botany—United States.
3. Botany—Canada.
I. Flora of North America Editorial Committee.
QK110.F55   2000   581.97   92-30459

9 8 7 6 5 4 3 2 1
Printed in the United States of America
on acid free paper

Contents

FOUNDING MEMBER INSTITUTIONS

Flora of North America Association

Arnold Arboretum
Jamaica Plain, Massachusetts

Biosystematics Research Institute
Ottawa, Ontario

Canadian Museum of Nature
(now part of Eastern Cereal and
 Oilseed Research Centre)
Ottawa, Ontario

Carnegie Museum of Natural
 History
Pittsburgh, Pennsylvania

Field Museum of Natural History
Chicago, Illinois

Fish and Wildlife Service
United States Department of the
 Interior
Washington, D.C.

Harvard University Herbaria
Cambridge, Massachusetts

Hunt Institute for Botanical
 Documentation
Carnegie Mellon University
Pittsburgh, Pennsylvania

Jacksonville State University
Jacksonville, Alabama

Jardin Botanique de Montréal
Montréal, Québec

Kansas State University
Manhattan, Kansas

Missouri Botanical Garden
St. Louis, Missouri

New Mexico State University
Las Cruces, New Mexico

New York State Museum
Albany, New York

Northern Kentucky University
Highland Heights, Kentucky

The New York Botanical Garden
Bronx, New York

The University of British Columbia
Vancouver, British Columbia

The University of Texas
Austin, Texas

Université de Montréal
Montréal, Québec

University of Alaska
Fairbanks, Alaska

University of Alberta
Edmonton, Alberta

University of California
Berkeley, California

University of California
Davis, California

University of Idaho
Moscow, Idaho

University of Illinois
Urbana-Champaign, Illinois

University of Iowa
Iowa City, Iowa

University of Kansas
Lawrence, Kansas

University of Michigan
Ann Arbor, Michigan

University of Oklahoma
Norman, Oklahoma

University of Ottawa
Ottawa, Ontario

University of Southwestern
 Louisiana
Lafayette, Louisiana

University of Western Ontario
London, Ontario

University of Wyoming
Laramie, Wyoming

Utah State University
Logan, Utah

For their support of the Flora of North America Project, we gratefully acknowledge and thank:

National Science Foundation (1989–1999)
The Pew Charitable Trusts (1989–1992)
The Caleb C. and Julia W. Dula Foundation (1991–1994)
The Surdna Foundation (1991–1994)
The David and Lucile Packard Foundation (1988–1996)
National Fish and Wildlife Foundation (1990; 1993–1994)
ARCO Foundation (1992–1995)
The William and Flora Hewlett Foundation (1994–)
Edward Chase Garvey Memorial Foundation (1990)
Waste Management, Inc./Environmental Affairs Department (1995)
The Andrew W. Mellon Foundation (1995–1999)
The Geraldine R. Dodge Foundation (1994–1996)
Chevron (1995; 1998–1999)
Union Pacific (1995–1996)
The Bellwether Foundation (1996–1998)
Enterprise Consulting Group (1998–1999)
Informix Corporation (1998–1999)

Friends of Missouri Botanical Garden made generous donations in honor of Peter H. Raven's twentieth and twenty-fifth anniversaries of service.

Project Staff

Gwen R. Ericson
Technical Editor (1998–1999)

Tricia L. Frye
Map Editor (1998–1999)

Helen K. Jeude
Senior Technical Editor

John Myers
Illustrator

Gina Otterson
Departmental Secretary (–1999)

Anne Keats Smith
Map Editor (–1997)

Judith Unger
Project Coordinator (–1999)

Alan T. Whittemore
Bryophyte and Vascular Plant Specialist (–1999)

Yevonn Wilson-Ramsey
Illustration Editor and Illustrator

James L. Zarucchi
Managing Editor

Eleanor Zeller
Clerk (–1999)

Acknowledgments

Flora of North America is fortunate to have a tremendously skilled and dedicated staff. Although every person in the Organizational Center has contributed in many ways to the completion of this volume, Helen Jeude and Gwen Ericson (technical editors), Tricia Frye and Anne Keats Smith (map editors), and Yevonn Wilson-Ramsey and John Myers (illustration editor and illustrators) deserve special recognition. In addition to authoring treatments of several families, Alan Whittemore worked extensively with the treatment of Juncaceae. James Zarucchi, Managing Editor, has overseen preparation of this volume and has assured a high accuracy level, especially in the nomenclatural citations.

John Thieret has done the lion's share of editing in this volume, having been responsible for 25 of the 30 families; in addition, Michael Moore edited Xyridaceae, Mayacaceae, Commelinaceae, and Eriocaulaceae; J. B. Phipps edited Juncaceae; Robert Kiger edited all literature citations. John McNeill provided his nomenclatural expertise. John Strother and Theodore Barkley edited all manuscripts at several stages.

Flora of North America owes its start to those who attended an organizing meeting on 30 April and 1 May 1982 at the Missouri Botanical Garden: G. Argus, D. Bates, W. Burger, S. Hatch, N. Holmgren, T. Jacobsen, M. Johnston, R. Kiger, J. Massey, J. McNeill, R. Mecklenburg, N. Morin, J. Phipps, D. Porter, J. Reveal, S. Shetler, F. Utech, and G. Webster.

The project has been designed, in large part, by the Editorial Committee. This committee has changed its makeup slightly over the years, but overall it has been a remarkably stable group. Since 1997, when Volume 3 was published, J. Estes and J. Fay have left the project. To our great sadness, Gerald Straley passed away in December 1997, and Michael Moore, in late 1998. Scott Peterson replaced J. Fay as liaison with governmental agencies, Paul Peterson joined as an editor for Poaceae, and Aaron Liston joined the project as Regional Coordinator for the Northwest. Theodore Barkley's move from Kansas State University, Manhattan to the Botanical Research Institute of Texas necessitated the appointment of a new Regional Coordinator for the North Central Region, and Craig Freeman, University of Kansas, has agreed to take on this assignment. Michael Moore has been replaced by Richard Wunderlin as Regional Coordinator for the Southeastern United States. Bryophyte editors John Engel, Marie Hicks, and Barbara Murray have left the project, and Sharon Bartholomew-Began, Paul Davison, Lloyd Stark, and Richard Zander have joined it.

One of the most exciting developments in the project has been the establishment of formal editorial centers in addition to the Flora of North America office at Missouri Botanical Garden.

Lead Editors at these centers will also serve on the Editorial Committee. Centers are currently located at the New York Botanical Garden for the Bryophytes under the direction of Barbara Thiers; the Hunt Institute for Botanical Documentation, Carnegie Mellon University, with Robert Kiger as Lead Editor; the Botanical Research Institute of Texas, with Theodore Barkley currently acting as Lead Editor; and the National Museum of Natural History, with Paul Peterson as Lead Editor. Mapping and distributions applications research is a new effort being done at the Harvard University Herbaria. We are grateful to these institutions for their work on the project.

ESRI generously donated software and training in support of Geographical Information System (GIS) applications employed in the production of this volume.

The University of Alabama Aquatic Biology GIS laboratory prepared the preliminary drafts of distribution maps for Butomaceae, Limnocharitaceae, Alismataceae, Hydrocharitaceae, Najadaceae, Potamogetonaceae, Juncaginaceae, Aponogetonaceae, Zannichelliaceae, Ruppiaceae, Zosteraceae, and Cymodoceaceae.

The Flora of North America project is coordinated from the Organizational Center, located at the Missouri Botanical Garden. Support by the Missouri Botanical Garden for this Center has been instrumental in the success of the project, and we are grateful to Peter H. Raven for committing institutional resources and for contributing his time and energy to Flora of North America. The institutions at which editors work provide more than half of the overall cost of the project through in-kind support. We are grateful to all of the foundations and individuals who have financially supported Flora of North America. In addition to the staff (listed on p. viii), undergraduate interns, some of whom were funded by the Research Experiences for Undergraduates program of the National Science Foundation, have contributed to the project.

We are deeply grateful to all of the people who have helped to create Flora of North America for their hard work and continuing support on behalf of the project.

Taxonomic Reviewers

W. P. Armstrong
Palomar College
San Marcos, California

Michael J. Balick
Institute of Economic
Botany
New York Botanical Garden
Bronx, New York

David Bar-Zvi
Fairchild Tropical Gardens
Miami, Florida

Paul A. Cox
Pacific Tropical Botanical
Garden
Lawai, Hawaii

Barbara J. Ertter
University of California
Berkeley, California

Neil A. Harriman
University of Wisconsin
Oshkosh, Wisconsin

Andrew Henderson
New York Botanical Garden
Bronx, New York

Nancy Hensold
Field Museum of Natural
History
Chicago, Illinois

Bruce Holst
The Marie Selby Botanical
Gardens
Sarasota, Florida

Walter S. Judd
University of Florida
Gainesville, Florida

Robert B. Kaul
University of Nebraska
Lincoln, Nebraska

James R. Massey
University of North
Carolina
Chapel Hill, North Carolina

Bruce Sorrie
Southern Pines, North
Carolina

Galen S. Smith
University of Wisconsin
Whitewater, Wisconsin

Gerald L. Smith
High Point University
High Point, North Carolina

Dennis Wm. Stevenson
New York Botanical Garden
Bronx, New York

Ronald L. Stuckey
Ohio State University
Columbus, Ohio

Natalie W. Uhl
Cornell University
Ithaca, New York

John H. Wiersema
U.S.D.A./Systematic Botany
& Mycology Lab
Beltsville, Maryland

J. W. Wooten
University of Southern
Mississippi
Hattiesburg, Mississippi

Richard P. Wunderlin
University of South Florida
Tampa, Florida

James C. Zech
Sul Ross State University
Alpine, Texas

Regional Reviewers

ALASKA

Alan Battan
University of Alaska Museum
Fairbanks, Alaska

Robert Lipkin
Alaska Natural Heritage Program
University of Alaska
Anchorage, Alaska

Mary Stensvold
U.S.D.A. Forest Service, Alaska Region
Sitka, Alaska

NORTHWEST

Edward R. Alverson
The Nature Conservancy
Eugene, Oregon

Kenton L. Chambers
Oregon State University
Corvallis, Oregon

Richard Old
XID Services, Inc.
Pullman, Washington

Cindy Roche
Asotin, Washington

Scott Sundberg
Oregon State University
Corvallis, Oregon

SOUTHWESTERN UNITED STATES

H. David Hammond
Flagstaff, Arizona

Charles T. Mason Jr.
University of Arizona
Tucson, Arizona

Donald J. Pinkava
Arizona State University
Tempe, Arizona

Teresa Prendusi
U.S.D.A. Forest Service, Southwestern
Region
Albuquerque, New Mexico

CALIFORNIA

Roxanne L. Bittman
California Department of Fish &
Game
Sacramento, California

Barbara J. Ertter
University of California
Berkeley, California

James R. Shevock
U.S.D.A. Forest Service
San Francisco, California

WESTERN CANADA

William T. Cody
Eastern Cereal and Oilseed Research
Centre
Ottawa, Ontario

Bruce A. Ford
University of Manitoba
Winnipeg, Manitoba

Vernon L. Harms
University of Saskatchewan
Saskatoon, Saskatchewan

ROCKY MOUNTAINS

Bonnie Heidel
Montana Natural Heritage Program
Helena, Montana

Tim Hogan
University of Colorado Museum
Boulder, Colorado

J. Stephen Shelly
U.S.D.A. Forest Service, Northern
Region
Missoula, Montana

Stanley L. Welsh
Brigham Young University
Provo, Utah

NORTH CENTRAL UNITED STATES

Anita F. Cholewa
University of Minnesota
St. Paul, Minnesota

Theodore S. Cochrane
University of Wisconsin
Madison, Wisconsin

Neil A. Harriman
University of Wisconsin
Oshkosh, Wisconsin

Robert B. Kaul
University of Nebraska
Lincoln, Nebraska

Gary E. Larson
South Dakota State University
Brookings, South Dakota

Deborah Q. Lewis
Iowa State University
Ames, Iowa

Ronald L. McGregor
University of Kansas
Lawrence, Kansas

Lawrence R. Stritch
U.S.D.A. Forest Service
W.F.R.P. Staff
Washington, D.C.

Connie Taylor
Southeastern Oklahoma State
University
Durant, Oklahoma

George Yatskievych
Missouri Botanical Garden
St. Louis, Missouri

Contributors

Ralph E. Brooks
Black & Veatch
Lake Oswego, Oregon

Gregory K. Brown
Department of Botany
University of Wyoming
Laramie, Wyoming

Kathleen Burt-Utley
Department of Biological Sciences
University of New Orleans
New Orleans, Louisiana

Steven E. Clemants
Herbarium
Brooklyn Botanic Garden
Brooklyn, New York

Robert B. Faden
Department of Botany
Smithsonian Institution
Washington, DC

Robert R. Haynes
Department of Biological Sciences,
  Biodiversity, and Systematics
University of Alabama
Tuscaloosa, Alabama

C. Barre Hellquist
Department of Biology
North Adams State College
North Adams, Massachusetts

Robert B. Kaul
School of Biological Sciences
University of Nebraska
Lincoln, Nebraska

Helen Kennedy
Botany Department
University of British Columbia
Vancouver, British Columbia

Robert Kral
Vanderbilt University
Nashville, Tennessee

W. John Kress
Department of Botany
Smithsonian Institution
Washington, DC

Elias Landolt
Geobotanisches Institüt ETH
Zürich, Switzerland

Harry E. Luther
The Marie Selby Botanical Gardens
Sarasota, Florida

Mark A. Nienaber
Department of Biology
Northern Kentucky University
Highland Heights, Kentucky

Linda M. Prince
Department of Biology
University of North Carolina
Chapel Hill, North Carolina

S. Galen Smith
University of Wisconsin
Whitewater, Wisconsin

Janice Coffee Swab
Department of Biology and Health
  Sciences
Meredith College
Raleigh, North Carolina

Sue A. Thompson
Section of Botany
Carnegie Museum of Natural History
Pittsburgh, Pennsylvania

John F. Utley
Department of Biological Sciences
University of New Orleans
New Orleans, Louisiana

Alan T. Whittemore
Flora of North America
Missouri Botanical Garden
St. Louis, Missouri

Scott A. Zona
Fairchild Tropical Garden
Miami, Florida

Introduction

Nancy R. Morin

Scope of the Work

Flora of North America North of Mexico is a synoptic floristic account of the plants of North America north of Mexico: the continental United States of America (including the Florida Keys and Aleutian Islands), Canada, Greenland (Kalâtdlit-Nunât), and St. Pierre and Miquelon. The flora is intended to serve both as a means of identifying plants within the region and as a systematic conspectus of the North American flora. Taxa and geographical areas in need of further study also are identified in the flora.

Flora of North America North of Mexico will be published in thirty volumes. Volume 1 contains background information that is useful for understanding patterns in the flora. Volume 2 contains treatments of ferns and gymnosperms. Families in volumes 3–26, the angiosperms, are arranged according to the classification system of A. Cronquist (1981). Bryophytes will be covered in volumes 27–29. Volume 30 will contain the cumulative bibliography and index.

The first two volumes were published in 1993; the third volume was published in 1997. The bibliographic citation for the flora is: Flora of North America Editorial Committee, eds. 1993+. *Flora of North America North of Mexico*. 4+ vols. New York and Oxford.

Volume 22 contains treatments of 30 families, 89 genera, and 423 species. For additional statistics, please refer to table 1.

Contents · General

The published flora includes accepted names, literature citations, selected synonyms, identification keys, summaries of habitats and geographic ranges, descriptions, chromosome numbers, phenological information, and other biological observations. Economic uses, weed status, and conservation status are provided from specified sources. Each volume contains a bibliography and an index to the taxa included in the volume. A comprehensive, consolidated bibliography and comprehensive index will be published in the last volume. The treatments, written and reviewed by experts from throughout the systematic botanical community, are based on original observations of herbarium specimens and, whenever possible, on living plants. These observations are supplemented by critical reviews of the literature.

Table 1. *Statistics for Volume 22 of Flora of North America.*

Family	Total Genera	Total Species	Endemic Genera	Endemic Species	Introduced Genera	Introduced Species	Conservation Taxa
Butomaceae	1	1	0	0	1	1	0
Limnocharitaceae	1	1	0	0	1	1	0
Alismataceae	4	34	0	17	0	4	5
Hydrocharitaceae	10	14	0	3	5	5	1
Aponogetonaceae	1	1	0	0	1	1	0
Scheuchzeriaceae	1	1	0	0	0	0	0
Juncaginaceae	2	5	0	1	0	0	1
Potamogetonaceae	2	37	0	18	0	1	6
Ruppiaceae	1	2	0	0	0	0	0
Najadaceae	1	8	0	2	0	3	2
Zannichelliaceae	1	1	0	0	0	0	0
Cymodoceaceae	2	2	0	0	0	0	0
Zosteraceae	2	5	0	1	0	1	0
Arecaceae	19	29	2	4	10	15	3
Acoraceae	1	2	0	1	0	1	0
Araceae	8	10	2	6	1	1	0
Lemnaceae	4	19	0	2	0	3	0
Xyridaceae	1	21	0	12	0	0	3
Mayaceae	1	1	0	0	0	0	0
Commelinaceae	6	51	0	23	2	17	0
Eriocaulaceae	3	17	0	12	0	1	5
Juncaceae	2	118	0	67	0	8	4
Sparganiaceae	1	9	0	2	0	0	0
Typhaceae	1	3	0	0	0	0	0
Bromeliaceae	4	19	0	1	0	0	0
Heliconiaceae	1	1	0	0	1	1	0
Musaceae	1	1	0	0	1	1	0
Zingiberaceae	4	4	0	0	4	4	0
Cannaceae	1	3	0	1	0	1	0
Marantaceae	2	3	0	1	1	1	0
Totals	89	423	4	174	28	71	30

Italic = introduced.

Basic Concepts

Our goal has been and continues to be to make the flora as clear, concise, and informative as practicable so that it can be an important resource for both botanists and nonbotanists. To this end, we are attempting to be consistent in style and content from the first volume to the last. Readers may assume that a term has the same meaning each time it appears and that, within groups, descriptions may be compared directly with one another. Any departures from consistent usage will be explicitly noted in the treatments (see also References).

Treatments are intended to reflect current knowledge of taxa throughout their ranges world-wide, and classifications are therefore based on all available evidence. Where notable differences of opinion about the classification of a group occur, appropriate references are mentioned in the discussion of the group.

Documentation and arguments supporting significantly revised classifications are published separately in botanical journals before publication of the pertinent volume of the flora. Similarly, all new names, names for new taxa, and new combinations are published prior to their use in the flora. No nomenclatural innovations will be published intentionally in the flora. Journals and series in which papers relevant to *Flora of North America North of Mexico* have been or may be published include *Brittonia, Canadian Journal of Botany, North American Flora, Novon, Systematic Botany, Systematic Botany Monographs,* and *Taxon,* among others.

Contents of Treatments

Taxa treated in full include native species, native species thought to be recently extinct, hybrids that are well established (or frequent), and waifs or cultivated plants that are found frequently outside cultivation and give the appearance of being naturalized. Taxa mentioned only in discussions include waifs or naturalized plants now known only from isolated old records and some nonnative, economically important or extensively cultivated plants, particularly when they are relatives of native species. Excluded names and taxa are listed at the ends of appropriate sections, e.g., species at the end of genus, genera at the end of family.

Treatments are intended to be succinct and diagnostic but adequately descriptive. Characters and character states used in the keys are repeated in the descriptions. Descriptions of related taxa at the same rank are directly comparable.

With few exceptions, taxa are presented in taxonomic sequence. If an author is unable to produce a classification, the taxa are arranged alphabetically, and the reasons are given in the discussion.

Treatments of hybrids follow that of one of the putative parents. Hybrid complexes are treated at the ends of their genera, after the descriptions of species.

We have attempted to keep terminology as simple as accuracy permits. Common English equivalents have been used in place of Latin or Latinized terms or other specialized terminology whenever the correct meaning could be conveyed in approximately the same space, e.g., "pitted" rather than "foveolate," but "striate" rather than "with fine longitudinal lines." Specialized terms that are used are defined in the generic or family descriptions and, in some cases, are illustrated.

References

Authoritative general reference works used for style are *Chicago Manual of Style,* ed. 14 (University of Chicago Press 1993); *Webster's New Geographical Dictionary* (Merriam-Webster 1988); and *The Random House Dictionary of the English Language,* ed. 2, unabridged (S. B. Flexner and L. C. Hauck 1987). *B-P-H/S. Botanico-Periodicum-Huntianum/ Supplementum* (G. D. R. Bridson and E. R. Smith 1991) has been used for abbreviations of titles of serials, and *Taxonomic Literature,* ed. 2 (F. A. Stafleu and R. S. Cowan 1976–1988) and its supplements by F. A. Stafleu and E. A. Mennega (1992+) have been used for abbreviations of titles of books.

Graphic Elements

All genera, and approximately one out of three species, are illustrated. Illustrated taxa are marked with an "F" (for figure) following the accepted name statement. The illustrations may be of typical or of unusual species, or they may show diagnostic traits or complex structures. Most illustrations have been drawn from herbarium specimens selected by the authors. In some cases living material or photographs have been used. Data on specimens that were used and parts that were illustrated have been recorded. This information, together with the archivally preserved original drawings, is deposited in the Missouri Botanical Garden Library and is available for scholarly study.

Specific Information in Treatments

Keys

A key to families of Magnoliophyta will be published separately. Keys are included in each volume for all ranks below families if two or more taxa are treated. For dioecious species, keys are designed for use with either staminate or pistillate plants. Keys are also designed to facilitate identification of taxa that flower before leaves appear. More than one key may be given, and for some groups tabular comparisons may be presented in addition to keys.

Nomenclatural Information

Basionyms, with author and literature citation, are given for accepted names. Synonyms in common use are listed in alphabetical order, without literature citations.

Common names in vernacular use are given in the appropriate language. In general, such names have not been created for use in the flora. Luc Brouillet and Richard Spellenberg assisted with the French and Spanish names, respectively.

The last names of authors of taxonomic names have been spelled out. The conventions of *Authors of Plant Names* (R. K. Brummitt and C. E. Powell 1992) have been used as a guide for including first initials to discriminate individuals who share surnames.

If only one infraspecific taxon within a species occurs in the flora area, nomenclatural information (literature citation, basionym with literature citation, relevant synonyms) is given for the species, as is information on the number of infraspecific taxa in the species and their distribution worldwide, if known. A description and detailed distributional information are given only for the infraspecific taxon.

Descriptions

Character states common to all taxa are treated in the description of the taxon at the next higher rank. For example, if corolla color is yellow for all species treated within a genus, that character state is given in the generic description. Characters used in keys are repeated in the descriptions. Characteristics are given as they occur in plants from the flora area. Notable characteristics that occur in plants from outside the flora area are given in square brackets or are included in a brief discussion at the end of the description. In families with one genus and one or more species, the family description is given as usual, the genus description is condensed, and the species description is as usual.

In reading descriptions of vascular plants, the reader may assume, unless otherwise noted, that: the plant is green, photosynthetic, and reproductively mature; a woody plant is perennial; stems are erect; roots are fibrous; leaves are simple and petiolate. Because measurements and elevations are almost always approximate, modifiers such as "about," "circa," or "±" are usually omitted.

Arrangements of elements within descriptions of taxa are from base to apex, proximal to distal, abaxial to adaxial. General features such as growth form, persistence, habit, and nutrition are given first. For a particular structure or organ system, description of parts follows the order: presence, number, position/insertion, arrangement, orientation, connation, adnation, coherence, adherence. Features of a whole organ follow the order: color, odor, symmetry, architecture, shape, dimensions (length, width, thickness, mass), texture, base, margin, peripheral region or sides, central area, apex, surface, vestiture, internal parts, exudates. Unless otherwise noted, dimensions are length × width. If only one dimension is given, it is length or height. All measurements are given in metric units. Measurements usually are based on dried specimens but these usually should not differ significantly from the measurements actually found in fresh or living material.

Chromosome numbers generally are given only if published, documented counts are available from North American material or from an adjacent region. No new counts are published intentionally in the flora. Chromosome counts from nonsporophyte tissue have been converted to the $2n$ form. A literature reference for each reported chromosome number is available in the Flora of North America database (see below). The base number ($x = $) is given for each genus. This represents the lowest known haploid count for the genus unless evidence is available that the base number differs.

Flowering time and often fruiting time are given by season, sometimes qualified by early, mid, or late or by months. Elevation generally is rounded to the nearest 100 m; elevations between 0 and 100 m are rounded to the nearest 10 m. Mean sea level is shown as 0 m, with the understanding that this is approximate. Elevation often is omitted from herbarium specimen labels, particularly for collections made where the topography is not remarkable, and therefore precise elevation is sometimes not known for a given taxon.

The term "introduced" is defined broadly to refer to plants that were released deliberately or accidentally into the flora and that now exist as wild plants in areas in which they were not recorded as native in the past. The distribution of nonnative plants is often poorly documented and may be ephemeral. [Introduced plants are flagged with an "I" after the accepted name and the nature of introduced populations is discussed as far as understood].

Taxa that occur only in the flora area are indicated as endemic by an "E" after the accepted name statement.

If a taxon is globally rare or if its continued existence is threatened in some way, the words "of conservation concern" appear before the statements of elevation and geographic range, and a "C" is shown after the accepted name statement. Criteria for taxa of conservation concern are based on The Nature Conservancy's designations of global rank (G-rank), G1 and G2:

G1   Critically imperiled globally because of extreme rarity (5 or fewer occurrences or fewer than 1000 individuals or acres) or because of some factor(s) making it especially vulnerable to extinction.

G2   Imperiled globally because of rarity (5–20 occurrences or fewer than 3000 individuals or acres) or because of some factor(s) making it very vulnerable to extinction throughout its range.

Taxa thought to have become extinct during the period of permanent European settlement, i.e., the past 500 years, are included in the flora. Treatments of such taxa have been reviewed by Larry Morse of The Nature Conservancy.

Range maps are given for each species or infraspecific taxon. The Nunavut boundary on the maps has been provided by the GeoAccess Division, Canada Center for Remote Sensing, Earth Science. The maps are generalized and, in order to represent the probable range of a taxon, parts of states or provinces may be shaded even though documentation of occurrence there may be lacking. Occurrences in states or provinces listed in distribution statements are documented by specimens. We have assumed that details such as "northeastern Florida" are apparent on the map; consequently, directional qualifiers are not given in the list of territories, provinces, and states. Authors are expected to have seen at least one specimen documenting each state, provincial, or territorial record and have been urged to examine as many specimens as possible from throughout the range of each taxon. Additional information about distribution may be given in the discussion.

Distributions are stated in the following order: Greenland; St. Pierre and Miquelon; Canada (provinces and territories in alphabetic order); United States (states in alphabetic order); Mexico (11 northern states may be listed specifically, in alphabetic order); West Indies; Bermuda; Central America (Guatemala, Belize, Honduras, El Salvador, Nicaragua, Costa Rica, Panama); South America; Europe, or Eurasia; Asia (including Indonesia); Africa; Pacific Islands; Australia; Antarctica.

Discussion

The discussion section includes information on taxonomic problems and interesting biological phenomena. Statements of economic uses were supplied and documented by the author(s). Weed status has been determined in consultation with weed specialist Robert H. Callihan and is given in order to make this information more easily available to users of the Flora. Weediness is indicated by code W at the end of the accepted name statement and is based on D. T. Patterson et al. (1989) and R. H. Callihan et al. (1995). Authors may provide additional discussions on the various aspects. Toxicity, if known, also is mentioned in the discussion, and pertinent literature is cited. Please see "CAUTION" below.

Selected References

Major references used in preparation of a treatment or containing critical information about a taxon are cited after the discussion. These, and other works that are referred to briefly in the discussion or elsewhere, are included in the bibliography at the end of the volume and in the consolidated bibliography in the last volume of the flora.

The Database

Data contained in printed volumes of the *Flora of North America North of Mexico*, additional supporting data, authority files, more precise maps, and other useful information are available online at http://www.fna.org.

CAUTION

The Flora of North America Editorial Committee **does not encourage, recommend, promote, or endorse** any of the folk remedies, culinary practices, or various utilizations of any plant described within these volumes. Information about medicinal practices and/or ingestion of plants, or of any part or preparation thereof, has been included only for historical background and as a matter of interest. Under no circumstances should the information contained in these volumes be used in connection with medical treatment. Readers are strongly cautioned to remember that many plants in the flora are toxic or can cause unpleasant or adverse reactions if used or encountered carelessly.

Key to letters after names:

C   of conservation concern
E   endemic
F   illustrated
I   introduced
W   weedy

Flora of North America

188. BUTOMACEAE Richard

• Flowering-rush Family

Robert R. Haynes

Herbs, perennial, rhizomatous, stemless, glabrous; sap clear. **Roots** not septate. **Leaves** basal, emersed or rarely submersed, sessile, sheathing somewhat proximally; blade with translucent markings absent, basal lobes absent; venation parallel. **Inflorescences** scapose umbels, erect, bracteate. **Flowers** bisexual, hypogynous, pedicellate; tepals persistent, 6, in 2 series, scarious; stamens 9, distinct; anthers 4-loculed, dehiscing longitudinally; pistils 6, coherent proximally, 1-loculed; placentation laminar; ovules 50. **Fruits** follicles. **Seeds:** embryo straight; endosperm absent in mature seed.

Genera 1, species 1: introduced, North America; native, temperate regions, Eurasia.

SELECTED REFERENCES   Anderson, L. C., C. D. Zeis, and S. F. Alam. 1974. Phytogeography and possible origins of *Butomus* in North America. Bull. Torrey Bot. Club 101: 292–296.   Stuckey, R. L. 1968. Distributional history of *Butomus umbellatus* (flowering-rush) in the western Lake Erie and Lake St. Clair region. Michigan Bot. 7: 134–142.   Stuckey, R. L. 1994. Map of the known distribution of *Butomus* (flowering-rush) in North America. In: W. R. Burk and R. L. Stuckey. 1994. Ronald L. Stuckey: His Role in the Ohio Academy of Science. Chapel Hill. P. 85.   Stuckey, R. L., G. Schneider, and M. L. Roberts. 1990. *Butomus umbellatus* L.: Notes from the German literature and North American field studies. Ohio J. Sci. 90(2): 5–6.   Tomlinson, P. B. 1982. Helobiae (Alismatidae). In: C. R. Metcalfe, ed. 1960+. Anatomy of the Monocotyledons. 8+ vols. Oxford. Vol. 7.

1. BUTOMUS Linnaeus, Sp. Pl. 1: 372. 1753; Gen. Pl. ed. 5, 174. 1754 • Flowering-rush

[Greek *butomos* / *butomon*, marsh plant; from Greek *bous*, cow, and *temno*, to cut, referring to sharp leaves, known or believed to cut mouths of cattle] [1]

Plants in fresh water to 2 m deep. **Leaves** emersed, submersed, or floating; blade triangular proximally, flattened distally. **Inflorescences** overtopping leaves; scape triangular; bracts 3, subtending umbel. **Flowers:** tepals light pink-purple with darker purple veins; stamens in 2 cycles, outer cycle of 3 pairs opposite outer tepals, inner cycle of 3 opposite inner tepals; anthers ovoid; pistils pink. **Fruits** leathery, beaked.

Species 1: introduced, North America; Eurasia.

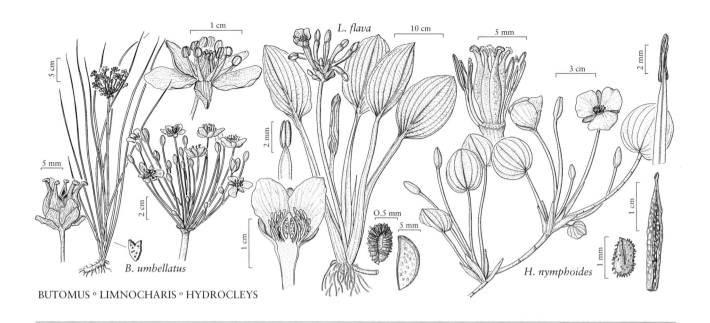

BUTOMUS ∘ LIMNOCHARIS ∘ HYDROCLEYS

1. Butomus umbellatus Linnaeus, Sp. Pl. 1: 372. 1753

F I W

Herbs, to 150 cm. **Leaves** linear, to 2.7 m. **Inflorescences** with 20–25 flowers; scape to 150 cm. **Flowers** 2–2.5 cm wide; pedicels 4–10 cm; outer tepals elliptic, 6–7.5 × 2–2.5 mm, apex acute, inner tepals oblanceolate, 9–11.5 × 4.5–6 mm, apex obtuse, erose; filaments 3–4.5 mm, anthers 1 mm. **Follicles** 1 cm.

Flowering summer–fall. Mud and shallow water of streams, lakes, and ditches; 0–700 m; introduced; Alta., Man., N.S., Ont., P.E.I., Que.; Conn., Idaho, Ill., Mich., Minn., Mont., N.Y., N.Dak., Ohio, Pa., S.Dak., Vt., Wis.; Eurasia.

The name *Butomus umbellatus* f. *vallisneriifolius* (Sagorski) Glück has been used for plants that grow totally submersed or have floating leaves. Field transplant experiments with North American plants (R. L. Stuckey et al. 1990) have demonstrated that the non-flowering submersed form can be converted to a flowering mudflat form, and that flowering terrestrial plants can be transformed into non-flowering submersed ones. Consequently, *B. umbellatus* f. *vallisneriifolius* is a deep-water growth form and should have no systematic status.

Two species, *Butomus umbellatus* and *B. junceus* Turczaninow, have been recognized in the natural range of the genus (L. C. Anderson et al. 1974), the former from Europe and western Asia, and the latter from east-

ern Asia. Reportedly, the distinguishing features are shorter scapes, fewer flowers, and straight stigmas for *B. junceus* as opposed to taller plants, more flowers, and curved stigmas for *B. umbellatus*.

Studies of *Butomus* in North America (L. C. Anderson et al. 1974) indicated that apparently the genus has become naturalized in North America at two separate locations, one near Detroit and another in the St. Lawrence River region. It is possible that plants naturalized in the St. Lawrence River region originated in eastern Asia, and those naturalized in the Detroit area originated in Europe or western Asia.

A map of *Butomus* in North America, prepared by R. L. Stuckey (1994), showed that he accepted two species. His map essentially had everything east of Niagara Falls as *B. junceus* and everything west of the Falls as *B. umbellatus*. At this time, I do not accept two species in the genus. Should two species be accepted, however, determinations would essentially follow the distribution given by Stuckey. He included dots for *B. umbellatus* from Indiana and British Columbia. I have not observed specimens from those two areas although the species is certainly to be expected in Indiana, and eventually in British Columbia if it does not already occur there.

Butomus umbellatus was first collected in North America near Laprairie on the St. Lawrence River in 1905; it was first observed in 1897 (R. L. Stuckey, pers. comm.). West of Niagara Falls, the taxon was first collected near Detroit in 1930 by O. A. Farwell, although he noted on the specimen, "Has been here since before 1918!!!" (R. L. Stuckey 1968).

189. LIMNOCHARITACEAE Takhtajan ex Cronquist
• Water-poppy Family

Robert R. Haynes

Herbs, perennial, rhizomatous or stoloniferous, caulescent, glabrous; sap milky. **Roots** not septate. **Leaves** basal [alternate], submersed and floating [emersed], sessile or petiolate, sheathing proximally; blade with translucent markings absent, basal lobes absent; venation reticulate, primary veins parallel from base of blade to apex, secondary veins reticulate. **Inflorescences** scapose umbels, floating [erect], bracteate. **Flowers** bisexual, hypogynous, pedicellate; sepals persistent, 3; petals deciduous, 3, usually delicate; stamens [6–]20–25, distinct; anthers 4-loculed, dehiscing longitudinally; pistils [3–]10–12, coherent proximally [distinct], 1-loculed; placentation laminar; ovules 50 or more. **Fruits** follicles. **Seeds:** embryo U-shaped; endosperm absent in mature seed.

Genera 3, species 8 (1 genus, 1 species in the flora): introduced; tropical and subtropical regions, North America, Central America, South America, Asia, Africa, Australia.

Another genus, *Limnocharis*, was reported from southern Louisiana and southern Florida (W. C. Muenscher 1944). The genus was not included for Louisiana by J. W. Thieret (1972); it was included by R. K. Godfrey and J. W. Wooten (1979) on the basis of Muenscher's report although the authors stated that they had seen no specimens from the United States. Also, the genus is not mapped in the "Florida Atlas" (available on the Internet). Richard Wunderlin, primary compiler of the "Atlas," has indicated that no specimens have been examined (pers. comm.).

SELECTED REFERENCE   Haynes, R. R. and L. B. Holm-Nielsen. 1992. The Limnocharitaceae. In: Organization for Flora Neotropica. 1968+. Flora Neotropica. 75+ nos. New York. No. 56.

1. HYDROCLEYS Richard, Mém. Mus. Hist. Nat. 1: 368. 1815 • Water-poppy [Greek *hydro*, water, and *clavis*, club-shaped, presumably from shape of pistils] Ⅰ

Stolons often present, terete. **Leaves:** submersed leaves phyllodia (expanded petioles resembling and functioning like leaves), sessile; floating leaves long-petiolate, petioles terete, septate; blade orbiculate to oblong-lanceolate, base cordate [rounded], apex mucronate to obtuse. **Inflores-cences** occasionally proliferating with stolons, leaves; scape septate; bracts subtending pedicel, distinct, elliptic to lanceolate, shorter than pedicel, delicate. **Flowers:** sepals persistent, erect, green, lanceolate, leathery, midvein absent [present], apex hoodlike; petals erect to spreading, yellow to white, orbiculate [oblong-obovate], longer than [shorter than] sepals; stamens in [1–] 4–5 series, outer often sterile; pistils terete, linear-lanceoloid, attenuate into style, style curved inward, apex papillose. **Fruits** ± terete, linear-lanceoloid, membranous, dehiscing along adaxial margins. **Seeds** 50 or more, sparsely to densely glandular-pubescent.

Species 5 (1 in the flora): introduced, North America; South America.

SELECTED REFERENCE Kenton, A. 1982. A Robertsonian relationship in the chromosomes of two species of *Hydrocleys* (Butomaceae sens. lat.). Kew Bull. 36: 487–492.

1. Hydrocleys nymphoides (Willdenow) Buchenau, Index Crit. Butom. Alism. Juncag., 2, 9. 1868 • Water-poppy [F] [I]

Stratiotes nymphoides Willdenow, Sp. Pl. 4(2): 821. 1806

Herbs, to 50 cm; stolons to 45 cm. **Leaves:** petioles 1.5–40 cm × 0.9–9 mm, sheathing base to 8.5 cm; blade broadly ovate to orbiculate, 1.4–11.9 × 0.9–10.6 cm, veins 5–9. **Inflorescences** with 1–6 flowers, proliferating with stolons, leaves; peduncles to 30 cm × 1.5–6 mm; bracts elliptic, 2–4.5 × 0.4–1 cm, apex obtuse; pedicels spreading, 3.5–17.5 cm × 1.5–6 mm. **Flowers** ca. 6.5 cm wide; sepals 13–28 × 7–13 mm; petals spreading, pale yellow to white with yellow base, 2.3–2.6 × 3.8–4.1 cm; stamens 20–25; staminodes 20+; pistils 5–8, 10 mm. **Fruits** 10–14.5 × 2–3.5 mm; beak 3.5–5.5 mm. **Seeds** ca. 1 mm, sparsely glandular-pubescent, glandular trichomes 0.15 mm, 150–200 µm apart, not present on every epidermal cell of seed coat.

Flowering summer. Margins of lakes and wet ditches; 0–100 m; introduced; Fla., Tex.; Central America; South America.

Hydrocleys nymphoides is cultivated, either in aquaria or in pools and ponds (D. S. Correll and M. C. Johnston 1970; R. K. Godfrey and J. W. Wooten 1979). The species apparently persists following cultivation or dumping of aquaria.

SELECTED REFERENCE Sattler, R. and V. Singh. 1973. Floral development of *Hydrocleis nymphoides*. Canad. J. Bot. 51: 2455–2458.

190. ALISMATACEAE Ventenat

• Water-plantain or Arrowhead Family

Robert R. Haynes

C. Barre Hellquist

Herbs, annual or perennial, rhizomatous, stoloniferous, or cormose, caulescent, glabrous to stellate-pubescent; sap milky. **Roots** septate or not septate. **Leaves** basal, submersed, floating, or emersed, sessile or petiolate, sheathing proximally; blade with translucent markings of dots or lines present or absent, basal lobes present or absent; venation reticulate, primary veins parallel from base of blade to apex, secondary veins reticulate. **Inflorescences** scapose racemes or panicles, rarely umbels, erect, rarely floating or decumbent, whorled (forming racemes) or whorls branching (forming panicles), bracteolate. **Flowers** bisexual or unisexual, if unisexual, staminate and pistillate on same or different plants, hypogynous, subsessile to long-pedicellate; sepals persistent, 3; petals deciduous, 3, delicate; stamens 0, 6, 9, or to 30, distinct; anthers 2-loculed, dehiscing longitudinally; pistils 0 or 6–1500 or more, distinct or coherent proximally, 1-loculed; placentation basal; ovules 1–2. **Fruits** achenes or follicles. **Seeds:** embryo U-shaped; endosperm absent in mature seed.

Genera 12, species ca. 80 (4 genera, 34 species in the flora): nearly worldwide, primarily tropical and subtropical regions.

SELECTED REFERENCES Argue, C. L. 1974. Pollen studies in the Alismataceae (Alismaceae). Bot. Gaz. 135: 338–344. Argue, C. L. 1976. Pollen studies in the Alismataceae with special reference to taxonomy. Pollen & Spores 18: 161–173. Beal, E. O. 1960b. The Alismataceae of the Carolinas. J. Elisha Mitchell Sci. Soc. 76: 68–79. Charlton, W. A. 1973. Studies in the Alismataceae. II. Inflorescences of Alismataceae. Canad. J. Bot. 51: 775–789. Correll, D. S. and H. B. Correll. 1972. Aquatic and Wetland Plants of Southwestern United States. Washington. Godfrey, R. K. and J. W. Wooten. 1979. Aquatic and Wetland Plants of Southeastern United States: Monocotyledons. Athens, Ga. Haynes, R. R. and L. B. Holm-Nielsen. 1985. A generic treatment of Alismatidae in the Neotropics. Acta Amazon. 15(suppl.): 153–193. Haynes, R. R. and L. B. Holm-Nielsen. 1994. The Alismataceae. In: Organization for Flora Neotropica. 1968+. Flora Neotropica. 75+ nos. New York. No. 64. Rogers, G. K. 1983. The genera of Alismataceae in the southeastern United States. J. Arnold Arbor. 64: 387–424.

1. Pistils weakly coherent proximally into starlike aggregation; petals erose. . . . . .2. *Damasonium*, p. 10
1. Pistils distinct, forming heads or rings; petals entire.
    2. Pistils arranged in ring around margin of flattened receptacle. . . . . 4. *Alisma*, p. 23
    2. Pistils spirally arranged on convex receptacle.

1. **ECHINODORUS** Richard ex Engelmann in A. Gray, Manual, 460. 1848 • [Greek *echius*, rough husk, and *doros*, leathern bottle, alluding to ovaries, which in some species are armed with persistent styles, forming prickly head of fruit]

Helianthium (Engelmann ex Hooker f.) J. G. Smith

Plants annual or perennial, emersed, floating-leaved, or rarely submersed, glabrous to stellate-pubescent; rhizomes present or absent; stolons absent; corms absent; tubers absent. **Roots** not septate. **Leaves** sessile or petiolate; petioles triangular, rarely terete; blade with translucent markings as dots or lines present or absent, linear to lanceolate to ovate, base attenuate to cordate, margins entire or undulating, apex obtuse to acute. **Inflorescences** racemes or panicles, rarely umbels, of 1–18 whorls, erect or decumbent, emersed; bracts coarse, apex obtuse to acute, surfaces smooth or papillose along veins. **Flowers** bisexual, subsessile to pedicellate; bracts subtending pedicels, subulate to lanceolate, shorter than to longer than pedicels, apex obtuse to acute; pedicels ascending to recurved; receptacle convex; sepals recurved to spreading, herbaceous to leathery, sculpturing absent; petals white, entire; stamens 9–25; filaments linear, glabrous; pistils 15–250 or more, spirally arranged on convex receptacle, forming head, distinct; ovules 1; style terminal or lateral. **Fruits** plump, often longitudinally ribbed, sometimes flattened, rarely abaxially keeled, abaxial wings absent, lateral wings absent, glands often present.

Species 26 (4 in the flora): Western Hemisphere.

SELECTED REFERENCES Fassett, N. C. 1955. *Echinodorus* in the American tropics. Rhodora 57: 133–156, 174–188, 202–212.   Haynes, R. R. and L. B. Holm-Nielsen. 1986. Notes on *Echinodorus* (Alismataceae). Brittonia 38: 325–332.   Rataj, K. 1975. Revizion [sic] of the Genus *Echinodorus* Rich. Prague.

1. Pistils 15–20; submersed leaves mostly present, sessile; emersed leaves narrowly lanceolate
   to ovate; plants relatively delicate. . . . . 1. *Echinodorus tenellus*
1. Pistils 45–250; submersed leaves mostly absent, when present, petiolate; emersed leaves
   lanceolate to elliptic to ovate; plants stout.
   2. Inflorescence decumbent to arching, proliferating; sepal veins papillate. . . . .
      . . . . 3a. *Echinodorus cordifolius* subsp. *cordifolius*
   2. Inflorescence erect, not proliferating; sepal veins not papillate.
      3. Stamens 9–15; plants to 70 cm. . . . . 2. *Echinodorus berteroi*
      3. Stamens 21; plants to 200 cm. . . . . 4. *Echinodorus floridanus*

1. **Echinodorus tenellus** (Martius) Buchenau, Index Crit. Butom. Alism. Juncag., 21. 1868

Alisma tenellum Martius in J. J. Roemer and J. A. Schultes, Syst. Veg. 7(2): 1600. 1830; *Echinodorus parvulus* Engelmann; *E. tenellus* var. *parvulus* (Engelmann) Fassett; *Helianthium parvulum* (Engelmann) Small

Herbs, annual, relatively slender, to 25 cm; rhizomes absent.

Leaves: submersed leaves mostly present, sessile; emersed leaves with petioles 4–5-ridged, 1.2–9.5 cm; blade with translucent markings absent, linear, 1–7.4 × 0.2–0.5 cm, base attenuate; if emersed, narrowly lanceolate to ovate. **Inflorescences** umbels, rarely racemes, of 1–2 whorls, each 4–6-flowered, erect, to 6 × 8 cm, not proliferating; peduncles terete, 1.2–4 cm; rachis absent or if present, terete; bracts connate ca. ½ length, deltate, 2.8–4.9 mm, coarse, margins delicate; pedicels spreading, 0.5–3 cm. **Flowers** 6–8 mm wide; sepals slightly appressed, 3–5-veined, veins not papillate; petals

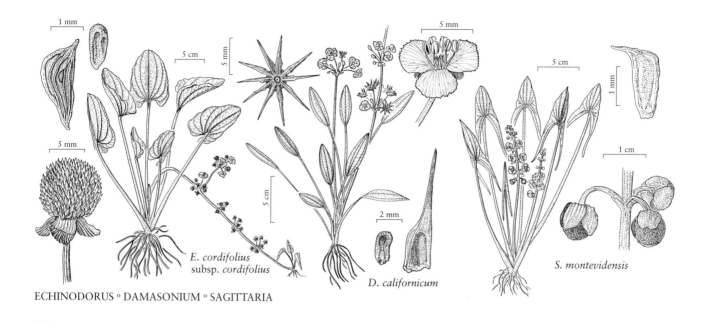

E. cordifolius
subsp. *cordifolius*

D. californicum

S. montevidensis

ECHINODORUS ∘ DAMASONIUM ∘ SAGITTARIA

clawed; stamens ca. 9; anthers basifixed; pistils 15–20. **Fruits** obovate, flattened, 0–3-ribbed, not abaxially keeled, 0.8–1.5 × 0.8–1 mm, glands absent; beak lateral, 0.1–0.2 mm. Chromosome number not available.

Flowering summer–fall. Sandy soil along margins of small streams or lakes; 0–1500 m; Ala., Conn., Fla., Ga., Ill., Kans., Mass., Mich., N.Y., S.C., Tex., Vt.; se Mexico; West Indies; Central America (Guatemala, Nicaragua); South America.

Echinodorus tenellus has often been separated into two species, one in North America and the other in the tropics. Whenever *Echinodorus* is studied throughout the range, differences between the two species break down, and a continuum exists from one species to the next. We recognize that overlap; therefore we are combining *E. parvulus* with *E. tenellus*.

2. Echinodorus berteroi (Sprengel) Fassett, Rhodora 57: 139. 1955

Alisma berteroi Sprengel, Syst. Veg. 2: 163. 1825; *Echinodorus berteroi* var. *lanceolatus* (Engelmann ex S. Watson & J. M. Coulter) Fassett; *E. rostratus* (Nuttall) Engelmann; *E. rostratus* var. *lanceolatus* Engelmann ex S. Watson & J. M. Coulter

Herbs, annual or perennial, stout, to 70 cm; rhizomes present. **Leaves** emersed or submersed; submersed leaves mostly absent; petiole terete to triangular, 2–36 cm; blade with translucent markings present as distinct lines, elliptic, lanceolate, or ovate, 2.6–15.5 × 0.5–20 cm, base truncate or occasionally cordate to tapering. **Inflorescences** racemes, rarely panicles, of 1–9 whorls, each 1–3(–4)-flowered, erect, 1.5–40 × 1.7–50 cm, not proliferating; peduncles 3–5-ridged, 2.1–57 cm; rachis triangular; bracts distinct, lanceolate, 0.3–2.5 cm, coarse, margins scarious; pedicels spreading to ascending, 0.6–2.8 cm. **Flowers** 6–11 mm wide; sepals spreading to recurved, 9–13-veined, veins not papillate; petals clawed; stamens 9–15; anthers versatile; pistils 45–200. **Fruits** oblanceolate, plump, 3–5-ribbed, abaxially 2-keeled, 0.9–3.2 × 0.6–2.5 mm; glands 1–2; beak terminal, 0.6–1.3 mm. $2n = 22$.

Flowering mid summer–fall. Clay soils of wet ditches, streams, and shallow ponds; 0–800 m; Ark., Calif., Fla., Ill., Iowa, Kans., La., Mo., Nebr., Nev., Ohio, Okla., S.Dak., Tex, Utah, Wis.; Mexico; West Indies; South America.

Echinodorus berteroi is an extremely easy species to recognize when in fruit. The elongated beaks of the fruits project upward, giving the fruiting head an echinate appearance. The generic name, in fact, came from the fruiting head of this species.

3. Echinodorus cordifolius (Linnaeus) Grisebach, Abh. Königl. Ges. Wiss. Göttingen 7: 257. 1857 [F]

Alisma cordifolia Linnaeus, Sp. Pl. 1: 343. 1753

Subspecies 2 (1 in flora): North America; Mexico; West Indies; South America.

3a. Echinodorus cordifolius (Linnaeus) Grisebach subsp. **cordifolius** [F]

Echinodorus radicans (Nuttall) Engelmann

Herbs, perennial, stout, to 100 cm; rhizomes present. **Leaves** emersed; petiole 5–6-ridged, 17.5–45 cm; blade with translucent markings distinct lines, ovate to elliptic, 6.5–32 × 2.5–19.1 cm, base truncate to cordate. **Inflorescences** racemes, of 3–9 whorls, each 3–15-flowered, decumbent to arching, to 62 × 8–18 cm, often proliferating; peduncles terete, 35–56 cm; rachis triangular; bracts distinct, subulate, 10–21 mm, coarse, margins coarse; pedicels erect to ascending, 2.1–7.5 cm. **Flowers** to 25 mm wide; sepals spreading, 10–12-veined, veins papillate; petals not clawed; stamens ca. 22; anthers versatile; pistils 200–250. **Fruits** oblanceolate, plump, 3–4-ribbed, abaxially 1-keeled, 2–3.5 × 0.9–1.5 mm; glands 3–4; beak terminal, 1–1.3 mm. $2n = 22$.

Flowering late summer–early fall. Clay soils of wet ditches, streams, and shallow ponds; 0–500 m; Ala., Ark., Fla., Ga., Ill., Ind., Kans., Ky., La., Miss., Mo., N.C., Okla., S.C., Tenn., Tex., Va.; Mexico (Tamaulipas); West Indies; South America.

Echinodorus cordifolius is very easily determined, as it is the only species of the genus with arching to decumbent inflorescences. In addition, it is the only one with papillate veins on the sepals.

4. Echinodorus floridanus R. R. Haynes & Burkhalter, Castanea 63: 180, fig. 1. 1998 [C][E]

Herbs, perennial, stout, to 200 cm; rhizomes present. **Leaves** emersed; petiole 4–5-ridged, to 115 cm; blade with translucent markings, dots and short distinct lines, absent in smaller leaves, elliptic to broadly ovate, 16–28 × 8–23 cm, base attentuate to cordate. **Inflorescences** panicles, of 4–9 whorls, each 9–16-flowered, erect, to 65 × 45 cm, not proliferating; peduncles ridged, to 131 cm; rachis 3–4-ridged; bracts distinct, lanceolate, 4–7 mm, coarse, margins coarse; pedicels spreading, 1.5–7.5 cm. **Flowers** 4 cm wide; sepals erect, 21-veined, veins not papillate; petals not clawed; stamens 21; anthers versatile; pistils ca. 200. **Fruits** oblanceolate, terete, 5-ribbed, not abaxially keeled, 2.5 × 0.6–1 mm; glands 4–5; beak terminal, 0.5 mm. Chromosome number not available.

Flowering summer–fall. Sandy soil at edge of broadleaf evergreen forest; of conservation concern; 0–10 m; Fla.

Echinodorus floridanus is a recently discovered taxon, known only from the type locality, which is within metropolitan Pensacola, Florida. The plants form a large colony that is bisected by a state highway. This species is by far the largest species of the genus in North America.

2. DAMASONIUM Miller, Gard. Dict. Abr. ed. 4, 1: 28. 1754 · [Greek, ancient name]

Machaerocarpus Small

Plants perennial, floating-leaved or emersed, glabrous; rhizomes usually absent; stolons absent; corms present; tubers absent. **Roots** not septate. **Leaves:** petiole triangular; blade with translucent markings absent, narrow-lanceolate to ovate, base rounded to attenuate, basal lobes absent, margins entire, apex acute. **Inflorescences** racemes [panicles], of 1–9 whorls, erect, emersed; bracts coarse, apex acute, surfaces smooth. **Flowers** bisexual, pedicellate; bracts subtending pedicels, lanceolate to ovate-lanceolate, much shorter than pedicels, apex acuminate; pedicels ascending to spreading; receptacle flat; sepals erect, without sculpturing, herbaceous; petals white or pink, erose; stamens 6; filaments dilated, glabrous; pistils 6–15, in single ring, radiating in starlike pattern, weakly coherent proximally; ovules 1–2; styles terminal. **Fruits** longitudinally ribbed, laterally compressed, not abaxially keeled, abaxial wings absent, lateral wings absent, glands absent. $x = 7$.

Species 5 (1 in the flora): North America, Europe, Australia.

SELECTED REFERENCE   Vuille, F.-L. 1987. Reproductive biology of the genus *Damasonium* (Alismataceae). Pl. Syst. Evol. 157: 63–71.

1. Damasonium californicum Torrey in G. Bentham, Pl. Hartw., 341. 1857 E F

Machaerocarpus californicus (Torrey) Small

Herbs, to 35 cm. **Leaves** erect; petiole 3–15 cm; blade absent or narrowly lanceolate to ovate, 3–9 × 0.5–3 cm, margins entire. **Inflorescences** racemes, rarely panicles; scapes erect or ascending; bracts broadly ovate-lanceolate, thin, scarious. **Flowers** 12–22 mm wide; sepals persistent, green, oblong or ovate, somewhat hooded, 4–5 mm; petals white with yellow patch near base, rarely pink, broadly cuneate, rhombic, 6–10 mm, erose distally; stamens 2–3 mm, 2 in front of each petal; pistils 6–10 in single ring on receptacle; ovules 1; styles erect in anthesis, soon spreading radially, elongate, stout. **Fruits** horned by enlargement of style, ribbed on angles, with prominent angular shoulder and depressed flat faces, 0.5 × 3–5.5 mm; beak 3–6 mm.

Flowering late summer. Vernal pools, margins of intermittent streams, or on mud; 100–1700 m; Calif., Idaho, Nev., Oreg.

3. SAGITTARIA Linnaeus, Sp. Pl. 2: 993. 1753; Gen. Pl. ed. 5, 429. 1754 • Arrowhead, sagittaire [Latin *sagitta*, arrow]

Lophotocarpus T. Durand

Plants perennial, rarely annual, submersed, floating-leaved, or emersed, glabrous to sparsely pubescent; rhizomes often present, occasionally terminated by tubers; stolons often present; corms absent; tubers white to brown, smooth. **Roots** septate. **Leaves** sessile or petiolate; petiole terete to triangular; blade with translucent markings absent, linear to obovate, base attenuate to hastate or sagittate, margins entire, apex round to acute. **Inflorescences** racemes, panicles, rarely umbels, of 1–17 whorls, erect, emersed or floating, rarely submersed; bracts coarse or delicate, apex obtuse to acute, surfaces smooth or papillose proximally to distally. **Flowers** unisexual, the proximal rarely with ring of sterile stamens; staminate flowers pedicellate, distal to pistillate flowers; pistillate flowers mostly pedicellate, rarely sessile; bracts subtending pedicels, lanceolate, shorter than pedicels, apex obtuse to acute; pedicels ascending to recurved; receptacle convex; sepals recurved in staminate flowers, recurved to erect in pistillate flowers, often sculptured, herbaceous to leathery; petals white, rarely with pink spot or tinge, entire; stamens 7–30; filaments linear to dilated, glabrous or pubescent; pistils to 1500 or more, spirally arranged, not radiating in starlike pattern, distinct; ovules 1; styles terminal. **Fruits** without longitudinal ribs, compressed, abaxially keeled or not, abaxial wings often present, lateral wing often present, 1, curved, glands present. $x = 11$.

Species ca. 30 (24 in the flora): mostly Western Hemisphere; Europe; Asia.

SELECTED REFERENCES Beal, E. O., J. W. Wooten, and R. B. Kaul. 1982. Review of *Sagittaria engelmanniana* complex (Alismataceae) with environmental correlations. Syst. Bot. 7: 417–432. Bogin, C. 1955. Revision of the genus *Sagittaria* (Alismataceae). Mem. New York Bot. Gard. 9: 179–233. Rataj, K. 1972. Revision of the genus *Sagittaria*. Part II. (The species of West Indies, Central and South America). Annot. Zool. Bot. 78. Wooten, J. W. 1973. Taxonomy of seven species of *Sagittaria* from eastern North America. Brittonia 24: 64–74.

1. Fruiting pedicels recurved or rarely spreading; pistillate sepals mostly erect and closely enclosing flower or fruiting head, occasionally spreading to recurved.
   2. Leaves, at least some, emersed.
      3. Fruiting heads 1.2–2.1 cm diam.; leaf blade hastate to sagittate. . . . . 3. *Sagittaria montevidensis*
      3. Fruiting heads 0.7–1.2 cm diam.; leaf blade linear-ovate to lance-elliptic.
         4. Achenes abaxially keeled, mostly more than 2 mm; faces not tuberculate, often with glands; California. . . . . 5. *Sagittaria sanfordii*
         4. Achenes not abaxially keeled, mostly less than 2 mm, faces tuberculate, without glands; e of Rocky Mountains. . . . . 6. *Sagittaria platyphylla*

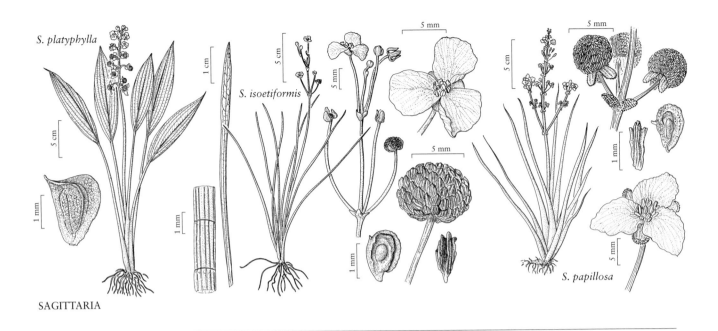

2. Leaves submersed, floating, rarely emersed, or plants stranded on shore.
    5. Achene faces tuberculate; floating leaves with sagittate blades present at least on
       some plants of population. . . . . 2. *Sagittaria guayanensis*
    5. Achene faces not tuberculate; floating leaves absent or, if present, with unlobed or
       hastate blades.
       6. Fruiting pedicels 0.2–1.1 cm; phyllodia lenticular (best observed in cross section);
          tidal muds; mostly brackish waters. . . . . 1. *Sagittaria subulata*
       6. Fruiting pedicels 1.5–6.5 cm; phyllodia flattened (best observed in cross section);
          rarely tidal muds; mostly fresh waters.
          7. Leaves with blades and petioles usually present on some plants; some plants
             often stranded along shore, these usually with expanded leaf blades. . . . . . . 7. *Sagittaria filiformis*
          7. Leaves all phyllodia; plants almost always submersed, rarely stranded, these
             without expanded leaf blades.
             8. Phyllodia 50–250 cm; Florida springs. . . . . 8. *Sagittaria kurziana*
             8. Phyllodia 12–53 cm; streams and lakes, New Mexico. . . . . 4. *Sagittaria demersa*
1. Fruiting pedicels spreading to ascending or absent; pistillate sepals mostly spreading to
  recurved, not enclosing flower.
    9. Filaments pubescent to tomentulose (except *S. fasciculata*).
      10. Leaves all phyllodia, nearly terete; ne United States. . . . . 9. *Sagittaria teres*
      10. Leaves with blades and petioles or if phyllodia, then flattened to triangular in cross
         section; mostly widespread.
         11. Pistillate flowers sessile to subsessile. . . . . 16. *Sagittaria rigida*
         11. Pistillate flowers obviously pedicellate.
           12. Filaments cylindric. . . . . 14. *Sagittaria lancifolia*
           12. Filaments dilated.
              13. Rhizomes present, coarse; stolons and corms absent.
                14. Abaxial wing of fruit ± entire; plants widespread. . . . . 13. *Sagittaria graminea*
                14. Abaxial wing of fruit scalloped or toothed; ne Alabama. . . . . . .
                . . . . 12. *Sagittaria secundifolia*
             13. Rhizomes absent or if present, then not coarse; corms and/or stolons
              present.
                15. Filaments exceeding anthers in length. . . . . 17. *Sagittaria cristata*

15. Filaments shorter than or equaling anthers in length.
  16. Blades of emersed leaves 0.5 cm or more wide; w Carolinas.
  . . . . 11. *Sagittaria fasciculata*
  16. Blades of emersed leaves, if present, 0.4(–0.5) cm or less wide;
  se coastal plain. . . . . 10. *Sagittaria isoetiformis*
9. Filaments glabrous.
  17. Emersed leaf blades linear to ovate.
    18. Bracts papillose. . . . . 15. *Sagittaria papillosa*
    18. Bracts not papillose.
      19. Emersed plants with erect to ascending petioles; leaf blades lanceolate to
      ovate. . . . . 18. *Sagittaria ambigua*
      19. Emersed plants with recurved petioles; leaf blades linear to sagittate. . . 19. *Sagittaria cuneata*
  17. Emersed leaf blades cordate, sagittate, or hastate.
    20. Bracts distinct or connate much less than ¼ total length.
      21. Flowers in 2–4 whorls; achenes with facial glands. . . . . 20. *Sagittaria engelmanniana*
      21. Flowers in 5–12 whorls; achenes without facial glands.
        22. Petiole winged in cross section; achene beak strongly recurved. . . . . 21. *Sagittaria australis*
        22. Petiole ridged in cross section; achene beak ascending apically. . . . 22. *Sagittaria brevirostra*
    20. Bracts connate at least ¼ total length.
      23. Achene beak 1–2 mm, horizontal. . . . . 24. *Sagittaria latifolia*
      23. Achene beak 0.1–0.6 mm, erect or incurved.
        24. Emersed plants with recurved petioles and linear to sagittate blades;
        basal lobes equal to or shorter than remainder of blade; submersed
        leaves phyllodial, floating leaves cordate to sagittate. . . . . 19. *Sagittaria cuneata*
        24. Emersed plants with ascending to erect petioles and sagittate blades;
        basal lobes longer than remainder of blade; submersed and floating
        leaves absent. . . . . 23. *Sagittaria longiloba*

1. **Sagittaria subulata** (Linnaeus) Buchenau, Abh. Naturwiss. Vereine Bremen 2: 490. 1871

Alisma subulatum Linnaeus, Sp. Pl. 1: 343. 1753 (as *subulata*); *Sagittaria lorata* (Chapman) Small; *S. subulata* var. *gracillima* (S. Watson) J. G. Smith; *S. subulata* var. *natans* (Michaux) J. G. Smith

Herbs, perennial, to 40 cm; rhizomes absent; stolons present; corms present. **Leaves** submersed or rarely floating, phyllodial, lenticular, 5–40 × 0.1–0.4 cm; petiole terete, 2.4–4 cm; blade linear-lanceolate to ovate, 1–2.5 × 0.3–1.5 cm. **Inflorescences** racemes, of 2–7 whorls, floating, 2–11 × 1.5–4.5 cm; peduncles 5–40 cm; bracts connate more than ¼ total length, subulate, 1.5–4.2 mm, delicate, not papillose; fruiting pedicels recurved, cylindric to club-shaped, 0.2–1.1 cm. **Flowers** 0.4–1.2 cm diam.; sepals spreading to recurved in staminate, erect in pistillate, enclosing flower or fruiting head; filaments dilated, longer than anther, glabrous; pistillate pedicellate, occasionally with ring of sterile stamens. **Fruiting heads** 0.55–0.8 cm diam.; achenes oblanceoloid, abaxially keeled, ca. 2 × 1.5 mm, beaked; faces not tuberculate, wings 1–2,

crenate, glands 0–1; beak lateral, erect, 0.2–0.4 mm. $2n = 22$.

Flowering summer–fall. Tidal muds, streams, and brackish bays; 0–100 m; Ala., Conn., Del., D.C., Fla., Ga., Md., Mass., Miss., N.J., N.Y., N.C., Pa., R.I., S.C., Va.; South America (Colombia).

Sagittaria subulata occurs in shallow brackish waters near the coast. The plants are especially common in areas that are exposed during low tides.

2. **Sagittaria guayanensis** Kunth in A. von Humboldt et al., Nov. Gen. Sp. 1: 250. 1816 [1]

Subspecies 2 (1 in the flora): introduced, North America; Mexico; West Indies; tropical regions, Asia, Africa.

Sagittaria guayanensis was divided into two subspecies, *S. guayanensis* subsp. *guayanensis* from the Neotropics and subsp. *lappula* (D. Don) Bogin from the Paleotropics. The two subspecies were separated by the shape and size of the fruit in addition to their distribution. *Sagittaria guayanensis* subsp. *lappula* has compressed achenes longer than 2.5 mm, whereas subsp. *guayanensis* has plump achenes shorter than 2.5 mm (C. Bogin 1955).

The name has often been spelled "guyanensis" (K. Rataj 1972), which is incorrect. *Sagittaria guayanensis* was the spelling in the protologue. The holotype was collected by Humboldt and Bonpland in Colombia, not Guyana. The type citation is "Colombia: Guainia: in wetlands near the sugar mill of Don Felix Farreras and the city of Bolivar." The type is supposedly at Paris, as with all HBK types; however, we (with the help of Alicia Lourteig) have been unable to locate it. MO has a fragment, however. Rataj did designate a neotype: Suriname, *Hostman 870* (TCD!). The fragment at MO takes precedence over this neotype.

SELECTED REFERENCE   Thieret, J. W. 1969b. *Sagittaria guayanensis* (Alismaceae) in Louisiana: New to the United States. Sida 3: 445.

2a. Sagittaria guayanensis Kunth subsp. **guayanensis**
I

Herbs, annual or perennial, to 50 cm; rhizomes absent; stolons absent; corms present. **Leaves** submersed or floating, phyllodial, flattened, 4–7 × 0.4–0.8 cm; petiole triangular, to 42 cm; blade sagittate, 3.5–10.5 × 1.5–8.5 cm, basal lobes shorter than to equal to length of remainder of blade. **Inflorescences** racemes, of 1–7 whorls, floating, 0.9–9 × 0.6–4 cm; peduncles 10.9–22 cm; bracts distinct or connate less than ¼ total length, broadly lanceolate to linear-lanceolate, 2.8–15 mm, delicate, not papillose; fruiting pedicels spreading to recurved, cylindric, 0.6–2.1 cm. **Flowers** 1.2–2 cm diam.; sepals spreading in staminate, erect in pistillate, enclosing flower or fruiting head; filaments cylindric, ± equaling anther length, glabrous; pistillate pedicellate, with ring of sterile stamens. **Fruiting heads** 0.5–1.8 cm diam.; achenes oblanceoloid, abaxially keeled, 1.7–2.2 × 1.2–1.5 mm, beaked; faces tuberculate, wings 1–3, ± dentate, glands absent; beak lateral, erect, 0.2–0.5 mm.

Flowering summer–fall. Rice fields, ephemeral pools, roadside ditches, and streams; 0–100 m; introduced; La.; c, s Mexico; West Indies; Central America; tropical South America.

3. Sagittaria montevidensis Chamisso & Schlechtendal, Linnaea 2: 156. 1827 · Sagittaire spongieuse   F W

Herbs, annual or perennial, to 100 cm; rhizomes present; stolons absent; corms present. **Leaves** submersed and emersed; submersed leaves absent or phyllodial, flattened in cross section, to 17 × 2 cm; emersed leaves with petiole triangular, 21–55 cm, blade hastate to sagittate, 2.5–17.5 × 0.6–22 cm, lobes longer than or equal to remainder of blade. **Inflorescences** racemes or panicles,

of 1–15 whorls, floating or emersed, 1.5–28 × 1.5–15 cm; peduncles 15–47 cm; bracts distinct or connate less than ¼ total length, lanceolate to elliptic, 4–34 mm, delicate, not papillose; fruiting pedicels recurved, club-shaped, 0.5–4.2 cm. **Flowers** 2–5 cm diam.; sepals spreading in staminate, erect in pistillate, enclosing flower or fruiting head; filaments cylindric, longer than anthers, glabrous; pistillate pedicellate, with or without ring of sterile stamens. **Fruiting heads** 1.2–2.1 cm diam.; achenes oblanceoloid, not abaxially keeled, 2–4.3 × 0.7–1.5 mm, beaked; faces not tuberculate, wings absent, glands 1; beak lateral, horizontal, 0.4–0.8 mm.

Subspecies 3 (3 in the flora): North America; Mexico; South America.

1. Plants of tidal mud flats, often stranded by low tide; emersed blades spatulate, rarely sagittate . . . . 3b. *Sagittaria montevidensis* subsp. *spongiosa*
1. Plants of fresh waters, not stranded by low tide; emersed blades hastate to sagittate.
  2. Pistillate flowers without ring of sterile stamens; petals with purple spot at base. . . . . 3a. *Sagittaria montevidensis* subsp. *montevidensis*
  2. Pistillate flowers with ring of sterile stamens; petals without purple spot at base. . . . . 3c. *Sagittaria montevidensis* subsp. *calycina*

3a. Sagittaria montevidensis Chamisso & Schlechtendal subsp. **montevidensis**   I

Herbs, perennial, not stranded by low tide. **Leaves** submersed and emersed; submersed leaves sessile, blade linear; emersed petiolate, blade hastate to sagittate. **Inflorescences** of 1–15 whorls, floating or emersed; bracts connate. **Flowers:** pistillate without ring of sterile stamens; petals with purple spot at base.

Flowering summer–fall (Jul–Oct). Wet ditches, mud flats of rivers; 0–10 m; introduced; Ala., Fla., Ga.; c, s South America.

Sagittaria montevidensis subsp. *montevidensis* apparently is native to South America. It has been introduced along the coasts of the Gulf of Mexico and the Atlantic Ocean.

3b. Sagittaria montevidensis Chamisso & Schlechtendal subsp. **spongiosa** (Engelmann) Bogin, Mem. New York Bot. Gard. 9: 198. 1955 • Tidal arrowhead [E]

Sagittaria calycina Engelmann var. *spongiosa* Engelmann in A. Gray, Manual ed. 5, 493. 1867; *Lophotocarpus spongiosus* (Engelmann) J. G. Smith; *Sagittaria spatulata* (J. G. Smith) Buchenau

Herbs, annual, often stranded by low tide. **Leaves** emersed, sessile, phyllodial, lanceolate, or petiolate with blade spatulate, rarely sagittate. **Inflorescences** of 1–4 whorls, emersed; bracts connate. **Flowers:** pistillate with ring of sterile stamens; petals without purple spot at base. $2n = 22$.

Flowering summer–fall (Jul–Oct). Brackish to freshwater tidal mud flats and salt marshes; 0–5 m; N.B.; Conn., Del., Maine, Md., Mass., N.H., N.J., N.Y., Pa.

3c. Sagittaria montevidensis Chamisso & Schlechtendal subsp. **calycina** (Engelmann) Bogin, Mem. New York Bot. Gard. 9: 197. 1955

Sagittaria calycina Engelmann in W. H. Emory, Rep. U.S. Mex. Bound. 2(1): 212. 1859; *Lophotocarpus calycinus* (Engelmann) J. G. Smith

Herbs, annual or perennial, not stranded by low tide. **Leaves** submersed and emersed; submersed leaves sessile, blade linear; emersed leaves petiolate, blade hastate to sagittate. **Inflorescences** of 1–15 whorls, floating or emersed; bracts distinct. **Flowers:** pistillate with ring of sterile stamens; petals without purple spot at base.

Flowering mid summer–fall. Mud flats of lakes and rivers; 5–2000 m; Ala., Ark., Calif., Colo., Del., Ga., Ill., Ind., Iowa, Kans., Ky., La., Mich., Minn., Miss., Mo., Nebr., N.Mex., N.C., Ohio, Okla., S.C., S.Dak., Tenn., Tex., Va., W.Va., Wis.; Mexico (Chihuahua, Coahuila, Sinaloa).

4. Sagittaria demersa J. G. Smith, N. Amer. Sagittaria, 32, plate 15, figs. 1–4. 1894

Herbs, annual, to 60 cm; rhizomes absent; stolons present; corms present. **Leaves** submersed, phyllodial, lenticular in cross section, to nearly terete, 12–53 × 0.3–0.7 cm; rare stranded plants without expanded leaf blades. **Inflorescences** racemes, of 2–7 whorls, floating or emersed, to 16 × 4 cm; peduncles 13.5–28 cm; bracts connate more

than ¼ total length, ovate to lanceolate, 1.5–2 mm, delicate, not papillose; fruiting pedicels spreading to reflexed in flower and fruit, cylindric, 1.5–6.5 cm. **Flowers** 1.5–5 cm diam.; sepals spreading in staminate, appressed to spreading in flower and fruit in pistillate, often enclosing flower or fruiting head; filaments dilated, longer than anthers, glabrous; pistillate pedicellate, without ring of sterile stamens. **Fruiting heads** 0.4–0.6 cm diam; achenes oblanceoloid to obovoid, not abaxially keeled, 1.5 × 1 mm, beaked; faces not tuberculate, wings absent, glands absent; beak lateral, erect, ca. 1.1 mm.

Flowering summer–fall. Streams and lakes; 1500–2000 m; N.Mex.; c Mexico.

Sagittaria demersa was known previously only from central Mexico. It is known in the United States from three recent collections taken in northern New Mexico.

5. Sagittaria sanfordii Greene, Pittonia 2: 158. 1890 [C] [E]

Herbs, perennial, to 130 cm; rhizomes present; stolons absent; corms present. **Leaves** emersed, sessile, phyllodial, lenticular, or with petiole triangular, to 53 cm, blade lance-elliptic, 11–18 × 1.5–4.5 cm. **Inflorescences** racemes, of 3–9 whorls, emersed, 2–13.5 × 3–6 cm; peduncles to 75 cm; bracts connate more than ¼ total length, lanceolate, 5–8 mm, delicate, not papillose; fruiting pedicels recurved, cylindric, 0.5–2.5 cm. **Flowers** to 3.5 cm diam.; sepals spreading in staminate, erect in pistillate, enclosing flower or fruiting head; filaments club-shaped, longer than anthers, pubescent; pistillate pedicellate, without ring of sterile stamens. **Fruiting heads** 0.7–1.2 cm diam.; achenes oblanceoloid, abaxially keeled, 2–3 × 1.4–2.1 mm, beaked; faces not tuberculate, wings absent, glands 0–1; beak lateral, erect, 0.2–0.6 mm. Chromosome number not available.

Flowering summer (Jul–Sep). Ditches, streams, and lake margins; of conservation concern; 10–300 m; Calif.

6. Sagittaria platyphylla (Engelmann) J. G. Smith, N. Amer. Sagittaria, 29. 1894 [F]

Sagittaria graminea Michaux var. *platyphylla* Engelmann in A. Gray, Manual ed. 5, 494. 1867; *S. mohrii* J. G. Smith

Herbs, perennial, to 150 cm; rhizomes absent; stolons present; corms present. **Leaves** submersed and emersed; submersed sessile, phyllodial, flattened, to 26 × 0.5

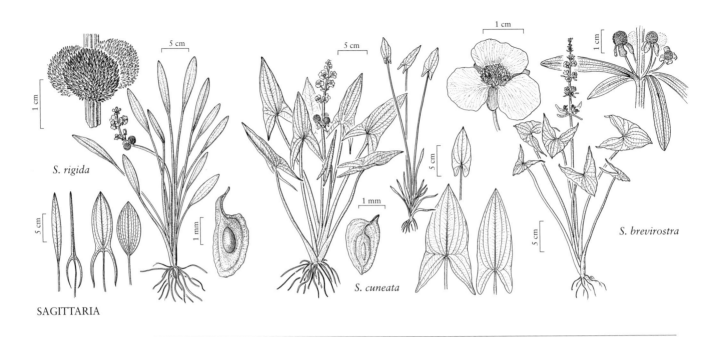

SAGITTARIA

S. rigida

S. cuneata

S. brevirostra

cm; emersed with petiole ± triangular, 21–70.5 cm, blade linear-ovate to ovate, 4.6–16.4 × 0.7–6.1 cm. **Inflorescences** racemes, of 3–9 whorls, emersed, 2.5–10 × 2–4.5 cm; peduncles 22–60 cm; bracts connate more than ¼ total length, lanceolate, 3–5.5 mm, delicate, not papillose; fruiting pedicels spreading to recurved, cylindric, 0.5–3 cm. **Flowers** to 1.8 cm diam.; sepals spreading to recurved, not enclosing flower or fruiting head; filaments dilated, longer than anthers, pubescent; pistillate pedicellate, without ring of sterile stamens. **Fruiting heads** 0.7–1.2 cm diam.; achenes oblanceoloid, not abaxially keeled, 1.2–2 × 0.8–1.2 mm, beaked; faces tuberculate, wings absent, glands absent; beak lateral, horizontal to erect, 0.3–0.6 mm. 2*n* = 22.

Flowering summer–fall. Streams and lakes; 0–900 m; Ala., Ark., Fla., Ga., Ill., Ky., La., Miss., Mo., Ohio, Okla., Pa., Tenn., Tex., W.Va.; Mexico (Nuevo León, sw Mexico); Central America (Panama).

Sagittaria platyphylla has been accepted at the varietal level, i.e., *Sagittaria graminea* var. *platyphylla* (C. Bogin 1955; J. W. Wooten 1973; E. O. Beal et al. 1982). After studying dozens of populations in the field from much of its range and hundreds of herbarium specimens, we have concluded that this taxon should be recognized at the specific level instead, a conclusion supported by cladistic analyses of morphologic characters (R. M. Kortright 1998)

SELECTED REFERENCE Kortright, R. M. 1998. The Aquatic Plant Genus *Sagittaria* (Alismataceae): Phylogeny Based on Morphology. M.S. thesis. University of Alabama.

7. **Sagittaria filiformis** J. G. Smith, N. Amer. Sagittaria, 20, plate 15, figs. 5–8. 1894 E

Sagittaria stagnorum Small; *S. subulata* (Linnaeus) Buchenau var. *gracillima* (S. Watson) J. G. Smith

Herbs, perennial, to 170 cm, mostly of fresh waters, some plants stranded along shore; rhizomes absent; stolons present; corms present. **Leaves** submersed or floating, rarely emersed; submersed phyllodial, flattened, 30–250 × 0.1–1.5 cm; floating with petiole flattened, to 40 cm, blade linear-ovate to ovate, rarely sagittate, to 3.5 × 0.5 cm; emersed with petiole 5–10 cm, blade linear-ovate to ovate, rarely hastate to sagittate, to 4 × 0.5 cm; stranded plants usually with expanded leaf blades. **Inflorescences** racemes, rarely panicles, of 4–10 whorls, floating to slightly emersed, 15–25 × 5–15 cm; peduncles 10–200 cm; bracts connate more than ¼ total length, lanceolate, 1–10 mm, delicate, not papillose; fruiting pedicels spreading to recurved, cylindric, 1.5–4.5 cm. **Flowers** to 3 cm diam.; sepals spreading in staminate, erect in pistillate, enclosing flower or fruiting head; filaments dilated, ± equaling anthers, glabrous; pistillate pedicellate, without ring of sterile stamens. **Fruiting heads** 0.7–1 cm diam.; achenes obovoid, abaxially keeled, 5 × 2.5 mm, beaked; faces not tuberculate, wings 0–3, ± entire, glands 0–1; beak lateral, erect, 1 mm. $2n = 22$.

Flowering summer–fall. Shallow, swift waters or deep streams in northern portion of range; ponds, lakes, drainage canals, and swamps in southern portion of range; 0–100 m; Ala., Conn., Fla., Ga., Maine, Mass., N.Y., N.C., Pa., R.I., S.C., Va.

8. Sagittaria kurziana Glück, Bull. Torrey Bot. Club 54: 257. 1927 [E]

Herbs, perennial, to 250 cm; rhizomes absent; stolons present; corms present. **Leaves** submersed and floating, sessile, phyllodial, flattened, 50–250 × 0.4–1.5 cm; rare stranded plants without expanded leaf blades. **Inflorescences** racemes, rarely panicles, of 4–10 whorls, floating to slightly emersed, 15–25 × 5–15 cm; peduncles to 200 cm; bracts connate less than or equal to ¼ total length, lanceolate, 0.4–0.8 mm, delicate, not papillose; fruiting pedicels spreading to recurved, cylindric, 1.5–4.5 cm. **Flowers** to 2.8 cm diam.; sepals erect to spreading in staminate, erect in pistillate, enclosing flower or fruiting head; filaments dilated, ± equaling anthers, glabrous; pistillate pedicellate, without ring of sterile stamens. **Fruiting heads** 0.7–1 cm diam.; achenes obovoid, abaxially keeled, 5 × 2.5 mm, beaked; faces not tuberculate, wings absent, glands 0–1; beak lateral, erect, 1 mm. $2n = 22$.

Flowering summer–fall. Large springs and their streams, in fresh to slightly brackish water; 0–100 m; Fla.

Sagittaria kurziana is a distinctive species found in springs and clear streams, often with high sulfur content. The plants form huge colonies, essentially filling the stream with their long flexible leaves that wave back and forth in the water current.

9. Sagittaria teres S. Watson in A. Gray et al., Manual ed. 6, 555. 1890 • Quill-leaved sagittaria [E]

Herbs, perennial, to 80 cm; rhizomes absent; stolons present; corms present. **Leaves** emersed or submersed, sessile, phyllodial, nearly terete; emersed, to 60 × 0.15–0.7 cm; submersed, 3.5–18.5 × 0.15–0.4 cm. **Inflorescences** racemes, of 1–4 whorls, emersed, 2.5–4 × 2.5–6 cm; peduncles 10–80 cm; bracts connate more than or equal to ¼ total length, subulate, 2–3 mm, delicate, not papillose; fruiting pedicels obliquely ascending, filiform, 1 cm. **Flowers** to 1.5 cm diam.; sepals recurved, not enclosing flower; filaments dilated, ± equaling anthers, pubescent; pistillate pedicellate, without ring of sterile stamens. **Fruiting heads** 0.6–1 cm diam.; achenes obovoid-cuneate, abaxially keeled, 2–3 × 1.2–1.5 mm, beaked; faces not tuberculate, wings absent, glands 1–2; beak erect to horizontal, 0.3–0.4 mm. $2n = 22$.

Flowering summer (Jul–Sep). Sandy pond shores and swamps of acid waters, mainly along Atlantic Coastal Plain; 0–100 m; Mass., N.J., N.Y., R.I.

10. Sagittaria isoetiformis J. G. Smith, Rep. (Annual) Missouri Bot. Gard. 6: 115, plate 53. 1895 [E] [F]

Herbs, perennial, to 65 cm; rhizomes absent; stolons present; corms present. **Leaves** emersed or submersed, phyllodial or rarely dilated apically, flattened, 4–40 cm × 0.05–0.4(–0.5) cm. **Inflorescences** racemes, of 1–5 whorls, emersed, 2–20 × 1.5–12 cm; peduncles 7–59 cm; bracts connate more than or equal to ¼ total length, oblanceolate, 0.2–0.3 mm, delicate, not papillose; fruiting pedicels spreading, cylindric, 0.5–6 cm. **Flowers** to 1.3 cm diam.; sepals recurved to spreading, not enclosing flower; filaments dilated, shorter than anthers, minutely tomentose; pistillate pedicellate, without ring of sterile stamens. **Fruiting heads** 0.5–1 cm diam.; achenes obovoid, abaxially keeled, 2.2–2.8 × 1.5–2 mm, beaked; faces not tuberculate, wings 1–2, ± entire, glands 3–5; beak lateral, incurved-erect, 0.2 mm.

Flowering summer–fall. Shores of sandy-bottomed lakes, se coastal plain; 0–100 m; Ala., Fla., Ga., N.C., S.C.

Sagittaria isoetiformis has often been misidentified as *Sagittaria teres* (E. O. Beal 1960b). The two species can be separated, however; *S. teres* has nearly terete phyllodia, whereas *S. isoetiformis* has flattened phyllodia (R. K. Godfrey and P. Adams 1964). A study of the genetics of the two species shows them to be genetically different, and the data indicate the two taxa actually should be considered at the specific level. The two species are capable of CAM photosynthesis, a process very uncommon in the genus and found among none of their supposedly closely related species (A. L. Edwards, pers. comm.).

SELECTED REFERENCE Godfrey, R. K. and P. Adams. 1964. The identity of *Sagittaria isoetiformis* (Alismataceae). Sida 1: 269–273.

11. Sagittaria fasciculata E. O. Beal, J. Elisha Mitchell Sci. Soc. 76: 76, fig. 3, map 5. 1960 [C] [E]

Herbs, perennial, to 35 cm; rhizomes absent; stolons present; corms present. **Leaves** submersed and emersed; submersed phyllodial, flattened, 9.5–16.5 × 0.6–1.5 cm; emersed with petiole nearly terete, 3–25 cm, blade linear-lanceolate to ovate, 5.5–8.5 × 0.5–2.1 cm. **Inflorescences** racemes, of 2–5 whorls, emersed, 4.5–15 × 2–6 cm; peduncles to 35 cm; bracts connate ¼ total length, lanceolate, 2–4.5 mm, delicate, not papillose; fruiting pedicels spreading to recurved, cylindric, 1.5–4.5 cm. **Flowers** to 1 cm diam.; sepals appressed to spreading, not enclosing flower or fruiting head; filaments dilated, ± equaling anthers, glabrous; pistillate pedicellate, without ring of sterile stamens. **Fruiting heads** 0.5–0.6 cm diam; achenes obovoid, abaxially keeled, 2.5–3 × 1.2–1.5 mm, beaked; faces not tuberculate, wings 1, ± entire, glands 0–1; beak lateral, horizontal, ca. 0.5 mm.

Flowering spring–summer. Swamps, bogs, and wet roadside ditches; of conservation concern; 250–1000 m; N.C., S.C.

12. Sagittaria secundifolia Kral, Brittonia 34: 12, fig. 1. 1982 [C] [E]

Herbs, perennial, to 50 cm; rhizomes coarse; stolons absent; corms absent. **Leaves** mostly submersed, rarely emersed, sessile, phyllodial, flattened to lenticular, 5–30 × 0.2–0.5 cm. **Inflorescences** racemes, of 2–5 whorls, emersed, 1.5–8 × 1–7 cm; peduncles 1–2.5 cm; bracts connate more than or equal to ¼ total length, lanceolate, 0.5–1.5 mm, delicate, not papillose; fruiting pedicels spreading, cylindric, 0.6–2.5 cm. **Flowers** to 2.5 cm diam.; sepals recurved to spreading, not enclosing flower or fruiting head; filaments dilated, shorter than anthers, minutely tomentose; pistillate pedicellate, without ring of sterile stamens. **Fruiting heads** 0.5 cm diam; achenes obovoid-triangular, abaxially keeled, 2 × 1 mm, beaked; faces tuberculate, wings 1, scalloped or toothed, glands 0–1; beak lateral, incurved-erect, 0.3 mm.

Flowering spring–fall. Riverine shoals and pools; of conservation concern; 300–400 m; Ala.

The rhizomes of *Sagittaria secundifolia* grow in crevices of bedrock along the Little River in Alabama. During lower water periods, the plant flowers, with peduncles reaching the water surface, and flowers more or less floating.

13. Sagittaria graminea Michaux, Fl. Bor.-Amer. 2: 190. 1803 • Sagittaire à feuilles de graminées [W]

Herbs, perennial, to 100 cm; rhizomes coarse; stolons absent; corms absent. **Leaves** submersed or emersed; submersed leaves phyllodial, angled abaxially, flattened adaxially, 6.4–35 × 0.5–4 cm; emersed with petiole triangular, 6.5–17 cm, blade linear to linear-oblanceolate, 2.5–17.4 × 0.2–4 cm. **Inflorescences** racemes or panicles, of 1–12 whorls, emersed, 2.5–21 × 1–8 cm; peduncles 6.5–29.7 cm; bracts connate more than to equal to ¼ total length, broadly subulate to lanceolate, 20–50 mm, coarse, not papillose; fruiting pedicels spreading, cylindric, 0.5–5 cm. **Flowers** to 2.3 cm diam.; sepals recurved to spreading, not enclosing flower; filaments dilated, shorter than anthers, pubescent; pistillate flowers pedicellate, without ring of sterile stamens. **Fruiting heads** 0.6–1.5 cm diam.; achenes oblanceoloid, without abaxial keel, 1.5–2.8 × 1.1–1.5 mm, beaked; faces not tuberculate, abaxial wings 0–1, ± entire, glands 1–2; beak lateral, erect, 0.2 mm.

Subspecies 3 (3 in the flora): North America; West Indies (Cuba).

Seven varieties of *Sagittaria graminea* have been recognized, i.e., var. *graminea*, var. *platyphylla* Engelmann, var. *teres* (S. Watson) Bogin, var. *weatherbiana* Fernald, var. *cristata* (Engelmann) Bogin, var. *chapmanii* J. G. Smith, and var. *macrocarpa* (J. G. Smith) Bogin (C. Bogin 1955). We accept only one infraspecific rank, i.e., subspecies. Consequently, we have made the appropriate combinations. We accept all of the taxa accepted by Bogin at the varietal level. At specific level we accept Bogin's var. *platyphylla*, var. *teres*, and var. *cristata* and at subspecific level his var. *graminea*, var. *chapmanii*, and var. *weatherbiana*.

Sagittaria graminea var. *macrocarpa* actually is synonymous with var. *graminea* (E. O. Beal 1960b). We therefore are following Beal in recognizing var. *macrocarpa* sensu Bogin as *S. fasciculata*. We also accept var. *platyphylla*, var. *teres*, and var. *cristata* at the specific level, leaving only three subspecies. These subspecies can be separated by the branching of the inflorescence and the length of pistillate pedicels.

1. Inflorescences panicles. . . . .
   . . . . 13b. *Sagittaria graminea* subsp. *chapmanii*
1. Inflorescences racemes.
   2. Pistillate pedicels 0.5–3 cm; phyllodia to 1 cm
      wide. . . . . 13a. *Sagittaria graminea* subsp. *graminea*
   2. Pistillate pedicels 2.1–5 cm; phyllodia more
      than 1 cm wide. . . . .
      . . . . . 13c. *Sagittaria graminea* subsp. *weatherbiana*

13a. Sagittaria graminea Michaux subsp. **graminea**

• Sagittaire graminoïde

Sagittaria cycloptera (J. G. Smith) C. Mohr; *S. eatonii* J. G. Smith; *S. macrocarpa* J. G. Smith

Phyllodia to 1 cm wide. **Inflorescences** racemes; pistillate pedicels 0.5–3 cm. 2*n* = 22.

Flowering summer–fall. Streams, lakes, and tidal areas; 0–700 m; N.B., Nfld. and Labr. (Nfld.), N.S., Ont., P.E.I., Que.; Ala., Ark., Conn., Del., Fla., Ga., Ill., Ind., Iowa, Kans., Ky., La., Maine, Md., Mass., Mich., Minn., Miss., Mo., Nebr., N.H., N.J., N.Y., N.C., Ohio, Okla., Pa., R.I., S.C., S.Dak., Tenn., Tex., Vt., Va., Wash., W.Va., Wis.; West Indies (Cuba).

13b. Sagittaria graminea Michaux subsp. **chapmanii** (J. G. Smith) R. R. Haynes & Hellquist, Novon 6: 370. 1996 [E]

Sagittaria graminea var. *chapmanii* J. G. Smith, N. Amer. Sagittaria, 26, plate 21. 1894 (as *chapmani*); *S. chapmanii* (J. G. Smith) C. Mohr

Phyllodia to 1 cm wide. **Inflorescences** panicles; pistillate pedicels 0.5–3 cm.

Flowering summer–fall. Margins of swamps, ponds, and small streams; 0–100 m; Ala., Fla., Ga., S.C.

13c. Sagittaria graminea Michaux subsp. **weatherbiana** (Fernald) R. R. Haynes & Hellquist, Novon 6: 371. 1996 [E]

Sagittaria weatherbiana Fernald, Rhodora 37: 387, plate 385, fig. 1. 1935; *S. graminea* var. *weatherbiana* (Fernald) Bogin

Phyllodia 1–2.5 cm wide. **Inflorescences** racemes; pistillate pedicels 2.1–5 cm. 2*n* = 22.

Flowering summer–fall. Swamps; 0–100 m; Fla., N.C., S.C., Va.

14. Sagittaria lancifolia Linnaeus, Syst. Nat. ed. 10, 2: 1270. 1759

Herbs, perennial, to 2 m; rhizomes coarse; stolons absent; corms absent. **Leaves** emersed; petiole terete, 44–58 cm; blade linear to ovate or elliptic, 20.3–35 × 0.7–16 cm. **Inflorescences** racemes or panicles, of 6–13 whorls, emersed, 16–72 × 19–60 cm; peduncles 75–125.5 cm; bracts connate more than or equal to ¼ total length, lanceolate to ovate-lanceolate, 3–32 mm, coarse, papillose or not; fruiting pedicels spreading, cylindric, 0.6–3.8 cm. **Flowers** to 3.3 cm diam.; sepals recurved to spreading, not enclosing flower or fruiting head; filaments cylindric, longer than anthers, pubescent; pistillate pedicellate, without ring of sterile stamens. **Fruiting heads** 0.5–1.2 cm diam.; achenes oblanceoloid, abaxially keeled, 1.6–2.5 × 0.8–1.1 mm, beaked; faces not tuberculate, wings absent, glands 1; beak lateral, erect, 0.3–0.7 mm.

Subspecies 2 (2 in the flora): North America; Mexico; West Indies; Central America; South America.

1. Bracts and sepals striate to ribbed; leaves linear to ovate. . . . . 14a. *Sagittaria lancifolia* subsp. *lancifolia*
1. Bracts and sepals papillose; leaves lanceolate to broadly ovate. . . 14b. *Sagittaria lancifolia* subsp. *media*

14a. Sagittaria lancifolia Linnaeus subsp. **lancifolia**

Sagittaria angustifolia Lindley

Leaves linear to ovate. **Inflorescences:** bracts striate to ribbed. **Flowers:** sepals striate to ribbed. 2*n* = 22.

Flowering summer–fall. Brackish marshes near coast; 0–50 m; Fla., Ga., S.C.; se Mexico; West Indies; Central America; tropical South America.

14b. Sagittaria lancifolia Linnaeus subsp. **media** (Micheli) Bogin, Mem. New York Bot. Gard. 9: 214. 1955

Sagittaria lancifolia var. *media* Micheli in A. L. P. de Candolle and C. de Candolle, Monogr. Phan. 3: 73. 1881; *S. falcata* Pursh

Leaves lanceolate to broadly ovate. **Inflorescences:** bracts papillose. **Flowers:** sepals papillose. 2*n* = 22.

Flowering summer–fall. Brackish marshes near coast; 0–50 m; Ala., Fla., Ga., La.,

Miss., N.C., S.C., Tex., Va.; coastal s Mexico; Central America; South America (Colombia and Ecuador).

15. Sagittaria papillosa Buchenau, Index Crit. Butom. Alism. Juncag., 27, 44. 1868 [E] [F]

Herbs, perennial, to 112 cm; rhizomes not coarse; stolons absent; corms absent. **Leaves** submersed and emergent; petiole triangular, 11.5–35 cm; blade elliptic to ovate, 25–56 cm. **Inflorescences** racemes or panicles, of 4–10 whorls, emersed, 8–25 × 5–36 cm; peduncles to 76 cm; bracts connate ¼ total length, lanceolate, 4–8 mm, papillose; fruiting pedicels ascending to spreading, cylindric, 1–4.5 cm. **Flowers** to 3 cm diam.; sepals appressed to spreading; filaments slightly dilated, shorter than anthers, glabrous; pistillate pedicellate, without ring of sterile stamens. **Fruiting heads** 0.7–1 cm diam.; achenes obovoid, abaxially keeled, 1.2–1.5 × 0.7–1.1 mm, beaked; faces not tuberculate, wings 1, ± entire, glands 0–1; beak lateral, erect, 0.1–0.3 mm.

Flowering spring–fall. Swamps, marshes, bogs, ditches, borders of lakes; 0–300 m; Ark., La., Okla., Tex.

16. Sagittaria rigida Pursh, Fl. Amer. Sept. 2: 397. 1814
· Sessile-fruited arrowhead, Sagittaire dressée [E] [F] [W]

Herbs, perennial, to 115 cm; rhizomes absent; stolons present; corms present. **Leaves** emersed or submersed; submersed leaves phyllodial, flattened, 30–70 cm, rarely widening into blade; emersed leaves phyllodial, flattened, or petiole triangular, 34–50 cm, blade linear to elliptic, rarely hastate to sagittate, 5–15 × 0.6–12 cm, basal lobes when present shorter than remainder of blade. **Inflorescences** racemes, of 2–8 whorls, emersed, 8–10 × 2–6 cm; peduncles 10–115 cm; bracts connate more than or equal to ¼ total length, ovate, 3–6 mm, delicate, not papillose; fruiting pedicels absent. **Flowers** to 3 cm diam.; sepals recurved, not enclosing flower; filaments dilated, shorter than anthers, pubescent; pistillate sessile to subsessile, without ring of sterile stamens. **Fruiting heads** 1–1.7 cm diam.; achenes obovoid to oblong, abaxially keeled, 2–3 × 1.3–1.6 mm, beaked; face not tuberculate, wings 1–2, ± entire, glands 1; beak lateral, recurved, 0.8–1.4 mm. 2*n* = 22.

Flowering summer (Jul–Sep). Calcareous or brackish shallow water and shores of ponds, swamps, and rivers, occasionally in deep water; 0–1000 m; Man., Ont., Que., P.E.I., Sask.; Ark., Calif., Conn., Del., D.C.,

Idaho, Ill., Ind., Iowa, Kans., Ky., Maine, Md., Mass., Mich., Minn., Mo., Nebr., N.H., N.J., N.Y., Ohio, Pa., R.I., Tenn., Vt., Va., Wash., W.Va., Wis.

17. Sagittaria cristata Engelmann, Proc. Davenport Acad. Nat. Sci. 4: 29. 1883 [E]

Sagittaria graminea Michaux var. *cristata* (Engelmann) Bogin

Herbs, annual or perennial, to 75 cm; rhizomes absent; stolons present; corms present. **Leaves** emersed, sessile, phyllodial, linear to lanceolate, flattened, 15–25 (–40) × 1.5–4 cm, or petiole triangular, 15–50 cm, blade linear to elliptic-lanceolate, 4–10 × 0.3–2 cm. **Inflorescences** racemes, of 3–6 whorls, 1.5 × 4 cm; peduncle 20–60 cm; bracts connate greater than or equal to ¼ total length, ovate, 4–10 mm, nearly scarious, not papillose; fruiting pedicels spreading, cylindric, 0.8–3 cm. **Flowers** to 0.8 cm diam.; sepals recurved to spreading, not enclosing flower; filaments dilated, exceeding anthers in length, pubescent; pistillate pedicellate, without ring of sterile stamens. **Fruiting heads** 1.2–2 cm diam.; achenes cuneate-obovoid, abaxially keeled, 2.5–3 × 1.4–1.8 mm, beaked; faces not tuberculate, wings 1, ± entire, glands absent; beak ascending to horizontal, 0.4–0.7 mm.

Flowering summer (Jul–Aug). Sandy margins and bottoms of lakes, ponds, and swamps; 100–1000 m; Ont.; Ill., Iowa, Mich., Minn., Mo., Nebr., Wis.

18. Sagittaria ambigua J. G. Smith, N. Amer. Sagittaria, 22, plate 17. 1894 [C] [E]

Herbs, perennial, to 90 cm; rhizomes absent; stolons present; corms present. **Leaves** emersed; petiole triangular, 15–54 cm; blade lanceolate to ovate, 5–20 × 1–10.5 cm. **Inflorescences** racemes, rarely panicles, of 2–11 whorls, emersed; bracts scarcely connate proximally, linear to lanceolate, 10–30 mm, delicate, not papillose; fruiting pedicels spreading to ascending, cylindric, 1.5–3.5 cm. **Flowers** to 23 mm diam.; sepals recurved, not enclosing flower or fruiting head; filaments linear, shorter than anthers, glabrous; pistillate pedicellate, without ring of stamens. **Fruiting heads** 0.8–1.2 cm diam; achenes cuneate-obovoid, abaxially keeled, 1.5–2.1 × 0.8–1.5 mm, beaked; faces not tuberculate, wings 0–1, ± entire, glands absent; beak lateral, horizontal or incurved, 0.1–0.2 mm.

Flowering spring–summer (Apr–Sep). Pond and lake shores, shallow water, ditches, and damp areas; of con-

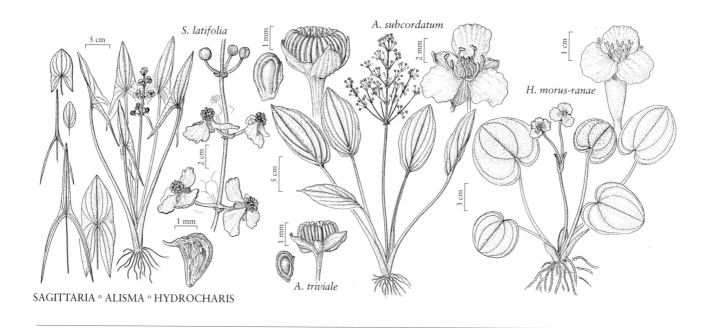

S. latifolia

A. subcordatum

H. morus-ranae

SAGITTARIA ∘ ALISMA ∘ HYDROCHARIS

A. triviale

servation concern; 100–1000 m; Ill., Ind., Kans., Mo., Okla.

19. Sagittaria cuneata E. Sheldon, Bull. Torrey Bot. Club 20: 283, plate 159. 1893 • Northern arrowhead, wapato, Sagittaria cunéaire E F

Sagittaria arifolia Nuttall ex J. G. Smith

Herbs, perennial, to 110 cm; rhizomes absent; stolons present; corms present. **Leaves** emersed, floating, and submersed; submersed phyllodial, flattened, to 45 cm; floating with petiole triangular, to 100 cm, blade cordate or sagittate, rarely linear or ovate, 7.5–9 × 3.5–4 cm; emersed with petiole recurved, 3.5–51 cm, blade linear to sagittate, 2.5–17 × 1.5–11 cm, basal lobes when present shorter than remainder of blade. **Inflorescences** racemes, rarely panicles, of 2–10 whorls, emersed, 14–21 × 2–10 cm, peduncle triangular, 10–50 cm; bracts connate more than or equal to ¼ total length, lance-attenuate or acute, mostly (4–)7–40 mm, membranous, not papillose; fruiting pedicels ascending, cylindric, 0.5–2 cm. **Flowers** to 25 mm diam.; sepals recurved, not enclosing flower or fruiting head; filaments not dilated, equal to or longer than anthers, glabrous; pistillate pedicellate, without ring of sterile stamens. **Fruiting heads** 0.8–1.5 cm diam.; achenes obovoid, abaxially keeled, 1.8–2.6 × 1.3–2.5 mm, beaked; face not tuberculate, wings 0–1, ± entire, glands 0–1; beak apical, erect, 0.1–0.4 mm. $2n = 22$.

Flowering late spring–summer (Jun–Sep). Calcareous and muddy shores and shallow waters of rivers, lakes, ponds, pastures, and ditches, occasional in tidal waters, or in deep flowing water with slow current; 100–2500 m; Alta., B.C., Man., N.B., Nfld. and Labr. (Labr.), N.W.T., N.S., Ont., Que., Sask., Yukon; Alaska, Ariz., Calif., Colo., Conn., Idaho, Ill., Ind., Iowa, Kans., Maine, Mass., Mich., Minn., Mont., Nebr., Nev., N.H., N.J., N.Mex., N.Y., N.Dak., Ohio, Okla., Oreg., Pa., S.Dak., Tex., Utah, Vt., Wash., Wis., Wyo.

Sagittaria cuneata is extremely variable. On emersed plants, the leaf petioles are often bent toward the ground. Submersed plants often grow from a basal rosette with long flexuous petioles and floating sagittate leaves. Plants in deep rivers often develop broad, straplike phyllodia.

The corms of *Sagittaria cuneata* were consumed by Native Americans.

20. Sagittaria engelmanniana J. G. Smith, Mem. Torrey Bot. Club 5: 25. 1894 • Acid-water arrowhead E

Herbs, perennial, to 70 cm; rhizomes absent; stolons present; corms present. **Leaves** emersed; petiole triangular, 10–40 cm; blade sagittate, 4.5–10 × 0.2–2 cm, basal lobes ± equal to remainder of blade. **Inflorescences** racemes, of 2–4 whorls, emersed, 5–14 × 2–4 cm; peduncles 20–38 cm; bracts distinct or if connate, then less than ¼ total length, lanceolate, 5–25 mm, not papillose; fruiting pedicels spreading, cylindric, 1.5–3.5 cm. **Flowers** to 30 mm diam.; sepals recurved to spreading, not enclosing flower or fruiting head; filaments linear, longer than to equal length of anther, glabrous; pistillate pedicellate, without ring of sterile stamens. **Fruiting heads** 1–1.8 cm diam.; achenes cuneate to obovoid, without abaxial keel, 2.4–4 × 1.5–3.8 mm, beaked; faces not tuberculate, wings 1–3, ± entire, glands 1–2; beak lateral, obliquely ascending, (0.7–)1–2.1 mm. $2n = 22$.

Flowering late spring–summer (Jun–Sep). Acid waters of ponds, lakes, bogs, and streams; 0–100 m; Ont.; Ala., Conn., Del., Fla., Ga., Md., Mass., Miss., N.J., N.Y., N.C., Pa., R.I., S.C., Va.

21. **Sagittaria australis** (J. G. Smith) Small, Fl. S.E. U.S., 45. 1903 • Appalachian arrowhead [E]

Sagittaria longirostra (Micheli) J. G. Smith var. *australis* J. G. Smith, Bull. Torrey Bot. Club 24: 20. 1897; *S. engelmanniana* J. G. Smith. subsp. *longirostra* (Micheli) Bogin

Herbs, perennial, to 130 cm; rhizomes absent; stolons present; corms present. **Leaves** emersed; petiole 5-winged, 19–85 cm; blade sagittate, 3–19 × 2.5–11 cm, basal lobes ± equal to remainder of blade. **Inflorescences** racemes, of 5–12 whorls, emersed, 10–29 × 3–5 cm; peduncles 25–105 cm; bracts distinct or if connate, then less than ¼ total length, lanceolate, 7–30 mm, papery, not papillose; fruiting pedicels spreading to ascending, cylindric, 0.3–2.3 cm. **Flowers** to 3 cm diam.; sepals recurved to spreading, not enclosing flower or fruiting head; filaments cylindric, longer than anthers, glabrous; pistillate pedicellate, without ring of sterile stamens. **Fruiting heads** 1–2.2 cm diam.; achenes obovoid, without abaxial keel, 2.1–3.2 × 1.4–2.3 mm, beaked; faces not tuberculate, wings 0–2, ± entire, glands absent; beak lateral, strongly recurved, 4–17 mm. $2n = 22$.

Flowering summer–early fall (Jul–Oct). Slightly basic to slightly acidic ponds, lakes, and swamps; 1–300 m; Ala., Ark., Fla., Ga., Ill., Ind., Iowa, Ky., La., Md., Miss., Mo., N.J., N.Y., N.C., Ohio, Pa., S.C., Tenn., Va., W.Va.

The name *Sagittaria longirostra* (Micheli) J. G. Smith has been misapplied to *S. australis* (J. G. Smith) Small (E. O. Beal et al. 1980).

SELECTED REFERENCE Beal, E. O., S. S. Hooper, and K. Rataj. 1980. Misapplication of the name *Sagittaria longirostra* (Micheli) J. G. Smith to *S. australis* (J. G. Smith) Small. Kew Bull. 35: 369–371.

22. **Sagittaria brevirostra** Mackenzie & Bush, Rep. (Annual) Missouri Bot. Gard. 16: 102. 1905 • Midwestern arrowhead [E] [F]

Sagittaria engelmanniana J. G. Smith subsp. *brevirostra* (Mackenzie & Bush) Bogin

Herbs, perennial, to 70 cm; rhizomes absent; stolons present; corms present. **Leaves** emersed; petiole terete-ridged, to 39 cm; blades sagittate, 5–20 × 2–8 cm, basal lobes equal to or less than remainder of blade. **Inflorescences** racemes or panicles, of 5–12 whorls, emersed, 25–30 × 6–15 cm; peduncles to 45 cm; bracts distinct or if connate, then less than ¼ total length, long-acuminate, 10–40 mm, firm, not papillose; fruiting pedicels spreading to ascending, cylindric, 1–2.5 cm. **Flowers** to 3.5 cm diam.; sepals recurved to spreading, not enclosing flower; filaments cylindric, longer than anthers, glabrous; pistillate pedicellate, without ring of sterile stamens. **Fruiting heads** (1.2–)1.5–2.5 cm diam.; achenes cuneate-obovoid, without abaxial keel, 2.1–3.1 × 1.4–2.2 mm, beaked; faces not tuberculate, wings 0–2, ± entire, glands absent; beak lateral, ascending apically, 0.4–1.7 mm.

Flowering summer (Jul–Sep). Slightly basic to slightly acidic to alkaline waters of ponds, lakes, and swamps; 100–1800 m; Sask.; Ala., Ark., Calif., Colo., Ill., Ind., Iowa, Kans., Ky., La., Mich., Minn., Miss., Mo., Nebr., Ohio, Okla., S.Dak., Tex., Va., Wis.

23. **Sagittaria longiloba** Engelmann ex J. G. Smith, N. Amer. Sagittaria, 16, plate 11. 1894 [W]

Sagittaria greggii J. G. Smith

Herbs, perennial, to 100 cm; rhizomes absent; stolons present; corms present. **Leaves** emersed; petiole 5-ridged, ascending to erect, 24.5–60 cm; blade sagittate, 11.5–26.5 × 0.8–15 cm, basal lobes longer than remainder of blade. **Inflorescences** racemes, rarely panicles, of 5–17 whorls, emersed, 20–37 × 5–27 cm; peduncles 25–96 cm; bracts connate more than or equal to ¼ total length, lanceolate, 6.5–15 mm, delicate, not papillose; fruiting pedicels spreading, cylindric, 1.5–4.4 cm. **Flowers** to 3 cm diam.; sepals recurved to spreading, not enclosing flower; filaments cylindric, shorter than anthers, glabrous; pistillate flowers pedicellate, without ring of sterile stamens. **Fruiting heads** 0.9–1.5 cm diam.; achenes oblanceoloid, abaxially keeled, 1.2–2.5 × 0.8–1.6 mm, beaked; faces tuberculate, wings absent, glands 0–1; beak lateral, erect, 0.1–0.6 mm.

Flowering summer–fall. Wet ditches, ephemeral pools, and margins of streams and lakes; 0–300 m; Ariz., Calif., Kans., Nebr., Okla., Tex.; Mexico; Central America (Nicaragua).

24. Sagittaria latifolia Willdenow, Sp. Pl. 4(1): 409. 1805 • Wapato, common arrowhead, duck-potato, sagittaire latifoliée F

Sagittaria latifolia var. *obtusa* (Muhlenberg) Wiegand; *S. latifolia* var. *pubescens* (Engelmann) J. G. Smith; *S. ornithorhyncha* Small; *S. planipes* Fernald; *S. pubescens* Muhlenberg; *S. viscosa* C. Mohr

Herbs, perennial, to 45 cm; rhizomes absent; stolons present; corms present. **Leaves** emersed; petiole triangular, erect to ascending, 6.5–51 cm; blade sagittate, rarely hastate, 1.5–30.5 × 2–17 cm, basal lobes equal to or less than remainder of blade. **Inflorescences** racemes, rarely panicles, of 3–9 whorls, emersed, 4.5–28.5 × 4–23 cm; peduncles 10–59 cm; bracts connate more than or equal to ¼ total length, elliptic to lanceolate, 3–8 mm, delicate, not papillose; fruiting pedicels spreading, cylindric, 0.5–3.5 cm. **Flowers** to 4 cm diam.; sepals recurved to spreading, not enclosing flower or fruiting head; filaments cylindric, longer than anthers, glabrous; pistillate pedicellate, without ring of sterile stamens. **Fruiting heads** 1–1.7 cm diam; achenes oblanceoloid, without abaxial keel, 2.5–3.5 × to 2 mm, beaked; faces not tuberculate, wings absent, glands (0–)1(–2); beak lateral, horizontal, 1–2 mm. $2n = 22$.

Flowering summer–fall. Wet ditches, pools, and margins of streams and lakes; 0–1500 m; Alta., B.C., Man., N.B., N.S., Ont., P.E.I., Que., Sask.; Ala., Ark., Calif., Colo., Conn., Del., D.C., Fla., Ga., Ill., Ind., Iowa, Kans., Ky., La., Maine, Md., Mass., Mich., Minn., Miss., Mo., Nebr., N.H., N.J., N.Y., N.C., N.Dak., Ohio, Okla., Pa., R.I., S.C., S.Dak., Tenn., Tex., Vt., Va., W.Va., Wis.; c, s Mexico; Central America (Guatemala); South America (Colombia, Ecuador, Venezuela).

Sagittaria latifolia has been divided into numerous species and varieties. It was divided into two varieties, based upon the presence of pubescence over the entire vegetative plant (C. Bogin 1955; K. Rataj 1972). We have examined numerous specimens and found that many from the southeastern United States are pubescent; we believe that this character alone is insufficient for recognition of the varieties.

The corms of *Sagittaria latifolia* were consumed by Native Americans.

4. ALISMA Linnaeus, Sp. Pl. 1: 343. 1753; Gen. Pl. ed. 5, 160. 1754 • Water-plantain [ancient Greek name, adopted by Linnaeus from Dioscorides]

Herbs, perennial, submersed, floating-leaved, emersed, glabrous; rhizomes often present; stolons absent; corms absent; tubers absent. **Roots** not septate. **Leaves** sessile or petiolate; petiole triangular; blade with translucent markings absent, linear to ovate, base attenuate to rounded, margins entire, apex obtuse to acute. **Inflorescences** panicles, of 2–10 whorls, erect, emersed, rarely submersed; bracts delicate, apex acuminate, surfaces smooth. **Flowers** bisexual; pedicels ascending; bracts subtending pedicels, lanceolate, shorter than pedicels, apex acuminate; receptacle flattened; sepals erect, not sculptured, herbaceous; petals pink or white, entire; stamens 6–9, filaments filiform, glabrous; pistils 15–20, in ring around margin of flattened receptacle, not radiating in starlike pattern, distinct; ovules 1; styles lateral. **Fruits** without longitudinal ribs, strongly laterally compressed, abaxial wings absent, lateral wings absent, abaxially 2–3-ribbed, abaxial keel absent, glands absent. $x = 7$.

Species 9 (5 in the flora): widely distributed, nearly worldwide.

Much controversy surrounds the treatment of *Alisma* in North America. At present three distinct native species in North America are generally recognized (P. Rubtzoff 1964) as well as the probable occurrence of two introduced species, one in California and the other in Alaska.

SELECTED REFERENCES Björkquist, I. 1968. Studies in *Alisma* L. II. Chromosome studies, crossing experiments and taxonomy. Opera Bot. 19: 1–138. Hendricks, A. J. 1957. A revision of the genus *Alisma* (Dill.) L. Amer. Midl. Naturalist 58: 470–493. Rubtzoff, P. 1964. Notes on the genus *Alisma*. Leafl. W. Bot. 10: 90–95. Voss, E. G. 1958. Confusion in *Alisma*. Taxon 7: 130–133.

1. Style longer than or equal to ovary; anthers ellipsoid.
    2. Petals with obtuse apex, purplish white to purplish pink; style ± straight; leaf blade
        ovate to broadly lanceolate; Alaska. . . . . 4. *Alisma plantago-aquatica*
    2. Petals with acute apex, purplish pink; style curved proximally; leaf blade narrowly
        elliptic to lanceolate; California, Oregon. . . . . 5. *Alisma lanceolatum*
1. Style shorter than or equal to ovary; anthers ovoid to rounded.
    3. Achenes with 2 abaxial grooves, 1 abaxial ridge; leaves submersed, ribbonlike, or if
        emersed, blades linear-lanceolate to narrowly elliptic. 1. *Alisma gramineum*
    3. Achenes with 1 abaxial groove; leaves emersed, rarely submersed or floating, blades
        ovate to elliptic.
        4. Fruiting heads 2–4 mm diam.; achenes 1.5–2.2 mm; petals 1–3 mm. . . . . 2. *Alisma subcordatum*
        4. Fruiting heads 4.1–7 mm diam.; achenes 2.1–3 mm; petals 3.5–6 mm. . . . . 3. *Alisma triviale*

1. **Alisma gramineum** Lejeune, Fl. Spa 1: 175. 1811
    • Grass-leaved water-plantain, alisma graminoïde

Alisma geyeri Torrey; *A. gramineum* var. *angustissimum* A. J. Hendricks; *A. gramineum* var. *wahlenbergii* Raymond & Kucyniak

Herbs, 0.5–30(–50) cm. **Leaves** submersed, floating, or emersed; submersed sessile, ribbonlike; floating linear, 15–100 × 0.2–2 (–3) cm, blade present or absent; emersed petiolate, rarely sessile, blade linear-lanceolate, lanceolate, or narrowly elliptic, 2–6 × 0.4–1.5 cm. **Inflorescences** to 50 cm. **Flowers** cleistogamous in submersed plants, chasmogamous in terrestrial; sepals 1.5–3 mm; petals faint purplish white, 2–4 mm, margin entire to slightly erose, apex obtuse; anthers ovoid, 0.3–0.6 mm; style ± coiled, 0.4–0.5 mm, ½ length ovary. **Fruiting heads** 3–6 mm diam.; achenes orbicular to orbicular-cuneate, 2–2.7 mm, with 2 shallow, abaxial grooves lateral to 1 abaxial ridge, beak erect. $2n = 14$.

Flowering summer–fall. Shallow fresh or brackish water or muddy shores; 30–1800 m; Alta., B.C., Man., Ont., Que., Sask.; Ariz., Calif., Colo., Idaho, Minn., Mont., Nebr., Nev., N.Y., N.Dak., Oreg., S.Dak., Utah, Vt., Wash., Wyo.; Eurasia; n Africa.

2. **Alisma subcordatum** Rafinesque, Med. Repos., hexade 2, 5: 362. 1808 • Southern water-plantain, alisma subcordé F

Alisma parviflorum Pursh; *A. plantago-aquatica* Linnaeus var. *parviflorum* (Pursh) Torrey

Herbs, to 6 dm. **Leaves** emersed, rarely floating, petiolate; blade ovate to elliptic, to 15 × 10 cm. **Inflorescences** to 1 m. **Flowers** chasmogamous; sepals 1.5–2.5 (–6) mm; petals white, 1–3 mm, margins slightly erose, apex mostly acuminate; anthers rounded, 0.4–0.6 mm; style ± curved, 0.2–0.4 mm, ¼ length ovary. **Fruiting heads** 2–4 mm diam.; achenes obliquely ovoid, 1.5–2.2 mm, abaxially rounded, with 1 abaxial groove; beak ascending. $2n = 14$.

Flowering spring–fall. Shallow ponds, stream margins, marshes, and ditches; 50–1500 m; N.B., Ont., Que.; Ala., Ark., Colo., Conn., Del., D.C., Ga., Ill., Ind., Iowa, Kans., Ky., La., Maine, Md., Mass., Mich., Minn., Miss., Mo., Nebr., N.H., N.J., N.Y., N.C., N.Dak., Ohio, Okla., Pa., R.I., S.C., S.Dak., Tenn., Tex., Vt., Va., W.Va., Wis.; Mexico.

3. **Alisma triviale** Pursh, Fl. Amer. Sept. 1: 252. 1814
    • Northern water-plantain, alisma commun F

Alisma plantago-aquatica Linnaeus var. *americanum* Schultes & Schultes f.

Herbs, to 1 m. **Leaves** emersed, petiolate; blade linear-lanceolate to broadly elliptic or oval, to 35 × 3–12 cm. **Inflorescences** to 1 m. **Flowers** chasmogamous; sepals 3–6 mm; petals white, 3.5–6 mm, margins ± erose, apex obtuse; anthers ovoid, 0.6–1 mm; style ± curved, 0.4–0.6 mm, equal to ovary length. **Fruiting heads** 4.1–7 mm diam.; achenes ovoid, 2.1–3 mm, abaxial keel broadly rounded with 1 median abaxial groove, beak erect or nearly erect. $2n = 28$.

Flowering spring–fall. Shallow muddy ponds, stream margins, marshes, and ditches; 0–2000 m; Alta., B.C., Man., N.B., Nfld. and Labr., N.W.T., N.S., Ont., P.E.I., Que., Sask.; Alaska, Ariz., Calif., Colo., Conn., Idaho, Ill., Ind., Iowa, Kans., Maine, Mass., Mich., Minn., Mont., Nebr., Nev., N.H., N.Mex., N.Y., N.Dak., Ohio, Okla., Oreg., Pa., S.Dak., Utah, Vt., Wash., Wis., Wyo.; n Mexico.

I have seen no specimens from West Virginia, but the species is expected to be there.

4. **Alisma plantago-aquatica** Linnaeus, Sp. Pl. 1: 342. 1753 [I]

Herbs, to 1 m. **Leaves** emersed, petiolate; blade linear-lanceolate to broadly elliptic to ovate, to 30 × 1–12 cm. **Inflorescences** to 1 m. **Flowers** chasmogamous; sepals 1.7–3.2 mm; petals purplish white to purplish pink, 3.4–6.4 mm, margins ± erose, apex obtuse; anthers ellipsoid, 0.7–1.4 mm; style ± straight, 0.6–1.5 mm, exceeding ovary length. **Fruiting heads** 4–6.5 mm diam; achenes ovoid, 1.7–3.1 mm, abaxial keels broadly rounded, with 1 median abaxial groove, rarely 2, beak erect or nearly erect. 2n = 14 (Eurasian material).

Flowering and fruiting late summer. Stream margins; 200 m; introduced; Alaska; Eurasia.

The name *Alisma plantago-aquatica* has been used in a variety of North American floras. We are following, however, the treatment of I. Björkqvist (1968), in which the native distribution of *A. plantago-aquatica* is restricted to Eurasia.

SELECTED REFERENCE Fernald, M. L. 1946. The North American representatives of *Alisma plantago-aquatica*. Rhodora 48: 86–88.

5. **Alisma lanceolatum** Withering, Arr. Brit. Pl. ed. 3, 2: 362. 1796 [I]

Herbs, to 70 cm. **Leaves** emersed, petiolate; blade lanceolate, 6–23 × 1–4.5 cm. **Inflorescences** to 70 cm. **Flowers** chasmogamous; sepals 1.6–3 mm; petals purplish pink, 4.4–6.5 mm, margins ± erose, apex acuminate; anthers ellipsoid, 0.6–1.1 mm; style ± curved, 0.4–0.6 mm, equaling ovary length. **Fruiting heads** 4–8 mm diam; achenes obovoid, 2–2.9 mm, with 1–2 abaxial grooves, beak erect. 2n = 26, 28 (Eurasian material).

Flowering and fruiting summer–fall. Streams and marshes; 0–500 m; introduced; Calif., Oreg.

191. HYDROCHARITACEAE Jussieu

• Tape-grass or Frog-bit Family

Robert R. Haynes

Herbs, annual or perennial, caulescent or without evident stem, glabrous or pubescent, entirely submersed, with both submersed and floating leaves, or with submersed stolons and emergent leaves, in fresh, brackish, or marine waters; turions rarely present. **Stems** rhizomatous, creeping, with abbreviated erect axis at nodes, or erect, leafy, elongate. **Leaves** basal, alternate, opposite, or whorled, sessile or petiolate; stipules sometimes present, forming tubular sheath around stem; blade margins entire or serrate; veins 1–many. **Inflorescences** axillary, terminal, or scapose, 1-flowered or cymose, subtended by spathe; spathe a 2-fid bract or pair of opposite bracts. **Flowers** unisexual, staminate and pistillate on same plants or on different plants, often with rudiments of opposite type, or bisexual, actinomorphic, rarely slightly zygomorphic; perianth epigynous, free, mostly 6-parted, then differentiated into sepals and petals, rarely 3-parted, then petals absent in *Thalassia* and *Halophila*; stamens (0–)2–many in 1 or more whorls (inner often staminodial), epigynous, distinct or ± connate; pollen spheric, in monads or tetrads or in slender chains; ovary 0–1, if present, inferior, 2–6[–16]-carpellate, 1-locular or falsely 6–9-locular; placentation parietal. **Fruits** berrylike. **Seeds** many, fusiform, ellipsoid, ovoid, or spheric; seed coat glabrous, papillose, or echinate.

Genera 17, species ca. 76 (10 genera, 14 species in the flora): nearly worldwide.

Hydrocharitaceae, like other members of the Alismatidae, have one or more (fewer than 20) scales (intravaginal squamules) in the axils of their leaves. These scales (or hairs in some taxa) secrete mucilage and are without any venation. The structures are often referred to as "squamulae intravaginales" or "intravaginal scales" in the literature.

SELECTED REFERENCES Ancibor, E. 1979. Systematic anatomy of vegetative organs of the Hydrocharitaceae. Bot. J. Linn. Soc. 78: 237–266. Catling, P. M. and W. G. Dore. 1982. Status and identification of *Hydrocharis morsus-ranae* and *Limnobium spongia* (Hydrocharitaceae) in northeastern North America. Rhodora 84: 523–545. Cook, C. D. K. 1982. Pollinating mechanisms in the Hydrocharitaceae. In: J.-J. Symoens et al., eds. 1982. Studies on Aquatic Vascular Plants. Brussels. Pp. 1–15. Godfrey, R. K. and J. W. Wooten. 1979. Aquatic and Wetland Plants of Southeastern United States: Monocotyledons. Athens, Ga. Hartog, C. den. 1970. The Sea-grasses of the World. Amsterdam.

1. Plants of marine waters; pollen in moniliform chains.
    2. Leaf-bearing branches arising from rhizomes at distances of several internodes; styles 6–8; fruits echinate, dehiscing into 6–8 irregular valves. . . . . .9. *Thalassia*, p. 35
    2. Leaf-bearing branches arising from rhizome at each node; styles 3–5; fruits smooth or ridged, not echinate, dehiscing by pericarp decay. . . . . .10. *Halophila*, p. 36
1. Plants of fresh or slightly brackish waters; pollen in monads or tetrads.
    3. Stems elongate (more than 3 cm), erect; leaves cauline, whorled.
        4. Leaves with prickles along abaxial surface of midvein; intravaginal squamules fringed with orange-brown hairs. . . . . .8. *Hydrilla*, p. 34
        4. Leaves without prickles along abaxial surface of midvein; intravaginal squamules entire or, if fringed, marginal hairs clear, not orange-brown.
            5. Whorls with 5 or more leaves per node. . . . . .6. *Egeria*, p. 31
            5. Whorls with 2–4(–7) leaves per node or leaves opposite at proximalmost nodes. . . . 7. *Elodea*, p. 32
    3. Stems short (less than 2 cm) or, if elongate, then stoloniferous; leaves basal.
        6. Stems floating on or suspended in water; some leaves with aerenchyma on abaxial surface; peduncles mostly short (less than 5 cm).
            7. Anthers oval; filaments distinct to base or nearly so; seeds minutely tuberculate or muricate; styles 6, 2-fid less than ½ length. .1. *Hydrocharis*, p. 27
            7. Anthers elongate; filaments connate at least ½ length; seeds echinate; styles 3–9, 2-fid nearly to base. . . . . .2. *Limnobium*, p. 28
        6. Stems rooted in substrate; leaves without aerenchyma on abaxial surface; peduncles mostly elongate (more than 5 cm).
            8. Leaves petiolate; spathe winged or ribbed. . . . . .3. *Ottelia*, p. 29
            8. Leaves sessile; spathe not winged or ribbed.
                9. Seeds glabrous; leaves with rows of lacunae on each side of midvein, giving blade 3-zoned appearance of middle, light-colored zone bordered on each side by darker zone; flowers unisexual. . . . . .4. *Vallisneria*, p. 30
                9. Seeds echinate; leaves with continuous intercellular spaces, blade of uniform color from margin to margin; flowers bisexual. . . . . .5. *Blyxa*, p. 31

1. HYDROCHARIS Linnaeus, Sp. Pl. 2: 1036. 1753; Gen. Pl. ed. 5, 458. 1754 • Frog-bit [Greek *hydr-*, water, and *chari*, grace] [I]

Plants perennial, of fresh waters. **Rhizomes** absent; stolons present. **Stems** floating on or suspended in water, rooted or not, unbranched, short. **Leaves** basal, emergent or floating, petiolate; blade cordate to reniform or orbiculate, base reniform or cordate, apex obtuse to almost truncate; midvein without rows of lacunae along sides, blade uniform in color throughout; abaxial surface without prickles, smooth on emergent leaves or with aerenchymous tissue on floating leaves; intravaginal squamules entire. **Inflorescences** 1-flowered or cymose, sessile or short-peduncluate; spathe not winged. **Flowers** unisexual, staminate and pistillate on different plants [on same plants], emersed, pedicellate; petals white to pinkish. **Staminate flowers:** filaments distinct or basally connate, distinct portion longer than connate; anthers oval; pollen in monads. **Pistillate flowers:** ovary 1-locular; styles 6, 2-fid less than ½ length. **Fruits** spheric, smooth to ridged, dehiscing irregularly. **Seeds** ellipsoid, minutely tuberculate or muricate.

Species 3 (1 in the flora): introduced, North America, Eurasia, Africa, Australia.

SELECTED REFERENCES Cook, C. D. K. and R. Lüönd. 1982. A revision of the genus *Hydrocharis* (Hydrocharitaceae). Aquatic Bot. 14: 177–204. Roberts, M. L., R. L. Stuckey, and R. S. Mitchell. 1981. *Hydrocharis morsus-ranae* (Hydrocharitaceae) new to the United States. Rhodora 83: 147–148.

1. Hydrocharis morsus-ranae Linnaeus, Sp. Pl. 2: 1036. 1753 • European frog-bit F I

Herbs, to 20 cm. **Stolon buds** with 1 root. **Leaves** floating or, in dense vegetation, emergent; blade 1.2–6 × 1.3–6.3 cm; primary veins forming 75–90° angle with midvein, broadly curving, aerenchyma confined to midvein region (not margin to margin as in *Limnobium*), individual aerenchyma space (located ca. 1 mm from either side of midvein) 0.1–0.5 mm across its longest axis. **Flowers:** staminate flowers 1–5 in each spathe, pedicel to 4 cm, stamens 9–12 in 4 whorls, filaments basally not obviously connate; pistillate flowers solitary, pedicels to 9 cm, styles 2-fid less than ½ length. **Seeds** 1–1.3 mm. $2n = 28$ (Netherlands).

Flowering spring–fall. Ponds, bays of rivers; 10–50 m; introduced; Ont., Que.; N.Y.; Eurasia.

Hydrocharis morsus-ranae was planted in ponds beside Dow's Lake in the Central Experimental Farm Arboretum at Ottawa in 1932 (P. M. Catling and W. G. Dore 1982). It apparently escaped from these ponds; by 1939 it was found in the Rideau Canal and by 1967 in the St. Lawrence River from Montreal as far as Lake St. Peter. It had spread into Lake Ontario, Lake Erie, and a couple of localities in New York.

2. LIMNOBIUM Richard, Mém. Cl. Sci. Math. Inst. Natl. France 12(2): 66. 1814

• American frog-bit [Greek *limnobios*, living in pools]

Plants perennial, of fresh waters. **Rhizomes** absent; stolons floating on or suspended in water, rooted or not, unbranched, short. **Leaves** basal, emergent or floating, petiolate; blade elliptic to orbiculate, base reniform or cordate, apex obtuse to acuminate; midvein without rows of lacunae along sides, blade uniform in color throughout; abaxial surface without prickles, smooth on emergent leaves or with aerenchymous tissue on floating leaves; intravaginal squamules entire. **Inflorescences** cymose, sessile or short-pedunculate; spathe not winged. **Flowers** unisexual, staminate and pistillate on same plants, emersed, pedicellate; petals greenish white to yellowish. **Staminate flowers:** filaments connate at least ½ their length; anthers elongate; pollen in monads. **Pistillate flowers:** ovary 1-locular, falsely 6–9-locular; styles 3–9, 2-fid nearly to base. **Fruits** ellipsoid to spheric, smooth to ridged, dehiscing irregularly. **Seeds** ellipsoid, echinate, covered with blunt cylindric hairs.

Species 2 (1 in the flora): North America, Central America, South America.

SELECTED REFERENCES   Cook, C. D. K. and K. Urmi-König. 1983. A revision of the genus *Limnobium* including *Hydromystria* (Hydrocharitaceae). Aquatic Bot. 17: 1–27.   Díaz-Miranda, D., D. Philcox, and P. Denny. 1981. Taxonomic clarification of *Limnobium* Rich. and *Hydromystria* G. W. F. Meyer (Hydrocharitaceae). Bot. J. Linn. Soc. 83: 311–323.   Hunziker, A. T. 1981. *Hydromystria laevigata* (Hydrocharitaceae) en el centro de Argentina. Lorentzia 4: 5–8.   Hunziker, A. T. 1982. Observaciones biológicas y taxonómicas sobre *Hydromystria laevigata* (Hydrocharitaceae). Taxon 31: 472–477.   Wilder, G. J. 1974. Symmetry and development of *Limnobium spongia* (Hydrocharitaceae). Amer. J. Bot. 61: 624–642.

1. Limnobium spongia (Bosc) Richard ex Steudel, Nomencl. Bot. ed. 2, 2: 45. 1841 E F

Hydrocharis spongia Bosc, Ann. Mus. Natl. Hist. Nat. 9: 396. 1807

Herbs, to 50 cm. **Roots** branched; stolon buds with 10 or more roots. **Leaves** floating or emersed in dense vegetation and when stranded; blade 1–10 × 0.9–7.8 cm; primary veins forming 30–80° angle with midvein, ascending, aerenchyma extensive, nearly margin to margin, individual aerenchyma space (located ca. 1 mm from either side of midvein) 0.4–1.6 mm across its longest axis. **Flowers:** staminate flowers with 9–12(–18) stamens; pistillate flowers with 3–4 petals, ovary 6–9-carpellate, ovules 200. **Fruits** 4–12 mm diam.

Flowering summer–fall. Floating on slow-moving water of streams, bayous, and lakes or stranded along shore; 0–100 m; Ala., Ark., Del., Fla., Ga., Ill., Ky., La., Md., Miss., Mo., N.Y., N.C., Okla., S.C., Tenn., Tex., Va.

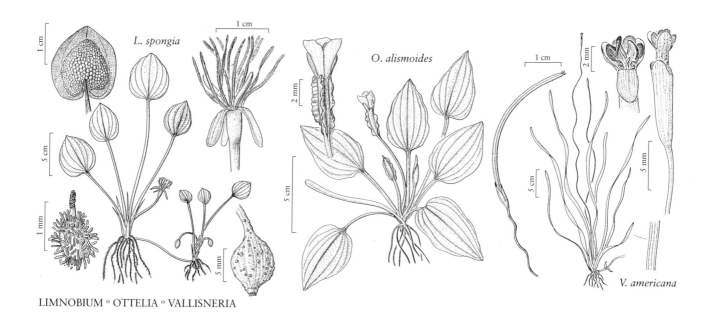

L. spongia

O. alismoides

V. americana

LIMNOBIUM ∘ OTTELIA ∘ VALLISNERIA

No specimens have been seen from New Jersey, but the species is to be expected there.

Limnobium spongia has two leaf forms, often on the same plant. The floating leaves have a thick layer of aerenchyma on the abaxial surface; the emersed leaves lack such tissue. Flowering and fruiting are predominantly on individuals with emersed leaves. Following pollination, the peduncle becomes recurved, forcing the developing fruit below the water surface.

3. OTTELIA Persoon, Syn. Pl. 1: 400. 1805 • [Malay *am ottelambel*, apparently from *otta*, to stick to, in reference to thin leaves that stick to body, and *am bel*, nymphaea] 1

Plants annual or perennial, of fresh waters. **Rhizomes and stolons** absent [present]. **Erect stems** rooted in substrate, unbranched, short. **Leaves** basal, submersed [floating], petiolate; blade lanceolate to reniform [linear], base cuneate to cordate, apex rounded to acute; midvein without rows of lacunae along sides, blade uniform in color throughout; abaxial surface without prickles or aerenchyma, smooth; intravaginal squamules entire. **Inflorescences** scapose, long-pedunculate; spathes winged or ribbed. **Flowers** bisexual [unisexual, staminate and pistillate on different plants], sessile [staminate pedicellate], floating or occasionally opening under water; petals white to pink or light violet, often yellow at base; filaments distinct; pollen in monads; ovary 1-locular; styles 3–9, not 2-fid. **Fruits** oblong, ridged, dehiscing irregularly. **Seeds** fusiform, covered with hairs.

Species 21 (1 in flora): introduced, North America, South America; Asia, Africa, Australia.

SELECTED REFERENCES   Cook, C. D. K., J.-J. Symoens, and K. Urmi-König. 1984. A revision of the genus *Ottelia* (Hydrocharitaceae). 1. Generic considerations. Aquatic Bot. 18: 263–274.   Cook, C. D. K. and K. Urmi-König. 1984. A revision of the genus *Ottelia* (Hydrocharitaceae). 2. The species of Eurasia, Australasia and America. Aquatic Bot. 20: 131–177.   Holmes, W. C. 1978. Range extension for *Ottelia alismoides* (L.) Pers. (Hydrocharitaceae). Castanea 43: 193–194.   Turner, C. E. 1980. *Ottelia alismoides* (L.) Pers. (Hydrocharitaceae)—U.S.A., California. Madroño 27: 177.

1. Ottelia alismoides (Linnaeus) Persoon, Syn. Pl. 1: 400. 1805 · Duck-lettuce [F] [I] [W]

Stratiotes alismoides Linnaeus, Sp. Pl. 1: 535. 1753

Leaves submersed; blade lanceolate to widely ovate, to 17 × to 20 cm, margins entire to crisped. **Inflorescences** 1-flowered; spathes 3–10-winged. **Flowers:** sepals 10–15 × 2–9 mm; stamens 3–12; ovary 1, 3–9-carpellate. **Fruits** 15–40 mm. **Seeds** to 2000.

Flowering spring–summer. Shallow water of bayous, pools, and lakes; 0–50 m; Calif., La.; Asia, Australia.

Ottelia alismoides is a quite distinctive member of our aquatic flora. It is the only species we have with flowers surrounded by a winged or ribbed spathe. The plant is entirely submersed except for the flower, which is projected to the water surface by the elongate peduncle. The flowers may open fully even when completely submersed.

First collected in North America in Calcasieu Parish, Louisiana, in 1939, *Ottelia alismoides* is a conspicuous weed of rice paddies and irrigation systems in the warmer countries of the Eastern Hemisphere. Propagules were probably inadvertently brought into North America with rice seed (D. H. Dike 1969).

SELECTED REFERENCE Dike, D. H. 1969. Contribution to the Biology of *Ottelia alismoides* (Hydrocharitaceae). M.S. thesis. University of Southwestern Louisiana.

4. VALLISNERIA Linnaeus, Sp. Pl. 2: 1015. 1753; Gen. Pl. ed. 5, 446. 1754 · Wild-celery [for Antonio Vallisneri, Italian botanist, 1661–1730]

Plants perennial, of fresh or brackish waters. **Rhizomes and stolons** present. **Erect stems** rooted in substrate, unbranched, short. **Leaves** basal, submersed, sessile; blade linear, base grading into sheath, apex obtuse to apiculate; midvein with 4–5 rows of lacunae on each side, blade appearing 3-zoned with light-colored middle zone bordered on each side by darker zone; abaxial surface without prickles or aerenchyma; intravaginal squamules entire. **Inflorescences** cymose, long-pedunculate; spathe not winged. **Flowers** unisexual, staminate and pistillate on different plants, submersed or floating, sessile (staminate) or pedicellate (pistillate); petals transparent. **Staminate flowers:** filaments distinct, released from spathe and floating to surface; anthers spheric; pollen in monads. **Pistillate flowers** floating; ovary 1-locular; styles 1, not 2-fid. **Fruits** cylindric to ellipsoid, ridged, dehiscing irregularly. **Seeds** ellipsoid, glabrous.

Species 2 (1 in the flora): North America, South America, Eurasia, Africa, Australia.

Vallisneria is generally considered to have 1-flowered pistillate inflorescences. A few populations in southern United States and Central America, however, have cymes with up to 30 flowers. The United States populations include those in Florida, Alabama, and Mississippi.

SELECTED REFERENCES Fernald, M. L. 1918. The diagnostic character of *Vallisneria americana*. Rhodora 20: 108–110. Lowden, R. M. 1982. An approach to the taxonomy of *Vallisneria* L. (Hydrocharitaceae). Aquatic Bot. 13: 269–298. Marie-Victorin, Frère. 1943. Les Vallisnéries Américaines. Contr. Inst. Bot. Univ. Montréal 46: 1–38.

1. Vallisneria americana Michaux, Fl. Bor.-Amer. 2: 220. 1803 · Eel-grass [F]

Vallisneria neotropicalis Victorin

Scapes: staminate scapes 30–50 mm, submersed; pistillate scapes elongate, projecting flowers to surface. **Leaves** 10–110 × 0.3–1.5 cm; leaf blade 3-zoned longitudinally, margins entire to serrate. **Flowers:** staminate flowers 1–1.5 mm wide, stamens 2, filaments basally connate; pistillate flowers solitary, rarely in umbel-like clusters. $2n = 20$.

Flowering summer–fall. Fresh to brackish waters of streams, lakes, rivers, and bays; 0–500 m; Man., N.B., N.S., Ont., Que.; Ala., Conn., Del., D.C., Fla., Ga., Ill., Ind., Iowa, Ky., La., Maine, Md., Mass., Mich., Minn., Miss., Mo., Nev., N.H., N.J., N.Y., N.C., Ohio, Oreg., Pa., R.I., S.C., S.Dak., Tenn., Tex., Vt., Va., Wash., W.Va., Wis.; Mexico; West Indies; Central America; Asia.

Vallisneria americana plus various species of *Sagittaria*, *Sparganium*, and *Blyxa aubertii* form usually sterile basal rosettes of long, linear leaves in shallow water in North America. *Vallisneria* can easily be separated from the others by the following combination of character states: base of leaves nearly flat in cross section, broad band of lacunae along each side of midvein, roots without cross septa, and absence of milky juice. The three other genera have different combinations for these characters.

Vallisneria spiralis Linnaeus has been reported in some of the older literature as being represented in North America. These reports are all based on a mis-application of the name *V. spiralis* and are actually *V. americana*. In warmer waters of southeastern United States are some populations of *Vallisneria* with much larger leaves that have been given the name *V. neotropicalis*. After considerable study of populations in the field, the plants formerly known as *V. neotropicalis* were determined to be just larger individuals of *V. americana* (R. M. Lowden 1982).

5. BLYXA Noroña ex Thouars, Gen. Nov. Madagasc., 4. 1806 • [Greek *blyxo*, to gush forth, spout out, bubble up] I

Plants annual [perennial], of fresh waters. **Rhizomes** absent [present]; stolons present. **Erect stems** rooted in substrate, unbranched, short. **Leaves** basal, submersed, sessile; blade linear, base grading into sheath, apex acute; midvein without rows of lacunae along sides, uniform in color from margin to margin, with continuous intercellular spaces; abaxial surface without prickles or aerenchyma; intravaginal squamules entire. **Inflorescences** 1-flowered, flowers rarely paired [cymose], long-pedunculate; spathe not winged. **Flowers** bisexual [unisexual, staminate and pistillate on different plants], emersed, pedicellate; petals white to reddish; filaments distinct; anthers fusiform; pollen in monads; ovary 1-locular; styles 1, not 2-fid. **Fruits** cylindric, ridged, dehiscing irregularly. **Seeds** ellipsoid or fusiform, echinate [without spines and with or without ridges or wings].

Species 9 (1 in flora): introduced, North America; Asia, Africa, Australia.

SELECTED REFERENCES   Cook, C. D. K. and R. Lüönd. 1983. A revision of the genus *Blyxa* (Hydrocharitaceae). Aquatic Bot. 15: 1–52.   Thieret, J. W., R. R. Haynes, and D. H. Dike. 1969. *Blyxa aubertii* (Hydrocharitaceae) in Louisiana: New to North America. Sida 3: 343–344.

1. **Blyxa aubertii** Richard, Mém. Cl. Sci. Math. Inst. Natl. France 12(2): 19. 1814   F I

Stems 0.5–3 cm. **Leaves** 2.5–60 × 0.2–1.2 cm, margins finely serrate, apex acute. **Inflorescences:** peduncles to 50 cm [sessile]; spathe flattened. **Flowers** mostly solitary, rarely paired, projected to water surface; stamens 3. **Fruits** terete, 20–80 × to 5 mm. **Seeds** 1.2–1.8 mm. $2n = 40$.

Flowering summer–fall. Shallow water of pools and lakes; 0–50 m; introduced; La.; native, Asia; Africa; Australia.

Blyxa aubertii is known in the Western Hemisphere only from a few collections in southern Louisiana.

6. EGERIA Planchon, Ann. Sci. Nat., Bot., sér. 3, 11: 79. 1849 • South American elodea [Latin *egeri*, a nymph, in reference to aquatic habitat] I

Plants perennial, of fresh waters. **Rhizomes and stolons** absent. **Erect stems** rooted in substrate, branched or unbranched, elongate. **Leaves** cauline, in whorls of 5 or more [opposite], submersed, sessile; blade linear, base sloping to stem, apex obtuse; midvein without rows of lacunae along side(s), blade uniform in color throughout; abaxial surface without prickles or aerenchyma; intravaginal squamules entire. **Inflorescences** 1-flowered, sessile; spathes not winged. **Flowers** unisexual, staminate and pistillate on different plants, projected to surface by

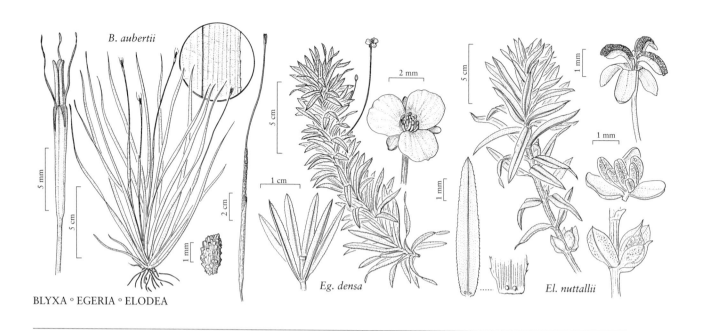

BLYXA ∘ EGERIA ∘ ELODEA

---

slender, elongate floral tube base, pedicellate; petals white. **Staminate flowers:** filaments distinct; anthers linear; pollen in monads. **Pistillate flowers:** ovary 1-locular; styles 3, not 2-fid. **Fruits** ovoid, smooth, dehiscing irregularly. **Seeds** fusiform, mucilaginous.

Species 2 (1 in the flora): introduced, North America; Central America; South America; Eurasia; Africa; Australia.

SELECTED REFERENCES   Cook, C. D. K. and K. Urmi-König. 1984b. A revision of the genus *Egeria* (Hydrocharitaceae). Aquatic Bot. 19: 73–96.   St. John, H. 1961. Monograph of the genus *Egeria* Planchon. Darwiniana 12: 293–307.

1. Egeria densa Planchon, Ann. Sci. Nat., Bot., sér. 3, 11: 80. 1849 • Water-weed [F] [I] [W]

Stems 1–3 mm diam. **Leaves** whorled, recurved, 10–40 × 1.5–4.5 mm. **Staminate spathes** 2–4-flowered, 7.5–12 mm; pedicel to 80 mm. $2n = 46$.

Flowers summer–fall. Shallow waters of lakes and streams; 0–500 m; introduced; Ala., Ark., Calif., Colo., Del., Fla., Ga., Ill., Ind., Kans., Ky., La., Md., Mass., Miss., Mo., N.J., N.Mex., N.Y., N.C, Okla., Oreg., Pa., S.C., Tenn., Tex., Vt., Va., W.Va.; Mexico; West Indies; Central America; native, South America; Europe; Africa; Asia; Australia.

Egeria densa is native to southeastern Brazil and has been widely sold in the aquarium trade, often becoming established in nature. Only staminate plants of *E. densa* have been observed outside its native range. Reproduction, then, occurs entirely by vegetative methods. No differentiated vegetative reproductive structures (turions, bulbils, etc.) are known (C. D. K. Cook and K. Urmi-König 1984b); however, the species is known to live temporarily under ice. The leaves of *Egeria densa*, which are only two cell-layers thick, are much used to demonstrate plant-cell structure and cytoplasmic streaming in introductory botany courses.

7. ELODEA Michaux, Fl. Bor.-Amer. 1: 20. 1803 • Water-weed [Greek *helodes*, marshy]

Plants perennial, of fresh waters. **Rhizomes and stolons** absent. **Erect stems** rooted in substrate, branched or unbranched, elongate. **Leaves** cauline, whorled, 3–7 at each node, or leaves opposite at 4+ proximalmost nodes, submersed, sessile; blade linear to linear-lanceolate, apex acute; midvein without lacunae along side(s), blade uniform in color throughout; abaxial surface without prickles or aerenchyma; intravaginal squamules entire. **Inflorescences** solitary, sessile; spathes not winged. **Flowers** unisexual, staminate and pistillate on different plants, or rarely bisexual, usually projected to surface of water by elongate floral tube base, sessile; petals white. **Staminate flowers:** filaments distinct or 3 inner connate ½ their length; anthers oval; pollen in monads or tetrads. **Pistillate flowers:** ovary 1-locular; styles 3, not 2-fid. **Fruits** ovoid to lance-ellipsoid, smooth, dehiscing irregularly. **Seeds** cylindric to fusiform, glabrous to hirsute.

Species 5 (3 in the flora): North America, Central America, South America, Europe.

An additional name, *Elodea schweinitzii* (G. Planchon) Caspary, has appeared in North American literature. Although it was accepted by some (H. St. John 1965), others stated, "*Elodea schweinitzii* is something of a taxonomist's nightmare" (C. D. K. Cook and K. Urmi-König 1985). The few collections referable to the name are exceedingly variable, so variable in fact that the authors added, "the main character is variability itself." I am following Cook and Urmi-König in not accepting the taxon.

SELECTED REFERENCES   Cook, C. D. K. and K. Urmi-König. 1985. A revision of the genus *Elodea* (Hydrocharitaceae). Aquatic Bot. 21: 111–156.   St. John, H. 1920. The genus *Elodea* in New England. Rhodora 22: 17–29.   St. John, H. 1962. Monograph of the genus *Elodea* (Hydrocharitaceae): Part I. The species found in the Great Plains, the Rocky Mountains, and the Pacific states and provinces of North America. Res. Stud. State Coll. Wash. 30(2): 19–44.   St. John, H. 1965. Monograph of the genus *Elodea*: Part 4 and summary: The species of eastern and central North America. Rhodora 67: 1–35, 155–180.   Shinners, L. H. 1956. *Elodea* correct without being conserved. Rhodora 58: 162.

1. Staminate spathes 4 mm or less; styles usually 2 mm or less; leaves usually less than 1.7 mm wide. . . . . 1. *Elodea nuttallii*
1. Staminate spathes 6 mm or more; styles usually more than 2 mm; leaves usually more than 1.8 mm wide.
   2. Pollen in tetrads; staminate pedicels detaching before or during anthesis; anthers 3 mm or less; leaves mostly in 3s; seeds 4.5–5.7 mm or more. . . . . 2. *Elodea canadensis*
   2. Pollen in monads; staminate pedicels remaining briefly following anthesis; anthers 3–4.5 mm; some leaves in 2s; seeds 4 mm or less. . . . . 3. *Elodea bifoliata*

1. Elodea nuttallii (Planchon) H. St. John, Rhodora 22: 29. 1920   F

Anacharis nuttallii Planchon, Ann. Mag. Nat. Hist., ser. 2, 1: 86. 1848

Leaves mostly in 3s, often recurved, linear to lanceolate, 4–15.5 × 0.9–1.7(–2.4) mm, margins folded. **Inflorescences:** staminate spathes subglobose to ovoid, 2.2–4 mm; peduncles abscissing in bud; pistillate spathes linear, 8.5–15 mm. **Flowers** unisexual; staminate flowers: stamens 9, pedicels briefly remaining attached following anthesis, inner 3 filaments connate proximally, forming column, anthers 1–1.4 mm, pollen in tetrads; pistillate flowers: styles mostly 1.2–2 mm. **Seeds** fusiform, 4–4.6 mm, base with long hairs. $2n = 48$ (Britain).

Flowering summer. Waters, mostly calcareous, of lakes and rivers; 50–1600 m; N.B., Ont., Que.; Ark., Calif., Colo., Conn., Del., Idaho, Ill., Ind., Iowa, Kans., Ky., Maine, Md., Mass., Mich., Minn., Mont., Mo., Nebr., Nev., N.H., N.J., N.Y., N.C., N.Dak., Ohio, Okla., Oreg., Pa., R.I., S.Dak., Tenn., Vt., Va., Wash., W.Va., Wis., Wyo.; Europe; Asia.

I know of no instance in North America where *Elodea nuttallii* or *E. canadensis* is weedy. Both are weedy in Europe, and *E. canadensis* is weedy also in Australia and New Zealand.

2. Elodea canadensis Michaux, Fl. Bor.-Amer. 1: 20. 1803

Elodea brandegeeae H. St. John

Leaves mostly in 3s, spreading to recurved, linear, oblong, to ovate, 5–13 × (1.1–)2–5 mm, flat. **Inflorescences:** staminate spathes cylindric, 8.2–13.5 mm, peduncles abscissing just before or during anthesis; pistillate spathes cylindric, 8.3–17.5 mm. **Flowers** unisexual; staminate flowers: stamens 7–9 (–18), pedicels detaching before or during anthesis, all filaments connate proximally, forming column, anthers 1.7–3 mm, pollen in tetrads; pistillate flowers: styles 2.6–4 mm. **Seeds** fusiform, 4–5.7 mm, basal hairs absent. $2n = 24$ (Britain).

Flowering summer. Waters, mostly calcareous, of lakes and rivers; 0–2000 m; B.C., Man., N.B., N.S., Ont., Que., Sask.; Ala., Ark., Calif., Colo., Conn., Del., Fla., Idaho, Ill., Ind., Iowa, Kans., Ky., Maine, Md., Mass., Mich., Minn., Mo., Mont., Nebr., Nev., N.H., N.J., N.Mex., N.Y., N.C., N.Dak., Ohio, Oreg., Pa., R.I., S.Dak., Tenn., Utah, Vt., Va., Wash., W.Va., Wis., Wyo.; Europe; Asia; Australia.

3. Elodea bifoliata H. St. John, Res. Stud. State Coll. Wash. 30(2): 23. 1962   [E]

Elodea longivaginata H. St. John

Leaves sometimes in 2s, spreading, linear to narrowly elliptic, 4.7–24.8 × (0.8–)1.8–4.3 mm, flat. **Inflorescences:** staminate spathes cylindric, 10.2–42 mm; peduncles abscissing following anthesis; pistillate spathes linear, 9–67 mm. **Flowers** unisexual; staminate flowers: stamens 7–9, pedicels remaining briefly following anthesis, inner 3 filaments connate proximally, forming column, anthers 3–4.5 mm, pollen in monads; pistillate flowers: styles 2.3–3 mm. **Seeds** ellipsoid, 2.8–3 mm, densely covered with long hairs.

Flowering summer–fall. Waters of rivers and reservoirs; 100–2500 m; Alta., B.C., Man., Sask.; Colo., Idaho, Kans., Minn., Mont., N.Mex., N.Dak., Oreg., Utah, Wyo.

8. **HYDRILLA** Richard, Mém. Cl. Sci. Math. Inst. Natl. France 12(2): 9, 61, 73, plate 2a–k. 1814 • [Greek *hydr-*, water, and *-illa*, diminutive]   [I]

Plants perennial, of fresh or brackish waters. **Rhizomes** present, stolons absent. **Erect stems** rooted in substrate, branched or unbranched, elongate. **Leaves** cauline, whorled, 3–8 per node, submersed, sessile; blade linear, rarely slightly elliptic, base tapering to stem, apex acute; midvein without lacunae along side(s), blade uniform in color throughout; abaxial surface with prickles along midvein, without aerenchyma; intravaginal squamules fringed with orange-brown hairs. **Inflorescences** 1-flowered, sessile to subsessile; spathe not winged. **Flowers** unisexual, staminate and pistillate on different plants or on same plants, submersed, sessile; petals whitish to reddish. **Staminate flowers:** filaments distinct, released under water, rising to surface; anthers oval; pollen in monads. **Pistillate flowers:** ovary 1-locular; floral tube long, styles 1, not 2-fid. **Fruits** linear, cylindric, smooth or with simple spiny processes, indehiscent. **Seeds** cylindric, glabrous.

Species 1: introduced, North America; Central America, South America, Eurasia, Africa, Australia.

SELECTED REFERENCES   Blackburn, R. D. and L. W. Weldon. 1970. Control of *Hydrilla verticillata*. Hyacinth Control J. 8: 4–9.   Cook, C. D. K. and R. Lüönd. 1982b. A revision of the genus *Hydrilla* (Hydrocharitaceae). Aquatic Bot. 13: 485–504.   Lakshmanan, C. 1951. A note on the occurrence of turions in *Hydrilla verticillata* Presl. J. Bombay Nat. Hist. Soc. 49: 802–804.   Shireman, J. V. and M. J. Maceina. 1981. The utilization of grass carp, *Ctenopharyngodon idella* Val., for *Hydrilla* control in Lake Baldwin, Florida. J. Fish Biol. 19: 629–636.

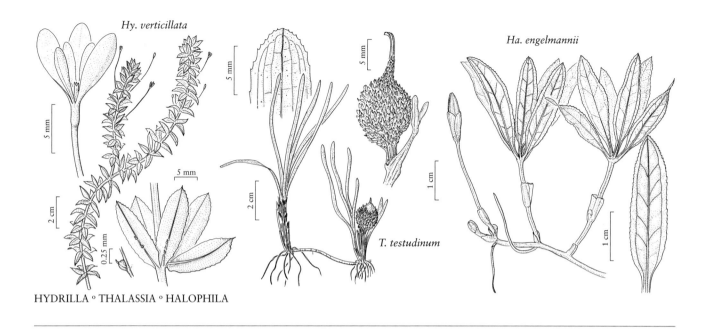

Hy. verticillata

Ha. engelmannii

T. testudinum

HYDRILLA ° THALASSIA ° HALOPHILA

1. Hydrilla verticillata (Linnaeus f.) Royle, Ill. Bot. Himal. Mts. 1: 376. 1839 [F] [I] [W]

Serpicula verticillata Linnaeus f., Suppl. Pl., 416. 1782

Rhizomes and erect stems with turions; subterranean turions cream-brown, appearing as tubers, surface smooth; turions from erect stems olive-green, covered with short, stiff scales. **Leaves** 8–15(–20) × 1.2–4 mm, margins serrulate. **Inflorescences:** spathe of 2 connate bracts. **Flowers** 1 per spathe; staminate pedicels 0.5 mm; pistillate flowers with floral tube 10–50 mm, ovary 1-locular. $2n = 32$.

Flowering summer–fall. Lakes, streams, rivers, bayous; 0–200 m; introduced; Ala., Calif., Conn., Del., D.C., Fla., Ga., La., Md., Miss., S.C., Tenn., Tex., Va.; Mexico; West Indies; Central America; South America (Venezuela); Eurasia; Africa; Australia.

Hydrilla verticillata apparently entered the United States in 1959 at Miami, Florida (D. F. Austin 1978). It was introduced as an aquarium plant, "star-vine" or "oxygen plant" (D. P. Tarver et al. 1978). By 1960 the species was reported as naturalized in Florida (G. E. Allen 1976).

Hydrilla verticillata is widely distributed in the Eastern Hemisphere but it is uncertain as to where it is truly native. It grows in a variety of aquatic habitats ranging from acidic to basic, oligotrophic to eutrophic, fresh to brackish, and from a few centimeters to a meter or more if light penetrates that deeply. Growth and spread often are rapid. Stem fragments become rooted by fine, unbranched adventitious roots and soon produce vegetative reproductive structures from both subterranean and erect stems. Tubers produced on subterranean stems are pale brown; those produced on erect stems are dark olive-green and covered with short, stiff scales. Both types germinate quickly to produce new stems.

9. **THALASSIA** Banks & Solander ex K. D. König, Ann. Bot. (König & Sims) 2: 96. 1805 • Turtle-grass [Greek *thalass*, sea]

Plants perennial, of marine waters. **Rhizomes** present; leaf-bearing branches arising from rhizomes at distances of several internodes; stolons absent. **Erect stems** rooted in substrate, unbranched, short. **Leaves** 2–6, basal, submersed, sessile; blade linear, base tapering to stem; apex obtuse; midvein without lacunae along side(s), blade uniform in color throughout; abaxial surface without prickles; intravaginal scales entire. **Inflorescences** 1-flowered to cymose, pedunculate; spathes not winged. **Flowers** unisexual, staminate and pistillate on different plants, submersed, short-pedicellate to nearly sessile; petals absent. **Staminate flowers:** filaments distinct; anthers linear; pollen embedded in gelatinous matrix, in moniliform chains. **Pistillate flowers:** ovary 1-locular; styles 6–8, not 2-fid. **Fruits** spheric, echinate, dehiscing into 6–8 irregular valves. **Seeds** pyriform; seed coat ephemeral.

Species 2 (1 in flora): North America, Central America, Africa, Asia, Australia.

SELECTED REFERENCES   Grey, W. F. and M. D. Moffler. 1978. Flowering of the seagrass *Thalassia testudinum* (Hydrocharitaceae) in the Tampa Bay, Florida area. Aquatic Bot. 5: 251–259.   Moffler, M. D., M. J. Durako, and W. F. Grey. 1981. Observations on the reproductive ecology of *Thalassia testudinum* (Hydrocharitaceae) in Tampa Bay, Florida. Aquatic Bot. 10: 183–187.   Moore, D. R. 1963. Distribution of the sea grass, *Thalassia*, in the United States. Bull. Mar. Sci. Gulf Caribbean 13: 329–342.   Orpurt, P. R. and L. L. Boral. 1964. The flowers and seeds of *Thalassia testudinum* König. Bull. Mar. Sci. Gulf Caribbean 14: 296–302.

1. Thalassia testudinum Banks ex K. D. König, Ann. Bot. (König & Sims) 2: 96. 1805 [F]

Rhizomes elongate, 3–6 mm thick. **Leaves** 10–60 × 0.4–1.2 cm, margins entire proximally, minutely serrulate near apex; veins 9–15. **Inflorescences:** staminate inflorescences 1–3-flowered, peduncles 3–7 cm, spathes connate on 1 side; pistillate inflorescences 1-flowered, peduncles 3–4 cm, spathes connate on both sides. **Flowers:** staminate flowers: pedicels 1.2–2.5 cm, stamens 9; pistillate flowers nearly sessile, styles 7–8. **Fruits** bright green to yellow-green or red, 1.5–2.5 cm diam., dehiscing in 5–8 valves; beak 4–7 mm.

Flowering spring–summer. Ocean floor consisting of organic matter, rock matter, coral sand, or dead reefs; -10–0 m; Fla., La., Tex.; Mexico; West Indies; Central America; South America (Colombia).

Thalassia testudinum is possibly the most important marine spermatophyte along the coasts of the Caribbean and Gulf of Mexico (C. den Hartog 1970). The species grows from the low-water mark to nearly 10 m depth in very clear water. Establishment occurs on a wide variety of substrates, including organic matter, rocky matter, coral sand, and dead reef-platforms. Once the species is established, the substrate type becomes less important, especially in areas of low current. Dead leaves and rhizomes accumulate among the erect living leaves for considerable periods of time. The beds are important not only in substrate development but also in substrate stabilization. Massive amounts of substrate are lost in areas without turtle-grass colonies during hurricanes, but only minimal loss occurs in turtle-grass beds. Substrate loss is lessened by roots and rhizomes binding the substrate, as well as by the leaves lowering water velocity.

Posidonia oceanica (Linnaeus) Delile was included in the Texas flora (D. S. Correll and M. C. Johnston 1970; D. S. Correll and H. B. Correll 1972) because of specimens washed ashore along the Gulf of Mexico. The specimens were later determined to be *Thalassia testudinum*, based upon comparative growth studies and upon flavonoid chemistry profiles (C. McMillan et al. 1975).

10. HALOPHILA Thouars, Gen. Nov. Madagasc., 2. 1806 • [Greek *halo*, sea, and *philein*, to love]

Plants of marine waters. **Rhizomes** present; leaf-bearing branches arising from rhizome at each node; stolons absent. **Erect stems,** if present, rooted in substrate, unbranched, short; scales 2, midway or higher on stem. **Leaves** 2–8, terminal pairs or pseudowhorls [distichous], submersed, sessile or petiolate; blade linear to ovate, base tapering to stem, apex obtuse; midvein without lacunae along side(s), blade uniform in color throughout; abaxial surface without prickles; intravaginal squamules entire. **Inflorescences** 1-flowered or cymose, sessile; spathes not winged. **Flowers** unisexual, staminate and pistillate on same plants or on different plants, submersed, sessile, nearly sessile (pistillate), or pedicellate (staminate); petals absent. **Staminate flowers:** filaments distinct; anthers linear to fusiform; pollen in moniliform chains. **Pistillate**

flowers: ovary 1-locular; styles 3–5, not 2-fid. **Fruits** ovoid to spheric, smooth or ridged, not echinate, dehiscing by decay of pericarp. **Seeds** spheric or nearly so, echinate to reticulate.

Species 10 (3 in the flora): North America, Central America, South America, s Europe, Asia, Africa, Australia.

Halophila baillonis Ascherson ex Dickie has been listed for the Florida Keys (R. K. Godfrey and J. W. Wooten 1979), but that species is restricted to areas south of the Keys (C. den Hartog 1970; N. J. Eiseman and C. McMillan 1980).

In addition to the three species treated here, three additional ones, *Halophila aschersonii* Ostenfeld, *H. baillonis* Ascherson ex Dickie, and *H. hawaiiana* Doty & Stone, have been credited to North America. These represent two taxa, neither of which approaches North America. *Halophila baillonis* occurs in the Caribbean Sea, and *H. aschersonii* is a synonym of that species. *Halophila hawaiiana* grows in the eastern Pacific Ocean and is known in the United States only from the Hawaiian Islands.

SELECTED REFERENCES Doty, M. S. and B. C. Stone. 1967. Typification for the generic name *Halophila* Thouars. Taxon 16: 414–418. Hartog, C. den. 1959. A key to the species of *Halophila* (Hydrocharitaceae), with descriptions of the American species. Acta Bot. Neerl. 8: 484–489. Sachet, M.-H. and F. R. Fosberg. 1973. Remarks on *Halophila* (Hydrocharitaceae). Taxon 22: 439–443.

1. Erect stems apparent; 2 scales on rhizome at each node, 2 scales on erect stem near middle; leaves in pseudowhorl at apex of erect shoot. . . . . 3. *Halophila engelmannii*
1. Erect stems very short to absent; 2 scales on rhizome at each node; leaves apparently attached to rhizome.
  2. Blade linear-lanceolate, margins entire, glabrous; staminate and pistillate flowers on different plants. . . . . 1. *Halophila johnsonii*
  2. Blade oblong-elliptic, margins serrulate, pubescent; staminate and pistillate flowers on same plants. . . . . 2. *Halophila decipiens*

1. Halophila johnsonii Eiseman, Aquatic Bot. 9: 16, fig. 1. 1980 C E

Rhizomes with 2 scales at each node. **Erect stems** absent. **Leaves** apparently attached to rhizome, 5–25 × 1–4 mm; blade linear-lanceolate, base cuneate, margins entire, glabrous, apex round to round-acute. **Inflorescences:** spathe 2-leaved, enclosing pistillate flowers. **Flowers** unisexual, staminate and pistillate on different plants; staminate flowers unknown; pistillate flowers sessile, ca. 5 mm. **Fruits** ovoid. **Seeds** not seen.

Flowering spring–summer. Sandy substrates of marine waters; -2–0 m; Fla.

SELECTED REFERENCE Eiseman, N. J. and C. McMillan. 1980. A new species of seagrass, *Halophila johnsonii*, from the Atlantic coast of Florida. Aquatic Bot. 9: 15–19.

2. Halophila decipiens Ostenfeld, Bot. Tidsskr. 24: 260, fig. s.n. [p. 261]. 1902

Rhizomes with 2 scales at each node. **Erect stems** absent. **Leaves** apparently attached to rhizome, 10–25 × 3–6 mm; blade oblong-elliptic, base cuneate, margins serrulate, apex obtuse to rounded, pubescent. **Inflorescences:** spathe enclosing both staminate and pistillate flowers. **Flowers** unisexual, staminate and pistillate on same plants; staminate flowers deciduous shortly after anthesis; pistillate flowers nearly sessile, 2 mm. **Fruits** ovoid. **Seeds** ca. 0.2 mm.

Flowering winter–spring. Sandy substrates of marine waters; -30–0 m; Fla.; West Indies; Central America (Panama, Costa Rica); South America (Venezuela, Colombia); Asia; Australia.

3. **Halophila engelmannii** Ascherson in G. B. von Neumayer, Anl. Wiss. Beobacht. Reisen, 368. 1875 [F]

Rhizomes with 2 scales at each node. **Erect stem** apparent, scales 2, near middle. **Leaves** in pseudowhorl at apex of erect stem, 10–30 × 3–6 mm; blade linear-oblong to oblong, base cuneate, margins serrulate, apex obtuse to apiculate. **Inflorescences:** spathes 1-flowered. **Flowers** unisexual, staminate and pistillate on different plants; staminate flowers unknown; pistillate flowers sessile to nearly sessile, ca. 4 mm. **Fruits** beaked. **Seeds** 0.3 mm.

Flowering spring. Sandy to coral substrates of marine waters; -40–0 m; Fla., La., Tex.; West Indies (Cuba, Bahama Islands).

SELECTED REFERENCE   Short, F. T. and M. L. Cambridge. 1984. Male flowers of *Halophila engelmanii*: Description and flowering ecology. Aquatic Bot. 18: 413–416.

192. APONOGETONACEAE J. Agardh

• Aponogeton or Cape-pondweed Family

C. Barre Hellquist

Robert R. Haynes

Herbs, perennial, rhizomatous, caulescent; turions absent. **Leaves** alternate, floating [submersed], petiolate [sessile]; sheath not persisting longer than blade, not leaving circular scar when shed, not ligulate, not auriculate; blade ovate to narrowly lanceolate [linear]; intravaginal squamules (i.e., minute appressed, planate trichomes attached at basal edge) scales, more than 2. **Inflorescences** terminal, spikes, subtended by spathe, pedunculate; peduncle following fertilization not elongating, not spiraling. **Flowers** bisexual [unisexual]; subtending bracts absent; perianth present [absent]; tepals 1[–6]; stamens 6–18[–50] in 2–3[–4] series, not epitepalous; anthers distinct, dehiscing longitudinally; pollen ellipsoid; pistils 2–6[–9], distinct, not stipitate; ovules basal-marginal, anatropous. **Fruits** follicles. **Seeds** 4; embryo straight.

Genera 1, species 52 (1 genus, 1 species in the flora): North America, s Africa, and tropical regions of Eastern Hemisphere.

SELECTED REFERENCES Bruggen, H. W. E. van. 1973. Revision of the genus *Aponogeton* (Aponogetonaceae): VI. The species of Africa. Bull. Jard. Bot. Natl. Belg. 43: 1–2, 193–233. Bruggen, H. W. E. van. 1985. Monograph of the Genus *Aponogeton* (Aponogetonaceae). Stuttgart.

1. APONOGETON Linnaeus f., Suppl. Pl., 32. 1782 • [Greek, from aquatic habitat] ⓘ

Herbs with milky sap. **Leaves** sheathing at base; axillary scales present. **Inflorescences** branched [unbranched], projected above water. **Flowers** bilaterally symmetric, sessile; stamens 6–18 [–50]; filaments distinct; anthers 2-locular; pistils 2–6[–9], sessile. **Fruits:** beak terminal [lateral], curved or straight. **Seeds:** endosperm absent; seed coat single or double. $x = 8$.

Species 47 (1 in the flora): North America; native, Africa; tropical regions, Eastern Hemisphere.

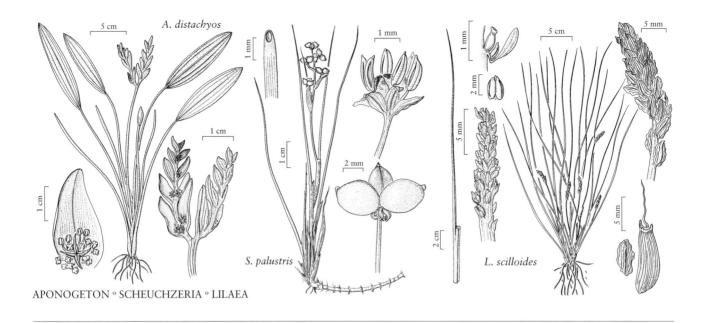

A. distachyos

S. palustris

L. scilloides

APONOGETON ° SCHEUCHZERIA ° LILAEA

1. **Aponogeton distachyos** Linnaeus f., Suppl. Pl., 32, 214. 1782 (as distachyon) • Cape-pondweed, water-hawthorne F I

Aponogeton distachyos var. *lagrangei* André

Leaves floating; petiole to 100 cm; blade ovate, narrowly oval, or narrowly lanceolate, 6–23 × 1.5–7.5 cm, base rounded to attenuate, apex obtuse to acute; veins 7–9. **Inflorescences:** spikes 1 or more, to 4.5 cm; spathe ca. 3 cm. **Flowers** in 2 rows, secund; tepals becoming green, enlarging, 10–15 × 3.5–6 mm, to 30 mm in fruit; veins 13 or more; stamens 3–4.5 mm; filaments expanded basally; anthers blackish purple; pollen yellow; pistils 2.5–3 × 0.7–1 mm; ovules usually 4. **Fruits** to 22 × 6 mm; beak 5 mm.

Flowering late winter–spring (Mar–Apr), fall (Oct–Nov). Quiet ponds; 0–550 m; introduced; Calif.; South America; Europe; Africa; Australia.

Aponogeton distachyos is native to temperate South Africa and has become widely established in Australia. It is very popular in water gardening because it blooms during the cooler periods of the year. The plant does not do well in the heat of the summer. It flowers early in the growing season, goes dormant during the summer, and flowers again late in the season. Where the weather is mild, it will flower all winter. Presently, it is known in North America only from central-coastal and southern California.

193. SCHEUCHZERIACEAE Rudolfi

• Scheuchzeria Family

Mark A. Nienaber

Herbs, perennial, rhizomatous, caulescent; turions absent. **Leaves** alternate, emergent, sessile; sheath with remains often persisting, auriculate; blade linear, nearly cylindric, with conspicuous round pore adaxially on leaf tip; intravaginal squamules hairs, numerous. **Inflorescences** terminal, bracteate racemes, not subtended by spathe; pedicels following fertilization elongating, not spiraling. **Flowers** bisexual; subtending bracts present; perianth present; tepals 6; stamens 6, epitepalous; anthers basifixed, distinct, dehiscing longitudinally, extrorse; pollen globose; pistils 3(–6), distinct to slightly connate at base; ovules basal-marginal, anatropous. **Fruits** follicles. **Seeds** 1–2(–3).

Scheuchzeriaceae have sometimes been included in Juncaginaceae. That *Scheuchzeria* represents a family by itself is accepted by the majority of post-1940 workers (J. W. Thieret 1988).

Genera 1, species 1: colder parts of the Northern Hemisphere.

SELECTED REFERENCES Britton, N. L. 1909. Scheuchzeriaceae. In: N. L. Britton et al., eds. 1905+. North American Flora. . . . 47+ vols. New York. Vol. 17, pp. 41–42.   Cody, W. J. 1975. *Scheuchzeria palustris* L. (Scheuchzeriaceae) in northwestern North America. Canad. Field-Naturalist 89: 69–71.   Fernald, M. L. 1923. The American variety of *Scheuchzeria palustris*. Rhodora 25: 177–179.   Thieret, J. W. 1988. The Juncaginaceae in the southeastern United States. J. Arnold Arbor. 69: 1–23.

1. SCHEUCHZERIA Linnaeus, Sp. Pl. 1: 338. 1753; Gen. Pl. ed. 5, 157. 1754 • [for Johann Jakob Scheuchzer, 1672–1733, Swiss botanist]

Stems erect, unbranched, usually sheathed at base by remains of old leaves. **Leaves** basal and cauline; cauline well separated, bases sheathing, dilated, ligules prominent, membranous; blade erect. **Flowers:** tepals persistent, distinct; ovaries broadly ellipsoid, 1-loculed, inflated at maturity; ovules (1–)2(–3) per pistil; stigmas abaxial, nearly apical. **Follicles** mostly 1–4 per aggregate, spreading.

Species 1: North America, Eurasia.

1. Scheuchzeria palustris Linnaeus, Sp. Pl. 1: 338. 1753

• Scheuchzérie palustre [F]

Scheuchzeria palustris var. *americana* Fernald; *S. palustris* subsp. *americana* (Fernald) Hultén; *S. americana* (Fernald) G. N. Jones

Herbs, glabrous. **Rhizomes** creeping, jointed, freely branching. **Stems** flexuous, zigzag, 1–4 dm. **Leaves** striate; cauline leaves gradually reduced to bracts; sheaths 1.5–10 cm, ligules 2–12 mm; hairs within leaf sheath 0.2–2 cm; blade 2–41 cm × 1–3 mm. **Inflorescences** 3–12-flowered, 3–10 cm; proximal bract foliaceous; distalmost sheaths sometimes bladeless; pedicels spreading, 5–25 mm. **Flowers:** tepals in 2 similar series of 3 each, white to yellow-green, lance-ovate to lanceolate, 1-veined, 2–3 mm, membranous, apex acute; filaments filiform; anthers linear, elongate; pistils 6–7 mm; styles oblong; stigmas papillose. **Follicles** light green to brown, ovoid, 4–10 mm, leathery, dehiscing on curved, adaxial side; beak 0.5–1 mm. **Seeds** brown to black, ovoid, 4–5 mm, smooth, hard; endosperm absent. $2n = 22$.

Flowering in late spring. Sphagnum bogs, marshes, and lake margins; 0–2000 m; Alta., B.C., Man., N.B., Nfld. and Labr., N.W.T., N.S., Ont., Que., Sask., Yukon; Alaska, Calif., Conn., Idaho, Ill., Ind., Iowa, Maine, Mass., Mich., Minn., Mont., N.H., N.J., N.Y., N.Dak., Ohio, Oreg., Pa., R.I., Vt., Wash., W.Va., Wis.; Eurasia.

North American representatives of this species have been regarded as being varietally distinct from Eurasian plants on the basis of follicle and stigma characters (M. L. Fernald 1923). Variability in those characters, in specimens from both hemispheres, vitiates their worth for varietal distinction.

194. JUNCAGINACEAE Richard

• Arrow-grass Family

Robert R. Haynes

C. Barre Hellquist

Herbs, perennial or annual, rhizomatous, evident stems absent; turions absent. **Leaves** basal, emersed, sessile; sheath persisting longer than blade, not leaving circular scar when shed, ligulate, auriculate with scarious lobes; blade linear; intravaginal squamules scales, more than 2. **Inflorescences** terminal or axillary, scapose spikes, spikelike racemes, rarely solitary flowers, without spathe, pedunculate; peduncle following fertilization not elongating, not spiraling. **Flowers** bisexual or unisexual, staminate and pistillate on same plant; subtending bracts absent; perianth present, rarely absent; tepals 1, or 6 in 1–2 series. **Bisexual and staminate flowers:** stamens 1, 4, or 6, epitepalous, when 4 or 6, then in 2–3 series; anthers distinct, dehiscing longitudinally; pollen globose. **Pistillate and bisexual flowers:** pistils 1, 3, or 6, not stipitate, when 3 or 6, coherent or weakly connate; ovules basal, anatropous. **Fruits** nutlets or schizocarps. **Seeds** 1; embryo straight.

Genera 4, species ca. 15 (2 genera, 5 species in the flora): nearly worldwide.

SELECTED REFERENCES   Haynes, R. R. and L. B. Holm-Nielsen. 1985. A generic treatment of Alismatidae in the Neotropics. Acta Amazon. 15(suppl.): 153–193.   Reveal, J. L. 1977. Juncaginaceae. In: A. Cronquist et al. 1972+. Intermountain Flora. 5+ vols. New York and London. Vol. 6, pp. 18–22.   Thieret, J. W. 1988. The Juncaginaceae in the southeastern United States. J. Arnold Arbor. 69: 1–23.   Thorne, R. F. 1993. Juncaginaceae. In: J. C. Hickman, ed. 1993. The Jepson Manual. Higher Plants of California. Berkeley, Los Angeles, and London. Pp. 1166–1168.

1. Inflorescences both scapose spikelike racemes with bisexual and unisexual flowers and sessile with pistillate flowers; pistils 1. . . . . .1. *Lilaea*, p. 44
1. Inflorescences all scapose spikelike racemes with bisexual flowers; fertile pistils 3 or 6. . . . . .2. *Triglochin*, p. 44

1. LILAEA Bonpland in A. von Humboldt and A. J. Bonpland, Pl. Aequinoct. 1: 221. 1808 • Flowering-quillwort [for French botanist Alire Raffeneau-Delile, 1778–1850]

Herbs, annual. **Roots** without terminal tubers. **Rhizomes** slender, short. **Leaves** erect, terete, with aerenchyma tissue; sheath with ligule free. **Inflorescences** both scapose and sessile; scapes shorter than leaves, terminated by spikes; spikes with pistillate flowers near base, bisexual medially, staminate proximally; sessile inflorescences 2 or more, with solitary pistillate flower only. **Flowers** bisexual and unisexual, to 5 types, sessile; tepals 0–1, green, linear; stamens 0–1; anthers sessile; pistils 1; ovules 1. **Bisexual and staminate flowers:** tepal present. **Pistillate flowers:** tepal present, style 0.5–2 mm or tepal absent, style 0.2–30 cm. **Fruits** nutlets, angular, beaked.

Species 1: North America, Central America, South America.

1. Lilaea scilloides (Poiret) Hauman, Publ. Inst. Invest. Geogr. Fac. Filos. Letras Univ. Buenos Aires, A 10: 26. 1925 [F]

Phalangium scilloides Poiret in J. Lamarck et al., Encycl. 5: 251. 1804; *Lilaea subulata* Humboldt & Bonpland

Herbs, to 30 cm. **Leaves** to 35 cm; sheath 2–5 cm; blade 15–30 cm × 0.9–3.1 mm. **Inflorescences** axillary, scapose, 0.3–4 cm × 1.5–8 mm; scape 8–10 cm × 0.5–1.5 mm. **Flowers** sessile; perianth 1–1.5 × 0.5 mm; anther 0.5 × 0.4 mm; ovary 1–2 mm; style 0.5 mm (bisexual), 30 cm (pistillate). **Fruits** 2–5 mm. $2n = 12$.

Flowering summer. Shallow water and adjacent mud flats; 0–1700 m; Alta., B.C., Sask.; Calif., Mont., Nev., Oreg., Wash.; Mexico; South America (Argentina, Chile, Ecuador, Peru); introduced, se Australia.

The taxonomic position of *Lilaea* is uncertain. The genus often is treated as a monotypic family (Lilaeaceae) that is closely allied to the Juncaginaceae. *Lilaea*, for example, possesses lactifers, unknown in other Juncaginaceae. The lactifers, plus uncertainty regarding the interpretation of floral morphology in *Lilaea*, prompted a decision to retain it in the Lilaeaceae pending further study (P. B. Tomlinson 1982). Embryologic, cytologic, and palynologic evidence, however, indicate a close relationship between *Lilaea* and other genera of the Juncaginaceae. Despite the presence of lactifers, *Lilaea* has been considered similar enough to warrant the inclusion of the genus in the Juncaginaceae (A. Cronquist 1981). Molecular evidence also corroborates the close relationship between *Lilaea* and *Triglochin* (D. H. Les and R. R. Haynes 1995).

2. TRIGLOCHIN Linnaeus, Sp. Pl. 1: 338. 1753; Gen. Pl. ed. 5, 157. 1754 • Arrowgrass, troscart [Greek *treis*, three, and *glochis*, a point]

Herbs, perennial. **Roots** occasionally with tubers. **Rhizomes** stout. **Leaves** erect, terete; sheath with ligule apically entire or 2-lobed. **Inflorescences** spikelike racemes, scapose; scapes shorter than to longer than leaves. **Flowers** bisexual, of 1 type, short-pedicellate; tepals 6, in 2 series, distinct, yellow-green, conchiform; stamens 4 or 6; anthers nearly sessile; pistils 6, 3 fertile, 3 sterile or 6 fertile, separating when mature; ovules 1 per locule; styles absent. **Fruits** schizocarps, globose to linear in fruit; mericarps 3 or 6. $x = 6$.

Species ca. 12 (4 in the flora): nearly all temperate areas, reaching tropics, especially in higher elevations.

The fruit type of *Triglochin* has been variously interpreted. We follow R. M. T. Dahlgren et al. (1985), who considered the fruits to be schizocarps with 1-seeded mericarps.

Although Linnaeus, in his original publication of the name, treated *Triglochin* as neuter, botanical tradition in North America and elsewhere has generally assigned feminine gender (International Code of Botanical Nomenclature, Art. 62.1); for this reason and because the

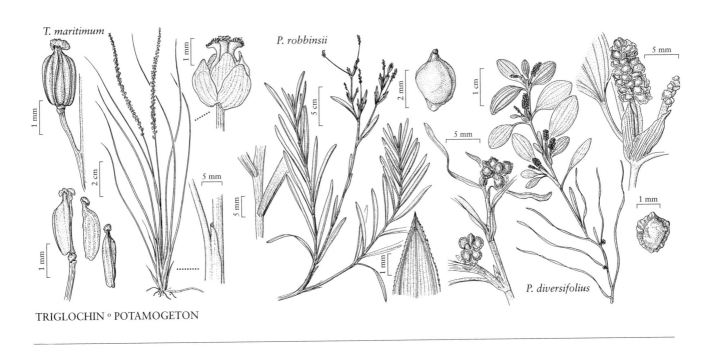

TRIGLOCHIN ∘ POTAMOGETON

Greek word *glochin* (γλωχιν) is feminine (ICBN, Art. 62.2), the feminine gender is the more correct under the Code and is adopted in the Flora. The use of the neuter gender in some recent works appears to reflect a pre-1987 wording of the Code that was held to require adoption of the gender assigned by the original author.

SELECTED REFERENCE   Looman, J. 1976. Biological flora of the Canadian prairie provinces IV. *Triglochin* L., the genus. Canad. J. Pl. Sci. 56: 725–732.

1. Fertile pistils 6; fruiting receptacle without wings; ligule apically 2-lobed.
  2. Leaves typically shorter than scape, slender to thickish, erect from sheath; racemes 6–45 cm. . . . . 2. *Triglochin maritima*
  2. Leaves usually equaling scape, very slender, curving outward from sheath at 30–50° angle; racemes 2–7 cm. . . . . 1. *Triglochin gaspensis*
1. Fertile pistils 3; fruiting receptacle with wings; ligule apically entire.
  3. Schizocarps linear; mericarps weakly ridged, more than 5 mm. . . . . 3. *Triglochin palustris*
  3. Schizocarps globose; mericarps strongly 3-keeled, less than 2 mm. . . . . 4. *Triglochin striata*

1. **Triglochin gaspensis** Lieth & D. Löve, Canad. J. Bot. 39: 1271, figs. 1, 2 Aa, 3, 4 Aa, 5 a, e, f, 6. 1961 (as gaspense) • Troscart de la Gaspésie [C] [E]

Plants with strands of old leaves at base, (5–)10–15(–20) cm. **Leaves** curving outward from sheath at 30–50° angle, slender, almost terete, equaling or slightly longer than scapes, (5–)10–15 (–20) cm; sheath 12.5–21 × 1.8–3 mm, ligule often hoodlike, apically 2-lobed; blade 0.5–1 mm wide, apex acute. **Inflorescences:** scapes often purple near base, 0.5–1(–1.5) mm thick; racemes (2–)3–5(–7) cm; pedicels 1.4–4 × 0.1 mm. **Flowers:** tepals somewhat rounded, 1.2–1.5 × 1.1–1.8 mm, apex obtuse; pistils (3–)6–(9–12), 6 fertile. **Fruits:** fruiting receptacle without wings; schizocarps linear, 3–4 × 1 mm; mericarps linear, weakly ridged, 3–4 × 1 mm, beak recurved, 0.9–1 mm. $2n = 96$.

Flowering summer (Jul–Aug). Tidal saltwater marshes, usually submerged daily; of conservation concern; 0 m; N.B., Nfld. and Labr. (Nfld.), N.S., P.E.I., Que.; Maine.

Plants of *Triglochin gaspensis* tend to form lawnlike patches in contrast to the clumped habit of other northern species of the genus.

SELECTED REFERENCE   Löve, D. and H. Lieth. 1961. *Triglochin gaspense*, a new species of arrow grass. Canad. J. Bot. 39: 1261–1272.

2. **Triglochin maritima** Linnaeus, Sp. Pl. 1: 339. 1753 (as maritimum) • Arrow-grass, troscart maritime F W

Triglochin concinna J. B. Davy; *T. concinna* var. *debilis* (M. E. Jones) J. T. Howell; *T. debilis* (M. E. Jones) Á. Löve & D. Löve; *T. elata* Nuttall

Plants with fibrous strands of old leaves at base, 3.5–61.5 cm. **Leaves** erect from sheath, mostly shorter than scape, 2.2–11.5 cm; sheath 0.7–2.5 cm × 1–1.8 mm, ligule occasionally hoodlike, apically 2-lobed; blade 0.9–1.4 mm wide, apex obtuse to round. **Inflorescences:** scape often purple near base, mostly exceeding leaves, 1–16.5 cm × 0.5–1 mm; racemes 6–45 cm × 1.5–7 mm; pedicel 1–4 × 0.2–0.3 mm. **Flowers:** tepals elliptic, 1.3–1.7 × 0.6–1.4 mm, apex acute; pistils 6, all fertile. **Fruits:** fruiting receptacle without wings; schizocarps linear to near globose, 2–4.5 × 1.5–2 mm; mericarps linear to linear-obovate, weakly ridged, 1.5–3.5 × 0.7–1 mm, beak erect to recurved, 0.2 mm. 2*n* = 12, 24, 36, 48, 120.

Flowering summer–fall. Coastal and mountain marsh areas and moist alkaline meadows; 0–4000 m; Alta., B.C., Man., N.B., Nfld. and Labr., N.W.T., N.S., Nunavut, Ont., P.E.I., Que., Sask., Yukon; Alaska, Ariz., Calif., Colo., Conn., Idaho, Ill., Ind., Iowa, Kans., Maine, Mass., Mich., Minn., Mont., Nebr., Nev., N.H., N.J., N.Mex., N.Y., N.Dak., Ohio, Oreg., Pa., R.I., S.Dak., Utah, Vt., Wash., Wis., Wyo.; Mexico; South America; n Europe; n Asia.

This taxon has been separated into *Triglochin concinna* and *T. maritima* based upon the lobing of the ligule and the smaller size of the plants of the former (e.g., J. L. Reveal 1977; R. F. Thorne 1993). On a local basis such a separation seems warranted. Examination of the *T. maritima* complex throughout the Americas, however, reveals continuous variation from small, widely spaced plants with 2-lobed ligules to large, tufted plants with unlobed ligules, including plants with all combinations of those characters.

Triglochin maritima is important in livestock management because it is quite toxic: it is a cyanide producer.

SELECTED REFERENCE   Löve, Á. and D. Löve. 1958. Biosystematics of *Triglochin maritimum* Agg. Naturaliste Canad. 85: 156–165.

3. **Triglochin palustris** Linnaeus, Sp. Pl. 1: 338. 1753 (as palustre) • Troscart des marais

Plants with fibrous strands of leaves at base, 9–42.5 cm. **Leaves** erect from sheath, shorter than scapes, 6–24.5 cm; sheath 3.5–5 cm × 1.5–5 mm, ligule not hoodlike, unlobed; blade 0.8–2.9 mm wide, apex acute. **Inflorescences:** scape often purple near base, mostly exceeding leaves, 5.5–27.2 cm × 1–2.1 mm; racemes 5.1–21.4 cm × 2–5 mm; pedicel 0.4–4.5 × 0.1–0.5 mm. **Flowers:** tepals elliptic, 1.1–1.6 × 0.7–0.9 mm, apex round; pistils 6, 3 fertile, 3 sterile. **Fruits:** fruiting receptacles with wings; schizocarps linear, 7–8.3 × 0.8–1.2 mm; mericarps linear, weakly ridged abaxially, 6.5–8.5 × 0.5–1.5 mm, beak erect, 0.3 mm. 2*n* = 24.

Flowering summer and early fall. Coastal and mountain marsh areas and moist alkaline meadows; 0–3700 m; Greenland; St. Pierre and Miquelon; Alta., B.C., Man., N.B., Nfld. and Labr., N.W.T., N.S., Nunavut, Ont., P.E.I., Que., Sask., Yukon; Alaska, Ariz., Calif., Colo., Idaho, Ill., Ind., Iowa, Maine, Mich., Minn., Mont., Nebr., Nev., N.H., N.Mex., N.Y., N.Dak., Ohio, Oreg., Pa., R.I., S.Dak., Utah, Wash., Wis., Wyo.; Mexico; South America; Eurasia.

4. **Triglochin striata** Ruiz & Pavón, Fl. Peruv. 3: 72. 1802 (as striatum)

Plants with fibrous strands of old leaves at base, 6–35 cm. **Leaves** erect from sheath, mostly longer than scapes, 4–35 cm; sheath 1.1–7.5 cm × 0.7–1 mm, ligule not hoodlike, unlobed; blade 0.2–5 mm wide, apex round-acute. **Inflorescences:** scape green to brown, mostly exceeded by leaves, 4.5–21.5 cm × 0.3–1.6 mm; racemes 0.6–20.3 × 0.4–1.3 cm; pedicels 0.4–2.1 × 0.1–0.3 mm. **Flowers:** tepals oval to elliptic, 0.6–1 × 0.8–0.9 mm, apex obtuse; pistils 6, 3 fertile, 3 sterile. **Fruits:** fruiting receptacle with wings; schizocarps globose to broader than long, 1–2 × 1.5–2.3 mm; mericarps obovate, strongly 3-keeled, 1–1.5 × 0.5–0.9 mm, beak reflexed, ca. 0.2 mm. Chromosome number not available.

Flowering summer–fall. Coastal alkaline marshes; 0–10 m; Ala., Calif., Fla., Ga., La., Miss., N.C., Oreg., S.C., Va.; Mexico; West Indies (Bahamas, Greater Antilles); South America (Argentina, Bolivia, Brazil, Chile, Peru).

195. POTAMOGETONACEAE Dumortier

• Pondweed Family

Robert R. Haynes

C. Barre Hellquist

Herbs, perennial or rarely annual, rhizomatous or not rhizomatous, caulescent; turions absent or present. **Leaves** alternate or nearly opposite, submersed or both submersed and floating, sessile or petiolate; sheath not persisting longer than blade, not leaving circular scar when shed, ligulate, not auriculate, or rarely auriculate; intravaginal squamules scales, more than 2. **Inflorescences** terminal or axillary, spikes, capitate spikes, or panicles of spikes, not subtended by spathe, pedunculate; peduncle not elongating, not spiraling following fertilization. **Flowers** bisexual; subtending bracts absent; tepals 4 in 1 series; stamens [2 or] 4, epipetalous, in 1 series; anthers distinct, dehiscing vertically; pollen spheric; pistils 1 or 4, mostly not stipitate, rarely short-stipitate; ovules marginal, orthotropous. **Fruits** drupaceous. **Seeds** 1; embryo curved.

Genera 3, species ca. 90 (2 genera, 37 species in the flora): nearly worldwide.

The family has historically been considered to consist of two genera, *Potamogeton* and *Groenlandia*. Recent molecular evidence (D. H. Les, unpublished), combined with existing morphologic evidence, indicates that *Potamogeton* in the broad sense actually represents two separate lineages. We recognize those lineages at the generic level, *Potamogeton* in the strict sense and *Stuckenia*. Consequently, we accept three genera in the family, *Potamogeton*, *Stuckenia*, and *Groenlandia*.

Members of Potamogetonaceae have been variously combined with members of Zosteraceae, Cymodoceaceae, Zannichelliaceae, and Najadaceae to compose Zosteraceae, Najadaceae, or Potamogetonaceae. Potamogetonaceae, as here interpreted, are separated from the other families by their bisexual flowers, the absence of spathelike bracts, and in some species, the presence of turions.

Aquatic vascular plants are known for their phenotypic plasticity. Plasticity may result from the varied environmental conditions in which the populations grow or from morphologic changes in individuals of a population during the growing season (R. R. Haynes 1975). Individuals in fruit have relatively consistent morphology within a species. Regardless of phenotypic plasticity, collections of Potamogetonaceae (and aquatic vascular plants in general) are often taken with little attention to the presence or absence of reproductive structures.

Reproductive features are most important in separating species of *Potamogeton* (R. R. Haynes 1978), and we include the entire family here. The keys may not always utilize reproductive features, but they are based on fruiting individuals. We strongly recommend that no one collect specimens of Potamogetonaceae that are lacking reproductive structures.

Leaves of Potamogetonaceae are stipulate. The stipules form a tubular sheath (stipular sheath) around the stem, free from or adnate to the base of the blade. In some species the leaf and sheath of submersed leaves are adnate for part of their length, and the leaf appears to have a sheathing base with an adaxial ligule at the junction of sheath and blade or petiole.

Fruits of Potamogetonaceae are drupaceous. The fruits do have endocarps but do not have fleshy mesocarps. Mesocarps exist but never become fleshy. Consequently, the fruits are not true drupes; they are drupaceous.

Many species of Potamogetonaceae undergo extensive vegetative reproduction either by turions or stem fragmentation. Turions are excellent modes of vegetative reproduction. The structures are produced at the stem tips and eventually fall to the substrate, either by a portion of the stem breaking off or by the stem itself falling to the substrate. The turions survive an unfavorable season, germinate, and grow into new plants during the next growing season. Because the unfavorable season is usually winter in North America, turions have been called "winter buds." At least one species, *Potamogeton crispus*, produces turions in early summer, and the turions survive the unfavorable season (summer, in this instance), germinating in the fall. The plant then survives the winter as a young individual, only a few centimeters long, even under ice, and begins growth as the water warms in the following spring. "Winter bud" is certainly not the correct term for *P. crispus*. The term "turions" designates all such structures, regardless of the unfavorable season.

SELECTED REFERENCES Hagström, J. O. 1916. Critical researches on the potamogetons. Kungl. Svenska Vetenskapsakad. Handl., n. s. 55(5): 1–281. Haynes, R. R. 1978. The Potamogetonaceae in the southeastern United States. J. Arnold Arbor. 59: 170–191. Les, D. H. 1983. Taxonomic implications of aneuploidy and polyploidy in *Potamogeton* (Potamogetonaceae). Rhodora 85: 301–323. Mason, H. L. 1957. A Flora of the Marshes of California. Berkeley. Reveal, J. L. 1977b. Potamogetonaceae. In: A. Cronquist et al. 1972+. Intermountain Flora. Vascular Plants of the Intermountain West, U.S.A. 5+ vols. New York and London. Vol. 6, pp. 24–42. Thorne, R. F. 1993b. Potamogetonaceae. In: J. C. Hickman, ed. 1993. The Jepson Manual. Higher Plants of California. Berkeley, Los Angeles, and London. Pp. 1304–1308.

1. Stipular sheaths of submersed leaves free from base of leaf blade, or if adnate, then adnate portion less than ½ length of stipule; leaves both submersed and floating or all submersed, submersed blades translucent, not channeled, flattened; peduncle stiff, if long enough then projecting inflorescence above surface of water. . . . . 1. *Potamogeton*, p. 48
1. Stipular sheaths of submersed leaves adnate to base of leaf blade for 2/3 or more length of stipule; leaves all submersed, blades opaque, channeled, turgid; peduncle flexible, not projecting inflorescence above surface of water. . . . . 2. *Stuckenia*, p. 70

1. POTAMOGETON Linnaeus, Sp. Pl. 1: 126. 1753; Gen. Pl. ed. 5, 61. 1754

• Pondweed, potamot [Greek *potamos*, river, and *geiton*, neighbor]

Herbs: rhizomes present or absent; tubers absent; turions present or absent. **Stems** terete or compressed, nodes occasionally with oil glands; turions with extremely shortened internodes, divided into outer and inner leaves; outer leaves 1–5 per side, similar to vegetative leaves or occasionally corrugate near base; inner leaves 1–10, rolled into fusiform structure, unmodified, or shortened and oriented at 90° angles to outer leaves. **Leaves** submersed or both submersed and floating, alternate to nearly opposite; stipules connate or not, if not, then convolute, tubular, sheathing stem and young inflorescences. **Submersed leaves** sessile or petiolate; stipules either free from or adnate to base of leaf blade for less than ½ length of stipule, if adnate, then extending past adnation as free ligule; blade translucent, linear to orbiculate, not channeled, flattened, base acute to perfoliate, margins entire or serrate, rarely crispate, apex subulate to obtuse; veins 1–35. **Floating leaves** petiolate, rarely nearly sessile; stipules free from base of leaf blade; blade elliptic to ovate, leathery, base cuneate to rounded or cordate, margins entire, apex acute to obtuse; veins 1–51. **Inflorescences** spikes or panicles of spikes, submersed or emersed, capitate or cylindric; peduncles stiff, if long enough then projecting inflorescence above surface of water. **Flowers:** pistils 1 or 4. **Fruits** abaxially rounded or keeled, flattened to turgid, beaked; embryo coiled 1 or more times. $x = 13$ or 14.

Species ca. 100 (33 in the flora): nearly worldwide.

Potamogeton is one of the most important genera in the aquatic environment, especially as food or habitat for aquatic animals (R. R. Haynes 1975). A few species become slightly weedy, but not significantly so. Plants of *Potamogeton* are important in stabilizing substrates and removing particulate matter from the water column.

The genus has been divided into several sections and numerous subsections (predominantly by J. O. Hagström 1916; see also R. R. Haynes 1975, 1985 for in-depth coverage of three subsections). After studying thousands of specimens over five continents, we believe that recognition of the many infrageneric categories is unwarranted. Consequently, we are not including infrageneric classification here.

Hybridization is common among members of the genus (J. O. Hagström 1916). Numerous hybrids were proposed, using intermediate stem anatomy as evidence of hybrid origin. We list all the hybrids that Hagström proposed for species that occur in North America. An additional 26 hybrids have been recognized for the British Isles (C. D. Preston 1995).

Vegetative and reproductive morphology varies considerably in the genus. Two types of stems occur, rhizomes and erect stems. Some species have both, others have only erect stems. Two types of leaves exist, submersed and floating. Floating leaves have well-developed epidermis abaxially and adaxially, and well-developed cuticle at least adaxially. Floating leaves may be similar in shape to that of the submersed, or they may differ considerably. Submersed leaves have no cuticle and do not have well-developed epidermis. All species of *Potamogeton* have submersed leaves; some also have floating leaves. Occasionally, individuals of floating-leaved species lose their submersed leaves because of decay or wave action. Leaves of *Potamogeton* may be sessile or petiolate and are divided into at least blade and stipule. The stipule may be adnate to the blade for 1/3 or less the length of the stipule. Venation in the stipule is parallel, and veins may appear coarse as distinct ridges on the stipule (fibrous), or they may be much less obvious, even difficult to observe (delicate). Stipular tissue between veins of fibrous stipules decays, leaving strands of fibers, whereas veins and the tissue between them decay in delicate stipules.

Many species have oil glands on the stem at the node of submersed leaves. These glands are especially common on species with sessile leaves. Circular and ranging from green to golden to white, they are present at most nodes, sometimes at all, or possibly only occasionally present. The glands (or nodal glands) are best observed with dried specimens, a good light source, and magnification of at least 15×, although they can be observed under less ideal conditions.

Inflorescences may be either emergent or submersed. Emergent inflorescences are elongate and almost always terminal on the stem, whereas submersed inflorescences are globular and axillary. Most species have either emergent inflorescences or submersed inflorescences, but not both (monomorphic). Other species have both types of inflorescences on one plant (dimorphic).

All specimens should be collected when in fruit. Fruiting characteristics are extremely important in the genus, although they are not always given in the key. Vegetative features during fruiting are distinctive for the species; consequently, they are included in the key. Important features of the fruit include presence or absence of lateral and abaxial wings, ribs, ridges, or keels. Here, "ribbed" indicates a raised "vein" on a rounded surface; "ridged," a ridge with an obtuse angle; "keeled," a ridge with an acute angle; and "winged," a ridge that appears to have a wing distally.

SELECTED REFERENCES   Fernald, M. L. 1932. The linear-leaved North American species of *Potamogeton* section *Axillaries*. Mem. Amer. Acad. Arts, n. s. 17: 1–183.   Haynes, R. R. 1975. A revision of North American *Potamogeton* subsection *Pusilli* (Potamogetonaceae). Rhodora 76: 564–649.   Haynes, R. R. 1985. A revision of the clasping-leaved *Potamogeton* (Potamogetonaceae). Sida 11: 173–188.   Ogden, E. C. 1943. The broad-leaved species of *Potamogeton* of North America north of Mexico. Rhodora. 45: 57–105, 119–163, 171–214.   Reznicek, A. A. and R. S. W. Bobbette. 1976. The taxonomy of *Potamogeton* subsection *Hybridi* in North America. Rhodora. 78: 650–673.

1. Stipular sheaths of submersed leaves adnate to base of leaf blade, tip projecting as ligule.
   2. Submersed leaves stiffish, conspicuously 2-ranked; blade lobed at junction with stipule, veins 20–60, fine. . . . . .1. *Potamogeton robbinsii*
   2. Submersed leaves lax, not conspicuously 2-ranked; blade without basal lobes, veins 20 or fewer.
      3. Submersed leaf tips obtuse to acute; floating leaf tips rounded.
         4. Apex of submersed leaf blade obtuse; fruits 1.3–2.4 mm wide, lateral wings with blunt tip, beak absent. . . . . .2. *Potamogeton spirillus*
         4. Apex of submersed leaf blade acute; fruits 0.9–2 mm wide, lateral wings with sharp tip, beak minute. . . . . .3. *Potamogeton diversifolius*
      3. Submersed leaf tips acute to long-tapering; floating leaf tips acute.
         5. Submersed leaf blade 0.1–0.4(–0.6) mm wide, without obvious lacunae; floating leaf blade 3–7-veined. . . . . .4. *Potamogeton bicupulatus*
         5. Submersed leaf blade 0.2–1(–2) mm wide, with abundant lacunae; floating leaves 9–23-veined. . . . . .5. *Potamogeton tennesseensis* (in part)
1. Stipular sheaths of submersed leaves free from base of leaf blade or with only ⅕ sheath adnate, ligule not obvious.
   6. Submersed leaf blades broadly linear-oblong to lanceolate to elliptic or nearly orbiculate, 3–58 mm wide (occasional stranded plants without submersed leaves).
      7. Leaf margins conspicuously serrate; stem flattened; fruit beak 2–3 mm; turions formed, hard. . . . . .23. *Potamogeton crispus*
      7. Leaf margins entire; stem terete; fruit beak 1 mm or less; turions rarely formed.
         8. Submersed leaves clasping stem; floating leaves absent.
            9. Leaf blade apex hoodlike, splitting when pressed; stipules persistent, conspicuous. . . . . .31. *Potamogeton praelongus*
            9. Leaf blade apex flat, not splitting when pressed; stipules deciduous and/or deteriorating into fibers.
               10. Leaf blade ovate-lanceolate to narrowly lanceolate, 1.6–13 cm, veins 3–35; stipules disintegrating to persistent fibers, even on proximal portion of stem. . . . . .32. *Potamogeton richardsonii*
               10. Leaf blade broadly lanceolate, orbiculate, or ovate, 0.9–7.6(–9.7) cm, veins 3–25; stipules deciduous and deteriorating, absent on proximal portion of stem. . . . . .33. *Potamogeton perfoliatus*
         8. Submersed leaves petiolate or sessile (not clasping); floating leaves absent or present.
            11. Floating leaf blade rounded to cordate at base; stem conspicuously rusty- or black-spotted; submersed leaf blades crispate, mostly arcuate.
               12. Submersed leaf blade with 19–49 veins; floating leaf blade with 27–49 veins. . . . . .24. *Potamogeton amplifolius*
               12. Submersed leaf blade with 7–19 veins; floating leaf blade with 15–19 veins. . . . . .25. *Potamogeton pulcher*

11. Floating leaf blade cuneate, tapering to petiole, rounded, or rarely cordate (*P. oblongus*); stems without spots; submersed leaf blades flat or rarely crispate (*P. illinoensis*), mostly not arcuate or occasionally arcuate (*P. illinoensis*).
   13. Submersed leaves with petioles 0.5–13 cm.
      14. Fruits (1.6–)2–2.5 mm, without well-developed abaxial keel; St. Pierre and Miquelon, Newfoundland, and Sable Island, Nova Scotia. . . . . .26. *Potamogeton oblongus*
      14. Fruits 2.5–4.3 mm, with well-developed abaxial keel; widespread.
         15. Larger submersed leaves acute at apex but without sharp awl-like tip; fruits red to reddish brown. . . . . .27. *Potamogeton nodosus*
         15. Larger submersed leaves acuminate at apex; fruits grayish green to olive-green. . . . . 30. *Potamogeton illinoensis* (in part)
   13. Submersed leaves sessile.
      16. Fruits plump, tawny olive-green, pedicellate; stipules blunt; submersed leaf blade 7–9-veined. . . . . .28. *Potamogeton alpinus*
      16. Fruits laterally compressed, greenish brown, grayish green to olive-green, sessile; stipules acute to obtuse; submersed leaf blade 3–19-veined.
         17. Fruits 1.9–2.3 × 1.8–2 mm; submersed leaf blade with 3–9 veins. . . . . .29. *Potamogeton gramineus*
         17. Fruits 2.5–3.6 × 2.1–3 mm; submersed leaf blade with 7–19 veins. . . . . 30. *Potamogeton illinoensis* (in part)
6. Submersed leaf blade linear, threadlike, or ribbonlike, 0.1–10 mm wide.
   18. Floating leaves often present; lacunae prominent in submersed leaves and rhizome present.
      19. Submersed leaves 1–3(–7)-veined; some with stipules partially adnate to leaf base; floating leaves acute at apex. . . . . .5. *Potamogeton tennesseensis* (in part)
      19. Submersed leaves 3–13-veined; stipules not adnate to leaf base; floating leaves rounded at apex. . . . . .6. *Potamogeton epihydrus*
   18. Floating leaves absent or present; lacunae sometimes present but not prominent in most species, if prominent then rhizome absent.
      20. Floating leaves present, at least in some plants in population.
         21. Floating leaf blade 0.6–1.5 cm; peduncle 0.5–3 cm; fruits 1.5–2.5 mm. . . . . .19. *Potamogeton vaseyi*
         21. Floating leaf blade 1.5–12 cm; peduncle 2.5–9.5 cm; fruits if formed 2.5–5 mm.
            22. Petiole junction with blade distinctly pale in color just proximal to apex; floating leaf blade ovate to oblong-ovate. . . . . .22. *Potamogeton natans*
            22. Petiole junction with blade continuous in color to apex; floating leaf blade elliptic, ovate-elliptic, or oblong-elliptic.
               23. Floating leaf blades 7–12 mm wide, tapering at both ends; no fruit produced; Florida. .20. *Potamogeton floridanus*
               23. Floating leaf blade 10–20(–30) mm wide, obtuse, rounded or tapering at base; fruit often produced; widespread n of Florida . . . . .21. *Potamogeton oakesianus*
      20. Floating leaves absent from all plants in population.
         24. Rhizome obvious; peduncle (3–)5–25 cm, ascending, not recurved; leaves threadlike, 0.1–0.5 mm wide. . . . . .7. *Potamogeton confervoides*
         24. Rhizomes absent or not apparent; peduncle 0.3–7 cm, erect to recurved; leaves not threadlike, 0.1–5 mm wide.
            25. Nodal glands absent.
               26. Leaves 15–35-veined, greater than 2 mm wide; stem conspicuously flattened; peduncles terminal. . . . . .17. *Potamogeton zosteriformis*
               26. Leaves 3–5-veined, less than 2 mm wide; stem terete; peduncles axillary.
                  27. Leaves bristle-tipped, occasionally apiculate to blunt, 3-

veined, 0.6–2.5(–4) mm wide; fruits 3-keeled, 2.3–4 mm. . .
. . . . 13. *Potamogeton hillii* (in part)

    27. Leaves acute, 1–3(–5)-veined, 0.3–2.3 mm wide; fruits 1-keeled, 1.4–2.3(–2.7) mm. . . . . 18. *Potamogeton foliosus* (in part)

25. Nodal glands present on at least some nodes.

    28. Stipules fibrous, often whitish.

        29. Leaf apex acute to apiculate; leaves 5–7(–9)-veined; turions with inner leaves at 90° angle to outer leaves. . . . . 8. *Potamogeton friesii*

        29. Leaf apex bristle-tipped, acute, rarely obtuse to apiculate; leaves 3–13-veined; turions flattened with inner and outer leaves in same plane.

            30. Fruits 1.9–2.1 mm; stipules connate; leaf blades 3–5(–7)-veined. . . . . 9. *Potamogeton strictifolius*

            30. Fruit 2.5–3 mm; stipules convolute; leaf blades 3–9(–13)-veined. . . . . 10. *Potamogeton ogdenii*

    28. Stipules not fibrous, green, brown, or white.

        31. Leaves 7–17-veined.

            32. Leaves 7–11-veined, 0.7–1.7 mm wide; Greenland. . . .
. . . . 11. *Potamogeton groenlandicus*

            32. Leaves 9–17-veined, 1.5–2 mm wide; Canada and Alaska. . . . . 12. *Potamogeton subsibiricus*

        31. Leaves 1–5-veined.

            33. Leaf blade apex bristle-tipped (rarely apiculate); peduncles recurved, 0.6–1.35 cm. . . . . 13. *Potamogeton hillii* (in part)

            33. Leaf blade apex blunt, acute, or apiculate, but not bristle-tipped; peduncles erect to ascending, rarely recurved, 0.5–6.6 cm.

                34. Fruits with abaxial surface rounded.

                    35. Fruits 2.5–3.6 mm; turions 3.5–7.8 cm × 2.3–5.1 mm; inner leaves undifferentiated. . . . . . .
. . . . 14. *Potamogeton obtusifolius* (in part)

                    35. Fruits 1.5–2.2 mm; turions 0.9–3.2 cm × 0.3–1.8 mm; inner leaves modified into fusiform structure. . . . . 16. *Potamogeton pusillus*

                34. Fruits with abaxial keel or ridge.

                    36. Abaxial keel undulate. . . . . . . 18. *Potamogeton foliosus* (in part)

                    36. Abaxial keel without undulations.

                        37. Inflorescences 8 mm or more; nodal glands 0.2–1 mm diam.; leaf blade light green to somewhat reddish, apex obtuse or round-apiculate; fruits without basal tubercles; northern plant. . . . . . 14. *Potamogeton obtusifolius* (in part)

                        37. Inflorescences 5.5–7.5 mm; nodal glands 0.2–0.3 mm diam.; leaf blade green, apex acute; fruits with basal tubercles; endemic to w Texas. . . . . 15. *Potamogeton clystocarpus*

1. **Potamogeton robbinsii** Oakes, Mag. Hort. Bot. 7: 180. 1841 • Robbins' pondweed, fern pondweed, potamot de Robbins [E] [F]

Rhizomes present. **Cauline stems** terete, without spots, to 100 cm; glands absent. **Turions** absent. **Leaves** submersed, conspicuously 2-ranked, sessile, stiffish; stipules persistent, conspicuous, adnate to base of blade ± ¼ length of stipule, connate, greenish brown to white, ligulate, 0.5–2 cm, fibrous, shredding at tip, apex obtuse; blade dark green to reddish green, linear to lanceolate, not arcuate, 2–7 (–12) cm × 3–4(–8) mm, base rounded, with basal lobes, not clasping, margins minutely spinulose to serrulate, not crispate, apex not hoodlike, acute, lacunae absent; veins 20–60, fine. **Inflorescences** often branched; peduncles not dimorphic, axillary, erect, cylindric, 3–5 (–7) cm; spikes not dimorphic, moniliform (i.e., beaded), 7–20 mm. **Fruits** stipitate, brown, obliquely obovoid, turgid, abaxially and laterally keeled, 3–4(–5) × 2–3.3 mm, lateral keels without points; beak erect, recurved at apex, 0.7–0.9 mm; sides without basal tubercles; embryo with less than 1 full spiral. $2n = 52$.

Flowering late summer–early fall. Shallow to deep water of ponds, lakes, and slow-flowing rivers; 0–3000 m; Alta., B.C., Man., N.B., N.S., Nunavut, Ont., P.E.I., Que., Sask.; Ala., Alaska, Calif., Conn., Idaho, Ill., Ind., Maine, Mass., Mich., Minn., Mont., N.H., N.J., N.Y., Ohio, Oreg., Pa., R.I., Utah, Vt., Wash., Wis., Wyo.

Potamogeton robbinsii is our most easily recognized species when it is fertile. It is the only species with branched inflorescences. The species, however, occurs in fairly deep water, forming large colonies that essentially cover the substrate. Only rarely do the plants flower. It also is the only species with truly auriculate leaves, the blades forming small lobes projecting past the stem on each side of the stem. Leaf blades of other *Potamogeton* species may have slightly rounded bases, but no others have lobes that actually protrude past the stem.

The species has a fairly large disjunction; primarily known from the northern part of the flora, it also occurs in the Tensas River area, Baldwin County, Alabama. The Alabama population has been collected on at least two occasions over 40 years, once as recently as 1970.

2. **Potamogeton spirillus** Tuckerman, Amer. J. Sci. Arts, ser. 2, 6: 228. 1848 • Northern snailseed pondweed, potamot spirillé [E]

Rhizomes present. **Cauline stems** compressed, without spots, 5–40 cm; glands absent. **Turions** absent. **Leaves** both submersed and floating or floating absent, ± spirally arranged. **Submersed leaves** sessile, lax; stipules persistent to deliquescent, inconspicuous, convolute, adnate to blade for ½ stipule length, reddish brown to light green, ligulate, 2–12 mm, not fibrous, not shredding at tip, apex obtuse; blade red-brown to light green, linear, not arcuate, 0.8–8 cm × 0.5–2 mm, base slightly tapering, without basal lobes, not clasping, margins entire, not crispate, apex not hoodlike, obtuse to acute, lacunae present, a broad band each side of midvein; veins 1–3. **Floating leaves** petiolate; petioles continuous in color to apex, 5–25 mm; blade adaxially light green, oblong to obovate, 0.7–3.5 cm × 2–13 mm, base tapered to rounded, apex obtuse, rounded; veins 5–15. **Inflorescences** unbranched; peduncles dimorphic, submersed axillary, recurved, clavate, 0.5–3 mm, emersed axillary or terminal, erect to recurved, slightly clavate, 4–27 mm; spikes dimorphic, submersed capitate, 2–5 mm, emersed ellipsoid to cylindric, 4–13 mm. **Fruits** sessile, greenish brown, somewhat orbicular, compressed, abaxially winged, laterally winged, 4–13 × 1.3–2.4 mm, lateral wing with blunt tips; beak absent; sides without basal tubercles; embryo with more than 1 full spiral.

Flowering mid summer–late fall. Neutral to acidic waters of ponds, lakes, and streams; 0–400 m; Man., N.B., Nfld. and Labr. (Nfld.), N.S., Ont., P.E.I., Que.; Conn., Iowa, Maine, Mass., Mich., Minn., Nebr., N.H., N.J., N.Y., Ohio, Pa., R.I., Vt., Wis.

This is one of three *Potamogeton* species in the flora area with dimorphic inflorescences. It can be separated from the other two species because its submersed leaf blades have broad lacunae, extending nearly from the midvein to the margin, and its fruits have lateral wings with blunt points along them.

3. **Potamogeton diversifolius** Rafinesque, Med. Repos., hexade 2, 5: 354. 1808 • Water-thread pondweed [F]

Potamogeton capillaceus Poiret

Rhizomes present. **Cauline stems** compressed, without spots, 10–35 cm; glands absent. **Turions** absent. **Leaves** both submersed and floating or floating absent, ± spirally arranged. **Submersed leaves** sessile, lax; stipules persistent to deliquescent, inconspicuous, convolute, adnate to blade ½ stipule length, light brown to red-brown, ligulate, 1.5–2.3 cm, not fibrous, not shredding at tip, apex obtuse; blade red-brown to light green, linear, often arcuate, 1–1.3 cm × 0.1 mm, base slightly tapering, without basal lobes, not clasping, margins entire, not crispate, apex not hoodlike, acute, lacunae present, 1–2 rows each side midrib; veins 1. **Floating leaves** petiolate; petioles continuous in color to apex, 0.7–0.8 cm; blade adaxially light green, obovate to elliptic, 0.8–1.6 cm × 3–8.5 mm, base acute, apex round to acute; veins 3–7. **Inflorescences** unbranched; peduncles dimorphic, submersed axillary, recurved, clavate, 3–5 mm, emersed axillary or terminal, erect to slightly recurved, clavate, 6–15 mm; spikes dimorphic, submersed capitate, 2–3 mm, emersed cylindric, 5–9.7 mm. **Fruits** sessile, greenish brown, orbicular, compressed, abaxially winged, laterally winged, 1–1.5 × 0.9–2 mm, lateral wings with sharp points; beak present, erect, 0.1 mm; sides without basal tubercles; embryo with more than 1 full spiral.

Flowering and fruiting spring–fall. Ponds, lakes, streams, and rivers; 5–2500 m; Ala., Ariz., Ark., Calif., Colo., Conn., Del., Fla., Ga., Idaho, Ill., Ind., Iowa, Kans., Ky., La., Md., Mass., Minn., Miss., Mo., Mont., Nebr., Nev., N.J., N.Y., N.C., N.Dak., Ohio, Okla., Oreg., Pa., S.C., S.Dak., Tenn., Tex., Vt., Va., W.Va., Wis.; Mexico.

Potamogeton diversifolius is very likely the most common species of the genus in the southeastern United States. It has been separated into two species, *P. diversifolius* in the strict sense and *P. capillaceus* (M. L. Fernald 1932). The species has also been divided into two varieties, var. *diversifolius* and var. *trichophyllus* Morong (D. S. Correll and M. C. Johnston 1970). *Potamogeton diversifolius* var. *trichophyllus* actually is misapplied, as the name really refers to the more northern *P. bicupulatus* Fernald. We are following E. J. Klekowski Jr. and E. O. Beal (1965) in accepting only one taxon, as we have studied the species over much of its range and reached similar conclusions.

SELECTED REFERENCE   Klekowski, E. J. Jr. and E. O. Beal. 1965. A study of variation in the *Potamogeton capillaceus–diversifolius* complex (Potamogetonaceae). Brittonia 17: 175–181.

4. **Potamogeton bicupulatus** Fernald, Mem. Amer. Acad. Arts, n. s. 17: 112. 1932 • Snail-seed pondweed [E]

Potamogeton diversifolius Rafinesque var. *trichophyllus* Morong

Rhizomes present. **Cauline stems** compressed, without spots, 10–25 cm; glands absent. **Turions** absent. **Leaves** both submersed and floating or floating absent, ± spirally arranged. **Submersed leaves** sessile, lax; stipules persistent to deliquescent, inconspicuous, convolute, adnate to blade for less than ½ stipule length, light green, ligulate, 0.2–1.2 cm, not fibrous, not shredding at tip, apex obtuse; blade light green to rarely brown, linear-setaceous, not arcuate, 1.5–11 cm × 0.1–0.4(–0.6) mm, base slightly tapering, without basal lobes, not clasping, margins entire, not crispate, apex not hoodlike, tapering, lacunae absent; veins 1. **Floating leaves** petiolate; petioles continuous in color to apex, 5–35 mm; blade adaxially light green, lanceolate-elliptic to broadly elliptic, 0.6–2.3(–2.8) cm × 1–11 mm, base tapering or rounded, apex acute to long tapering; veins 3–7. **Inflorescences** unbranched; peduncles dimorphic, submersed axillary, somewhat recurved, clavate, 1–10 mm, emersed axillary or terminal, erect to slightly recurved, slightly clavate, 3.5–22 mm; spikes dimorphic, submersed globular to ellipsoid, 1.5–7 mm, emersed ellipsoid to cylindric, 3–14 mm. **Fruits** sessile, greenish brown, somewhat orbicular, compressed, abaxially keeled, laterally keeled, 1.1–2.1 × 1.1–2 mm, lateral keel without points; beak absent; sides without basal tubercles; embryo with more than 1 full spiral.

Flowering early summer–fall. Acidic waters of ponds, lakes, and streams; 0–300 m; Ont.; Conn., Del., Ind., Maine, Mass., Mich., N.H., N.Y., Pa., R.I., Vt., Wis.

Potamogeton bicupulatus is an uncommon species of the acid lakes and streams of northeastern United States and southern Canada. It is the third species we have with dimorphic inflorescences and embryos with more than one full spiral. It can be separated from the other two, *Potamogeton spirillus* and *P. diversifolius*, because it has very narrow submersed leaves without lacunae and fruits with lateral keels without sharp points.

5. Potamogeton tennesseensis Fernald, Rhodora 38: 167, plate 412. 1936 • Tennessee pondweed C E

Rhizomes present. **Cauline stems** terete, without spots, 10–35 cm; glands absent. **Turions** absent. **Leaves** both submersed and floating or floating absent, ± spirally arranged. **Submersed leaves** sessile, lax; stipules persistent, inconspicuous, convolute, adnate to blade for ¼ or less stipule length, light brown to dark green, ligulate, 0.5–1.5 cm, not fibrous, not shredding at tip, apex acute; blade red-brown to dark green, linear-filiform, not arcuate, 2.5–10.5 cm × 0.2–1(–2) mm, base slightly tapering, not clasping, without basal lobes, margins entire, not crispate, apex not hoodlike, long tapering, lacunae abundant, broad, filling area between margin and midvein; veins 1–3. **Floating leaves** petiolate; petioles continuous in color to apex, 2.5–6 cm; blade greenish brown adaxially, lance-oblong, 2–4(–5.5) cm × 5–13 mm, base acute, apex acute; veins 9–23. **Inflorescences** unbranched, emersed; peduncle not dimorphic, axillary, ascending, cylindric, 3–8 cm; spikes not dimorphic, cylindric, 10–22 mm. **Fruits** sessile, greenish brown, quadrate-orbicular, slightly compressed, abaxially keeled, laterally ridged, 2.5–3 × 2–2.5 mm; lateral ridges without points; beak present, erect, 0.5 mm; sides without basal tubercles; embryo with less than 1 full spiral.

Flowering mid spring–fall. Slow- to fast-moving streams and rivers; of conservation concern; 200–1000 m; Ky., N.C., Ohio, Pa., Tenn., Va., W.Va.

No specimens have been seen from Maryland although the species is to be expected there.

6. Potamogeton epihydrus Rafinesque, Med. Repos., hexade 2, 5: 354. 1808; hexade 3, 2: 409. 1811 • Ribbon-leaf pondweed, potamot émergé E F

Potamogeton epihydrus subsp. *nuttallii* (Chamisso & Schlechtendal) Calder & Roy L. Taylor; *P. epihydrus* var. *ramosus* (Peck) House

Rhizomes present. **Cauline stems** flattened, without spots, 10–90 cm; glands absent. **Turions** absent. **Leaves** both submersed and floating or floating absent, ± spirally arranged. **Submersed leaves** sessile, lax; stipules persistent, inconspicuous, convolute, free from blade, red-brown, not ligulate, 1–3 cm, not fibrous, not shredding at tip, apex obtuse; blade red-brown to light green, linear, not arcuate, 5–22 cm × 1–10 mm, base not clasping, without basal lobes, margins entire, not crispate, apex not hoodlike, blunt to acute, lacunae present, broad band each side of midvein; veins 3–13. **Floating leaves** petiolate; petioles continuous in color to apex, 2–12.5 cm; blade adaxially light green, narrowly oblong-oblanceolate to elliptic, 2–8 cm × 4–20 mm, base acute, apex rounded or bluntly cuspidate; veins 11–41. **Inflorescences** unbranched, emersed; peduncles not dimorphic, axillary, erect, cylindric, 1.5–5(–16) cm; spikes not dimorphic, cylindric, 0.8–4 cm. **Fruits** sessile, greenish brown, round-obovoid, flattened, abaxially and laterally keeled, 2.5–4.5 × 2–3.6 mm, lateral keels without sharp points; beak erect, 0.5 mm; sides without basal tubercles; embryo with less than 1 full spiral. $2n = 26$.

Flowering summer–fall. Still or flowing waters of lakes, ponds, streams, and rivers; 10–1900 m; St. Pierre and Miquelon; B.C., Man., N.B., Nfld. and Labr., N.S., Ont., P.E.I., Que., Sask.; Ala., Alaska, Calif., Conn., Colo., Del., Fla., Ga., Idaho, Ind., Iowa, La., Maine, Md., Mass., Mich., Minn., Miss., Mont., N.H., N.J., N.Y., N.C., Ohio, Oreg., Pa., R.I., S.C., S.Dak., Tenn., Vt., Va., Wash., W.Va., Wis., Wyo; Europe.

No specimens have been seen from Maryland, but the species is to be expected there.

Two varieties, *Potamogeton epihydrus* var. *epihydrus* and var. *ramosus*, have been recognized. These prove not to be distinct. Both varieties often grow in the same body of water in the same population. The wider-leaved plants often occur in more alkaline waters. Two hybrids, *P. epihydrus* var. *nuttallii* × *P. gramineus* and *P. epihydrus* × *P. nodosus* (=*P.* × *subsessilis* Hagström), have been described.

Potamogeton epihydrus is a common species of lakes and streams of northern United States and southern Canada. It extends southward in the eastern United States to Louisiana and Alabama. *Potamogeton epihydrus* is one of our more easily recognized species: it has floating leaves, linear submersed leaves, and fruits with an embryo with one full spiral or less. The only other North American pondweed with a similar set of characteristics is *P. tennesseensis*, which differs from *P. epihydrus* by the former having long tapering apices in the submersed leaves whereas the latter has blunt to acute apices.

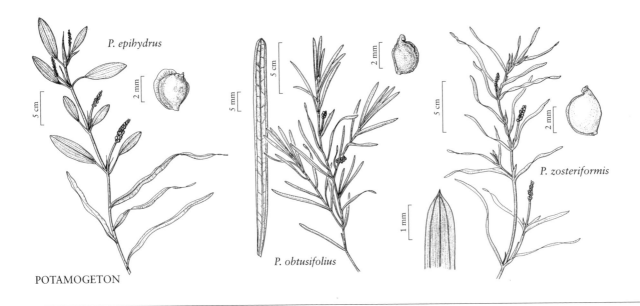

P. epihydrus

P. obtusifolius

P. zosteriformis

POTAMOGETON

---

7. **Potamogeton confervoides** Reichenbach in H. G. L. Reichenbach et al., Icon. Fl. Germ. Helv. 7: 13. 1845 • Alga pondweed, Tuckerman's pondweed, potamot confervoïde E

Potamogeton tuckermanii J. W. Robbins (as tuckermani)

Rhizomes obvious. **Cauline stems** terete, without spots, 10–80 cm. **Turions** present, in axils of old leaves and from disintegrating branches, fusiform, 0.7–2 cm, leaves spreading to ascending. **Leaves** submersed, ± spirally arranged, flaccid, sessile; stipules deliquescent, inconspicuous, convolute, free from blade, pale green, not ligulate, 0.5–1.2 cm, not fibrous, not shredding at tip, apex obtuse; blade pale green, linear, not arcuate, 1.8–6.5 cm × 0.1–0.5 mm, base slightly tapering, without basal lobes, not clasping, margins entire, not crispate, apex not hoodlike, extremely attenuate, bristly, lacunae present, each side of midvein to margins; veins 1. **Inflorescences** unbranched, emersed; peduncles not dimorphic, terminal, ascending, somewhat clavate, (3–)5–25 cm; spikes not dimorphic, capitate, 5–12 mm. **Fruits** sessile, light green, round-obovoid or nearly orbicular, compressed, abaxially and laterally keeled, 2–3 × 1.7–2.8 mm, lateral keels without sharp point; beak erect, 0.5 mm; sides without basal tubercles; embryo with less than 1 full spiral. Chromosome number not available.

Flowering early–late summer. Acidic waters of bogs, ponds, and lakes, often at higher elevation in e portion of range; 0–1500 m; St. Pierre and Miquelon; N.B.,

Nfld. and Labr., N.S., Ont., Que.; Conn., Maine, Mass., Mich., N.H., N.J., N.Y., N.C., Pa., R.I., S.C., Vt., Wis.

Potamogeton confervoides is most uncommon and found only in fairly acidic waters. It is easily recognized by its linear, bristly leaves and the unusually long peduncle that seems out of place on a plant with such fine leaves. The leaves are so fine that they almost appear as greenish colored hair in the water. When the plant is removed from the water, the extremely flaccid leaves collapse onto each other.

8. **Potamogeton friesii** Ruprecht, Hist. Stirp. Fl. Petrop., 43. 1845 • Potamot de Fries

Rhizomes absent. **Cauline stems** compressed, without spots, 10–135 cm; glands green, greenish brown, or gold, to 0.7 mm diam. **Turions** terminal or lateral, common, 1.5–5 cm × 1.5–4 mm, soft; leaves ± 4-ranked; outer leaves 2–3 per side, base corrugate, apex apiculate to acute; inner leaves reduced, arranged into fan-shaped structure and oriented at 90° angles to outer leaves. **Leaves** submersed, ± spirally arranged, delicate to rigid, sessile; stipules not persistent, inconspicuous, convolute, free from blade, white, not ligulate, 0.55–2.1 cm, fibrous, coarse, shredding at tip, apex obtuse; blade light green, rarely olive-green to somewhat reddish, linear, not arcuate, 2.3–6.5 cm × 1.2–3.2 mm, base slightly tapering, without basal lobes, not clasping, margins entire, not crispate, apex not hoodlike, acute to apiculate, lacunae absent or 1 narrow row each side of midrib; veins 5–7(–9). **Inflorescences** unbranched, emersed; peduncles not dimorphic, terminal or axillary, erect or rarely recurved, slightly clavate, 1.2–4.1(–7) cm; spike not dimorphic, cylindric, 7–16 mm. **Fruits** sessile, olive-green to brown, obovoid, turgid, not abaxially or laterally keeled, 1.8–2.5 × 1.2–2 mm; beak erect, 0.3–0.7 mm; sides without basal tubercles; embryo with 1 full spiral. $2n = 26$.

Flowering and fruiting summer–fall. Calcareous to brackish waters of lakes and slow-flowing streams; 0–3100 m; Alta., B.C., Man., N.B., Nfld. and Labr. (Nfld.), N.W.T., N.S., Ont., P.E.I., Que., Sask., Yukon; Alaska, Conn., Idaho, Ill., Ind., Iowa, Maine, Mass., Mich., Minn., Mont., Nebr., N.H., N.Y., N.Dak., Ohio, Pa., R.I., S.Dak., Utah, Vt., Wash., Wis., Wyo.; Eurasia.

Potamogeton friesii is a fairly common linear-leaved species, especially of calcareous waters of lakes and streams of the upper Midwest. Whenever turions are present, the species is easily identified, as it is the only one with the outer leaves of the turions having corrugate bases and the inner leaves turned at right angles to the outer leaves.

Two hybrids, *Potamogeton friesii* × *P. pusillus* (= *P.* ×*pusilliformis* Fischer [*P.* ×*intermedius* Fischer]) and *P. friesii* × *P. obtusifolius* (= *P.* ×*semifructus* A. Bennett ex Ascherson & Graebner), have been described.

9. Potamogeton strictifolius A. Bennett, J. Bot. 40: 148. 1902 • Potamot à feuilles raides E

Potamogeton strictifolius var. *rutiloides* Fernald

Rhizomes absent. **Cauline stems** terete, without spots, 27–95 cm; glands white, green, greenish brown, or gold, to 0.3 mm diam. **Turions** terminal or lateral, common, 2.5–4.8 cm × 0.8–2.2 mm, soft; leaves ± 2-ranked, flattened with outer and inner leaves in same plane; outer leaves 3–4 per side, base not corrugate, or rarely corrugate, apex acute; inner leaves undifferentiated. **Leaves** submersed, ± spirally arranged, rigid, sessile; stipules disintegrating, inconspicuous, connate, free from blade, white, not ligulate, 0.6–1.6 cm, fibrous, shredding at tip, apex obtuse; blade green to olive-green, linear, not arcuate, 1.2–6.3 cm × 0.6–2 mm, base slightly tapering, without basal lobes, not clasping, margins entire, not crispate, apex not hoodlike, acute to nearly bristle-tipped, rarely obtuse to apiculate, lacunae absent; veins 3–5(–7). **Inflorescences** unbranched, emersed; peduncles not dimorphic, terminal, erect, rarely recurved, cylindric, rarely slightly clavate, 1–4.5 cm; spike not dimorphic, cylindric, 0.6–1.3 cm. **Fruits** sessile, green-brown,

ovoid, turgid, not abaxially or laterally keeled, 1.9–2.1 × 1.3–1.8 mm; beak erect, 0.5–0.8 mm; sides without basal tubercles; embryo with 1 full spiral. $2n = 26$.

Flowering and fruiting summer–fall. Alkaline waters of lakes and slow-moving streams; 50–2000 m; Alta., Man., N.B., Ont., Que., Sask., Yukon; Conn., Ill., Ind., Maine, Mass., Mich., Minn., Nebr., N.Y., N.Dak., Ohio, Pa., S.Dak., Utah, Vt., Va., Wis., Wyo.

Potamogeton strictifolius is a relatively uncommon species found in alkaline waters. Fairly rigid leaves of the species make floating onto paper unnecessary in the collecting process. The leaves have a tendency to become revolute during the growing season. The species superficially resembles several other species of linear-leaved pondweeds. Consequently, many specimens of this species have been misidentified as other species and vice versa. Thus, literature records are often suspect.

One hybrid, *Potamogeton strictifolius* × *P. zosteriformis* (=*P.* ×*haynesii* Hellquist & G.E. Crow), has been described.

10. Potamogeton ogdenii Hellquist & R. L. Hilton, Syst. Bot. 8: 88, plates 1–2, figs. 1–3. 1983 • Ogden's pondweed, potamot d'Ogden C E

Rhizomes absent. **Cauline stems** compressed-filiform, without spots, to 50 cm; glands green, golden brown to dark brown, 0.2–0.6 mm diam. **Turions** terminal or lateral, uncommon, 3.7–9.9 × 2.6–6 cm, soft to hard; leaves flattened with outer and inner leaves in same plane; outer leaves 1–2 per side, base not corrugate, apex apiculate; inner leaves undifferentiated or rolled into hardened, fusiform structure. **Leaves** submersed, ± spirally arranged, sessile, rigid; stipules persistent, inconspicuous, convolute, free from blade, brown or rarely white, not ligulate, 0.9–2.1 cm, slightly fibrous, partially shredding at tip, apex obtuse; blade somewhat reddish to olive-green, linear, not arcuate, 1.5–10 cm × 1.2–2.9 mm, base slightly tapering, without basal lobes, not clasping, margins entire, not crispate, apex not hoodlike, cuspidate to bristle-tipped, lacunae present or absent, in 0–3 rows each side of midvein; veins 3–9(–13). **Inflorescences** unbranched, emersed; peduncles not dimorphic, terminal or occasionally axillary, erect or rarely recurved, slightly clavate, 1–3 cm; spikes not dimorphic, cylindric, 5–11 mm. **Fruits** sessile, dark green, orbicular, turgid, abaxial keel obscure, lateral keels obscure or absent, 2.5–3 × 2.2–3 mm, lateral keels if present without points; beak erect, 0.5 mm; sides without basal tubercles; embryo with 1 full spiral. Chromosome number not available.

Flowering mid summer–fall. Alkaline waters of ponds and lakes; of conservation concern; 100–300 m; Ont.; Conn., Mass., N.Y., Vt.

Potamogeton ogdenii is an extremely local species, probably known from fewer than a dozen localities. The species is here reported for the first time from Canada, being known from that country by a single collection made in 1987.

11. Potamogeton groenlandicus Hagström, Kongl. Svenska Vetenskapsakad. Handl., n. s. 55(5): 127. • 1916 [C] [E]

Potamogeton pusillus Linnaeus subsp. *groenlandicus* (Hagström) Boch

Rhizomes absent. **Cauline stems** terete, without spots, 20–50 cm; glands white, 0.1–0.7 mm diam. **Turions** terminal or axillary, common, 4.2–8.4 cm × 1.4–2.5 mm, soft; leaves ± 4-ranked; outer leaves 2 per side, base without corrugations, apex acute to apiculate; inner leaves unmodified. **Leaves** submersed, ± spirally arranged, sessile, delicate; stipules deliquescent to persistent, inconspicuous, convolute, free from blade, brown, not ligulate, 0.45–1.91 cm, not fibrous, not shredding at tip, apex obtuse; blade pale to deep green, linear, not arcuate, 2.6–8.8 cm × 0.7–1.7 mm, base slightly tapering, without basal lobes, not clasping, margins entire, not crispate, apex not hoodlike, acute to apiculate, lacunae absent or present, 0–3 rows each side of midrib; veins 7–11. **Inflorescences** unbranched, emersed; peduncles not dimorphic, terminal, erect, cylindric, 20 mm; spikes not dimorphic, cylindric, 5 mm. **Fruits** not laterally keeled, 2.6–2.7 × 2 mm; beak erect, 0.3 mm; sides without basal tubercles; embryo with 1 full coil. $2n = 26$.

Flowering late summer. Ponds and lakes; of conservation concern; 10–450 m; Greenland.

12. Potamogeton subsibiricus Hagström, Kongl. Svenska Vetenskapsakad. Handl., n. s. 55(5): 84. 1916 • Yenissei River pondweed, potamot de l'Ienisseï

Potamogeton porsildiorum Fernald

Rhizome absent. **Cauline stems** compressed-filiform, without spots, to 50 cm; glands white, 0.3–0.5 mm diam. **Turions** lateral, common, 3.5–9.5 cm × 2–5 mm, soft; leaves ± 2-ranked; outer leaves 3–4 per side, base not corrugate; inner leaves undifferentiated. **Leaves** submersed, ± spirally arranged, sessile, flaccid; stipules deliquescent, inconspicuous, con-

volute, free from blade, pale brown, not ligulate, 1–2 cm, not fibrous, not shredding, apex obtuse; blade dark green, linear, not arcuate, 3.5–9.5 cm × 1.5–2 mm, base slightly tapering, without basal lobes, not clasping, margins entire, not crispate, apex not hoodlike, rounded or nearly acute to mucronate, lacunae in 1–2 rows each side of midvein; veins 9–17. **Inflorescences** unbranched, emersed; peduncles not dimorphic, axillary, erect, cylindric, 1.7–3.5 cm; spikes not dimorphic, cylindric, 10–30 mm. **Fruits** sessile, reddish brown, oblong-obovoid, compressed, abaxially ridged, not laterally keeled, 3–4 × 1.5–2 mm; beak nearly erect, 0.3–0.5 mm; sides without basal tubercles; embryo with 1 full coil.

Fruiting early summer–late summer. Shallow water of ponds and lakes; 0–915 m; Man., N.W.T., Nunavut, Ont., Que., Yukon; Alaska; Siberia.

13. Potamogeton hillii Morong, Bot. Gaz. 6: 290, fig. 3. 1881 [E]

Potamogeton porteri Fernald

Rhizomes absent. **Cauline stems** slightly compressed, without spots, 30–60 cm; glands rare, when present, brown to green, 0.1–0.3 mm diam. **Turions** terminal, rare, 2.8–3 cm × 1.5–3 mm, soft; leaves ± 2-ranked; outer leaves 3–4 per side, base not corrugate, apex acute to apiculate; inner leaves undifferentiated. **Leaves** submersed, ± spirally arranged, sessile, delicate; stipules persistent, inconspicuous, convolute, free from blade, white to light brown, not ligulate, 0.7–1.6 cm, slightly fibrous, rarely shredding at tip, apex obtuse; blade pale green to olive-green, linear, not arcuate, 2–6 cm × 0.6–2.5(–4) mm, base slightly tapering, without basal lobes, not clasping, margins entire, not crispate, apex not hoodlike, apiculate to bristle-tipped or rarely blunt, lacunae in 1–2 rows each side of midrib; veins 3. **Inflorescences** unbranched, emersed; peduncles not dimorphic, axillary and/or terminal, erect to ascending, rarely recurved, slightly clavate, 6–13.5 mm; spikes not dimorphic, globose, (2–) 4–7 mm. **Fruits** brown to light greenish brown, ovoid to orbicular, turgid, sessile, abaxially and laterally keeled (3-keeled), 2.3–4 × 2–3.2 mm, lateral keels without points; beak erect, 0.3–0.7 mm; sides without basal tubercles; embryo with 1 full spiral. Chromosome number not available.

Flowering and fruiting summer. Alkaline waters of marshes, ponds, lakes, and slow-moving streams; 50–400 m; Ont.; Conn., Mass., Mich., N.Y., Ohio, Pa., Vt., Va., Wis.

Potamogeton hillii is an easily recognized species either in fruit or when sterile. The leaf blade has a bristle tip and five or fewer veins. Those characters combined with the usual absence of nodal glands will separate this species from all other North American linear-leaved species. Ecologically, it is consistently found in more alkaline waters than any other North American pondweed. A study of 35 localities established the mean to be 124.1 mg/l $CaCO_3$ (C. B. Hellquist 1984).

SELECTED REFERENCE Hellquist, C. B. 1984. Observations of *Potamogeton hillii* Morong in North America. Rhodora 86: 101–111.

14. **Potamogeton obtusifolius** Mertens & W. D. J. Koch in J. C. Röhling et al., Deutschl. Fl. ed. 3, 1: 855. 1823 • Potamot à feuilles obtuses F

Rhizomes absent. **Cauline stems** slightly compressed, without spots, 35–90 cm; glands yellow-green to gold, 0.2–1 mm diam. **Turions** terminal, abundant, 3.5–7.8 cm × 2.3–5.1 mm, soft; leaves ± 2-ranked; outer leaves 3–4 per side, base not corrugate, apex apiculate to obtuse; inner leaves undifferentiated. **Leaves** submersed, ± spirally arranged, sessile, flaccid; stipules persistent, inconspicuous, convolute, free from blade, white, not ligulate, 0.6–1.8 cm, fibrous, rarely shredding at tip, apex obtuse; blade light green to somewhat reddish, linear, not arcuate, 3–8.2 cm × 1–3.5 mm, base slightly tapering, without basal lobes, not clasping, margins entire, not crispate, apex not hoodlike, obtuse or round-apiculate, lacunae in 1–3 rows each side of midrib; veins 3. **Inflorescences** unbranched, emersed; peduncles not dimorphic, axillary, erect, rarely recurved, cylindric, 0.8–1.9 (–4.2) cm; spike not dimorphic, cylindric, 8–13 mm. **Fruits** sessile, olive-green to brown, obovoid, turgid, abaxially keeled or not, laterally keeled or not, 2.5–3.6 × 1.7–2.4 mm, lateral keels when present without points; beak erect, 0.8–1 mm; sides without basal tubercles; embryo with 1 full spiral. $2n = 26$.

Flowering and fruiting summer–fall. Medium- to low-alkaline waters of lakes and slow-flowing streams; 50–2000 m; Alta., B.C., Man., N.B., Nfld. and Labr. (Nfld.), N.W.T., N.S., Ont., Que., Sask., Yukon; Alaska, Conn., Maine, Mass., Mich., Minn., Mont., N.H., N.J., N.Y., Pa., R.I., Vt., Wash., Wis., Wyo.; Eurasia.

Potamogeton obtusifolius is a distinctive linear-leaved species with the leaf blades round at the apex, especially when fruiting inflorescences 5–7 mm wide are present. This is unusually wide for one of the linear-leaved species.

Two hybrids, *Potamogeton obtusifolius* × *P. pusillus*

(=*P.* ×*saxonicus* Hagström) and *P. friesii* × *obtusifolius* (=*P.* ×*semifructus* A. Bennett ex Ascherson & Graebner), have been described.

15. **Potamogeton clystocarpus** Fernald, Mem. Amer. Acad. Arts, n. s. 17: 79, plates 15, 30, fig. 5. 1932 C E

Rhizomes absent. **Cauline stems** terete to slightly compressed, without spots, to 57 cm; glands white to gold, 0.2–0.3 mm diam. **Turions:** unknown. **Leaves** submersed, ± spirally arranged, sessile, delicate; stipules persistent, inconspicuous, convolute, free from blade, brown, not ligulate, to 0.62 cm, not fibrous, not shredding at tip, apex obtuse; blade green, linear, not arcuate, 3.2–7.8 cm × 0.7–1.7 mm, base slightly tapering, without basal lobes, not clasping, margins entire, not crispate, apex not hoodlike, acute, lacunae rarely absent, in 0–4 rows each side of midrib; veins 3(–5). **Inflorescences** unbranched, emersed; peduncles not dimorphic, axillary or terminal, erect, cylindric, 3.2–4.8 cm; spike not dimorphic, capitate to cylindric, 5.5–7.5 mm. **Fruits** sessile, brown to yellow-green, obovoid, compressed, abaxially and laterally keeled, 2–2.2 × 1.7–1.8 mm, lateral keels without points; beak erect, 0.5–0.6 mm; sides with 1–3 tubercules near base; embryo with 1 full spiral.

Flowering and fruiting spring–summer. Small streams; of conservation concern; 1800 m; Tex.

Potamogeton clystocarpus is an extremely limited species known from only one canyon in west Texas.

16. **Potamogeton pusillus** Linnaeus, Sp. Pl. 1: 127. 1753 (as pusillum) W

Rhizomes absent. **Cauline stems** terete to slightly compressed, without spots, 18–150 cm; glands present on at least some nodes, green, gold, brown, or rarely white, to 0.5 mm diam. **Turions** common, lateral or terminal, 0.9–3.2 cm × 0.3–1.8 mm, soft; leaves ± 2-ranked; outer leaves 1–3 per side, base not corrugate, apex subulate to obtuse; inner leaves rolled into hardened fusiform structure. **Leaves** submersed, ± spirally arranged, sessile, delicate to coarse; stipules persistent, inconspicuous, connate or convolute, free from base of blade, brown to green or white, not ligulate, 0.31–0.92 cm, rarely appearing fibrous, not shredding at tip, apex obtuse; blade pale green to olive-green, rarely somewhat reddish, linear, not arcuate, 0.9–6.5 cm × 0.2–2.5 mm, base slightly tapering, without basal lobes, not clasping, margins entire, not crispate, apex not hoodlike, subulate to obtuse, lacunae absent or present, in 0–5 rows each side of midrib; veins 1–3(–5). **Inflorescences** unbranched, submersed or emersed; peduncles not dimorphic, axillary or terminal, erect, rarely recurved, filiform to slightly clavate, 0.5–6.2(–6.6) cm; spikes not dimorphic, capitate to cylindric, 1.5–10.1 mm. **Fruits** sessile, green to brown, ovoid to obovoid, turgid to concave, not abaxially or laterally keeled, 1.5–2.2 × 1.2–1.6 mm; beak erect, 0.1–0.6; sides without basal tubercles; embryo with less than 1 full spiral.

Subspecies 3 (3 in the flora): nearly worldwide.

Three hybrids, *Potamogeton perfoliatus* × *P. pusillus* (=*P. ×mysticus* Morong), *P. friesii* × *P. pusillus* (=*P. ×pusilliformis* Fisher [*P. ×intermedius* Fischer]), and *P. obtusifolius* × *P. pusillus* (=*P. ×saxonicus* Hagström), have been described.

1. Leaf blade 1-veined, subulate, 0.2–0.7 mm wide; New England and s Quebec. . . . .
. . . . . . 16c. *Potamogeton pusillus* subsp. *gemmiparus*
1. Leaf blade 1–5-veined, acute to obtuse, 0.2–2.5 mm wide; widespread throughout North America.
  2. Mature fruits obovoid, sides concave; beak toward adaxial edge, rarely median; peduncles filiform to cylindric, 1–3 per plant; inflorescences interrupted; leaf blade with 0–2 rows of lacunae along midrib, apex acute, rarely apiculate, rarely with bristle; stipules connate. . . . . 16a. *Potamogeton pusillus* subsp. *pusillus*
  2. Mature fruits widest at middle or ovoid, sides rounded; beak median, not toward adaxial edge; peduncles cylindric, more than 3 per plant; inflorescences continuous; leaf blade with 1–5 rows of lacunae along midrib, apex acute to obtuse; stipules convolute. . . . . . .
. . . . . 16b. *Potamogeton pusillus* subsp. *tenuissimus*

16a. Potamogeton pusillus Linnaeus subsp. **pusillus**
   • Potamot nain

Potamogeton panormitanus Bivona-Bernardi; *P. pusillus* var. *minor* (Bivona-Bernardi) Fernald & B. G. Schubert

Leaves: stipules connate; blade 1.4–6.5 cm × 0.5–1.9 mm, apex acute or rarely apiculate, rarely with bristle, lacunae present or absent, 0–2 rows each side of midrib; veins 1–3. **Inflorescences:** peduncles 1–3 per plant, filiform to cylindric; spikes cylindric, interrupted. **Fruits** obovoid, sides centrally concave; beak toward adaxial edge, rarely median. $2n = 26$ (Eurasia).

Flowering and fruiting spring–fall. Streams, lakes, or marshes; 0–3300 m; Alta., B.C., Man., N.B., N.W.T.,

N.S., Ont., P.E.I., Que., Sask., Yukon; Ala., Ariz., Ark., Calif., Colo., Conn., Del., D.C., Fla., Ga., Idaho, Ill., Ind., Iowa, Kans., Ky., La., Maine, Md., Mass., Mich., Minn., Miss., Mo., Mont., Nebr., Nev., N.H., N.J., N.Mex., N.Y., N.C., N.Dak., Ohio, Okla., Oreg., Pa., R.I., S.C., S.Dak., Tenn., Tex., Utah, Vt., Va., Wash., W.Va., Wis., Wyo.; South America; Eurasia; Africa.

Potamogeton pusillus subsp. *pusillus* is nearly worldwide. When it is in fruit, the inflorescence is interrupted. That character combined with its narrow, linear, 1–3-veined leaves makes this taxon easily recognized. The nodal glands are green, essentially the color of the stems. Often appearing only as bumps on the stem at the nodes, they are difficult to see. Also, because the glands frequently occur at only a few nodes per plant, one can easily overlook them.

16b. Potamogeton pusillus Linnaeus subsp. **tenuissimus** (Mertens & W. D. J. Koch) R. R. Haynes & Hellquist, Novon 6: 370. 1996 • Potamot très ténu

Potamogeton pusillus var. *tenuissimus* Mertens & W. D. J. Koch in J. C. Röhling, Deutschl. Fl. ed. 3, 1: 857. 1823; *P. berchtoldii* Fieber; *P. berchtoldii* var. *colpophilus* (Fernald) Fernald; *P. berchtoldii* var. *lacunatus* (Hagström) Fernald; *P. berchtoldii* var. *polyphyllus* (Morong) Fernald;
P. berchtoldii var. *tenuissimus* (Mertens & W. D. J. Koch) Fernald

Leaves: stipules convolute; blade 0.9–5.4 cm × 0.2–2.5 mm; apex acute to obtuse, lacunae in 1–5 rows each side of midrib; veins 1–3(–5). **Inflorescences:** peduncles more than 3 per plant, cylindric to slightly clavate; spikes capitate to cylindric, continuous. **Fruits** ovoid, sides rounded, rarely concave; beak median, rarely toward adaxial edge. $2n = 26$.

Flowering and fruiting summer–fall. Shallow waters of lakes and streams; 0–2100 m; Alta., B.C., Man., N.B., Nfld. and Labr., N.W.T., N.S., Nunavut, Ont., P.E.I., Que., Sask., Yukon; Ala., Alaska, Ariz., Ark., Calif., Colo., Conn., D.C., Fla., Idaho, Ill., Ind., Iowa, Ky., La., Maine, Md., Mass., Mich., Minn., Miss., Mont., Nev., N.H., N.J., N.Y., N.C., N.Dak., Ohio, Okla., Oreg., Pa., R.I., S.C., Vt., Va., Wash., Wis., Wyo.; Eurasia.

Although Delaware lies within the mapped area, we know of no collections from that state.

Potamogeton pusillus subsp. *tenuissimus* is the most common linear-leaved taxon of the family in temperate North America. Whenever one finds a linear-leaved pondweed with 1–5 rows of lacunae on each side of the midvein, chances are that it is subsp. *tenuissimus*. Only

Potamogeton obtusifolius could be confused with the taxon, and it can be separated by its cylindric inflorescence, whereas subsp. *tenuissimus* has a capitate inflorescence.

16c. Potamogeton pusillus Linnaeus subsp. **gemmiparus** (J. W. Robbins) R. R. Haynes & Hellquist, Novon 6: 370. 1996 • Potamot gemmipare E

Potamogeton pusillus var. *gemmiparus* J. W. Robbins in A. Gray, Manual ed. 5, 489. 1867; *P. gemmiparus* (J. W. Robbins) Morong

Leaves: stipules convolute; blade 1.1–6 cm × 0.2–0.7 mm, apex subulate, lacunae present or absent, 0–2 rows on each side of midrib; veins 1. **Inflorescences:** peduncles 1–3 per plant, cylindric; spikes cylindric, continuous. **Fruits** obovoid, sides centrally concave; beak median.

Flowering and fruiting summer–fall. Acid waters of lakes and streams; 0–100 m; Que.; Conn., Maine, Mass., N.H., Vt.

Potamogeton pusillus subsp. *gemmiparus* is an uncommon taxon that superficially resembles subsp. *pusillus*. It can be separated, however, by its continuous inflorescences, whereas those of subsp. *pusillus* are interrupted.

17. Potamogeton zosteriformis Fernald, Mem. Amer. Acad. Arts, n. s. 17: 36. 1932 • Flatstem pondweed, potamot zostériforme E F

Potamogeton zosterifolius Schumacher subsp. *zosteriformis* (Fernald) Hultén; *P. zosterifolius* var. *americanus* A. Bennett

Rhizomes absent. **Cauline stems** conspicuously flattened ("wing-flattened"), without spots, 60–120 cm; glands absent or rarely present, when present, gold, 0.3 mm diam. **Turions** common, terminal or lateral, 4–7.5 × 2–4.5 cm, firm; leaves ± 2-ranked; outer leaves 3–4 per side, base not corrugate, apex acute; inner leaves undifferentiated. **Leaves** submersed, ± spirally arranged, sessile, rigid; stipules persistent, conspicuous, convolute, free from blade, white, not ligulate, 1.5–3.5 cm, fibrous, shredding at tip, apex obtuse to acuminate; blade light green, linear, not arcuate, 10–20 cm × 2–5 mm, base rounded, without basal lobes, not clasping, margins entire, not crispate, apex not hoodlike, blunt, acuminate, or bristle-tipped, lacunae absent; veins 15–35. **Inflorescences** unbranched, emersed; peduncles not

dimorphic, terminal, erect to ascending, occasionally recurved, cylindric, 2–5 cm; spikes not dimorphic, cylindric, 15–30 mm. **Fruits** light green to olive-green, quadrate-oblong or nearly orbicular, turgid, sessile, abaxially keeled, not laterally keeled, 4–5 × 3–3.5 mm, keel winglike; beak erect, 0.6–1 mm; sides without basal tubercles; embryo with 1 full spiral. $2n = 52$.

Flowering summer–fall. Lakes, ponds, and slow streams; 0–1500 m; Alta., B.C., Man., N.B., Nfld. and Labr. (Nfld.), N.W.T., N.S., Ont., Que., Sask., Yukon; Alaska, Calif., Conn., Idaho, Ill., Ind., Iowa, Kans., Maine, Mass., Mich., Minn., Mont., Nebr., N.H., N.J., N.Y., N.Dak., Ohio, Oreg., Pa., S.Dak., Utah, Vt., Wash., Wis., Wyo.

One hybrid, *Potamogeton zosteriformis* × *P. strictifolius*, has been described and has been given the name *P.* ×*haynesii* Hellquist & G. E. Crow and is known from northern Michigan, Minnesota, Vermont, and southern Canada.

SELECTED REFERENCE Hellquist, C. B. and G. E. Crow. 1986. *Potamogeton* ×*haynesii* (Potamogetonaceae), a new species from northeastern North America. Brittonia 38: 415–419.

18. Potamogeton foliosus Rafinesque, Med. Repos., hexade 2, 5: 354. 1808 F W

Rhizomes absent. **Cauline stems** slightly compressed, without spots, 4–75 cm; glands rarely present, black to gold, to 0.5 mm diam. **Turions** uncommon, lateral or terminal, 0.9–2.5 cm × 0.6–2 mm, soft; outer leaves 1–3 per side, base not corrugate, apex acute to apiculate; inner leaves rolled into hardened fusiform structure. **Leaves** submersed, ± spirally arranged, sessile, delicate; stipules persistent, inconspicuous, convolute, free from blade, greenish to brown or rarely white, not ligulate, 0.2–2.2 cm, delicate to fibrous, rarely shredding at tip, apex obtuse; blade pale green to olive-green, rarely somewhat reddish, linear, not arcuate, 1.3–8.2 cm × 0.3–2.3 mm, base slightly tapering, without basal lobes, not clasping, margins entire, not crispate, apex not hoodlike, acute to apiculate, rarely with bristle, lacunae rarely present, 0–2 rows each side of midrib; veins 1–3(–5). **Inflorescences** unbranched, emersed; peduncles not dimorphic, in axils of proximal, rarely distal leaves, recurved, clavate, 0.3–1.1(–3.7) cm; spikes not dimorphic, capitate to cylindric, 1.5–7 mm. **Fruits** pale green to olive-green or brown, obovate to nearly orbicular, turgid to concave, sessile, abaxially keeled, not laterally keeled, 1.4–2.7 × 1.1–2.2 mm, abaxial keel undulate, winglike; beak erect, 0.2–0.6 mm; sides without basal tubercles; embryo with 1 full spiral.

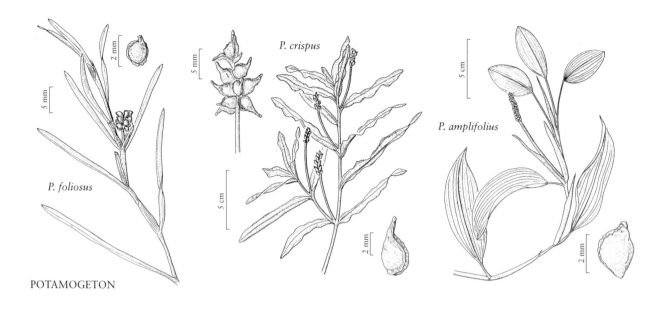

POTAMOGETON

Subspecies 2 (2 in the flora): North America, Central America.

1. Spike rarely interrupted; fruits olive to green-brown, 1.5–2.7 × 1.2–2.2 mm; keel 0.2 mm or more high; beak 0.2–0.6 mm; stipular veins decaying with age. . . . . 18a. *Potamogeton foliosus* subsp. *foliosus*
1. Spike interrupted; fruits pale green, 1.4–1.7 × 1.1–1.2 mm; keel less than 0.2 mm high; beak 0.2 or less; stipular veins with age remaining as fibers. . . . 18b. *Potamogeton foliosus* subsp. *fibrillosus*

18a. Potamogeton foliosus Rafinesque subsp. foliosus
• Leafy pondweed, potamot feuillé

Potamogeton curtissii Morong; *P. foliosus* var. *macellus* Fernald

Stems: glands rarely present, less than or equal to 0.3 mm diam. **Leaves:** stipules greenish to brown, delicate to slightly fibrous, veins decaying with age. **Inflorescences:** spikes continuous, rarely interrupted. **Fruits** olive-green to green-brown, 1.5–2.7 × 1.2–2.2 mm; keel greater than or equal to 0.2 mm high; beak 0.2–0.6 mm. $2n = 28$.

Flowering and fruiting spring–fall. Wide variety of waters of lakes and streams, either slow- or rapid-flowing; 0–2300 m; Alta., B.C., Man., N.B., N.W.T., N.S., Ont., P.E.I., Que., Sask., Yukon; Ala., Alaska, Ariz., Ark., Calif., Colo., Conn., Del., D.C., Fla., Ga., Idaho, Ill., Ind., Iowa, Kans., Ky., La., Maine, Md., Mass.,

Mich., Minn., Miss., Mo., Mont., Nebr., Nev., N.H., N.J., N.Mex., N.Y., N.C., N.Dak., Ohio, Okla., Oreg., Pa., R.I., S.C., S.Dak., Tenn., Tex., Utah, Vt., Va., Wash., W.Va., Wis., Wyo.; Mexico; West Indies; Central America (Guatemala, Costa Rica).

Potamogeton foliosus subsp. *foliosus* is probably the most common linear-leaved taxon of the family in North America, and it is probably the easiest to determine. Any linear-leaved *Potamogeton* specimen with fruits having an undulating winglike abaxial keel most likely is this taxon.

18b. Potamogeton foliosus Rafinesque subsp. fibrillosus (Fernald) R. R. Haynes & Hellquist, Novon 6: 370. 1996 Ⓒ Ⓔ

Potamogeton fibrillosus Fernald, Mem. Amer. Acad. Arts, n. s. 17: 51, plate 28, figs. a–c, plate 32, fig. 5. 1932; *P. foliosus* var. *fibrillosus* (Fernald) R. R. Haynes & Reveal

Stems: glands common, to 0.5 mm diam. **Leaves:** stipules brown or rarely white, fibrous, shredding at tip, veins in age remaining as fibers. **Inflorescences:** spikes interrupted. **Fruits** pale green, 1.4–1.7 × 1.1–1.2 mm; keel less than or equal to 0.2 mm high; beak less than or equal to 0.2 mm. Chromosome number not available.

Flowering and fruiting summer–fall. Warm waters of shallow lakes, springs, streams, and rivers; of conservation concern; 600–2700 m; Calif., Idaho, Oreg., Utah, Wash., Wyo.

Potamogeton foliosus subsp. *fibrillosus* is known from the warm waters of the northwestern United States. It differs from subsp. *foliosus* by the stipular tissue between the veins decomposing, leaving only strands formed by the fibrous veins. In addition, nodal glands are quite common.

19. **Potamogeton vaseyi** J. W. Robbins in A. Gray, Manual ed. 5, 485. 1867 • Vasey's pondweed, potamot de Vasey [E]

Rhizomes absent. **Cauline stems** terete, without spots, 2–5 cm; glands absent. **Turions** common, axillary, 0.5–2 cm × 0.5–1.2 mm, soft; leaves ± 2-ranked; outer leaves 2–3 per side, base not corrugate, apex acute; inner leaves undifferentiated or rolled into tight, hardened structure. **Leaves** submersed, or both submersed and floating, ± spirally arranged. **Submersed leaves** sessile, delicate; stipules persistent, inconspicuous, convolute, free from blade, green to brown, not ligulate, 0.4–1.2 cm, not fibrous, not shredding at tip, apex attenuate; blade light green, linear-filiform, not arcuate, 2–8 cm × 0.1–1 mm, base slightly tapering, without basal lobes, not clasping, margins entire, not crispate, apex not hoodlike, acute to almost bristle-tipped, lacunae present, rarely absent, 0–2 rows each side of midvein; veins 1(–3). **Floating leaves:** petioles continuous in color to apex, 5–25 mm; blade adaxially greenish brown, elliptic, spatulate, or obovate, 0.6–1.5 cm × 3–8 mm, base acute, apex obtuse; veins 5–9. **Inflorescences** unbranched, emersed; peduncles not dimorphic, terminal, ascending in flower, recurved in fruit, cylindric, 5–30 mm; spikes not dimorphic, cylindric or moniliform, 6–8 mm. **Fruits** sessile, green to brown, obliquely round-obovoid, compressed, abaxially keeled, not laterally keeled, 1.5–2.5 × 1.2–1.6 mm; beak erect, 0.3–0.5 mm; sides without basal tubercles; embryo with 1 full spiral. $2n = 28$.

Flowering and fruiting summer–fall. Quiet waters of lakes, ponds, and rivers; 50–500 m; N.B., Ont., Que.; Conn., Ill., Ind., Iowa, Maine, Mass., Mich., Minn., N.H., N.Y., Ohio, Pa., R.I., Vt., Wis.

All of the original material of *Potamogeton lateralis* Morong, including the collection designated as the lectotype, has been studied (C. B. Hellquist et al. 1988). Every specimen, was a mixed collection of *P. pusillus* and *P. vaseyi*. Based on the results of the study, *P. lateralis* is nomenclaturally invalid and should be rejected.

Potamogeton vaseyi is an uncommon species that has submersed leaves very similar to *P. pusillus* subsp. *gem-*

miparus. Floating leaves apparently are present only when the species is fertile, and the species often grows intermixed with that subspecies. Collections are consequently often a mixture of the two taxa. Also, sterile collections of either taxon can easily be mistaken for the other.

SELECTED REFERENCE Hellquist, C. B., C. T. Philbrick, and R. L. Hilton. 1988. The taxonomic status of *Potamogeton lateralis* Morong (Potamogetonaceae). Rhodora 90: 15–20.

20. **Potamogeton floridanus** Small, Fl. S.E. U.S., 37, 1326. 1903 [C] [E]

Rhizomes present. **Cauline stems** terete, without spots, ca. 50 cm. **Turions** unknown. **Leaves** submersed and floating, ± spirally arranged. **Submersed leaves** sessile, lax; stipules persistent, inconspicuous, convolute, free from blade, whitish, not ligulate, 2.5–4 cm, delicately fibrous, eventually shredding, apex acute; blade dark green to olive-green, narrowly linear, not arcuate, 15–20 cm × 0.7 mm, base attenuate, not clasping, not lobed, margins entire, not crispate, apex not hoodlike, acute, lacunae absent; veins 1. **Floating leaves:** petioles continuous in color to apex, 4–6.5 cm; blade adaxially dark brown, elliptic, 4.5–6.6 cm × 7–12 mm; base acute, apex acute; veins 7–11. **Inflorescences** unknown. **Fruits** unknown.

Flowering and fruiting unknown. Shallow waters of flowing streams; of conservation concern; 0–10 m; Fla.

Potamogeton floridanus is an extremely localized species of the panhandle of Florida. No one has treated the species recently (E. C. Ogden 1943; R. R. Haynes 1978). Four populations of the taxon were reported from Santa Rosa County, Florida (G. S. Wilhelm and R. H. Mohlenbrock 1986). Those populations clearly match the protologue and the original collections. One population has recently been found fertile (J. R. Burkhalter, pers. comm.), but we have not examined the specimens. Because the taxon persists and has a vegetative morphology unlike any other pondweed, we believe it must be recognized. Until a better understanding of the taxon is developed, we prefer to recognize it at the specific level.

SELECTED REFERENCE Wilhelm, G. S. and R. H. Mohlenbrock. 1986. The rediscovery of *Potamogeton floridanus* Small (Potamogetonaceae). Sida 11: 340–346.

21. Potamogeton oakesianus J. W. Robbins in A. Gray, Manual ed. 5, 485. 1867 • Oakes pondweed, potamot d'Oakes E

Rhizomes present. Cauline stems terete, with red spots, 7–75 cm; nodal glands absent. Turions absent. Leaves both submersed and floating, ± spirally arranged. Submersed leaves sessile, lax; stipules persistent, conspicuous, convolute, free from blade, whitish, not ligulate, 1–3 cm, delicately fibrous, not shredding at tip, apex obtuse; blade pale green, linear to phyllodial, not arcuate, 5–16 cm × (0.25–)0.3–1 mm, base slightly tapering, without basal lobes, not clasping, margins entire, not crispate, apex acute, not hoodlike, lacunae absent, setaceous; veins 3. Floating leaves: petioles continuous in color to apex, 3.2–7.5 cm; blade adaxially light to dark green, lanceolate to elliptic or ovate, (1.5–)2–4(–5.5) cm × 10–20(–30) mm, base rounded or tapering, apex acute; veins (7–)9–19(–23). Inflorescences unbranched, emersed; peduncles not dimorphic, terminal, ascending to spreading, cylindric, 2.5–8 cm; spikes not dimorphic, cylindric, 10–35 mm. Fruits sessile, greenish brown, obovoid, turgid, abaxially keeled, laterally keeled, 2.5–3.5(–3.7) × (1.6–)2–2.4 mm, lateral keels without points; beak erect, 0.4–0.8 mm; sides without basal tubercles; embryo with 1 full spiral. Chromosome number apparently unknown.

Flowering summer–fall. Quiet acidic waters of bogs, ponds, and lakes; 50–1500 m; St. Pierre and Miquelon; B.C., N.B., Nfld. and Labr., N.S., Ont., Que.; Conn., Maine, Mass., Mich., Minn., Mont., N.H., N.J., N.Y., Pa., R.I., Vt., Va., W.Va., Wis.

Potamogeton oakesianus, along with *Potamogeton floridanus* and *P. natans*, has floating leaves and phyllodial submersed leaves. The petioles of this species and of *P. floridanus* lack a short area of light-colored tissue immediately proximal to the blade.

22. Potamogeton natans Linnaeus, Sp. Pl. 1: 126. 1753 • Floating-leaf pondweed, potamot flottat

Rhizomes present. Cauline stems terete, often rust-spotted, 30–90 cm; nodal glands absent. Turions absent. Leaves both submersed and floating, ± spirally arranged. Submersed leaves sessile, rigid; stipules persistent, conspicuous, convolute, free from blade, whitish, not ligulate, 4.5–1 cm, fibrous, not shredding at tip, apex obtuse; blade light to dark green, phyllodial, not arcuate, 9–20 cm × 0.7–2.5 mm, base slightly tapering, without basal lobes, not clasping, margins entire, not crispate, apex not hoodlike, obtuse, lacunae absent; veins 3–5, obscure. Floating leaves: petioles lighter green immediately proximal to apex, 5.5–29 cm; blade adaxially light green, elliptic to ovate, 3.5–11 cm × 15–60 mm, base cordate, apex acute to rounded; veins 17–37. Inflorescences unbranched, emersed; peduncles not dimorphic, terminal, erect to ascending, cylindric, 4.5–9.5 cm; spikes not dimorphic, cylindric, 25–50 mm. Fruits sessile, green to greenish brown, obovoid, turgid, not abaxially or laterally keeled, 3.5–5 × 2–3 mm; beak erect to apically recurved, 0.4–0.8 mm; sides without basal tubercles; embryo with 1 full spiral. 2n = 52.

Flowering summer–fall. Quiet or slow-flowing waters of ponds, lakes, and streams; 0–3100 m; Alta., B.C., Man., N.B., Nfld. and Labr. (Nfld.), N.W.T., N.S., Ont., P.E.I., Que., Sask., Yukon; Alaska, Ariz., Calif., Colo., Conn., Idaho, Ill., Ind., Iowa, Kans., Maine, Mass., Mich., Minn., Mont., Nebr., Nev., N.H., N.J., N.Mex., N.Y., N.Dak., Ohio, Oreg., Pa., R.I., S.Dak., Utah, Vt., Wash., W.Va., Wis., Wyo.; Eurasia.

Potamogeton natans is the common floating-leaved pondweed of the north temperate areas. It is essentially circumboreal and can easily be identified by floating leaves that are almost always cordate at the base of the blade, the petiole with a short band of light tissue at its apex, and the submersed phyllodial leaves. Also, the apex of the petiole usually is bent so that the blade appears oriented in the opposite direction from which the petiole appears to be oriented.

Two hybrids, *Potamogeton natans* × *P. nodosus* (=*P.* ×*schreberi* Fischer [*P.* ×*perplexus* A. Bennett]) and *P. gramineus* × *P. natans* (=*P.* ×*sparganiifolius* Laestadius ex Fries), have been described.

23. Potamogeton crispus Linnaeus, Sp. Pl. 1: 126. 1753
(as crispum) • Curled pondweed, potamot crépu
F W

Rhizomes absent. **Cauline stems** flattened, without spots, to 100 cm; nodal glands absent. **Turions** common, axillary or terminal, 1.5–3 × ca. 2 cm, hard; leaves ± 2-ranked; outer leaves 1–4 per side, base not corrugate, apex rounded; inner leaves rolled into linear, terete structure, oriented parallel to outer leaves. **Leaves** submersed, ± spirally arranged, sessile, lax; stipules persistent to deliquescent, inconspicuous, convolute, free from blade, brownish, not ligulate, to 0.5 cm, not fibrous, not shredding at tip, apex obtuse; blade light to dark green, linear, not arcuate, 1.2–9 cm × 4–10 mm, base obtuse to rounded, without basal lobes, not clasping to nearly clasping, margins conspicuously serrate, not crispate, apex not hoodlike, round to round-acute, lacunae in 2–5 rows each side of midrib; veins 3–5. **Inflorescences** unbranched, emersed; peduncles not dimorphic, terminal or rarely axillary, erect to ascending, cylindric, 2.5–4 cm; spikes not dimorphic, cylindric, 10–15 mm. **Fruits** sessile, red to reddish brown, obovoid, turgid to slightly concave, not abaxially or laterally keeled, 6 × 2.5 mm; beak apically recurved, 2–3 mm; sides without basal tubercles; embryo with 1 full spiral. $2n = 52$ (Europe).

Flowering spring–summer. Quiet waters, especially brackish, alkaline, or eutrophic waters of ponds, lakes, and streams; 0–2000 m; introduced; Alta., B.C., Ont., Que., Sask.; Ala., Ariz., Ark., Calif., Conn., Del., D.C., Fla., Ga., Ill., Ind., Iowa, Kans., Ky., La., Maine, Md., Mass., Mich., Minn., Mo., Nebr., Nev., N.H., N.J., N.Y., N.C., Ohio, Okla., Oreg., Pa., R.I., S.Dak., Tenn., Tex., Utah, Vt., Va., Wash., W.Va., Wis., Wyo.; Central America (Costa Rica); South America (Colombia, Argentina); Eurasia; Australia.

No specimens have been seen from New Brunswick, but the species is to be expected there.

Potamogeton crispus, an introduced species, has spread throughout much of North America. The expansion of this species's range from its original collection in North America, apparently about 1840, has been discussed (R. L. Stuckey 1979). This is the only species of pondweeds in North America with serrate leaves and consequently it is easily recognized.

Life history of *Potamogeton crispus* is unusual as it flowers and fruits in late spring and early summer, at which time it also produces turions. The plants decay shortly after those structures develop, leaving only fruits and turions, which survive the summer. No one has observed any seed germination, but the turions (referred to as dormant apices) germinate in late summer or fall, and the plants overwinter as small plants only a few centimeters in size, even under the ice in northern climates (R. L. Stuckey et al. 1978). Growth then continues as the water begins warming in the spring.

One hybrid, *Potamogeton crispus* × *P. praelongus* (= *P.* ×*undulatus* Wolfgang ex Schultes & Schultes f.), has been described.

SELECTED REFERENCES Sastroutomo, S. S. 1981. Turion formation, dormancy and germination of curly pondweed, *Potamogeton crispus* L. Aquatic Bot. 10: 161–173. Stuckey, R. L. 1979. Distributional history of *Potamogeton crispus* (curly pondweed) in North America. Bartonia 46: 22–42. Stuckey, R. L., J. R. Wehrmeister, and R. J. Bartolotta. 1978. Submersed aquatic vascular plants in ice-covered ponds of central Ohio. Rhodora 80: 203–208.

24. Potamogeton amplifolius Tuckerman, Amer. J. Sci. Arts, ser. 2, 6: 225. 1848 • Broad-leaved pondweed, potamot à grandes feuilles E F

Rhizomes present. **Cauline stems** terete, often rusty spotted, 6–110 cm; nodal glands absent. **Turions** absent. **Leaves** both submersed and floating or floating absent, ± spirally arranged. **Submersed leaves** petiolate, lax; stipules persistent, conspicuous, convolute, free from blade, light brown, not ligulate, 1.5–11.7 cm, not fibrous, not shredding at tip, apex acute; petioles terete, 0.9–11.5 cm; blade light to dark green, ovate to oblanceolate, distinctly arcuate, 5–12.5 cm × 15–58 mm, base rounded to acute, without basal lobes, not clasping, margins entire, crispate, apex not hoodlike, acute to round-apiculate, lacunae absent; veins 19–49. **Floating leaves:** petioles continuous in color to apex, 2.3–22.6 cm; blade adaxially light green, lanceolate to round elliptic, 4.3–9.2 cm × 25–38 mm, base rounded to cordate, apex acute to rounded; veins 27–49. **Inflorescences** unbranched, emersed; peduncles not dimorphic, terminal or axillary, erect, cylindric, 4.5–22.3 cm; spikes not dimorphic, cylindric, 34–65 mm. **Fruits** sessile, reddish brown, obovoid, turgid, abaxially keeled, laterally ridged, 5–6.7 × 4.5–5.2 mm, lateral ridges without points; beak erect, 0.5–0.8 mm; sides without basal tubercles; embryo with full spiral. $2n = 52$.

Flowering summer–fall. Waters of lakes, ponds, streams, and rivers; 0–1900(–2900) m; B.C., Man., N.B., Nfld. and Labr. (Nfld.), N.S., Ont., Que., Sask.; Ala., Ark., Calif., Conn., Ga., Idaho, Ill., Ind., Iowa, Kans., Maine, Md., Mass., Mich., Minn., Mo., Mont., Nebr., N.H., N.J., N.Y., N.C., Ohio, Okla., Oreg., Pa., R.I., S.Dak., Tenn., Vt., Va., Wash., W.Va., Wis., Wyo.

No specimens have been seen from Kentucky or Texas, but the species is to be expected there.

Potamogeton amplifolius is common throughout much of North America. Its submersed leaves are larger than those of most other species of *Potamogeton*, are arcuate, and have more veins than do any other species.

One hybrid, *Potamogeton amplifolius* × *P. illinoensis* (=*P.* ×*scoliophyllus* Hagström), has been described.

25. Potamogeton pulcher Tuckerman, Amer. J. Sci. Arts 45: 38. 1843 • Spotted pondweed, potamot gracieux [E]

Rhizomes present. **Cauline stems** terete, conspicuously spotted, 8–95 cm; nodal glands absent. **Turions** absent. **Leaves** both submersed and floating, ± spirally arranged. **Submersed leaves** petiolate, lax; stipules deliquescent, inconspicuous, convolute, free from blade, light to dark brown, not ligulate, 0.7–1.2 cm, not fibrous, not shredding at tip, apex obtuse; petioles 0.5–4.5 cm; blade dark green, linear-lanceolate to lanceolate, often arcuate, 3.5–13.8 cm × 60–165 mm acute, base acute to rounded, without basal lobes, margins entire, crispate, apex not hoodlike, acute to obtuse, lacunae in 2–5 rows each side of midrib; veins 7–19. **Floating leaves:** petioles continuous in color to apex, 1–16.5 cm; blade adaxially light to dark green, lanceolate to round-ovate, 2.5–8.5 cm × 11–44 mm, base rounded to cordate, apex acute to rounded; veins 15–21. **Inflorescences** unbranched, emersed; peduncles not dimorphic, terminal or axillary, erect to ascending, cylindric, 3.3–9.4 cm; spikes not dimorphic, cylindric, 17–36 mm. **Fruits** sessile, dark green to dark brown, ovoid to obovoid, turgid, abaxially keeled, laterally ridged, 5–6.5 × 4.1–5 mm, lateral ridges without points; beak erect, 0.5 mm; embryo with 1 full spiral. Chromosome number not available.

Flowers summer–fall. Stagnant to slow-flowing waters of streams, lakes, ponds, and small rivers; 0–700 m; N.S., Ont.; Ala., Ark., Conn., Del., Fla., Ga., Ill., Ind., Ky., La., Maine, Md., Mass., Mich., Minn., Miss., Mo., N.H., N.J., N.Y., N.C., Ohio, Okla., Pa., R.I., S.C., Tenn., Tex., Va., W.Va., Wis.

Potamogeton pulcher is similar in morphology to *P. amplifolius* and occurs in similar habitats. *Potamogeton pulcher* differs from *P. amplifolius* by the former having lanceolate to linear-lanceolate submersed leaves with fewer than 19 veins, whereas the latter has ovate submersed leaves with more than 19 veins.

26. Potamogeton oblongus Viviani, Ann. Bot. (Genoa) 1(2): 102. 1802 • Potamot oblong [I]

Rhizome present. **Cauline stems** terete, without spots, 5–30 cm; nodal glands absent. **Turions** absent. **Leaves** both submersed and floating or only submersed, ± spirally arranged. **Submersed leaves** petiolate, lax; stipules persistent, conspicuous, convolute, free from blade, greenish, not ligulate, to 3 cm, not fibrous, not shredding at tip, apex acute; petioles 1–3(–8) cm; blade dark green, lanceolate to occasionally ovate, occasionally arcuate, 3–13 cm × 4–15(–44) mm, base acute, without basal lobes, not clasping, margins entire, not crispate, apex not hoodlike, acute, lacunae in 2–4 rows each side of midvein; veins 7–11. **Floating leaves:** petioles continuous in color to apex, 1–17 cm; blade adaxially reddish brown, ovate, 3–8(–10) cm × 10–46 mm, base rounded or slightly cordate, apex obtuse, acute, or somewhat apiculate; veins (11–)15–19(–21). **Inflorescences** unbranched, emersed; peduncles not dimorphic, erect to slightly recurved, cylindric, 4–12 cm; spikes not dimorphic, cylindric, 20–40 mm. **Fruits** sessile, dark reddish brown, obovoid to orbicular, compressed, not abaxially or laterally keeled, (1.6–)2–2.5 × (1.2–)1.5–2.1 mm; beak minute or obsolete, recurved; sides without basal tubercles; embryo with 1 full spiral. 2*n* = 26.

Flowering Jul–Aug. Acid to neutral waters of ponds, lakes, and occasionally streams; 10–100 m; introduced; St. Pierre and Miquelon; Nfld. and Labr. (Nfld.), N.S.; Europe; n Africa; Atlantic Islands (Azores).

27. Potamogeton nodosus Poiret in J. Lamarck et al., Encycl., Suppl. 4(2): 535. 1816 • Long-leaf pondweed, potamot noeuex [W]

Rhizomes present. **Cauline stems** terete, without spots, to 100 cm; nodal glands absent. **Turions** absent. **Leaves** both submersed and floating, or floating absent, ± spirally arranged. **Submersed leaves** petiolate, lax; stipules persistent to deliquescent, conspicuous, convolute, free from blade, light brown, not ligulate, 3–9 cm, not fibrous, not shredding at tip, apex round to acute; petioles 2–13 cm; blade light to dark green, linear-lanceolate to lance-elliptic, not arcuate, 9–20 cm × 10–35 mm, base acute, without basal lobes, not clasping, margins entire, not crispate, apex not hoodlike, acute, without sharp awl-like tip, lacunae in 2–5 rows each side of midrib; veins 7–15. **Floating leaves:** petioles 3.5–26 cm; blade adaxially light green, lenticular to elliptic, 3–11 cm × 15–45 mm, base cuneate to rounded, apex acute to rounded; veins 9–21. **Inflorescences** unbranched, emersed; peduncles not dimorphic, terminal, erect to ascending, cylindric, 3–15 cm; spikes not dimorphic, cylindric, 20–70 mm. **Fruits** sessile, red to reddish brown, obovoid, abaxially keeled, laterally ridged, 2.7–4.3 × 2.5–3 mm, keel well developed, lateral ridges with blunt to sharp tips; beak erect; sides without basal tubercles; embryo with 1 full spiral. $2n = 52$.

Flowering summer–fall. Clear to turbid waters of lakes, streams, rivers, and sloughs; 0–3300 m; Alta., B.C., N.B., Ont., Que., Sask.; Ala., Ariz., Ark., Calif., Colo., Conn., D.C., Fla., Ga., Idaho, Ill., Ind., Iowa, Kans., Ky., La., Maine, Md., Mass., Mich., Minn., Miss., Mo., Mont., Nebr., Nev., N.H., N.J., N.Mex., N.Y., N.C., N.Dak., Ohio, Okla., Oreg., Pa., R.I., S.C., S.Dak., Tenn., Tex., Utah, Vt., Va., Wash., W.Va., Wis., Wyo.; Mexico; West Indies; Central America; South America; Eurasia.

Six hybrids, *Potamogeton gramineus* × *P. nodosus* (=*P.* ×*argutulus* Hagström), *P. illinoensis* × *P. nodosus* (=*P.* ×*faxonii* Morong), *P. nodosus* × *P. richardsonii* (=*P.* ×*rectifolius* A. Bennett), *P. natans* × *P. nodosus* (=*P.* ×*schreberi* Fischer [*P.* ×*perplexus* A. Bennett]), *P. alpinus* × *P. nodosus* (=*P.* ×*subobtusus* Hagström), and *P. epihydrus* × *P. nodosus* (=*P.* ×*subsessilis* Hagström), have been described.

Potamogeton nodosus is a common floating-leaved species throughout much of the United States and southern Canada. When both submersed and floating leaves are present, it is very easily recognized by the petioles of the submersed leaves being longer than 5 cm.

28. **Potamogeton alpinus** Balbis, Misc. Bot., 13. 1804
   • Potamot alpin [F]

Potamogeton alpinus subsp. *tenuifolius* (Rafinesque) Hultén; *P. alpinus* var. *subellipticus* (Fernald) Ogden; *P. alpinus* var. *tenuifolius* (Rafinesque) Ogden; *P. tenuifolius* Rafinesque; *P. tenuifolius* var. *subellipticus* Fernald

Rhizome present. **Cauline stems** terete, without spots, to 200 cm; nodal glands absent. **Turions** absent. **Leaves** submersed and floating or floating absent, ± spirally arranged. **Submersed leaves** sessile, lax; stipules persistent, inconspicuous, convolute, free from blade, light brown to reddish, not ligulate, 1.2–)1.5–2.5(–4) cm, not fibrous, not shredding at tip, apex blunt; blade reddish green, oblong-linear to linear-lanceolate, not arcuate, 4.5–18(–25) cm × 5–20 mm, base rounded, without basal lobes,

slightly clasping, margins entire, not crispate, apex not hoodlike, obtuse or acute, lacunae in 0–6 rows each side of midvein; veins 7–9. **Floating leaves:** petioles continuous in color to apex, 0.1–1.2 cm; blade reddish green, elliptic or oblanceolate to obovate or oblong-linear, 4–7(–10) cm × 10–25(–40) mm, base gradually tapering into petiole, apex obtuse or acute; veins (7–)9–13(–15). **Inflorescences** unbranched, emersed; peduncles not dimorphic, terminal or axillary, erect, cylindric, 3–10(–16) cm; spikes not dimorphic, cylindric, 10–35 mm. **Fruits** pedicellate, tawny olive-green, obovoid, plump, turgid, abaxially keeled, laterally keeled or not, (2.5–)3–3.5 × (1.7–)2–2.4 mm, lateral keels when present without points; beak abaxially curved, 0.5–0.9 mm; sides without basal tubercles; embryo with 1 full spiral. $2n = 52$.

Flowering early summer–fall. Ponds, lakes, and slow-moving streams; 400–2500 m; Greenland; Alta., B.C., Man., N.B., Nfld. and Labr., N.W.T., N.S., Nunavut, Ont., P.E.I., Que., Sask., Yukon; Alaska, Calif., Colo., Idaho, Maine, Mass., Mich., Minn., Mont., Nev., N.H., N.J., N.Y., Oreg., Pa., S.Dak., Utah, Vt., Wash., Wis., Wyo.; Eurasia.

Plants of *Potamogeton alpinus* often are red whenever taken from the water, a feature that makes this species quite distinctive.

Two varieties, *Potamogeton alpinus* var. *tenuifolius* and var. *subellipticus*, have been recognized in North America, based mainly on submersed leaf shape. Plants bearing both leaf types have been observed in the same population; hence the varieties are not recognized.

Four hybrids, *Potamogeton alpinus* × *P. nodosus* (=*P.* × *subobtusus* Hagström), *P. alpinus* × *P. gramineus* (=*P.* ×*nericius* Hagström), *P. alpinus* × *P. praelongus* (=*P.* ×*griffithii* A. Bennett), and *P. alpinus* × *P. perfoliatus* (=*P.* ×*prussicus* Hagström), have been described.

29. **Potamogeton gramineus** Linnaeus, Sp. Pl. 1: 127. 1753 (as gramineum)

Potamot à feuilles de graminées

Potamogeton gramineus var. *maximus* Morong; *P. gramineus* var. *myriophyllus* J. W. Robbins

Rhizomes present. **Cauline stems** terete to flattened, without spots, to 150 cm; nodal glands absent. **Turions** absent. **Leaves** both submersed and floating or submersed only, ± spirally arranged. **Submersed leaves** sessile or rarely petiolate, lax; stipules persistent, inconspicuous, convolute, free from blade, pale green to brown, not ligulate, 1.3–1.6 cm, not fibrous, not shredding at tip, apex acute to obtuse; peti-

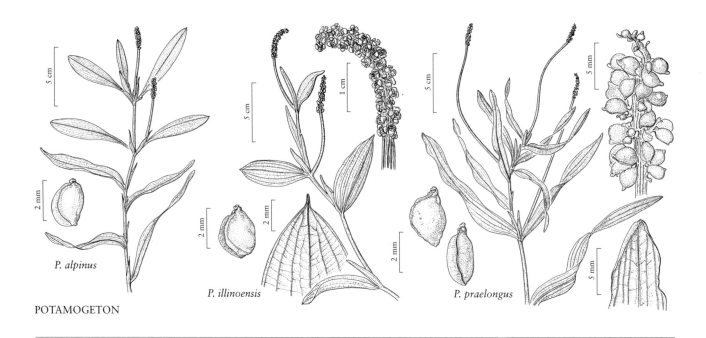

P. alpinus

POTAMOGETON

P. illinoensis

P. praelongus

oles to 3 cm; blade light green to brownish green, elliptic, not arcuate, 3.1–9.1 cm × 3–27 mm, base
attenuate, without basal lobes, not clasping, margins
entire, rarely crispate, apex not hoodlike, acuminate,
lacunae in 1–2 rows each side of midvein; veins 3–9.
Floating leaves: petioles continuous in color to apex, 3–
4.5 cm; blade yellow-green to dark green, elliptic to
ovate, 3.5–4 cm × 16–20 mm, base rounded, apex acuminate; veins 11–13. **Inflorescences** emersed, unbranched; peduncles not dimorphic, both axillary and
terminal, erect to ascending, cylindric, 3.2–7.7 cm;
spikes not dimorphic, cylindric, 15–35 mm. **Fruits** sessile, greenish brown, ovoid, laterally compressed, abaxially and laterally keeled, 1.9–2.3 × 1.8–2 mm, lateral
keels without points; beak erect, 0.3–0.5 mm; sides
without basal tubercles; embryo with less than 1 full
spiral. $2n = 52$.

Flowering summer–fall. Ponds, lakes, streams, and
rivers; 0–3500 m; Greenland; St. Pierre and Miquelon;
Alta., B.C., Man., N.B., Nfld. and Labr., N.W.T., N.S.,
Nunavut, Ont., P.E.I., Que., Sask., Yukon; Alaska,
Ariz., Calif., Colo., Conn., Idaho, Ill., Ind., Iowa, Kans.,
Ky., Maine, Md., Mass., Mich., Minn., Mont., Nebr.,
Nev., N.H., N.J., N.Y., N.Dak., Ohio, Oreg., Pa., R.I.,
S.Dak., Utah, Vt., Wash., Wis., Wyo.; Eurasia.

Seven hybrids, *Potamogeton gramineus* × *P. nodosus* (=*P.* ×*argutulus* Hagström), *P. gramineus* × *P.
richardsonii* (=*P.* ×*hagstroemii* A. Bennett [as hags-
tromii]), *P. alpinus* × *P. gramineus* (=*P.* ×*nericius*
Hagström), *P. gramineus* × *P. perfoliatus* (=*P.* ×*nitens*
Weber [*P.* ×*subnitens* Hagström]), *P. gramineus* × *P.
natans* (=*P.* ×*sparganiifolius* Laestadius ex Fries), *P.*

gramineus × *P. illinoensis* [=*P.* ×*spathuliformis* (J. W.
Robbins) Morong], and *P. gramineus* × *P. praelongus*
(=*P.* ×*vilnensis* Galinis), have been described.

Three varieties were recognized (E. C. Ogden 1943)
and treated (M. L. Fernald 1950). These varieties, *Potamogeton gramineus* var. *gramineus*, var. *myriophyllus*, and var. *maximus*, were said to be separated by the
shape and size of the submersed leaves. We have studied
many populations of this species in the field and have
observed on several occasions that a single population
has leaf morphology variable enough to include all
three varieties. We have therefore chosen not to recognize any infraspecific categories for this species.

30. **Potamogeton illinoensis** Morong, Bot. Gaz. 5: 50.
1880 • Illinois pondweed, potamot de l'Illinois [F]

Rhizomes present. **Cauline stems**
terete, without spots, 28–120 cm;
nodal glands absent. **Turions** absent. **Leaves** both submersed and
floating or the floating absent, ±
spirally arranged. **Submersed
leaves** sessile or petiolate, lax;
stipules persistent, conspicuous,
convolute, free from blade, light
brown to red-brown, not ligulate, 1–8 cm, not fibrous,
not shredding at tip, apex acute; petioles if present 0.5–
4 cm; blade red-brown to light green, elliptic to lanceolate or rarely linear, often arcuate, 5–20 cm × 2–45
mm, base acute, margins entire, often crispate, apex not
hoodlike, acute-mucronate, lacunae in 2–5 rows each

side midrib; veins 7–19. **Floating leaves:** petioles continuous in color to apex, 2–9 cm; blade adaxially light green, elliptic to oblong-elliptic, 4–19 cm × 20–65 mm, base cuneate, apex round-mucronate; veins 13–29. **Inflorescences** emersed, unbranched; peduncles not dimorphic, terminal or axillary, erect to ascending, cylindric, 4–30 cm; spikes not dimorphic, cylindric, 25–70 mm. **Fruits** sessile, grayish green to olive green, obovoid to ovoid, laterally compressed, abaxially keeled, laterally ridged, 2.5–3.6 × 2.1–3 mm, abaxial keel well developed, lateral ridges without points; beak erect to slightly recurved, 0.5–0.8 mm; sides without basal tubercles; embryo with 1 full spiral. $2n = 104$.

Flowering and fruiting summer–fall. Alkaline waters of streams, rivers, lakes, ponds, and sloughs; 0–2700 m; B.C., Man., N.B., N.W.T., Ont., Que.; Ala., Ark., Calif., Colo., Conn., Fla., Ga., Ill., Ind., Iowa, Kans., Ky., Md., Mass., Mich., Minn., Mo., Mont., Nebr., N.H., N.J., N.Y., N.C., Okla., Ohio, Oreg., Pa., S.C., S.Dak., Tex., Utah, Vt., Va., Wash., W.Va., Wis., Wyo.; Mexico; West Indies; Central America; South America.

Potamogeton illinoensis and *P. gramineus* are often difficult to separate. Certainly, in the extreme of each they are easily separated, but they continually grade into each other. Features to look for are the acute-mucronate apex of the submersed leaves of *P. illinoensis* and the acuminate apex for *P. gramineus*. Also, the number of veins seems to work as well.

Three hybrids, *Potamogeton illinoensis* × *P. nodosus* (=*P.* ×*faxonii* Morong), *P. amplifolius* × *P. illinoensis* (=*P.* ×*scoliophyllus* Hagström), and *P. gramineus* × *P. illinoensis* [=*P.* × *spathuliformis* (J. W. Robbins) Morong], have been described.

31. **Potamogeton praelongus** Wulfen, Arch. Bot. (Leipzig) 3: 331. 1805 • White-stemmed pondweed, potamot à longs pédoncules [F]

Rhizome present. **Cauline stems** terete, without spots, to 210 cm; nodal glands absent. **Turions** absent. **Leaves** submersed, ± spirally arranged, sessile, lax; stipules persistent, conspicuous, convolute, free from blade, white to green, not ligulate, 3–8.1 cm, fibrous, shredding at apex; blade pale green, rarely olive green, linear-lanceolate, not arcuate, 0.8–2.8 cm × 11–46 mm, base clasping, without basal lobes, margins entire, rarely crispate, apex hoodlike, splitting when pressed, obtuse, lacunae absent; veins 11–33. **Inflorescences** emersed, unbranched; peduncles not dimorphic, terminal or axillary, erect to spreading, cylindric, 9.5–53 cm; spikes not dimorphic, cylindric, 34–75 mm. **Fruits** sessile, greenish brown, ob-

ovoid, turgid, abaxially keeled, occasionally laterally keeled, 4–5.7 × 3.2–4 mm, lateral keels when present without points; beak erect, 0.6–1 mm; sides without basal tubercles; embryo with 1 full spiral. $2n = 52$.

Flowering summer–fall. Neutral to alkaline waters of lakes, rivers, and streams; 100–3000 m; Greenland; Alta., B.C., Man., N.B., Nfld. and Labr. (Nfld.), N.W.T., N.S., Nunavut, Ont., P.E.I., Que., Sask., Yukon; Alaska, Calif., Colo., Conn., Idaho, Ind., Iowa, Maine, Mass., Mich., Minn., Mont., Nebr., N.H., N.J., N.Y., Oreg., Pa., S.Dak., Utah, Vt., Wis., Wyo.; Mexico; Eurasia.

Potamogeton praelongus is one of the easiest pondweeds to identify with its submersed leaves only clasping the more or less zigzagged stem. The persistent, large, white stipules provide another clue to this species.

Four hybrids, *Potamogeton perfoliatus* × *P. praelongus* (=*P.* ×*cognatus* Ascherson & Graebner), *P. alpinus* × *P. praelongus* (=*P.* ×*griffithii* A. Bennett), *P. crispus* × *P. praelongus* (=*P.* ×*undulatus* Wolfgang ex Schultes & Schultes f.), and *P. gramineus* × *P. praelongus* (=*P.* ×*vilnensis* Galinus), have been described.

32. **Potamogeton richardsonii** (A. Bennett) Rydberg, Bull. Torrey Bot. Club 32: 599. 1905 • Potamot de Richardson [E] [F] [W]

Potamogeton perfoliatus Linnaeus var. *richardsonii* A. Bennett, J. Bot. 27: 25. 1889; *P. perfoliatus* subsp. *richardsonii* (A. Bennett) Hultén

Rhizome present. **Cauline stems** terete, without spots, to 100 cm; nodal glands absent. **Turions** absent. **Leaves** submersed, ± spirally arranged, sessile, lax; stipules persistent, conspicuous, convolute, free from blade, white, not ligulate, 0.12–0.17 cm, fibrous, disintegrating to persistent fibers, even on proximal portion of stem, shredding at apex, apex obtuse; blade olive green, ovate-lanceolate to narrowly lanceolate, not arcuate, 1.6–13 cm × 5–28 mm, base rounded, without basal lobes, clasping, margins entire to crispate, apex not hoodlike, not splitting when pressed, acute to obtuse, lacunae absent; veins 3–35. **Inflorescences** emersed, unbranched; peduncles not dimorphic, terminal or axillary, erect to rarely recurved, clavate, 1.5–14.8 cm; spikes not dimorphic, cylindric, 13–37 mm. **Fruits** sessile, greenish brown, obovoid, turgid to concave, not or rarely abaxially keeled, not laterally keeled, 2.2–4.2 × 1.7–2.9 mm; beak erect, 0.4–0.7 mm; sides without basal tubercles; embryo with 1 full spiral. $2n = 52$.

Flowering summer–fall. Alkaline waters of lakes, streams, and rivers; 0–3000 m; Alta., B.C., Man., N.B., N.W.T., N.S., Ont., Que., Sask., Yukon; Alaska, Calif.,

Colo., Conn., Idaho, Ill., Ind., Iowa, Maine, Mass., Mich., Minn., Mont., Nebr., Nev., N.H., N.Y., N.Dak., Ohio, Oreg., Pa., S.Dak., Utah, Vt., Wash., Wis., Wyo.

Potamogeton richardsonii is quite similar to *P. perfoliatus*. Specific characteristics to separate the two species are the shape of the leaf blade apex, acute in *P. richardsonii* and obtuse in *P. perfoliatus*, and the condition of the stipules, disintegrating between the veins leaving fibrous strands in *P. richardsonii*, and the entire stipule, including the veins, disintegrating in *P. perfoliatus*.

Two hybrids, *Potamogeton gramineus* × *P. richardsonii* (=*P.* ×*hagstroemii* A. Bennett [as *hagstromii*]) and *P. nodosus* × *P. richardsonii* (=*P.* ×*rectifolius* A. Bennett), have been described.

33. **Potamogeton perfoliatus** Linnaeus, Sp. Pl. 1: 126. 1753 (as *perfoliatum*) · Potamot perfolié

Rhizomes present. **Cauline stems** terete, without spots, to 250 cm; nodal glands absent. **Turions** absent. **Leaves** submersed, ± spirally arranged, sessile, lax; stipules deteriorating and deciduous, inconspicuous (absent on proximal portion of stem), convolute, free from blade, light brown to green, not ligulate, 3.5–6.5 cm, not fibrous, not shredding at apex; blade olive-green, broadly lanceolate, orbiculate, or ovate, not arcuate, 0.9–7.6(–9.7) cm × 7–40 mm, base rounded, without basal lobes, clasping, margins entire, often crispate, apex not hoodlike, not splitting when pressed, round, rarely acute; veins 3–25. **Inflorescences** unbranched, emersed; peduncles not dimorphic, terminal or axillary, erect to rarely recurved, cylindric, 1–7.3 cm; spike not dimorphic, cylindric, 0.4–4.8 cm. **Fruits** sessile, greenish brown to olive-green, obovoid, turgid or rarely concave, not abaxially or laterally keeled, 1.6–3 × 1.3–2.2 mm; beak erect, 0.4–0.6 mm; sides without basal tubercles; embryo with 1 full spiral. 2*n* = 52.

Flowering summer–fall. Waters, often brackish, of lakes, streams, rivers, and bays; 0–100 m; St. Pierre and Miquelon; N.B., Nfld. and Labr., N.S., Ont., P.E.I., Que.; Ala., Conn., Del., D.C., Fla., La., Maine, Md., Mass., N.H., N.J., N.Y., N.C., Ohio, Pa., R.I., Vt., Va.; Central America (Guatemala); Eurasia; Africa; Australia.

Four hybrids, *Potamogeton perfoliatus* × *P. praelongus* (=*P.* ×*cognatus* Ascherson & Graebner), *P. perfoliatus* × *P. pusillus* (=*P.* ×*mysticus* Morong), *P. gramineus* × *P. perfoliatus* (=*P.* ×*nitens* Weber [*P.* ×*subnitens* Hagström]), and *P. alpinus* × *P. perfoliatus* (=*P.* ×*prussicus* Hagström), have been described.

2. STUCKENIA Börner, Abh. Naturwiss. Vereine Bremen 21: 258. 1912 · Potamot

Coleogeton (Reichenbach) Les & R. R. Haynes

Herbs: rhizomes present; turions absent; tubers absent or present. **Stems** terete, nodes without oil glands. **Leaves** submersed, alternate, opaque, sessile, linear, channeled, turgid, base acute, margins entire, apex obtuse to acute, veins 1–5; stipules not tubular, adnate to base of leaf blades for ⅔ or more length of stipule, extending past adnation as free ligule. **Inflorescences** spikes, capitate or cylindric, submersed; peduncles flexible, not projecting inflorescence above surface of water. **Flowers:** pistils 4. **Fruits** abaxially rounded, beaked or not, turgid; embryo with less than 1 full coil. *x* = 13.

Species ca. 6 (4 in the flora): nearly worldwide.

The stipules of *Stuckenia* are adnate to the blade for two-thirds to nearly the entire length of the stipule. The few species of *Potamogeton* with adnate stipules have the adnation less than half the length of the stipule, in fact, less than 4 mm. Submersed leaves of *Potamogeton* are translucent, flat, and without grooves or channels, whereas those of *Stuckenia* are opaque, channeled, and turgid.

A proposal to elevate *Potamogeton* subg. *Coleogeton* to the generic level, retaining the name *Coleogeton*, was presented (D. H. Les and R. R. Haynes 1996). *Potamogeton pectinatus* Linnaeus was chosen as the nomenclatural type. The name *Stuckenia* had been previously published, however, and *P. pectinatus* cited (C. Börner 1912), making the generic name *Coleogeton* superfluous. *Stuckenia* is the correct name, and the appropriate specific combinations have been made (J. Holub 1997).

SELECTED REFERENCES Börner, C. 1912. Botanisch-systematische Notizen. Abh. Naturwiss. Vereine Bremen 21: 245–282. Holub, J. 1997. *Stuckenia* Börner 1912—the correct name for *Coleogeton* (Potamogetonaceae). Preslia 69: 361–366. Les, D. H. and R. R. Haynes. 1996. *Coleogeton* (Potamogetonaceae), a new genus of pondweeds. Novon 6: 389–391. St. John, H. 1916. A revision of the North American species of *Potamogeton* of the section *Coleophylli*. Rhodora 18: 121–138.

1. Leaf apex acute, apiculate, cuspidate, rarely round; proximal stipular sheaths not inflated; stems abundantly branched, especially on distal portions; fruits distinctly beaked.
  2. Leaves 0.4–1.5–3(–8.5) mm wide; leaf apex apiculate, cuspidate, or rarely round; plants from restricted range in w United States. . . . . 1. *Stuckenia striata*
  2. Leaves 0.2–1 mm wide; leaf apex acute to mucronate or apiculate; widespread throughout United States and Canada. . . . . 2. *Stuckenia pectinata*
1. Leaf apex notched, obtuse, or round, rarely apiculate; proximal stipular sheaths often inflated; stems sparsely branched on distal portions; fruits without beak.
  3. Stipules with distinct ligules to 20 mm, especially on distal stipules; summit of midstem stipules tight to stem, ± same width as stem; fruit 2–3 mm. . . . . 3. *Stuckenia filiformis*
  3. Stipules without ligules or these to 2 mm on distal stipules; summit of midstem stipules inflated at least 2 times width of stem; fruit 3–3.8 mm. . . . . 4. *Stuckenia vaginata*

1. Stuckenia striata (Ruíz & Pavon) Holub, Preslia 68: 364. 1997 • Nevada-pondweed

Potamogeton striatus Ruíz & Pavon, Fl. Peruv. 1: 70, fig. 1066. 1798; *P. latifolius* (J. W. Robbins) Morong

Stems branched distally, terete to 5-ridged, to 200 cm. **Leaves:** length and width of those on main stem 2 times or more those on branches; stipules: stipular sheaths not inflated, 1.2–3.4 cm, ligule 0.2–1.1 cm; blade linear, 5–21 cm × 0.4–5.1 (–8.5) mm, apex apiculate, cuspidate, or rarely round; veins 3–5. **Inflorescences:** peduncles axillary, rarely terminal, erect to ascending, cylindric, 1.2–5.2 cm; spikes cylindric, rarely moniliform, 13–45 mm; verticels 4–9. **Fruits** brown to reddish brown, obovoid to oblanceoloid, 3–3.9 × 2.8–3 mm; beak toward abaxial margin, erect or rarely recurved to apex, 0.2–0.3 mm. Chromosome number not available.

Flowering summer–fall. Waters of alkaline rivers, canals, and ponds; 800–2000 m; Ariz., Calif., Colo., Idaho, N.Mex., Nev., Oreg., Tex., Utah; Mexico; Central America; South America.

Stuckenia striata is widespread, although not common, in the western United States, extending southward into Argentina and Chile. *Potamogeton latifolius* [no combination in *Stuckenia* has been proposed] was accepted by R. F. Thorne (1993b). After examining specimens throughout the range and studying dozens of populations in the field, we have determined that the two names represent the same taxon. We are placing *P. latifolius* in synonymy.

2. Stuckenia pectinata (Linnaeus) Börner, Fl. Deut. Volk, 713. 1912 • Sago-pondweed, potamot pectiné [F]

Potamogeton pectinatus Linnaeus, Sp. Pl. 1: 127. 1753 (as pectinatum)

Stems branched, especially distally, terete to slightly compressed, to 75 cm. **Leaves:** those of main stem only slightly larger than those of branches; stipules: stipular sheaths not inflated, 0.8–1.1 cm, ligule 0.8 mm; blade linear, 5.6–9.2 cm × 0.2–1 mm, apex acute to mucronate or apiculate; veins 1–3. **Inflorescences:** peduncles terminal or axillary, erect to ascending, cylindric, 4.5–11.4 cm; spikes moniliform to cylindric, 14–22 mm; verticels 3–5. **Fruits** yellow-brown to brown, oblanceoloid, 3.8–4 × 2.5–3.1 mm; beak toward abaxial margin, erect, 0.5–1.1 mm. $2n = 78$.

Flowering summer–fall. Brackish to alkaline waters of lakes, streams, rivers, and estuaries; 0–2400 m; St. Pierre and Miquelon; Alta., B.C., Man., N.B., Nfld. and Labr. (Nfld.), N.W.T., N.S., Ont., P.E.I., Que., Sask., Yukon; Ala., Alaska, Ariz., Ark., Calif., Colo., Conn., Fla., Idaho, Ill., Ind., Iowa, Kans., Ky., La., Maine, Md., Mass., Mich., Minn., Mo., Mont., Nebr., Nev., N.H., N.J., N.Mex., N.Y., N.C., N.Dak., Ohio, Okla., Oreg., Pa., R.I., S.Dak., Tenn., Tex., Utah, Vt., Va., Wash., W.Va., Wis., Wyo.; Mexico; Central America; South America; Eurasia; Australia.

No specimens have been seen from Delaware, but the species is to be expected there.

The sago-pondweed is among the most important

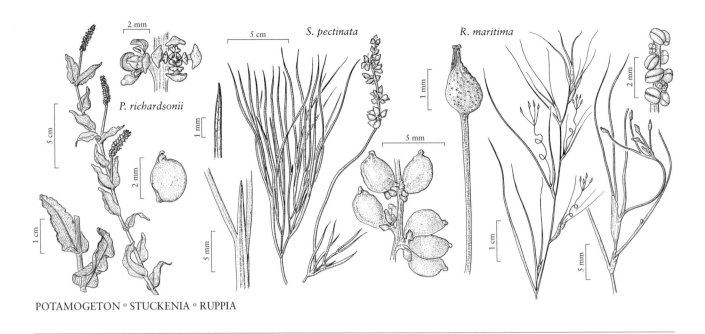

POTAMOGETON ∘ STUCKENIA ∘ RUPPIA

species as food for waterfowl (E. Moore 1913). The species reproduces vegetatively by underground tubers and is spread by various duck species, especially canvasbacks. In a study of food for ducks, a population of canvasbacks was observed feeding in aquatic vegetation comprised of several genera, including sago-pondweed. When the stomach contents were examined, they were found to contain essentially 100% tubers of sago-pondweed (E. Moore 1913).

Two hybrids with this species as a putative parent have been described under the genus *Potamogeton*. These are *P. pectinatus* × *P. vaginatus* (=*P.* ×*bottnicus* Hagström) and *P. filiformis* × *P. pectinatus* (=*P.* ×*suecicus* K. Richter).

3. **Stuckenia filiformis** (Persoon) Börner, Fl. Deut. Volk, 713. 1912 • Threadleaf-pondweed

Potamogeton filiformis Persoon, Syn. Pl. 1: 152. 1805

Stems freely branching proximally, sparsely branching distally, subterete, (10–)20–60(–100) cm. **Leaves:** length and width of those on main stem only slightly larger than those on branches; stipules: stipular sheaths often inflated on proximal portion of stem, 1–4(–9.5) cm, summit of midstem stipules tight to stem, ± same width as stem, ligule 2–20 mm, distinct, especially on distal stipules; blade filiform or slenderly linear, 1–15 cm × 0.2–2(–3.7) mm, apex notched, blunt, or short-apiculate;

veins 1–3. **Inflorescences:** peduncles terminal, erect, filiform to slender, 2–10(–15) cm; spikes cylindric to moniliform, 5–55 mm; verticels 2–6(–9). **Fruits** dark brown, obovoid, 2–3 × 1.5–2.4 mm; beak inconspicuous.

Subspecies ca. 5 (3 in the flora): nearly worldwide.

Three distinct subspecies apparently occur in North America. They are separated mainly by the size of the plants and the peduncle characteristics. *Stuckenia filiformis* subsp. *occidentalis* typically grows in cold deep water, standing or with a strong current. This variety tends to become robust and is easily confused with *Stuckenia vaginata* and *S. striata*. *Stuckenia filiformis* subsp. *alpina* is a much smaller plant typically growing in standing waters. *Stuckenia filiformis* subsp. *filiformis* is restricted to the far north. In the intermountain region of western United States is a more robust form that is quite similar to *S. filiformis* subsp. *alpina* although it has previously been recognized as *Potamogeton* [*Stuckenia*] *filiformis* var. *macounii* Morong (J. L. Reveal 1977b).

1. Plants 20–100 cm; stipules on proximal portion of stem inflated, disintegrating with age; fruits often absent. . . . . .3c. *Stuckenia filiformis* subsp. *occidentalis*
1. Plants 10–30 cm; stipules on proximal portion of stem tightly clasping or slightly enlarged, persistent; fruits common.
    2. Peduncles with flowers and/or fruits 4 cm or more apart; leaves 0.2–0.5 mm wide. . . . . .3a. *Stuckenia filiformis* subsp. *filiformis*

2. Peduncles with flowers and/or fruits less than
   4 cm apart; leaves to 1 mm wide. . . . .
   . . . . .3b. *Stuckenia filiformis* subsp. *alpina*

3a. Stuckenia filiformis (Persoon) Börner subsp.
filiformis

Stems 10–30 cm. **Leaves:** stipules persistent, those on proximal portion of stem tightly clasping or slightly enlarged, 0.2–0.5 mm wide. **Inflorescences:** peduncles with flowers and/or fruits 4 cm or more apart. **Fruits** common.

Flowering summer. Calcareous waters of ponds, lakes, and streams; 0–915 m; Greenland; N.W.T., Yukon; Alaska; Eurasia.

No specimens have been seen from British Columbia or Nunavut, but the subspecies is to be expected there.

3b. Stuckenia filiformis (Persoon) Börner subsp. **alpina**
(Blytt) R. R. Haynes, Les, & M. Král, Novon 8: 241.
1998 • Potamot filiforme alpin

Potamogeton marinus Linnaeus f. *alpinus* Blytt in M. N. Blytt and A. G. Blytt, Norges Fl. 1: 370. 1861; *P. borealis* Rafinesque; *P. filiformis* Persoon var. *alpinus* (Blytt) Ascherson & Graebner; *P. filiformis* var. *borealis* (Rafinesque) H. St. John; *P. filiformis* var. *macounii* Morong; *P. interior* Rydberg; *P. marinus* var. *alpinus* (J. W. Robbins) Morong; *P. marinus* var. *macounii* Morong

Stems 10–30 cm. **Leaves:** stipules persistent; those on proximal portion of stem tightly clasping or slightly enlarged, to 1 mm wide. **Inflorescences:** peduncles with flower and/or fruits adjacent or separated by less than 2 cm. **Fruits** common. $2n = 78$.

Flowering spring–early fall. Calcareous, saline, or brackish shallow to deep waters of ponds, lakes, streams, ditches, and coastal inshore waters; 0–3280 m; Greenland; Alta., B.C., Man., N.B., Nfld. and Labr., N.W.T., N.S., Nunavut, Ont., P.E.I., Que., Sask., Yukon; Alaska, Ariz., Calif., Colo., Idaho, Maine, Mich., Minn., Mont., Nebr., Nev., N.H., N.J., N.Mex., N.Y., N.Dak., Ohio, Oreg., Pa., S.Dak., Utah, Vt., Wash., Wis., Wyo.; Asia.

No specimens have been seen from Maryland, but the subspecies is to be expected there.

One hybrid, *P. filiformis* × *P. pectinatus* (=*P.* ×*suecicus* K. Richter), with this species as a putative parent has been described under the genus *Potamogeton*.

3c. Stuckenia filiformis (Persoon) Börner subsp.
occidentalis (J.W. Robbins) R. R. Haynes, Les, & M.
Král, Novon 8: 241. 1998 • Potamot filiforme
occidental E

Potamogeton marinus Linnaeus var. *occidentalis* J. W. Robbins in S. Watson, Botany (Fortieth Parallel), 339. 1871

Stems 20–100 cm. **Leaves:** stipules disintegrating with age, those on proximal portion of stem inflated, 0.2–5 mm wide. **Inflorescences:** peduncles with flower and/or fruits adjacent or separated by less than 2 cm. **Fruits** often absent.

Flowering spring–early fall. Calcareous waters, mainly in cold, slow- to fast-flowing streams and rivers, occasionally in standing waters of ponds and lakes; 0–2195 m; Alta., B.C., Man., Nfld. and Labr., N.W.T., N.S., Nunavut, Ont., Que., Sask., Yukon; Alaska, Colo., Idaho, Iowa, Maine, Mich., Minn., Mont., Nev., N.H., N.Mex., N.Y., N.Dak., Oreg., Utah, Vt., Wash., Wis., Wyo.

In the Great Lakes region, the name *Potamogeton* [*Stuckenia*] *vaginatus* has often been misapplied to this taxon. The distribution of *S. vaginatus* is further to the north and west than the Great Lakes region, however. *Stuckenia filiformis* subsp. *occidentalis* rarely produces fruit in the eastern portion of its range, while to the north and west it commonly fruits. This subspecies is very similar to *Potamogeton rostratus* Hagström (the combination does not exist within the genus *Stuckenia*) of Asia and may prove to be the same taxon. Both grow in similar habitats. The proximal stipules are much enlarged and inflated, and the leaves are generally wider than the other subspecies.

4. Stuckenia vaginata (Turczaninow) Holub, Folia
Geobot. Phytotax. 19: 215. 1984 • Bigsheath-
pondweed, potamot engainé

Potamogeton vaginatus Turczaninow, Bull. Soc. Imp. Naturalistes Moscou 27(2): 66. 1854

Stems freely branching proximally to distally, terete, 20–50 (–70) cm. **Leaves:** length and width of those on main stem only slightly larger than those on branches; stipules: stipular sheaths inflated 3–5 times stem thickness, 2–9 cm, ligule absent, obscure, or to 0.2 mm; blade narrowly filiform to linear, 1–10(–15) cm × 0.2–2.9 mm, apex rounded, obtuse, or slightly notched;

veins 1(–3). **Inflorescences:** peduncles terminal, erect, slender, 3–15 cm; spikes moniliform, 10–80 mm; verticels 3–12. **Fruits** brown, obliquely obovoid, 3–3.8 mm × 2–2.9 mm; beak inconspicuous. $2n = 78$.

Flowering late summer–fall. Deep lakes and ponds; 0–2300 m; Alta., B.C., Man., N.W.T., Nunavut, Ont., Que., Sask., Yukon; Alaska, Colo., Idaho, Minn., Mont., N.Dak., Oreg., S.Dak., Utah, Wash., Wis., Wyo.; Eurasia.

One hybrid, *P. pectinatus* × *P. vaginatus* (=*P.* ×*bottnicus* Hagström), with this species as a putative parent has been described under the genus *Potamogeton*.

196. RUPPIACEAE Hutchinson

• Ditch-grass Family

Robert R. Haynes

Herbs, annual or rarely perennial, not rhizomatous, caulescent; turions absent [present]. **Leaves** alternate to subopposite, submersed, sessile; sheath not persisting longer than blade, not leaving circular scar when shed, not ligulate, not auriculate; blade linear; intravaginal squamules scales, 2. **Inflorescences** terminal, capitate spikes, with subtending spathe, pedunculate; peduncle following fertilization often elongating, often spiraling. **Flowers** bisexual; subtending bracts absent; perianth absent; stamens 2, in 1 series; anthers distinct, dehiscing longitudinally; pollen arcuate; pistils 4–16, distinct, stipitate; ovules parietal, campylotropous. **Fruits** drupaceous. **Seeds** 1; embryo straight.

Genera 1, species ca. 10 (2 species in the flora): nearly worldwide.

Ruppiaceae are found submersed in brackish or saline waters or fresh waters with very high calcium or sulfur ion concentrations.

SELECTED REFERENCE Haynes, R. R. 1978. The Potamogetonaceae in the southeastern United States. J. Arnold Arbor. 59: 170–191.

1. RUPPIA Linnaeus, Sp. Pl. 1: 127. 1753; Gen. Pl. ed. 5, 61. 1754 • Ditch-grass, widgeon grass, ruppia [for Heinrich Bernhard Ruppius, 1689–1719, German botanist]

Herbs, rooting at proximal nodes. **Leaves:** blade entire proximally, minutely serrulate distally, apex ± obtuse to acute; veins 1. **Inflorescences** fewer than 20-flowered, at first enclosed by sheathing leaf bases. **Flowers** bisexual; anthers 2-loculed, locules separated by broad connective; stipe elongating after anthesis. **Fruits** beaked, long-stipitate; beak erect or slightly recurved.

Species ca. 10 (2 in the flora): almost worldwide.

Considerable confusion exists about North American taxa of *Ruppia*. Two distinct forms, along with several intermediates, are known: one with short peduncles with four or fewer spirals and another with long peduncles with five or more coils (often many more). Usually, those with few coils are in brackish waters near the coast, whereas those with many coils are inland, often in lakes that have high mineral contents. The forms have been considered variants of one species, *R. maritima* (M. L. Fernald and K. M. Wiegand 1914); more recently they have been accepted at species level (R. F. Thorne 1993). I am adopting Thorne's concepts.

SELECTED REFERENCES   Fernald, M. L. and K. M. Wiegand. 1914. The genus *Ruppia* in eastern North America. Rhodora 16: 119–127.   Jacobs, S. W. L. and M. A. Brock. 1982. A revision of the genus *Ruppia* (Potamogetonaceae) in Australia. Aquatic Bot. 14: 325–337.   Reese, G. 1962. Zur intragenerischen Taxonomie der Gattung *Ruppia* L. Z. Bot. 50: 237–264.   Thorne, R. F. 1993b. Potamogetonaceae. In: J. C. Hickman, ed. 1993. The Jepson Manual. Higher Plants of California. Berkeley, Los Angeles, and London. Pp. 1304–1310.   Verhoeven, J. T. A. 1979. The ecology of *Ruppia*-dominated communities in western Europe. I. Distribution of *Ruppia* representatives in relation to their autecology. Aquatic Bot. 6: 197–268.

1. Peduncle with 5 or more coils in fruit, longer than 30 mm; plants mostly from inland localities. . . . . 1. *Ruppia cirrhosa*
1. Peduncle with fewer than 5 coils in fruit, shorter than 25 mm; plants mostly from coastal localities. . . . . 2. *Ruppia maritima*

1. Ruppia cirrhosa (Petagna) Grande, Bull. Orto Bot. Regia Univ. Napoli 5: 58. 1918

Buccaferrea cirrhosa Petagna, Inst. Bot. 5: 1826. 1787; *Ruppia cirrhosa* subsp. *occidentalis* (S. Watson) Á. Löve & D. Löve; *R. occidentalis* S. Watson

Stems to 55 cm × 0.1–0.3 mm. **Leaves** 3.2–45.1 cm; blade 0.2–0.5 mm wide, apex acute. **Inflorescences:** peduncles with 5–30 coils, 30–300 × 0.5 mm. **Flowers:** pistils 4–6. **Fruits** 1.5–2 × 1.1–1.5 mm; gynophore 2–3.5 cm; beak lateral, erect, 0.5–1 mm. $2n = 40$ (Europe).

Flowering summer–fall. Shallow to deep fresh waters of lakes with high concentrations of sulfur or calcium; 300–2500 m; Alta., B.C., Man., N.W.T., Sask., Yukon; Alaska, Ariz., Calif., Colo., Ill., Kans., Mich., Minn., Mo., Mont., Nebr., Nev., N.Mex., N.Dak., Ohio, Okla., Oreg., S.Dak., Tex., Utah, Wash., Wyo.; West Indies; Central America; South America; Europe.

Ruppia spiralis Linnaeus has occasionally been used for this taxon in North America. Linnaeus, however, had never published that name (J. T. A. Verhoeven 1979). *Ruppia cirrhosa* is, indeed, the correct name for the taxon (J. C. Gamerro 1968).

An implication that the North American material with long, spiraling peduncles is different from the European material of *Ruppia cirrhosa* stems from the statement that *R. occidentalis* occurs in inland lakes (J. T. A. Verhoeven 1979). No differences between European *R. cirrhosa* and the North American material are listed, however. In fact, a comparison of Verhoeven's descriptive discussions of *R. cirrhosa* with the North American material shows that the two are the same. The genus should be studied on a worldwide basis. Until further studies indicate otherwise, I am considering North American and European material to be conspecific.

SELECTED REFERENCE   Gamerro, J. C. 1968. Observaciones sobre la biología floral y morfología de la Potamogetonácea *Ruppia cirrhosa* (Petagna) Grande (= *R. spiralis* L. ex Dum.). Darwiniana 14: 575–608.

2. Ruppia maritima Linnaeus, Sp. Pl. 1: 127. 1753

• Ruppie maritime   F

Ruppia maritima var. *brevirostris* Agardh; *R. maritima* var. *exigua* Fernald & Wiegand; *R. maritima* var *intermedia* (Thedenius) Ascherson & Graebner; *R. maritima* var. *longipes* Hagström; *R. maritima* var. *obliqua* (Schur) Ascherson & Graebner; *R. maritima* var. *rostrata* Agardh; *R. maritima* var. *subcapitata* Fernald & Wiegand

Stems to 50 cm × 0.1–0.7 mm. **Leaves** 6–10.5 cm; blade 0.3–0.5 mm, apex ± obtuse. **Inflorescences:** peduncles with 0–3[–4] coils, 0.5–16.5 × 0.5 mm. **Flowers:** pistils 4–8. **Fruits** 1.8–2 × 0.7–1.5 mm; gynophore 1.2–1.9 cm; beak terminal, slightly recurved, 0.6–1 mm. $2n = 16$.

Flowering spring–fall. Shallow waters of brackish streams, ditches, and lakes along ocean shore; 0–200 m; St. Pierre and Miquelon; B.C., N.B., Nfld. and Labr. (Nfld.), N.S., Ont., P.E.I., Que., Sask.; Ala., Alaska, Calif., Conn., Del., Fla., Ga., La., Maine, Md., Mass., Miss., N.H., N.J., N.Y., N.C., Oreg., R.I., S.C., Tex., Va., Wash.; coastal Mexico; West Indies; Bermuda; Central America; South America; Eurasia; Africa; Australia.

No specimens have been seen from the District of Columbia, but the species is to be expected there.

SELECTED REFERENCES   Graves, A. H. 1908. The morphology of *Ruppia maritima*. Trans. Connecticut Acad. Arts 14: 59–170.   Richardson, F. D. 1980. Ecology of *Ruppia maritima* L. in New Hampshire (U.S.A.) tidal marshes. Rhodora 82: 403–439.

197. NAJADACEAE Jussieu

• Naiad or Water-nymph Family

Robert R. Haynes

Herbs, annual, rarely perennial, not rhizomatous, caulescent; turions absent. **Leaves** nearly opposite or appearing whorled, submersed, sessile; sheath not persisting longer than blade, not leaving scar when shed, not ligulate, not auriculate or rarely auriculate; intravaginal squamules scales, 2, linear. **Inflorescences** axillary, solitary flowers or cymes, often subtended by spathe as involucre, sessile or short-pedunculate. **Flowers** unisexual, staminate and pistillate on same or separate plants; subtending bracts absent; perianth absent; stamens 1; anthers dehiscing irregularly; pollen spheric; pistils 1, not stipitate; ovules basal, anatropous. **Fruits** achenelike. **Seeds** 1; embryo straight.

Genera 1, species 40 (8 species in the flora): nearly worldwide.

The testa of *Najas* seeds contains little-known structures, the endotegmen tubercules, which are found, other than in *Najas*, only in the testa of a few species of Hydrocharitaceae (M. Shaffer-Fehre 1991, 1991b); this was the basis of Shaffer-Fehre's proposal to merge Najadaceae with Hydrocharitaceae. A summary of the systematics of the subclass (D. H. Les and R. R. Haynes 1995) shows that molecular data support those conclusions. Although the genus would best be treated under the Hydrocharitaceae (R. F. Thorne 1993c), neither the morphologic data nor the molecular data were available to A. Cronquist (1981) at the time of his study. Consequently, he accepted Najadaceae. The Flora of North America project basically follows Cronquist's system of classification.

The structure of the flowers of *Najas* has been controversial. Results of a study of the development of *N. flexilis* flowers (U. Posluszny and R. Sattler 1976) are summarized here. Early development of both staminate and pistillate flowers is so similar that they are essentially indistinguishable. Both are domelike at inception and each develops a girdling primordium. The girdling primordium of the staminate flower develops into an outer envelope (here "involucre," as in the literature). Eventually a second girdling primordium develops, producing the inner envelope. The inner envelope so closely adheres to the developing stamen, it becomes virtually indistinguishable.

Pistillate flower initiation is exactly terminal on the floral axis rather than lateral, as is the usual case. The girdling primordium of the pistillate flower gives rise to the ovary wall, which is two cell layers thick. As the ovary wall overgrows the floral apex, the apex is transformed into an ovule primordium. Two integuments are initiated, after which the ovule turns down, eventually becoming anatropous. Because the pistillate flower is initiated in an exact terminal position, the carpel concept cannot be applied. Consequently, the flower is clearly acarpellate (U. Posluszny and R. Sattler 1976).

At maturity, the seed is closely enveloped by the extremely delicate ovary wall. Because the flower is acarpellate, the term fruit does not accurately apply. In fact, no term applies to that situation. Consequently, I am using "fruit" to indicate the seed plus the closely enveloping ovary wall. This wall is extremely thin and slightly transparent. To examine the seeds clearly, remove the wall, best done by carefully scraping it with a dissecting needle. With care, it will come off the seed partially intact. Occasionally it shreds into numerous fine pieces, however, and cannot be removed completely.

SELECTED REFERENCES   Haynes, R. R. 1977. The Najadaceae in the southeastern United States. J. Arnold Arbor. 58: 161–170.   Shaffer-Fehre, M. 1991. The endotegmen tuberculae: An account of little-known structures from the seed coat of the Hydrocharitoideae (Hydrocharitaceae) and *Najas* (Najadaceae). Bot. J. Linn. Soc. 107: 169–188.   Shaffer-Fehre, M. 1991b. The position of *Najas* within the subclass Alismatidae (Monocotyledones) in the light of new evidence from seed coat structures in the Hydrocharitoideae (Hydrocharitales). Bot. J. Linn. Soc. 107: 189–209.   Thorne, R. F. 1993c. Hydrocharitaceae. In: J. C. Hickman, ed. 1993. The Jepson Manual. Higher Plants of California. Berkeley, Los Angeles, and London. Pp. 1150–1151.

1. NAJAS Linnaeus, Sp. Pl. 2: 1015. 1753; Gen. Pl. ed. 5, 445. 1754 • Bushy-pondweed, naid, water-nymph, naïade [Greek *naias*, a water-nymph]

Herbs, aquatic, glabrous, submersed in fresh or brackish waters. **Stems** slender, much branched, rooting at proximal nodes, sometimes armed with prickles on internodes. **Leaves:** sheaths variously shaped, margins usually toothed with 1–15 teeth per side, teeth similar in size and structure to those of blades; blade linear, 1-veined, sometimes armed with prickles on midvein abaxially, margins usually serrate to minutely serrulate with 5–100 teeth per side, apex acute to acuminate, with 1–3 teeth, teeth multicellular, formed by layers of cells decreasing in cell number distally, terminated by large, sharp-tipped cell, or teeth unicellular. **Inflorescences:** involucres mostly present in staminate flowers, rare in pistillate, clear, bronze, brown, light green, purple, or red-purple. **Staminate flowers** subtended by membranous involucre, involucre rarely absent; peduncle short, elongating at anthesis, pushing flower through involucre; anther sessile, 1- or 4-loculed. **Pistillate flowers** sessile; ovary 1-loculed; ovules basal, 1; style terminal (arising off-center at apex of ovary and fruit in *Najas gracillima*), 2–4-branched. **Fruits** dehiscing by decay of ovary wall; ovary wall extremely delicate, closely enveloping seed. **Seeds** fusiform to obovoid, apex occasionally asymmetric or recurved, areolate; raphe basal; testa 3 or 10–15 cell layers thick, hard, brittle; areoles formed by outer two cell layers of testa, irregularly arranged or in 15–60 longitudinal rows, end walls often raised, giving testa papillose appearance; endosperm absent. $x = 6$.

Species 40 (8 in the flora): nearly worldwide.

Seeds are necessary for certain determination of species of *Najas*.

SELECTED REFERENCES   Chase, S. S. 1947. Preliminary Studies in the Genus *Najas* in the United States. Ph.D. thesis. Cornell University.   Chase, S. S. 1947b. Polyploidy in an immersed aquatic angiosperm. [Abstract.] Amer. J. Bot. 34: 581–582.   Haynes, R. R. 1979. Revision of North and Central American *Najas* (Najadaceae). Sida 8: 34–56.   Lowden, R. M. 1986. Taxonomy of the genus *Najas* L. (Najadaceae) in the Neotropics. Aquatic Bot. 24: 147–184.   Triest, L. 1988. A revision of the genus *Najas* L. (Najadaceae) in the Old World. Mém. Acad. Roy. Sci. Belgique, Cl. Sci. (8°), n. s. 22: 1–172, 29 plates.   Wentz, W. A. and R. L. Stuckey. 1971. The changing distribution of the genus *Najas* (Najadaceae) in Ohio. Ohio J. Sci. 71: 292–302.

1. Staminate and pistillate flowers on different plants; testa pitted, 10–15 cell layers thick; midrib of leaf blade abaxially with prickles; internodes of stem usually with prickles; seeds 1.2 mm wide or wider. . . . . 1. *Najas marina*
1. Staminate and pistillate flowers on same plants; testa smooth or pitted, 3 cell layers thick; midrib of leaf blade abaxially without prickles; internodes of stem without prickles; seeds 1.2 mm or less wide, if 1.2 mm wide, then with smooth testa.
    2. Style offset to one side of center at seed apex. . . . . 5. *Najas gracillima*
    2. Style situated at center of seed apex.
        3. Seeds recurved; leaves usually recurved with age.
            4. Areoles of testa broader than long, ladderlike; widespread in a variety of habitats . . . . 7. *Najas minor*
            4. Areoles of testa longer than broad, never ladderlike; plants of Florida and Georgia in sandy-bottomed lakes. . . . . 6. *Najas filifolia*
        3. Seeds not recurved; leaves spreading to ascending with age.
            5. Teeth of leaf blade multicellular, 8–22 per side. . . . . 8. *Najas wrightiana*
            5. Teeth of leaf blade unicellular, 18–100 per side.
                6. Testa smooth, glossy; seeds narrowly to broadly obovoid, deep brown to yellow; anthers 1-loculed. . . . . 2. *Najas flexilis*
                6. Testa pitted, dull; seeds fusiform, yellowish white to greenish or reddish brown or with purple tinge; anthers 1- or 4-loculed.
                    7. Leaf sheath deeply auriculate at apex; restricted in North America to California. . . . . 4. *Najas graminea*
                    7. Leaf sheath rounded to slightly auriculate at apex; widespread in the flora. . . . . 3. *Najas guadalupensis*

1. Najas marina Linnaeus, Sp. Pl. 2: 1015. 1753 • Spiny naiad [F]

Najas gracilis (Morong) Small

Stems branched distally, 6–45 cm × 0.5–4 mm; internodes 0.3–11 cm, usually with prickles. **Leaves** spreading to ascending with age, 0.5–3.9 cm, stiff in age; sheaths 2–4.4 mm wide, apex acute; blade 0.4–4.5 mm wide, margins coarsely serrate, teeth 8–13 per side, apex acute, with 1 tooth, teeth multicellular; midvein with prickles abaxially. **Flowers** 1 per axil, staminate and pistillate on different plants. **Staminate flowers** in distal to proximal axils, 1.7–3 mm; involucral beaks 2-lobed, 0.3–0.7 mm; anthers 4-loculed, 1.7–3 mm. **Pistillate flowers** in distal to proximal axils, 2.5–5.7 mm; styles 1.2–1.7 mm; stigmas 3-lobed. **Seeds** not recurved, reddish brown, ovoid, 2.2–4.5 × 1.2–2.2 mm, apex with style situated at center; testa dull, 10–15 cell layers thick, pitted; areoles irregularly arranged, not in distinctive rows, not ladderlike, 3–4-angled, longer than broad, end walls slightly raised. $2n = 12$ (Europe).

Flowering summer–fall. Brackish or highly alkaline waters of ponds and lakes; 0–1000 m; Ariz., Calif., Fla., Ind., Mich., Minn., Nev., N.Y., N.Dak., Ohio, Pa., S.Dak., Tex., Utah, Wis.; Mexico; West Indies; Central America (El Salvador, Panama); South America; Eurasia.

With its prickly internodes and prickles along the abaxial surface of the leaves, *Najas marina* is the easiest of our *Najas* to recognize. Over its entire range, the species displays considerable morphologic variability (L. Triest et al. 1986), giving cause for ten subspecies to be recognized (L. Triest 1988). In North America, however, variability is relatively minor, so I am recognizing the taxon at the specific level (R. R. Haynes 1979). The species should be studied over its entire range, utilizing a variety of approaches, to determine adequately what, if any, infraspecific taxa should be recognized.

SELECTED REFERENCES Stuckey, R. L. 1985. Distributional history of *Najas marina* (spiny naiad) in North America. Bartonia 51: 2–16. Triest, L., J. van Geyt, and V. Ranson. 1986. Isozyme polymorphism in several populations of *Najas marina* L. Aquatic Bot. 24: 373–384. Viinikka, Y. 1973. The occurrence of B chromosomes and their effect on meiosis in *Najas marina*. Hereditas (Lund) 75: 207–212.

2. Najas flexilis (Willdenow) Rostkovius & W. L. E. Schmidt, Fl. Sedin., 382. 1824 [F]

Caulinia flexilis Willdenow, Mém. Acad. Roy. Sci. Hist. (Berlin) 1798: 89, plate1, fig. 1. 1801; *Najas caespitosus* (Maguire) Reveal; *N. canadensis* Michaux

Stems often profusely branched distally, 2.5–5 cm × 0.2–0.6 mm; internodes 0.16–6.8 cm, without prickles. **Leaves** spreading to ascending with age, 0.2–0.6 cm, lax in age; sheath 0.7–1.6 mm wide, apex rounded; blade 0.2–0.6 mm wide,

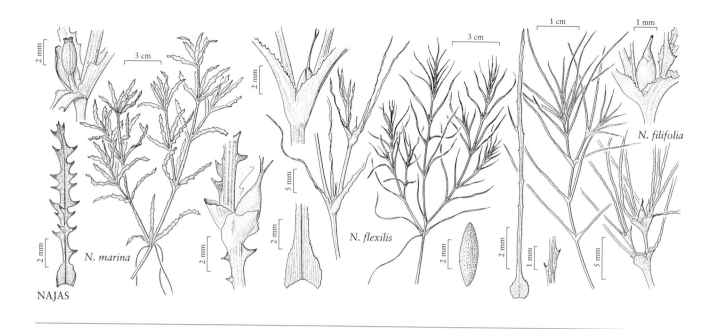

NAJAS

margins minutely serrulate, teeth 35–80 per side, apex acute, with 1–2 teeth, teeth unicellular; midvein without prickles abaxially. **Flowers** 1(–2) per axil, staminate and pistillate on same plants. **Staminate flowers** in distal axils, 1.1–2.7 mm; involucral beaks 3-lobed, 0.7–1.2 mm; anther 1-loculed, 1.1–2.7 mm. **Pistillate flowers** in distal to proximal axils, 2.5–4.7 mm; styles 1.5–1.7 mm; stigmas 3-lobed. **Seeds** not recurved, deep brown to yellow, narrowly to broadly obovoid, (1.2–)2.5–3.7 × 0.2–1.2 mm, apex with style situated at center; testa glossy, 3 cell layers thick, smooth; areoles regularly arranged in ca. 50 longitudinal rows, not ladderlike, 3–4-angled, longer than broad, end walls not raised. $2n = 12, 24$.

Flowering summer–fall. Lakes and rivers; 0–1500 m; Alta., B.C., Man., N.B., Nfld. and Labr. (Nfld.), N.S., Ont., P.E.I., Que., Sask.; Calif., Conn., Del., Idaho, Ill., Ind., Iowa, Ky., Maine, Md., Mass., Mich., Minn., Mo., Mont., N.H., N.J., N.Y., N.Dak., Ohio, Oreg., Pa., R.I., S.Dak., Utah, Vt., Va., Wash., W.Va., Wis.; Eurasia.

In habit, *Najas flexilis* is most similar to *N. guadalupensis*. When seeds are present, *N. flexilis* can be separated easily from the latter species by the glossy, smooth, yellowish seeds that are widest above the middle. In the northern United States and in Canada, *N. flexilis* is by far the most common species of *Najas*, although in Ohio and surrounding areas, it is disappearing as eutrophication (depletion of oxygen from lakes) continues (W. A. Wentz and R. L. Stuckey 1971).

SELECTED REFERENCE Posluszny, U. and R. Sattler. 1976. Floral development of *Najas flexilis*. Canad. J. Bot. 54: 1140–1151.

3. **Najas guadalupensis** (Sprengel) Magnus, Beitr. Kenntn. Najas, 8. 1870 [W]

Caulinia guadalupensis Sprengel, Syst. Veg. 1: 20. 1824

Stems often profusely branched distally, 11–90 cm × 0.1–2 mm; internodes 0.1–9 cm, without prickles. **Leaves** spreading with age, 0.3–3.3 cm, lax in age; sheath 1–3.4 mm wide, apex rounded to truncate; blade 0.2–2.1 mm wide, margins minutely serrulate, teeth 18–100 per side, apex rounded to slightly auriculate, with 1–3 teeth, teeth unicellular; midvein without prickles abaxially. **Flowers** 1–3 per axil, staminate and pistillate on same plants. **Staminate flowers** in distal axils, 1.5–3 mm; involucral beaks 4-lobed, 0.2–1.3 mm; anther 1- or 4-loculed, 1–1.7 mm. **Pistillate flowers** in proximal axils, 1.5–4 mm; styles 0.3–1.5 mm; stigmas 4-lobed. **Seeds** not recurved, yellowish white with purple tinge, fusiform, 1.2–3.8 × 0.4–0.8 mm, apex with style situated at center; testa dull, 3 cell layers thick, pitted; areoles regularly arranged in 20–60 longitudinal rows, not ladderlike, 4–6-angled, longer than broad, end walls not raised. $2n = 12, 36, 42, 48, 54, 60$.

Subspecies 4 (4 in the flora): widespread.

By having unicellular teeth on the leaf margin, *Najas guadalupensis* resembles *N. flexilis* and *N. graminea*. The seeds of *N. guadalupensis* are pitted; those of *N. flexilis* are smooth. The sheaths of *N. guadalupensis* are rounded to slightly auriculate; those of *N. graminea* are deeply auriculate.

1. Seeds 3.3–3.8 mm; areoles of testa arranged in 50–60 longitudinal rows. . . . . 3d. *Najas guadalupensis* subsp. *muenscheri*
1. Seeds 2.5 mm or less; areoles of testa in fewer than 50 longitudinal rows.
    2. Teeth of leaf blade evident to unaided eye; anthers 1-loculed. . . . . 3c. *Najas guadalupensis* subsp. *floridana*
    2. Teeth of leaf blade invisible to unaided eye; anthers 4-loculed.
        3. Leaf blade with 50–100 teeth per side; stems 0.8 mm or less diam. . . . . 3a. *Najas guadalupensis* subsp. *guadalupensis*
        3. Leaf blade with 20–40 teeth per side; stem 1 mm or more diam. . . . . 3b. *Najas guadalupensis* subsp. *olivacea*

3a. Najas guadalupensis (Sprengel) Magnus subsp. **guadalupensis** W

Stems 11–75 cm × 0.1–0.8 mm. **Leaves** 0.3–2.8 mm; sheath 1–1.9 mm wide, apex rounded; blade 0.2–1.8 mm wide, teeth 50–100 per side, invisible to unaided eye, apex acute to mucronate. **Flowers** 1–3 per axil. **Staminate flowers** 1.5–2.5 mm; anther 4-loculed. **Pistillate flowers** 1.5–2 mm. **Seeds** 1.2–2.5 × 0.4–0.6 mm; areoles of testa in 20 longitudinal rows. 2*n* = 24.

Flowering early summer–fall. Lakes, rivers, and canals; 0–1500 m; Alta., Ont.; Ala., Ariz., Ark., Calif., Colo., Conn., Del., D.C., Fla., Ga., Ill., Ind., Iowa, Kans., Ky., La., Maine, Md., Mass., Mich., Minn., Miss., Mo., Mont., Nebr., Nev., N.H., N.J., N.Y., N.C., Ohio, Okla., Oreg., Pa., R.I., S.C., S.Dak., Tenn., Tex., Utah, Vt., Va., Wash., W.Va., Wis.; Mexico; West Indies; Central America; South America.

3b. Najas guadalupensis (Sprengel) Magnus subsp. **olivacea** (Rosendahl & Butters) R. R. Haynes & Hellquist, Novon 6: 371. 1996 E W

Najas olivacea Rosendahl & Butters, Rhodora 37: 347. 1935

Stems 15–40 cm × 1–2 mm. **Leaves** 0.9–1.8 cm; sheath 2.5–3.4 mm wide, apex rounded to truncate; blade 1.5–2 mm wide, teeth 20–40 per side, invisible to unaided eye, apex acute. **Flowers** 1 per axil. **Staminate flowers** 2.3–2.8 mm; anther 4-loculed. **Pistillate flowers** 2.7–3.1 mm. **Seeds** 2.3–2.5 × 0.6–0.8 mm; areoles of testa in 20–40 longitudinal rows. Chromosome number not available.

Flowering late summer. Lakes and rivers; 0–100 m; Man., Ont., Que.; Ind., Iowa, Mich., Minn., N.Y., Wis.

3c. Najas guadalupensis (Sprengel) Magnus subsp. **floridana** (R. R. Haynes & Wentz) R. R. Haynes & Hellquist, Novon 6: 371. 1996 E W

Najas guadalupensis var. *floridana* R. R. Haynes & Wentz, Sida 5: 262. 1974

Stems 7–51 cm × 0.1–1.7 mm. **Leaves** 0.9–3.3 cm; sheath 1.2–2.5 mm wide, apex rounded; blade 0.5–2.1 mm wide, teeth 18–42 per side, visible to unaided eye, apex round obtuse to acuminate. **Flowers** 1 per axil. **Staminate flowers** 1.5–2.4 mm; anthers 1-loculed. **Pistillate flowers** 1–3.4 mm. **Seeds** 1.6–2.2 × 0.3–0.8 mm; areoles of testa in 20 longitudinal rows. 2*n* = 48.

Flowering spring–fall. Lakes, streams, and canals; 0–100 m; Ala., Fla., Ga.

3d. Najas guadalupensis (Sprengel) Magnus subsp. **muenscheri** (R. T. Clausen) R. R. Haynes & Hellquist, Novon 6: 371. 1996 C E

Najas muenscheri R. T. Clausen, Rhodora 39: 59. 1937

Stems 30–90 cm × 0.8–1 mm. **Leaves** 0.9–1.3 cm; sheath 1–1.5 mm wide, apex rounded; blade 0.5–1.6 mm wide, teeth 50–100 per side, invisible to unaided eye, apex acute. **Flowers** 1 per axil. **Staminate flowers** 2–3 mm; anther 1-loculed. **Pistillate flowers** 2.9–4 mm. **Seeds** 3.3–3.8 × 0.5–0.7 mm; areoles of testa in 50–60 longitudinal rows. Chromosome number not available.

Flowering late summer. Shallow water of rivers; of conservation concern; 0–100 m; N.Y.

4. Najas graminea Delile, Descr. Egypte, Hist. Nat., 282, plate 50, fig. 3. 1813 I W

Stems sparingly branched distally, to 35 cm × 0.2–0.5 mm; internodes 0.4–1.9 cm, without prickles. **Leaves** spreading to ascending with age, 0.8–2 cm, lax in age; sheath 1–1.5 mm wide, apex deeply auriculate; blade 0.5–1 mm wide, margins minutely serrulate, teeth 40 per side, apex acute with 2–3 teeth, teeth unicellular; midvein without prickles abaxially. **Flowers** 1–2 per axil, staminate and pistillate on same plants. **Staminate flowers** in distal axils, 2–3 mm; involucre beaks 4-lobed, 1–1.5 mm; anther 4-loculed, 1.5 mm. **Pistillate flowers** distal to proximal, to 3.5 mm; styles to 1 mm; stigmas 2-lobed. **Seeds** not recurved, greenish brown, fusiform, 1.7–2.5 × 0.4–6 mm, apex with style situated at center; testa dull, 3 cell layers thick, pitted; areoles regularly arranged in 35 longitudinal rows, not ladderlike, 4-angled, as long as broad, end walls slightly raised. 2*n* = 24, 36, 48 (Asia).

Flowering summer. Rice fields; 200 m; introduced; Calif.; Eurasia; Africa; Australia.

Because of the deeply auriculate sheaths, *Najas graminea* is one of the easiest of the North American and Central American *Najas* to recognize. *Najas graminea* often becomes weedy in rice fields of southeast Asia.

5. **Najas gracillima** (A. Braun ex Engelmann) Magnus, Beitr. Kenntn. Najas, 23. 1870 [E]

Najas indica (Willldenow) Chamisso var. *gracillima* A. Braun ex Engelmann in A. Gray, Manual ed. 5, 681. 1867

Stems slightly branched distally, 4.5–48 cm × 0.2–0.7 mm; internodes 0.1–3.2 cm, without prickles. **Leaves** spreading to ascending with age, 0.6–2.8 cm, lax in age; sheath 0.5–1.5 mm wide, apex truncate; blade 0.1–0.5 mm wide, margins minutely serrulate, teeth 13–17 per side, apex acute with 2–3 teeth, teeth unicellular; midvein without prickles abaxially. **Flowers** 1–3 per axil, staminate and pistillate on same plants. **Staminate flowers** in distal axils, 1.5–2 mm; involucral beaks 2-lobed, 0.8 mm; anthers 1-loculed, 1.3 mm. **Pistillate flowers** distal to proximal on plant, 0.5–2.7 mm; styles 0.3–1.5 mm; stigmas 2-lobed. **Seeds** not recurved, light brown, fusiform, 2–3.2 × 0.4–0.7 mm, apex with style situated off center; testa dull, 3 cell layers thick, pitted; areoles regularly arranged in 40 longitudinal rows, not ladderlike, 4-angled, longer than broad, end walls raised. 2*n* = 24, 36.

Flowering summer–fall. Soft water lakes; 0–200 m; Ont., Nfld. and Labr. (Nfld.), N.S.; Ala., Calif., Conn., Del., D.C., Ill., Ind., Ky., Maine, Md., Mass., Mich., Minn., Mo., N.H., N.J., N.Y., N.C., Ohio, Pa., R.I., Tenn., Vt., Va., Wis.; Eurasia.

No specimens were seen from Alberta, Manitoba, or Quebec, but the species is to be expected there.

Najas gracillima is most similar to *N. minor*, especially in vegetative condition. *Najas gracillima*, however, can be separated from the latter species by its style arising off-center at the apex of the ovary wall and by its areoles being longer than broad. Late in the growing season, the leaves of *N. minor* become recurved; those of *N. gracillima* do not.

6. **Najas filifolia** R. R. Haynes, Brittonia 37: 392. 1985 [C] [E] [F]

Stems profusely branched distally to proximally, 7–22 cm × 0.2–1 mm; internodes 0.2–4 cm, without prickles. **Leaves** usually recurved with age, 0.8–2.6 cm, stiff in age; sheath 1–1.3 mm wide, apex round to slightly auriculate; blade 0.1–0.7 mm wide, margins conspicuously serrulate, teeth 5–9 per side, apex acute, with 1 tooth, teeth multicellular; midvein without prickles abaxially. **Flowers** 1 per axil, staminate and pistillate on same plants. **Staminate flowers** in distal axils, 2–3 mm; involucral beaks 4-lobed, 0.4–0.6 mm; anthers 4-loculed, 1.5–2 mm. **Pistillate flowers** in proximal axils, to 2.5 mm; styles less than 0.4 mm; stigmas 4-lobed. **Seeds** strongly recurved, greenish brown, fusiform, 2.5–3 × 0.5–0.7 mm, apex with style central; testa dull, 3 cell layers thick, pitted; areoles regularly arranged in ca. 20 longitudinal rows, not ladderlike, 4-angled, longer than broad, end walls raised. Chromosome number not available.

Flowering late summer. Sandy-bottomed lakes; of conservation concern; 0–100 m; Fla., Ga.

Najas filifolia is known from four localities: Open Pond and Cane Water Pond in Decatur County, Georgia; Milton in Santa Rosa County, Florida; and Lake Jackson in Leon County, Florida.

The vegetative organs of *Najas filifolia* resemble those of *N. minor* and *N. wrightiana* by the presence of large teeth scattered along the margin of quite narrow leaves. *Najas filifolia* is easily distinguished from the latter two species, however, by its recurved fruits (sometimes crescent-shaped). I know of no other *Najas* with such curved fruits.

The name *Najas conferta* A. Braun has been misapplied to this taxon.

7. **Najas minor** Allioni, Fl. Pedem. 2: 221. 1785 [I] [W]

Stems profusely branched near apex distally, 11–120 cm × 0.2–1 mm; internodes 0.5–5.8 cm, without prickles abaxially. **Leaves** usually recurved with age, 0.5–3.4 cm, stiff in age; sheath 1–3 mm wide, apex truncate to auriculate; blade 0.1–1.2 mm wide, margins conspicuously serrulate, teeth 7–15 per side, apex acute, with 1–2 teeth, teeth multicellular; midvein without prickles abaxially. **Flowers** 1–2 per axil, staminate and pistillate on same plant. **Staminate flowers** in distal axils, 1.9–2.2 mm; involucre 2-lobed, beaks 0.4–2.1 mm; anther 1-loculed, 0.3 mm.

Pistillate flowers in distal to proximal axils, 2.2 mm; styles 1–1.2 mm; stigmas 2-lobed. **Seeds** slightly recurved, purplish, fusiform, 1.5–3 × 0.5–0.7 mm, apex with style situated at center; testa dull, 3 cell layers thick, pitted; areoles regularly arranged in ca. 15 longitudinal rows, ladderlike, 4-angled, broader than long, end walls not raised. $2n = 24$.

Flowering summer–fall. Ponds, lakes, and slow-moving streams; 0–500 m; introduced; Ala., Del., Fla., Ga., Ill., Ind., Ky., La., Mass., Mich., Miss., Mo., N.Y., N.C., Ohio, Okla., Pa., S.C., Tenn., Vt., Va., W.Va.; Eurasia; Africa.

Najas minor, with its mature leaves recurved and with its areoles broader than long and arranged in longitudinal rows like the rungs of a ladder, is one of the more distinctive species of *Najas*. Young sterile individuals resemble *N. gracillima*, however.

8. **Najas wrightiana** A. Braun, Sitzungsber. Ges. Naturf. Freunde Berlin 1868: 17. 1868 [I] [W]

Stems profusely branched proximally to distally, 9–45 cm × 0.3–1 mm; internodes 0.3–5.2 cm, without prickes. **Leaves** ascending with age, 0.5–2.6 cm, stiff in age; sheath 0.7–2.8 mm wide, apex rounded; blade 0.2–1.3 mm wide, margins conspicuously serrulate, teeth 8–22 per side, apex acute, with 1 tooth, teeth multicellular; midvein without prickles abaxially. **Flowers** 1–2 per axil, staminate and pistillate on same plant. **Staminate flowers** in distal axils, 1.2–1.9 mm; involucral beaks 2-lobed, 0.3–0.5 mm; anther 4-loculed, 1 mm. **Pistillate flowers** in proximal axils, 2 mm; styles to 0.7 mm; stigmas 4-lobed. **Seeds** not recurved, whitish, fusiform, 0.7–1.5 × 0.3–0.5 mm, apex with style situated at center; testa dull, 3 cell layers thick, pitted; areoles regularly arranged in 20 longitudinal rows, not ladderlike, 4–5-angled, longer than broad, end walls raised. Chromosome number not available.

Flowering late spring–fall. Slow-moving streams and ponds; 0–100 m; introduced; Fla.; Mexico; Central America (Guatemala, Belize, Honduras); West Indies (Cuba).

198. ZANNICHELLIACEAE Dumortier

• Horned-pondweed Family

Robert R. Haynes

C. Barre Hellquist

Herbs, annual [perennial], not rhizomatous, caulescent; turions absent. **Leaves** alternate, opposite, or pseudowhorled, often all on same plant, submersed, sessile; sheath not persisting longer than blade, not leaving circular scar when shed, ligulate, not auriculate; blade linear; intravaginal squamules scales, 2. **Inflorescences** axillary, cymes, without subtending spathe, sessile. **Flowers** unisexual, staminate and pistillate flowers on same plant; subtending bracts absent; perianth absent. **Staminate flowers:** stamens 1; anther dehiscing longitudinally; pollen globose. **Pistillate flowers:** pistils (1–)4–5(–8), distinct, short-stipitate; ovules pendulous, anatropous. **Fruits** drupaceous. **Seeds** 1; embryo curved.

Genera 4, species 10–12 (1 genus, 1 species in the flora): nearly worldwide.

SELECTED REFERENCES Campbell, D. H. 1897. A morphological study of *Naias* and *Zannichellia*. Proc. Calif. Acad. Sci., ser. 3, 1: 1–70. Haynes, R. R. and L. B. Holm-Nielsen. 1987. The Zannichelliaceae in the southeastern United States. J. Arnold Arbor. 68: 259–268. Posluszny, U. and P. B. Tomlinson. 1977. Morphology and development of floral shoots and organs in certain Zannichelliaceae. Bot. J. Linn. Soc. 75: 21–46. Taylor, N. 1909. Zannichelliaceae. In: N. L. Britton et al., eds. 1905+. North American Flora. . . . 47+ vols. New York. Vol. 17, pp. 13–27. Tomlinson, P. B. and U. Posluszny. 1976. Generic limits in the Zannichelliaceae (sensu Dumortier). Taxon 25: 273–279.

1. ZANNICHELLIA Linnaeus, Sp. Pl. 2: 969. 1753; Gen. Pl. ed. 5, 416. 1754

• Horned-pondweed [for Gian Girolamo Zannichelli, 1662–1729, Venetian apothecary and botanist]

Roots single or in pairs at nodes. **Leaves:** blade entire, veins 1[–3], nearly terete. **Inflorescences** usually of 2 flowers, 1 staminate, 1 pistillate. **Flowers** short-pedicellate. **Staminate flowers:** anthers 4-loculed. **Pistillate flowers:** pistils surrounded proximally by membranaceous envelope; stigma asymmetrically funnel-shaped. **Fruits:** convex margin minutely dentate; endocarp often coarsely papillose (visible after decay of outer fruit wall); stipe developing into podogyne (stalk supporting pistil).

Species ca. 5 (1 in the flora): nearly worldwide.

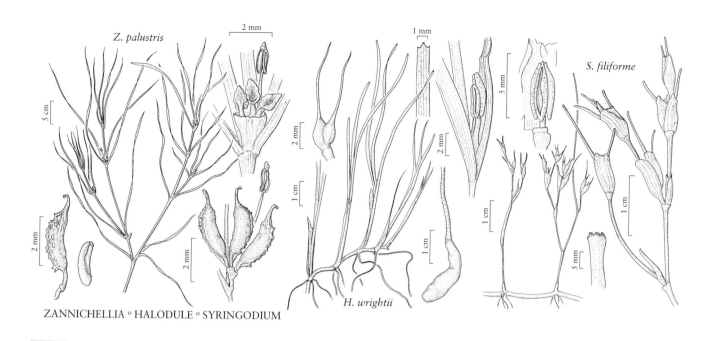

Z. palustris

S. filiforme

H. wrightii

ZANNICHELLIA ∘ HALODULE ∘ SYRINGODIUM

SELECTED REFERENCE Van Vierssen, W. 1982. The ecology of communities dominated by *Zannichellia* taxa in western Europe. I. Characterization and autecology of the *Zannichellia* taxa. Aquatic Bot. 12: 103–155.

1. **Zannichellia palustris** Linnaeus, Sp. Pl. 2: 969. 1753 F

Herbs entirely submersed. **Stems** to 50 cm × 0.2–0.6 mm. **Leaves** 3.5–4.2 cm × 0.2–1 mm, apex acute. **Staminate flowers:** filament 1.5–2 mm, connective prolonged into blunt tip 0.1 mm. **Pistillate flowers:** pistils 4–5; style 0.4–0.7 mm. **Fruits** 1.7–2.8 × 0.6–0.9 mm; rostrum 0.7–2 mm; podogyne 0.1–1.5 mm; pedicel 0.3–1.2 mm.

Flowering spring–summer. Brackish or fresh waters of streams, lakes, or estuaries; 0–2900 m; Alta., B.C., Man., N.B., Nfld. and Labr. (Nfld.), N.S., Ont., P.E.I., Que., Sask.; Ala., Alaska, Ariz., Ark., Calif., Colo., Conn., Del., D.C., Fla., Idaho, Ill., Ind., Iowa, Kans., Ky., La., Maine, Md., Mass., Mich., Minn., Miss., Mo., Mont., Nebr., Nev., N.H., N.J., N.Mex., N.Y., N.C., N.Dak., Ohio, Okla., Oreg., Pa., R.I., S.Dak., Tenn., Tex., Utah, Vt., Va., Wash., W.Va., Wis., Wyo.; Mexico; West Indies; Central America; South America; Eurasia; Africa; Australia.

Outside of Europe most *Zannichellia* are considered to be *Z. palustris* (W. Van Vierssen 1982). In Europe three species have been recognized based on stamen length, fruit length, podogyne length, and the rostrum to fruit length ratio. For *Z. palustris* in Europe the mean rostrum length is 0.78 ± 0.2 mm, the mean podogyne length is 0.4 ± 0.19 mm, and the rostrum to fruit ratio is less than 0.5 (W. Van Vierssen 1982). North American *Zannichellia* does not match any of these figures exactly. In North America *Zannichellia* has been considered historically to comprise only one species, which has been called *Z. palustris*. Until further research determines the range of *Zannichellia* and species delimitations, we are continuing to consider all North American material to be monospecific and are applying the name *Z. palustris* to that material.

199. CYMODOCEACEAE N. Taylor

• Manatee-grass Family

Robert R. Haynes

Herbs, perennial, rhizomatous, caulescent; turions absent. **Leaves** submersed, alternate or nearly opposite, sessile; sheath persisting longer than blade, leaving circular scar when shed, not ligulate, auriculate, lobes not scarious; blade linear; intravaginal squamules scales, more than 2. **Inflorescences** axillary or terminal, solitary or cymes, without spathe, sessile or pedunculate; peduncle, when present, not elongating following fertilization, not spiraling. **Flowers** unisexual, staminate and pistillate on separate plants; subtending bracts absent; perianth absent. **Staminate flowers:** stamens 2, in 1 series; anthers adaxially connate, dehiscing vertically; pollen linear. **Pistillate flowers:** pistils 2, distinct, not stipitate; ovules pendulous, orthotropous. **Fruits** achenelike or drupaceous. **Seeds** 1; embryo straight.

Genera 5, species 16 (2 genera, 2 species in the flora): widespread in warm oceanic waters worldwide.

Cymodoceaceae comprise one of three families of flowering plants in North America that inhabit oceanic waters. Individuals of this family form carpetlike vegetation over sandy to muddy substrates of the tropical and subtropical waters along the southern Atlantic and Gulf of Mexico coasts of the United States.

SELECTED REFERENCES Hartog, C. den. 1970. The Sea-grasses of the World. Amsterdam. Haynes, R. R. and L. B. Holm-Nielsen. 1985. A generic treatment of Alismatidae in the Neotropics. Acta Amazon. 15(suppl.): 153–193. Johnson, E. A. and S. L. Williams. 1982. Sexual reproduction in seagrasses: Reports for five Caribbean species with details for *Halodule wrightii* Aschers. and *Syringodium filiforme* Kutz. Caribbean J. Sci. Math. 18: 61–70.

1. Leaves flattened; flowers solitary; styles not divided into stigmas; anthers attached at different levels on axis. . . . . 1. *Halodule*, p. 87
1. Leaves terete or semi-terete; flowers in cymes; styles divided into 2 stigmas; anthers attached at same point on axis. . . . . 2. *Syringodium*, p. 88

1. HALODULE Endlicher, Gen. Pl., suppl. 1(2): 1368. 1841 • [Greek *halos*, salt]

Diplanthera Thouars, Gen. Nov. Madagasc., 3. 1806, not Gleditsch 1764

Stems erect, subtended by elliptic or ovate scales. **Leaves** 1–4; blade flattened, apex 2–3-toothed, lacunae absent; veins 3, midvein conspicuous, widened distally, lateral veins inconspicuous, each ending in tooth. **Inflorescences** solitary. **Flowers** axillary to leafy bract, inflated sheath absent. **Staminate flowers:** anthers attached to axis at different levels. **Pistillate flowers** nearly sessile, styles not divided into stigmas. **Fruits** achenelike, slightly compressed, nearly globose to ovoid.

Species 6 (1 in the flora): warm oceanic waters of both hemispheres.

The taxonomy of *Halodule* is based almost entirely on the shape of the vegetative leaf tip; many of the species are unknown in sexual reproductive condition. Features of the leaf tip important in systematics include the presence or absence of lateral teeth, the presence or absence of a middle tooth, and the length of the middle tooth relative to the two lateral ones.

For a discussion of pollination in *Halodule pinifolia* (Miki) Hartog, see P. A. Cox (1988) and P. A. Cox and R. B. Knox (1989). The flowers are enclosed in a vegetative shoot that closely resembles a perianth (E. A. Johnson and S. L. Williams 1982). Staminate flowers are borne at the surface of the substrate, and pistillate flowers are produced below the substrate (P. A. Cox 1988). During low tide, the filament elongates, projecting the anther to the water surface. Following exposure to air, the anthers dehisce, releasing a cottonlike mass of threadlike pollen grains approximately 1000 μm in length. These pollen grains rapidly assemble themselves into "search vehicles" that are relatively large, ca. 5000 μm. This large size apparently increases the possibility of contacting a stigma. Apparently most search vehicles are composed of pollen from one anther. Submarine dehiscence does occur (P. A. Cox 1988); no explanation is given for what occurs afterwards.

The pistillate flower is produced below the substrate surface (P. A. Cox 1988). The elongate style projects the stigma to the water surface during low tide. Possibly a relatively large "search vehicle" floating on the water surface will contact a stigma and pollination is accomplished. The fruit develops in the substrate, remaining there until released from the plant. Fruits remaining in the substrate possibly are the reason so few collections have fruits on them. They could fall off during the collection and removal of the substrate.

SELECTED REFERENCES Cox, P. A. 1988. Hydrophilous pollination. Annual Rev. Ecol. Syst. 19: 261–280. Cox, P. A. and R. B. Knox. 1989. Two-dimensional pollination in hydrophilous plants: Convergent evolution in the genera *Halodule* (Cymodoceaceae), *Halophila* (Hydrocharitaceae), *Ruppia* (Ruppiaceae), and *Lepilaena* (Zannichelliaceae). Amer. J. Bot. 76: 164–175. Hartog, C. den. 1964. An approach to the taxonomy of the sea-grass genus *Halodule* Endl. (Potamogetonaceae). Blumea 12: 289–312. McMillan, C. 1991. Isozyme patterning in marine spermatophytes. Opera Bot. Belg. 4: 193–200. Phillips, R. C. 1967. On species of the seagrass, *Halodule*, in Florida. Bull. Mar. Sci. 17: 672–676.

1. Halodule wrightii Ascherson, Sitzungsber. Ges. Naturf. Freunde Berlin 1868: 19. 1868 [F]

Diplanthera beaudettei Hartog; *Halodule beaudettei* (Hartog) Hartog

Rhizomes: internodes 0.5–4 cm; scales 4–10 mm. **Leaves** dark red-brown; sheath 1.5–6 cm × 1–2 mm; blade 5–20 cm × 0.3–1.5 mm, apex notched or with very prominent, acute median tooth; veins ending in teeth or not. **Inflorescences** solitary. **Staminate flowers:** peduncles 10–25 mm; distal anthers ca. 0.5 mm higher than the proximal. **Pistillate flowers:** styles lateral to terminal. **Fruits** ovoid to globose, 1.5–2 × 1.2–1.8 mm.

Flowering summer. Intertidal zone of marine waters with sandy or muddy substrates; -2–0 m; Ala., Fla., La., Miss., N.C.; e Mexico; West Indies; Central America (Guatemala, Belize, Nicaragua); South America (Venezuela).

Halodule wrightii occupies the shallowest waters in

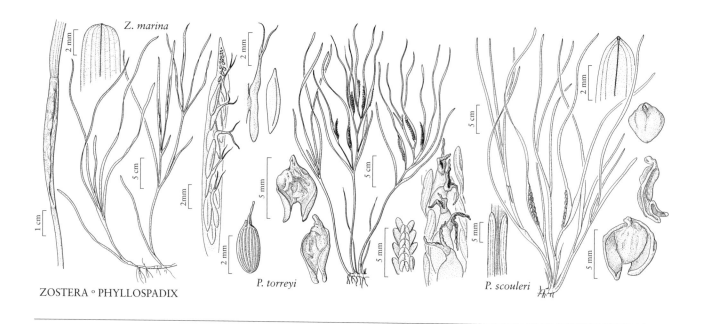

ZOSTERA ∘ PHYLLOSPADIX   *P. torreyi*   *P. scouleri*

the Gulf of Mexico. In fact, the plants are often exposed during low tides.

All *Halodule* along the North American coast have been considered to be *H. beaudettei* (C. den Hartog 1964, 1970; concept accepted by D. S. Correll and H. B. Correll 1972 and R. K. Godfrey and J. W. Wooten 1979). A study of the morphology of *Halodule* in one large population in the northern Gulf of Mexico showed the leaf tips in one population to range from that of *H. beaudettei* to that of *H. wrightii* (R. C. Phillips 1967). A later study of the isozymes of the two morphologic types showed no difference between the two (C. McMillan 1991). I am following both Phillips and McMillan in accepting one species of *Halodule* in the Gulf of Mexico, i.e., *H. wrightii*. *Halodule beaudettei*, therefore, is a synonym of *H. wrightii*.

2. SYRINGODIUM Kützing in R. F. Hohenacker, Algae Marinae Exsiccatae 9: 426. 1860 • Manatee-grass

Stems erect, subtended by ovate scales. **Leaves** 2–3; blade terete or semi-terete, apex acute to obtuse, lacunae 2; veins 1–2[–7–10], midrib inconspicuous, not obviously widened distally, lateral veins inconspicuous, not ending in tooth. **Inflorescences** cymose. **Flowers** axillary to bract, inflated sheath present. **Staminate flowers:** anthers attached at same height on axis. **Pistillate flowers** sessile, styles divided into 2 long stigmas. **Fruits** drupaceous, obovoid [ellipsoid], quadrangular in cross section.

Species 2 (1 in the flora): warm oceanic waters of both hemispheres.

SELECTED REFERENCES   Cox, P. A., T. Elmqvist, and P. B. Tomlinson. 1990. Submarine pollination and reproductive morphology in *Syringodium filiforme* (Cymodoceaceae). Biotropica 22: 259–265.   Tomlinson, P. B. and U. Posluszny. 1978. Aspects of floral morphology and development in the seagrass *Syringodium filiforme* (Cymodoceaceae). Bot. Gaz. 139: 333–345.

1. **Syringodium filiforme** Kützing in R. F. Hohenacker, Algae Marinae Exsiccatae 9: 426. 1860 ⟦F⟧

Cymodocea filiformis (Kützing) Correll; *C. manatorum* Ascherson

Leaves: sheath 1.5–7 cm; blade 5–30 cm × 0.3–2 mm; peripheral veins 2. **Inflorescences:** proximal branches dichasial, the distal monochasial, 1.5–5 × 1–4.5 cm; bracts: sheath 5–9 × 2–4 mm, blade 0.5–1 × 0.1–0.3 mm. **Staminate flowers:** peduncles 5–10 mm; anthers ovate to elliptic, 3–5 mm. **Pistillate flowers:** ovary ellipsoid. **Fruits** obovoid, 5–8 × 3.5–5 mm.

Flowering and fruiting winter–summer (Feb–Jun). Sublittoral zone of marine waters with sandy or muddy substrates; -25–0 m; Fla., La., Miss., Tex.; e Mexico; West Indies; Bermuda; Central America (Belize, Nicaragua, Costa Rica, Panama); South America (Colombia, Venezuela).

Syringodium filiforme has been listed as *Cymodocea filiformis* (D. S. Correll and H. B. Correll 1972). *Cymodocea* was separated from *Syringodium*, however, by the flattened leaves and solitary flowers of *Cymodocea* and terete leaves and cymose inflorescences of *Syringodium* (C. den Hartog 1970). *Syringodium* occurs in the Caribbean Sea and Gulf of Mexico, as well as subtropical oceans of the Old World, whereas *Cymodocea* is restricted to the tropical and subtropical oceans of the Old World (C. den Hartog 1970).

200. ZOSTERACEAE Dumortier

• Eel-grass Family

Herbs, perennial, rarely annual, rhizomatous, caulescent; turions absent. **Leaves** alternate, submersed, sessile; sheath persisting longer than blade or decaying with age into bundles of woolly fibers, not leaving circular scar when shed, not ligulate, auriculate with scarious lobes; blade linear; intravaginal squamules scales, more than 2. **Inflorescences** axillary or terminal, spadices, surrounded by spathe, pedunculate; peduncle following fertilization not elongating, not spiraling. **Flowers** unisexual, staminate and pistillate on same plant or different plants; subtending bracts (retinacula) often present; perianth absent. **Staminate flowers:** stamens 1; anthers dehiscing longitudinally; pollen linear. **Pistillate flowers:** pistils 1, not stipitate; ovules pendulous, orthotropous. **Fruits** achenelike. **Seeds** 1; embryo straight.

Genera 3, species 18 (2 genera, 5 species in the flora): widely distributed in temperate oceans worldwide.

Substantial differences occur in the literature concerning the morphology of the Zosteraceae, especially the reproductive structures. For a discussion of those differences, see C. den Hartog (1970). In summary, the flowers are unisexual, embedded in a spadix, and may or may not each be subtended by a bract or retinaculum. The spadix is subtended by a spathe, and it may or may not be retained within that spathe following fruit production. Spathes are produced on specialized branches (generative shoots), rather than occurring over the entire plant. The generative shoots may or may not contain normal vegetative leaves, but they do contain one or more spathes. The term inflorescence is ambiguous for Zosteraceae, and den Hartog used 'rhipidium' instead. In this treatment 'inflorescence' is used to include the spathe, spadix, and flowers.

SELECTED REFERENCES Hartog, C. den. 1970. The Sea-grasses of the World. Amsterdam. Haynes, R. R. and L. B. Holm-Nielsen. 1985. A generic treatment of Alismatidae in the Neotropics. Acta Amazon. 15(suppl.): 153–193. Phillips, R. C. and C. P. McRoy, eds. 1980. Handbook of Seagrass Biology: An Ecosystem Perspective. New York.

1. Staminate and pistillate flowers on same plant; spadix always enclosed within spathe sheath. . . . . 1. *Zostera*, p. 91

1. Staminate and pistillate flowers on different plants; mature spadix projecting from spathe
   sheath. . . . . 2. *Phyllospadix*, p. 93

1. ZOSTERA Linnaeus, Sp. Pl. 2: 968. 1753; Gen. Pl. ed. 5, 415. 1754 • Eel-grass, zostère [Greek *zoster*, belt]

Herbs, perennial or rarely annual, rooted in substrate. **Roots** 5–20, nodal. **Leaves:** sheath tubular or open, persisting longer than blade, often rupturing with age but not remaining as bundles of woolly fibers; blade entire or slightly denticulate distally; veins 3–11. **Generative shoot** lateral or terminal. **Inflorescences:** peduncle partially adnate to stem; spadix lanceolate, enclosed permanently within spathe sheath; staminate and pistillate flowers alternating on same spadix. **Staminate flowers** often subtended by bract. **Pistillate flowers** without subtending bract, pistil elliptic. **Fruits** achenelike, ovoid to ellipsoid.

Species 12 (2 in the flora): temperate waters worldwide.

Zostera, especially *Z. marina,* is very possibly the most important taxon of marine angiosperms in the Northern Hemisphere. Much has been written about the biology and economic importance of *Z. marina,* and the literature was summarized by C. den Hartog (1970). The species functions in sediment deposition, substrate stabilization, as substrate for epiphytic algae and micro-invertebrates, and as nursery grounds for many species of economically important marine vertebrates and macro-invertebrates. In addition to those functions, the species once was the principal material for the Dutch dikes, and it has been utilized as packing material and as stuffing for mattresses and cushions. Finally, the species has been used for food and recreation by the Seri tribe of Native Americans (R. Felger and M. B. Moser 1973).

During the 1930s, *Zostera marina* of North America began to develop large brown spots on the leaves and rhizomes. Slowly the plants of a population died, the dieback eventually spreading throughout the North Atlantic, until there was very little eel-grass remaining there. This slow dieback of eel-grass became known as "wasting disease." Biologists began studying the disease and eventually determined the causative agent to be *Labyrinthula zosterae,* a slime mold. That dieback resulted in huge decreases in the population sizes of most fauna that depended on *Zostera.* Slowly, eel-grass returned to most areas that it had occupied prior to the disease.

SELECTED REFERENCES Harrison, P. G. 1979. Reproductive strategies in intertidal populations of two co-occurring seagrasses (*Zostera* spp.). Canad. J. Bot. 57: 2635–2638. Nomme, K. M. and P. G. Harrison. 1991. A multivariate comparison of the seagrasses *Zostera marina* and *Zostera japonica* in monospecific versus mixed populations. Canad. J. Bot. 69: 1984–1990.

1. Leaf sheath tubular, without 2 membranous flaps, rupturing with age; bract absent or rarely
   1, subtending lowermost staminate flower. . . . . 1. *Zostera marina*
1. Leaf sheath open, with 2 membranous flaps, persisting without rupturing; bract subtending
   each staminate flower. . . . . 2. *Zostera japonica*

1. Zostera marina Linnaeus, Sp. Pl. 2: 968. 1753
• Zostére marine

Zostera marina var. *stenophylla* Ascherson & Graebner

Herbs, perennial. **Rhizomes** 2–6 mm thick; roots 5–20 at each node. **Leaves:** sheath tubular, rupturing with age, 5–20 cm, membranous flaps absent; blade to 110 cm × 2–12 mm, apex round-obtuse or slightly mucronate; veins 5–11. **Generative shoots** terminal, repeatedly branched, each branch with 1–5 spathes. **Inflorescences:** peduncles with adnate portion 15–100 mm, free portion 2–3 cm; spathes 10 or more, sheath 4–8 cm × 2–5 mm; blade 5–20 cm; spadix linear, apex acute or mucronate; staminate flowers 1–20; pistillate flowers 1–20. **Staminate flowers:** bracts absent or rarely 1, subtending most proximal flower; pollen sacs 4–5 × 1 mm. **Pistillate flowers:** ovary 2–3 mm, style 1–3 mm. **Fruits** ellipsoid to ovoid, 2–5 mm, often beaked.

Flowering late spring–summer. Intertidal to sublittoral of marine waters; -10–0 m; B.C., N.B., Nfld. and Labr. (Nfld.), Nunavut, N.S., Ont., P.E.I., Que., Yukon; Alaska, Calif., Conn., Del., Maine, Md., Mass., N.H., N.Y., N.C., Oreg., R.I., Va., Wash.; Mexico (Baja California, Sinaloa, Sonora); Eurasia.

Zostera marina is adapted to the cold waters of the North Atlantic and North Pacific. It extends southward to North Carolina in the Atlantic and Baja California in the Pacific. At the southern limits of its range, active growth mostly is in the cooler months of autumn and spring, with flowering and fruiting mostly in the spring and the plants dying in the hotter summer months, the vegetation becoming dislodged from the substrate and floating to the water surface. The fruits apparently remain in the floating vegetation for a period of time, eventually falling from the shoots to the substrate. Movement in dislodged vegetative material is the only adaptation the fruits have for dispersal (C. den Hartog 1970).

The species is found mostly in the sublittoral region, only rarely being exposed at low tide. It occurs in more or less sheltered areas on soft mud or firm sand. Plants of sandy substrates have narrower leaves than plants growing on muddy substrates (C. H. Ostenfeld 1905). Fruits fall from the floating vegetation to the substrate and settle on the substrate ripple marks, which run more or less perpendicular to the direction of current. Seedling establishment is parallel with the ripple marks, forming vegetated ridges separated by depressions, which gradually fill with sediments, and the plants then grow laterally into them, forming a meadow (C. den Hartog 1970). The vegetation lowers the velocity of current flow, causing some suspended particles to settle out and accumulate around the base of the plants, slowly building the substrate. As more particles accumulate, the substrate gets deeper over the rhizomes, since the rhizomes grow horizontally, not vertically. Eventually, the rhizomes are too deep, and the plants begin to die back, a phenomenon followed by erosion.

SELECTED REFERENCES Backman, T. W. H. 1991. Genotypic and phenotypic variability of *Zostera marina* on the west coast of North America. Canad. J. Bot. 69: 1361–1371. Cock, A. W. A. M. de. 1980. Flowering, pollination and fruiting in *Zostera marina* L. Aquatic Bot. 9: 201–220. Phillips, R. C. 1964. Comprehensive Bibliography of *Zostera marina*. Washington. [U.S.D.I. Fish Wildlife Serv., Special Sci. Rep. Wildlife 79.]

2. Zostera japonica Ascherson & Graebner in H. G. A. Engler, Pflanzenr. 31[IV,11]: 32. 1907 [I]

Zostera americana Hartog

Herbs, annual, rarely perennial. **Rhizomes** 0.5–1.5 mm thick; roots 2 at each node. **Leaves:** sheath open, persistent, 3–5.5 cm, membranous flaps 2; blade to 15 cm × 1–1.5 mm, apex obtuse or retuse; veins 3. **Generative shoots** lateral, mostly unbranched, with 1–5 spathes. **Inflorescence:** peduncles with adnate portion 20–90 mm, free portion 1.2–3.1 cm; spathes 2–5, sheath 2–4 cm × 2–2.5 mm; blade 5–10 cm; spadix lanceolate, apex obtuse; staminate flowers 5–7; pistillate flowers 5–8. **Staminate flowers:** bracts subtending every flower; pollen sacs 0.5 × 0.7 mm. **Pistillate flowers:** ovary 1.5–2 mm; style 1–2 mm. **Fruits** ellipsoid, ca. 2.5 mm, beaked.

Flowering early fall. Intertidal marine waters; -2–0 m; introduced; B.C.; Wash.; Asia.

The name *Zostera americana* was proposed for some of the collections made by Neil Hotchkiss from Pacific County, Washington (C. den Hartog 1970). Because *Z. americana* resembled a previously published species, it was suggested the name should be placed in synonymy, at least until further study could be undertaken of at least the ecology and genetics of the complex (R. C. Phillips and R. F. Shaw 1976; P. G. Harrison 1976). A proposal that *Z. americana* was synonymous with *Z. noltii* was based upon the identical or overlapping ranges of most characteristics (R. C. Phillips and R. F. Shaw 1976). *Zostera noltii* is native to the Atlantic coasts of Europe and Africa and to the Mediterranean Sea area. Therefore, the suggestion implies that *Z. noltii* has been introduced into North America. No mode of introduction was discussed, however.

Similarly P. G. Harrison (1976) suggested an introduction of an exotic species, but he suggested *Zostera japonica* instead. A study of populations of *Z. ameri-*

cana from Boundary Bay, south of Vancouver, British Columbia revealed no "obvious differences" between those plants and individuals of *Z. japonica* and *Z. noltii*. A comparison of the British Columbia specimens with illustrations by C. den Hartog (1970) of both *Z. japonica* and *Z. noltii* indicated the British Columbia plants resembled more the illustrations of *Z. japonica* than those of *Z. noltii*. A discussion of possible modes of introduction noted that a brown alga, *Sargassum muticum*, was introduced into the North American Pacific coast area with seed oysters. *Zostera japonica* occurs in areas where the oysters were obtained in Japan,

and oysters were packed in *Zostera* species during shipment. Such shipments were possibly the means by which the species was introduced into North America. Harrison's explanation is quite plausible, and I am accepting it until further research solves the problem.

SELECTED REFERENCES Bigley, R. E. and J. L. Barreca. 1982. Evidence for synonymizing *Zostera americana* den Hartog with *Zostera japonica* Aschers. & Graebn. Aquatic Bot. 14: 349–356. Harrison, P. G. 1976. *Zostera japonica* (Aschers. & Graebn.) in British Columbia, Canada. Syesis 9: 359–360. Phillips, R. C. and R. F. Shaw. 1976. *Zostera noltii* Hornem. in Washington, U.S.A. Syesis 9: 355–358.

2. PHYLLOSPADIX Hooker, Fl. Bor.-Amer. 2(10): 171. 1838 • Surf-grass [Greek *phyllon*, leaf, and *spadix*, spadix]

Herbs, perennial, attached to rocks. **Roots** 2 or in 2 rows of 3–5, internodal. **Leaves:** sheath open, decaying with blade, remaining as bundles of woolly fibers; blade irregularly toothed proximally to distally; veins 3–7. **Generative shoot** lateral. **Inflorescences:** peduncle free from stem; spadix linear, enclosed by spathe sheath when young, projecting from sheath when mature; staminate flowers and pistillate flowers on different spadices on different plants. **Staminate flowers** subtended by bract. **Pistillate flowers** subtended by bract, pistil crescent-shaped. **Fruits** drupaceous, crescent-shaped.

Species 5 (3 in the flora): temperate waters of the northern Pacific.

Plants of *Phyllospadix* grow attached to rocks, many of which are exposed at low tide. The ecology and importance of *Phyllospadix* is not known nearly as well as that of *Zostera*. In summary *Phyllospadix* vegetation protects the rocky substrate from erosion, and by accumulating sand in and between the tussocks, transforms the rocky substrate into sandy beaches or sublittoral sand flats. Rejuvenation of the *Phyllospadix* vegetation, however, is then no longer possible on the sand-covered rocks. The plants eventually die, exposing the sand-covered rocks to wave action, which results in erosion of the sand, again exposing the rocks (C. den Hartog 1970).

SELECTED REFERENCES Phillips, R. C. 1979. Ecological notes on *Phyllospadix* (Potamogetonaceae) in the northeast Pacific. Aquatic Bot. 6: 159–170. Soros-Pottruff, C. L. and U. Posluszny. 1994. Developmental morphology of reproductive structures of *Phyllospadix* (Zosteraceae). Int. J. Pl. Sci. 155: 405–420.

1. Leaves serrulate distally; veins 5–7; nodes with 2 roots. . . . . 3. *Phyllospadix serrulatus*
1. Leaves entire distally; veins 3; nodes with 2 rows of 3–5 roots.
  2. Spathes 1–5; pistillate bract distinctly narrowed at base; leaves 0.5–1.5 mm wide. . . . 1. *Phyllospadix torreyi*
  2. Spathes 1(–2); pistillate bract not narrowed at base; leaves 1–4 mm wide. . . . . . . 2. *Phyllospadix scouleri*

1. Phyllospadix torreyi S. Watson, Proc. Amer. Acad.
   Arts 14: 303. 1879 • Torrey's surf-grass F

Herbs; nodes with 2 rows of 3–5 roots. **Leaves:** sheath 6.5–55 cm, margins overlapping; blade to 2 m × 0.5–1.5 mm, margins entire, apex notched to rounded or slightly apiculate; veins 3. **Generative shoots** 50–60 cm, nodes 4–6, proximal 2 nodes sterile, each with 1 leaf, distal 2–4 nodes each bearing 1 leaf and 1–5 spathes. **Inflorescences:** peduncles 8–45 × 1–2 mm; staminate bract 2.5–4.3 × 2–2.7 mm; pistillate bract 3.5–6 × 1–1.5 mm, base distally narrowed, apex obtuse to acute. **Fruits** 2.8–3.2 × 2.9–3.5 mm.

Flowering and fruiting spring–fall. Intertidal zone; -15 m; B.C.; Calif., Oreg.; Mexico (Baja California).

SELECTED REFERENCE Williams, S. L. 1995. Surfgrass (*Phyllospadix torreyi*) reproduction: Reproductive phenology, resource allocation, and staminate rarity. Ecology 76: 1953–1970.

2. Phyllospadix scouleri Hooker, Fl. Bor.-Amer. 2(10):
   171. 1838 • Scouler's surf-grass F

Herbs; nodes with 2 rows of 3–5 roots. **Leaves:** sheath 4–35 cm, margins not overlapping; blade to 2 m × 1–4 mm, margins entire, apex obtuse to truncate or rarely slightly notched; veins 3. **Generative shoots** 0.2–11 cm, nodes 1–2, proximal node when present sterile, with 1 leaf, distal node with 1 leaf and 1(–2) spathes. **Inflorescences:** peduncles 11–60 × 1–2 mm; staminate bract 4–5.5 × 2–3 mm; pistillate bract 4–8 × 1.5–3 mm, base not narrowed, apex truncate to acute. **Fruits** 4–5 × ca. 5.5 mm.

Flowering and fruiting late spring–summer. Intertidal and upper part of sublittoral; -2 m; B.C.; Alaska, Calif., Oreg., Wash.; Mexico (Baja California).

SELECTED REFERENCE Barnabas, A. D. 1994. Anatomical, histochemical and ultrastructural features of the seagrass *Phyllospadix scouleri* Hook. Aquatic Bot. 49: 167–182.

3. Phyllospadix serrulatus Ruprecht ex Ascherson,
   Linnaea 35: 169. 1868 E

Herbs; nodes with 2 roots. **Leaves:** sheath 4–20 cm, margins not overlapping; blade 30–60 cm × 1–4 mm, margins serrulate distally, apex obtuse to truncate or rarely slightly notched; veins 5–7. **Generative shoot** 1–6 cm, nodes 1, without leaves, of 1(–2) spathes. **Inflorescences:** peduncles 11–40 × 1–2 mm; staminate bract 4–5.5 × 2–3 mm; pistillate bract 4–8 × 1.5–2 mm, base not narrowed, apex obtuse to truncate. **Fruits** 4–5 × ca. 5 mm.

Flowering and fruiting summer. Upper tidal to subtidal, attached to rocks or rarely on deep clay substrates; -6 m; B.C.; Alaska, Wash.

201. ARECACEAE Schultz-Schultzenstein
• Palm Family

Palmae Jussieu

Scott A. Zona

Trees or shrubs [lianas], perennial, branched or unbranched, solitary or clustered. **Roots** adventitious, thick. **Stems** woody, subterranean or terrestrial, creeping or erect [climbing], slender or massive, sometimes conspicuously enlarged and storing starch and water, smooth or covered with fibrous or prickly remains of leaf bases. **Leaves** spirally arranged; sheaths tubular, often forming crownshaft, sometimes with ligular appendages; petioles terete, channeled, or ridged, unarmed or bearing prickles or marginal teeth; hastula (flap of tissue from petiole apex at junction with surface of blade) absent or present adaxially, rarely present abaxially. **Leaf blade** palmate, costapalmate (intermediate between palmate and pinnate), pinnate, or 2-pinnate [undivided]; plication (folding lengthwise into pleats or furrows) ∧- or tent-shaped (reduplicate, splitting along abaxial ridges) or V-shaped (induplicate, splitting along adaxial ridges); segments lanceolate, linear, or cuneate [rhombic], glabrous or variously scaly, unarmed or bearing prickles (proximal segments modified into spines in *Phoenix*). **Inflorescences** from solitary [clustered] axillary buds, borne within, below, or above crown of leaves, paniculate, rarely spicate, usually branched to 1–5 orders; prophyll (1st bract on main inflorescence axis) 2-keeled; peduncular bract(s) (empty bract[s] between 1st prophyll and 1st bract subtending branch) present [absent]; flowers bisexual, unisexual with staminate and pistillate on same plants or on different plants, or both bisexual and unisexual on same plant. **Flowers** solitary or variously clustered along rachillae of inflorescence, radially symmetric; perianth 1–2-seriate; sepals [2–]3[–4], distinct or connate; petals [2–]3[–4], distinct or variously connate; stamens [3–]6–34[–1000]; filaments distinct or connate or basally adnate to petals; anthers basifixed or dorsifixed, dehiscing latrorsely or introrsely; staminodes in pistillate flowers distinct or variously connate or adnate to pistil or petals; pistils 1 or 3, distinct or partially connate, each bearing 1 ovule and 1 stigma, or 1 pistil bearing 1–3 ovules and 3 stigmas; styles distinct or connate, short; stigmas dry; pistillode in staminate flower present or absent. **Fruits** drupaceous or berrylike; stigmatic remains basal or apical; exocarp smooth, warty, prickly, or hirsute [corky or scaly]; mesocarp fleshy or dry and fibrous; endocarp papery, leathery, or bony, sometimes with 3 germination pores. **Seeds** 1(–2+), free or adhering to endocarp; seed coat thin; endosperm homogeneous or ruminate, sometimes penetrated by seed coat; embryo basal, lateral, or apical, peglike, minute; eophyll (1st seedling leaf with blade) undivided and lanceolate or 2-cleft [pinnate].

Genera 191, species ca. 2500 (19 genera, 29 species in the flora): worldwide, especially abundant in Central America, South America, se Asia.

Although palms appeared in various taxonomic schemes since the time of Linnaeus, the first attempt at a modern phylogenetic classification of the palms was published by H. E. Moore Jr. (1973). Moore left his "major groups" unranked, and his untimely death in 1980 prevented his completing a formal synthesis. J. Dransfield and N. W. Uhl (1986) gave formal ranks to Moore's groups and divided the family into six subfamilies and numerous tribes and subtribes. Their *Genera Palmarum* (N. W. Uhl and J. Dransfield 1987, 1999) is a model of accuracy and completeness and will long serve the needs of the scientific, horticultural, and resource-management communities. With the advent of molecular techniques and a resurgence in palm research, however, realignments in the classification may be expected, and indeed additional data already require some changes in the current scheme (J. Dransfield and H. J. Beentje 1995; N. W. Uhl and J. Dransfield 1999; N. W. Uhl et al. 1990, 1995).

Modern cladistic analyses place the palms as the sister group to the Commelinanae clade (M. W. Chase et al. 1993; J. I. Davis 1995; M. R. Duvall et al. 1993b), with which they share ultraviolet-fluorescent phenolic compounds in their cell walls and *Strelitzia*-type epicuticular wax morphology (W. Barthlott and D. Frölich 1983; P. J. Harris and R. D. Hartley 1980). Palms are currently treated as the sole representative of the superorder Arecanae, order Arecales (R. M. T. Dahlgren et al. 1985; R. F. Thorne 1992b).

Morphologically the family is diverse and complex (see especially P. B. Tomlinson 1990). The majority of palms produce a single indeterminate stem with axillary inflorescences; several noteworthy departures, however, also occur in numbers of vegetative and floral axes, position of inflorescence, and displacement of terminal bud. Stems may be solitary (monopodial) or clustered (sympodial), erect, prostrate, or lianoid. A majority of palms have unbranched vegetative axes, although aerial branching, sometimes dichotomous, is known in a variety of unrelated genera (e.g., *Korthalsia* Blume, *Nannorrhops* H. Wendland). Branching may also be nonaxiallary in some genera (J. B. Fisher et al. 1989).

Studies of pollination (F. Borchsenius 1997; F. Ervik and J. P. Feil 1997; A. Henderson 1986; C. Listabarth 1992, 1993, 1993b, 1994; A. O. Scariot et al. 1991) indicate that insect pollination, especially by beetles (Coleoptera), bees and wasps (Hymenoptera), and flies (Diptera), is apparently more common than wind pollination. Bats (Chiroptera) play a role in the pollination of some species (S. A. Cunningham 1995).

Dispersal of seeds is generally by means of animals for fleshy-fruited palms (S. Zona and A. Henderson 1989). Many species of mammals include palm fruits in their diets (S. H. Bullock 1980; R. F. Harlow 1961; W. D. Klimstra and A. L. Dooley 1990; D. S. Maehr 1984; D. S. Maehr and J. R. Brady 1984), but birds also play a significant role. In the Eastern Hemisphere, *Cocos* Linnaeus and *Nypa* Steck have achieved a wide distribution as the result of dispersal by water. For the relationship between palms and seed-eating bruchid beetles (Bruchidae: Pachymerinae: Pachmerini), see C. D. Johnson et al. (1995).

SELECTED REFERENCES Dransfield, J. and N. W. Uhl. 1986. An outline of a classification of palms. Principes 30: 3–11. Henderson, A. 1986. A review of pollination studies in the Palmae. Bot. Rev. 52: 221–259. Henderson, A., G. Galeano, and R. G. Bernal. 1995. Field Guide to the Palms of the Americas. Princeton. McClintock, E. 1993. Arecaceae [Palmae]. In: J. C. Hickman, ed. 1993. The Jepson Manual. Higher Plants of California. Berkeley, Los Angeles, and London. P. 1105.   Moore, H. E. Jr. 1973. The major groups of palms and their distribution. Gentes Herb. 11: 27–141.   Tomlinson, P. B. 1990. The Structural Biology of Palms. Oxford.   Uhl, N. W. and J. Dransfield. 1987. Genera Palmarum. Lawrence, Kans.   Zona, S. 1997. The genera of Palmae (Arecaceae) in the southeastern United States. Harvard Pap. Bot. 2: 71–107.

1. Leaf blades palmate to costapalmate.
  2. Pistils 3, distinct, tomentose; leaf sheath armed with coarse needlelike fibers. . . 3. *Rhapidophyllum*, p. 100
  2. Pistils 3(-4), partially connate, or apparently 1-carpellate, glabrous; leaf sheath producing soft fibers.
    3. Perianth 1-seriate.
      4. Petiole split at base; exocarp white; seed excavate or perforate, not furrowed. . . . . 1. *Thrinax*, p. 98
      4. Petiole not split at base; exocarp purplish; seed irregularly brain-shaped. . . 2. *Coccothrinax*, p. 99
    3. Perianth 2-seriate.
      5. Petioles entirely unarmed; rachillae of inflorescence glabrous. . . . . 8. *Sabal*, p. 107
      5. Petioles, especially basal portions, coarsely to obscurely armed; rachillae of inflorescence pubescent or glabrous.
        6. Stems procumbent or ascending; hastula present on both surfaces of leaf. . . . . 6. *Serenoa*, p. 104
        6. Stems erect; hastula present on adaxial surface of leaf
          7. Stems less than 15 cm diam., clustered. . . . . 5. *Acoelorraphe*, p. 103
          7. Stem more than 20 cm diam., solitary.
            8. Petiole split at base; leaf segments bearing fibers. . . . . 7. *Washingtonia*, p. 105
            8. Petiole not split at base; leaf segments not bearing fibers. . . . . 4. *Livistona*, p. 102
1. Leaf blade pinnate or 2-pinnate.
  9. Leaf segment plication induplicate (V-shaped).
    10. Leaves pinnate; leaf segment apices acute; basal segments modified into spines. . . . 9. *Phoenix*, p. 110
    10. Leaves 2-pinnate (in mature plants); leaf segment apices jagged and irregular; basal segments not modified into spines. . . . . 12. *Caryota*, p. 114
  9. Leaf segment plication reduplicate (∧-shaped).
    11. Plant (especially leaves) armed.
      12. Petiole margins unarmed, trunk and blades armed. . . . . 19. *Acrocomia*, p. 122
      12. Petiole margins armed (base of midvein lignified and persisting as spine), trunk and blades unarmed. . . . . 18. *Elaeis*, p. 121
    11. Plant with unarmed trunks, leaf bases, and leaf rachises.
      13. Flowers mostly bisexual (distal flowers with reduced gynoecia), borne singly on rachillae on pedicel-like stalks. . . . . 10. *Pseudophoenix*, p. 111
      13. Flowers unisexual, sessile.
        14. Staminate and pistillate flowers on different plants; stems clustering. . . . . 11. *Chamaedorea*, p. 113
        14. Staminate and pistillate flowers on same plants; stems various.
          15. Crownshaft absent.
            16. Fruits dry, greater than 10 cm diam. . . . . 16. *Cocos*, p. 119
            16. Fruits fleshy, less than 4 cm diam. 17. *Syagrus*, p. 120
          15. Crownshaft present.
            17. Stems solitary, greater than 20 cm diam. . . . . 14. *Roystonea*, p. 116
            17. Stems solitary or clustered, less than 15 cm diam.
              18. Leaf segment apices jagged and irregular; leaf sheath green; endocarp 5-lobed in cross section. . . . . 15. *Ptychosperma*, p. 118
              18. Leaf segment apices acuminate; leaf sheath yellowish; endocarp terete in cross section. . . . . 13. *Dypsis*, p. 115

201a. ARECACEAE Schultz-Schultzenstein subfam. CORYPHOIDEAE Griffith, Calcutta J. Nat. Hist. 5: 311. 1844

201a.1. ARECACEAE Schultz-Schultzenstein (subfam. CORYPHOIDEAE) tribe CORYPHEAE Martius in S. L. Endlicher, Gen. Pl. 4: 252. 1837

201a.1a. ARECACEAE Schultz-Schultzenstein (tribe CORYPHEAE) subtribe THRINACINAE Beccari, Webbia 2: 9. 1907

1. THRINAX Swartz, Prodr. 4, 57. 1788 • Thatch palm [Greek *thrinax*, trident or winnowing fork; presumably in reference to shape of leaf]

Hemithrinax Hooker f.

Plants small to moderate. **Stem** solitary, erect, ± smooth. **Leaves:** sheath producing soft fibers; petiole split at base, unarmed; abaxial hastula a small crescent-shaped ridge or absent; adaxial hastula irregularly semicircular to nearly cylindric; blade palmate; plication induplicate; segments lanceolate, basally connate; cross veins conspicuous [obscure]. **Inflorescences** interfoliar, emerging through split leaf bases, arching beyond leaves, with 2 orders of pendent branches; prophyll short; peduncular bracts many, tubular. **Flowers** bisexual, borne singly along rachillae, short- to long-pedicellate; perianth 1-seriate, shallowly cupulate, lobes 5–7, apiculate; stamens 6–12, erect [inflexed in bud]; filaments acute; anthers dorsifixed; pistils 1, 1-carpellate, glabrous; stigma funnelform. **Fruits** globose; stigmatic scar apical; exocarp white, smooth, slightly warty or rugose when dry; mesocarp mealy; endocarp membranaceous. **Seeds** oblate-globose, hilum deeply intruded into seed, forming cylindric depression or perforation; endosperm homogeneous, bony; embryo nearly apical; eophyll undivided, lanceolate. $x = 18$.

Species 7 (2 in the flora): regions along shores, North America (Florida), Mexico, West Indies (Bahamas), and Central America.

Thrinax lacks a showy perianth and is wind pollinated, although many insects visit the inflorescence (D. W. Roubik, pers. comm.) and may transfer some pollen once the lateral lobes of the stigma have opened to expose the stigmatic surfaces. Fruits are taken by birds (R. W. Read 1975), including the red-bellied woodpecker (*Melanerpes carolinus*), lizards (J. B. Iverson 1979), and gray squirrels, which probably disperse the seeds. Key deer consume *T. morrisii* fruits, but they are probably not seed dispersers (W. D. Klimstra and A. L. Dooley 1990; W. D. Klimstra, pers. comm.).

SELECTED REFERENCE   Read, R. W. 1975. The genus *Thrinax* (Palmae: Coryphoideae). Smithsonian Contr. Bot. 19: 1–98.

1. Leaf sheath liguliform opposite petiole; hastula silky-pubescent adaxially; leaf blades glaucous; segment apices stiff; pedicel inconspicuous in fruit. . . . . 1. *Thrinax morrisii*
1. Leaf sheath with V-shaped cleft opposite petiole; hastula glabrous adaxially; leaf blades green; segment apices lax; pedicel conspicuous in fruit (usually exceeding 1 mm). . . . . 2. *Thrinax radiata*

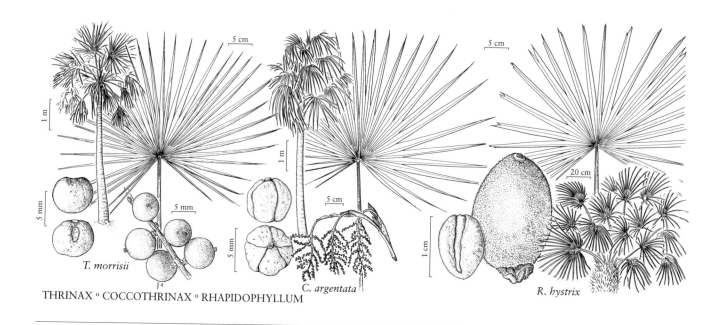

T. morrisii

THRINAX ° COCCOTHRINAX ° RHAPIDOPHYLLUM

C. argentata

R. hystrix

1. **Thrinax morrisii** H. Wendland, Gard. Chron., ser. 3, 11: 104. 1892 • Brittle-thatch F

Thrinax bahamensis O. F. Cook; *T. keyensis* Sargent; *T. microcarpa* Sargent; *T. ponceana* O. F. Cook; *T. praeceps* O. F. Cook

Leaves: sheath liguliform opposite petiole; hastula adaxially silky-pubescent; blade glaucous, with dark, minute dots on abaxial surface; segment apices stiff. **Fruits:** pedicel inconspicuous in fruit. $2n = 36$.

Flowering spring–summer. Rocky, calcareous soil of coastal hammocks and scrub; 0–10 m; Fla.; West Indies (Bahamas, Cuba, Hispaniola, Puerto Rico, Lesser Antilles).

See *Coccothrinax* for discussion of hybridization with that genus.

2. **Thrinax radiata** Loddiges ex Schultes & Schultes f. in J. J. Roemer et al., Syst. Veg. 7(2): 1301. 1830

Coccothrinax radiata (Schultes & Schultes f.) Sargent ex K. Schumann; *Thrinax floridana* Sargent; *T. martii* Grisebach & H. Wendland ex Grisebach

Leaves: sheath with V-shaped cleft opposite petiole; hastula adaxially glabrous; blade green, with minute, whitish dots on abaxial surface; segment apices lax. **Fruits:** pedicel conspicuous in fruit (usually exceeding 1 mm). $2n = 36$.

Flowering summer. Pinelands and littoral hammocks and scrub on limestone soils; 0–10 m; Fla.; se Mexico; West Indies.

2. COCCOTHRINAX Sargent, Bot. Gaz. 27: 87. 1899 • Silver palm [Greek *coccos*, berry, and *thrinax*, trident or winnowing fork]

Haitiella L. H. Bailey; *Thrincoma* O. F. Cook; *Thringis* O. F. Cook

Plants small to moderate. **Stems** solitary or occasionally cespitose, erect, slender, at first covered with leaf sheaths. **Leaves:** sheath fibers soft, netlike, eventually sloughing off to reveal smooth to fissured stem; petiole not split at base, unarmed; abaxial hastula a low ridge or absent; adaxial hastula a crescent-shaped ridge to semi-cylindric excrescence; blade palmate; plication induplicate; segments lanceolate, basally connate; cross veins obscure [conspicuous]. **Inflores-**

cences interfoliar, downcurved [long and arching], not extending beyond leaves, with 2 orders of branching; prophyll short, peduncular bracts several, sterile, tubular, distally expanded, silky-pubescent; primary branches subtended by smaller peduncular bracts. **Flowers** bisexual, borne singly along rachillae, short-pedicellate; perianth 1-seriate, shallowly cupulate, lobes 5–7, apiculate; stamens 7–12; filaments acute; anthers dorsifixed, twisted when dry; pistils 1, 1-carpellate, glabrous; style slender; stigma funnelform. **Fruits** globose; stigmatic scar apical; exocarp purplish, smooth, slightly warty when dry; mesocarp fleshy; endocarp membranaceous. **Seeds** globose, irregularly brain-shaped; endosperm homogeneous, bony; embryo apical [superior]; eophyll undivided, lanceolate.

Species 14–50 (1 in the flora): Caribbean Basin.

Coccothrinax shares a similar floral morphology with *Thrinax*, and like *Thrinax* it is wind pollinated. Fruits of *C. argentata* are one of the most important foods of Florida's Key deer (W. D. Klimstra and A. L. Dooley 1990), but seeds are not excreted intact (W. D. Klimstra, pers. comm.).

Coccothrinax includes a great number of species with ornamental potential, and many of the cultivated ones are discussed by C. E. Nauman and R. W. Sanders (1991). Because of their generally small and slender stature and their predictable growth form, they make elegant horticultural subjects.

This genus is in dire need of systematic study.

SELECTED REFERENCES   Nauman, C. E. 1990. Intergeneric hybridization between *Coccothrinax* and *Thrinax* (Palmae: Coryphoideae). Principes 34: 191–198.   Nauman, C. E. and R. W. Sanders. 1991. An annotated key to the cultivated species of *Coccothrinax*. Principes 35: 27–46.

1. **Coccothrinax argentata** (Jacquin) L. H. Bailey, Gentes Herb. 4: 223. 1939 [F]

Palma argentata Jacquin, Fragm. Bot., 39, plate 43, fig. 1. 1801; *Coccothrinax garberi* (Chapman) Sargent; *C. jucunda* Sargent; *T. garberi* Chapman

Stems generally solitary, 0–3 m, occasionally with vegetative sprouts forming on trunk. **Leaves** less than 1 m wide; segments silvery abaxially, stiff or lax. **Inflorescences** compact, not exceeding leaves; rachillae stiff; pedicels conspicuous. **Fruits** dark purple-black, globose, 6.5–12 mm diam. 2*n* = 36.

Flowering spring–fall. Rocky, calcareous soil of coastal hammocks and scrub; 0–10 m; Fla.; West Indies (Bahama Islands).

Palms of this species from Dade and Broward counties, Florida, are typically smaller in stature than those from the Florida Keys. It is not known if this dwarfism is purely the result of edaphic conditions or if it is a genetic trait.

Coccothrinax argentata is found in the Florida Keys and extreme southern Florida as far north as the vicinity of Boca Raton, Palm Beach County. It also occurs in the Bahamas. It is usually a small, single-trunked palm but occasionally develops multiple stems. It is most easily distinguished from *Thrinax morrisii*, with which it is sometimes confused in Florida, by the obscure transverse venation in the leaves and the purple-black fruits on long pedicels (versus conspicuous transverse venation and white fruits on short pedicels in *T. morrisii*).

This species has been reported (C. E. Nauman 1990) to produce sterile natural hybrids with *Thrinax morrisii* in the Florida Keys where the two species are sympatric.

3. RHAPIDOPHYLLUM H. Wendland & Drude, Bot. Zeitung (Berlin) 34: 803. 1876

• Needle palm, palmier à aiguilles [Greek *rhapidos*, a rod, and *phyllon*, leaf, in reference to the spines of the leafbases, or perhaps meaning having the leaf of *Rhapis* (a genus of small Asian palms)]   [E]

Plants small, shrubby. **Stems** solitary or cespitose, compact, erect or procumbent. **Leaves:** sheath armed with sharp, stout, needlelike fibers; petiole unarmed; abaxial hastula absent; adaxial hastula a crescent-shaped ridge or flange; blade palmate or weakly costapalmate; plication induplicate, but segments separating between folds to form plications appearing reduplicate; segments lanceolate, basally connate; cross veins obscure to conspicuous. **Inflorescences** axillary, within crown of leaves, closely appressed to trunk, with 1 or 2 orders of branching; prophyll tubular, expanded distally, membranaceous; peduncular bracts 2, expanded distally. **Flowers** unisexual or bisexual, borne singly along rachillae or densely in clusters of 2 or 3, short-pedicellate; sepals 3, distinct to slightly connate at base; petals 3, distinct, orbiculate to acute, fleshy; stamens 6; filaments linear; anthers dorsifixed; staminodes in pistillate flowers 6, with poorly developed anthers; pistils 3, distinct, tomentose; pistillode in staminate flowers minute. **Fruits** drupes, ovoid, ellipsoid, or globose; exocarp brown, hirsute; mesocarp mealy; endocarp bony. **Seeds** ellipsoid, with prominent raphe; endosperm homogeneous; embryo lateral; eophyll undivided, lanceolate. $x = 18$.

Species 1: North America.

Rhapidophyllum is probably a relictual genus endemic to the southeastern United States. It is most closely related to *Trachycarpus* H. Wendland of the Himalayan region of southeastern Asia (N. W. Uhl and J. Dransfield 1987). *Trachycarpus* is the sister group to *Rhapidophyllum*, which is itself the sister group to the clade containing *Guihaia* J. Dransfield, S. K. Lee, & F. N. Wei and *Rhapis* Linnaeus f. ex Aiton, two genera from southeastern Asia (N. W. Uhl et al. 1995). Those genera most likely evolved from a widespread common ancestor with a Laurasian distribution. This group of genera is perhaps another example of the well-known eastern North America–eastern Asia disjunction pattern.

SELECTED REFERENCES Clancy, K. E. and M. J. Sullivan. 1988. Some observations on seed germination, the seedling, and polyembryony in the needle palm, *Rhapidophyllum hystrix*. Principes 32: 18–25. Shuey, A. G. and R. P. Wunderlin. 1977. The needle palm: *Rhapidophyllum hystrix*. Principes 21: 47–59.

1. Rhapidophyllum hystrix (Pursh) H. Wendland & Drude, Bot. Zeitung (Berlin) 34: 803. 1876 [E] [F]

Chamaerops hystrix Pursh, Fl. Amer. Sept. 1: 240. 1814; *Rhapis caroliniana* Hort ex Kunth; *Sabal hystrix* (Pursh) Nuttall

Stems erect or procumbent, short, 1–7 dm, cespitose shoots often present. **Leaves** less than 70 cm wide; leaf sheath bearing stout emergent spinelike fibers, ca. 50 cm. **Flowers** yellowish. **Fruits** brown, ellipsoid, length ca. 2 cm, diam. 1.5 cm. $2n = 36$.

Flowering spring. Rich soil over limestone in shady mesic to wet woodlands, and along banks of ravines and streams; 20–100 m; Ala., Fla., Ga., Miss., S.C.

The palm usually grows procumbently with adventitious roots emerging from the trunk where it contacts moist soil (A. G. Shuey and R. P. Wunderlin 1977). In the wild, *Rhapidophyllum* forms suckers along its stem, and it is that vegetative reproduction, more than seedling reproduction, that maintains most populations (K. E. Clancy and M. J. Sullivan 1988).

Flowers are protandrous and most likely to be pollinated by a species of *Notolomus* (Coleoptera: Curculionidae). The evil-smelling and curiously hirsute fruits are taken by black bears (D. S. Maehr and J. R. Brady 1984) and other mammals (A. G. Shuey and R. P. Wunderlin 1977).

201a.1b. A R E C A C E A E Schultz-Schultzenstein (tribe CORYPHEAE) subtribe LIVISTONINAE Saakov, Palmy S.S.S.R., 193. 1954

4. L I V I S T O N A R. Brown, Prodr., 267. 1810 • Cabbage-palm [for Patrick Murray, Baron of Livistone (d. 1671), whose collections formed the nucleus of the Edinburgh Botanic Garden] □

Saribus Blume; *Wissmania* Burret

Plants small to large. **Stems** solitary, erect, slender (rarely) to robust (more than 20 cm), at first covered by persistent leaf sheaths, later becoming bare or covered with persistent petiole bases, ringed conspicuously or obscurely with leaf scars. **Leaves:** petiole not split at base, strongly armed [unarmed]; abaxial hastula minute or absent; adaxial hastula conspicuous; blade palmate or costapalmate; plication induplicate; segments basally connate, lanceolate, not producing fibers between segments. **Inflorescences** axillary within crown of leaves, paniculate with 3–5 orders of branching, ± as long as leaves; prophyll membranaceous; peduncular bracts many, obscuring rachis; rachillae pubescent. **Flowers** bisexual [unisexual], borne singly along rachillae [in small groups]; perianth 2-seriate; calyx cupulate, 3-lobed; corolla 3-lobed, valvate; stamens 6, connate in short tube; pistils 3, distinct basally, glabrous; styles connate, slender. **Fruits** drupes; exocarp blackish, smooth; mesocarp fleshy; endocarp bony. **Seeds** globose or ellipsoid; endosperm homogeneous; embryo lateral; eophyll undivided, lanceolate. $x = 18$.

Species 25+ (2 in the flora): introduced in Florida, West Indies, Bermuda; native to Asia, Afric, Pacific Islands, and Australia.

1. Inflorescences with 3 orders of branching; fruits oblong (rarely globose), ripening from green to blue-green. . . . . .1. *Livistona chinensis*
1. Inflorescences with 5 orders of branching, immediately 3-forking to 3 equal main axes; fruits globose, ripening from green through orange to black. . . . . .2. *Livistona rotundifolia*

1. Livistona chinensis (Jacquin) R. Brown ex Martius in C. F. P. von Martius et al., Hist. Nat. Palm. 3: 240, plate 146. 1838 • Chinese fan palm, palmier évantail de Chine F □

Latania chinensis Jacquin, Fragm. Bot., 16, plate 11, fig. 1. 1801; *Saribus chinensis* (Jacquin) Blume.

Leaves: segment apices lax. **Inflorescences** with single primary axis and 3 orders of branching. **Fruits** usually oblong or olive-shaped (rarely globose), ripening from green to blue-green. $2n = 36$.

Flowering spring–summer. Disturbed hammocks and mesic woods; 0–10 m; introduced; Fla.; native, Asia (s China, Bonin Islands).

Livistona chinensis is naturalized in Florida (E. H. Butts 1959).

SELECTED REFERENCE Butts, E. H. 1959. *Livistona chinensis* naturalized in Florida. Principes 3: 133.

2. Livistona rotundifolia (Lamarck) Martius in C. F. P. von Martius et al., Hist. Nat. Palm. 3: 241, plate 102. 1838 • Footstool palm □

Corypha rotundifolia Lamarck in J. Lamarck et al., Encycl. 2: 131. 1786; *Saribus rotundifolia* (Lamarck) Blume

Leaves: segment apices stiff. **Inflorescences** dividing into 3 main axes, each axis bearing up to 4 orders of branching. **Fruits** globose, ripening from green through orange to black. $2n = 36$.

Flowering spring–summer. Disturbed hammocks and mesic woods; 0–10 m; introduced; Fla.; native to Malay Peninsula, Java.

5. ACOELORRAPHE H. Wendland, Bot. Zeitung (Berlin) 37: 148. 1879 • Paurotis palm, palmier des Everglades [Greek *a-*, without, *coelo*, hollow, and *raphe*, in reference to shape of the seed]

Paurotis O. F. Cook

Plants moderate, clustered palms. **Stems** cespitose, erect, slender, less than 15 cm diam. **Leaves:** sheath fibers soft, eventually sloughing off to reveal smooth to fissured stem; petiole not split at base, armed with stout teeth, occasionally obscurely armed with minute teeth; abaxial hastula absent; adaxial hastula conspicuous, acute to acuminate; costa minute to absent; blade palmate; plication induplicate; segments lanceolate, stiff, basally connate. **Inflorescences** axillary within crown of leaves, paniculate, arching, longer than leaves, with 3(–4) orders of branching; prophyll short; peduncular bracts 2, tubular, 2-keeled, long; rachillae pendulous, pubescent. **Flowers** bisexual, borne singly or in clusters of 2–3 along rachillae, sessile; perianth 2-seriate; sepals 3, imbricate, briefly connate, margins ciliate; petals 3, basally connate, valvate, triangular with adaxial crest; stamens 6, filaments basally connate, acuminate; anthers dorsifixed, versatile; dehiscence latrorse-introrse; pistils 3(–4), distinct basally, glabrous; styles connate, filiform; stigmas minute. **Fruits** globose, stigmatic scar apical; exocarp blackish, smooth, slightly rugose when dry; mesocarp thin, fleshy; endocarp membranaceous. **Seeds** 1 per fruit, globose, with circular raphe; endosperm bony, homogeneous; embryo nearly basal; eophyll undivided, lanceolate. $x = 18$.

Species 1: widespread, North America, se Mexico, West Indies, Central America, and n South America (Colombia–Isla de Providencia).

Acoelorraphe is a widespread monotypic genus. The taxa *A. wrightii* forma *inermis* Hadač and *A. wrightii* var. *novogeronensis* Beccari probably have no biological significance.

Recent molecular studies indicated that *Serenoa* is the sister genus to *Acoelorraphe* (N. W. Uhl et al. 1995).

SELECTED REFERENCE Galeano-Garcés, G. 1986. Primer registro de dos géneros de palmas para la flora Colombiana. Mutisia 66: 1–4.

1. **Acoelorraphe wrightii** (Grisebach & H. Wendland) H. Wendland ex Beccari, Webbia 2: 109. 1907 C F

Copernicia wrightii Grisebach & H. Wendland in A. H. R. Grisebach, Cat. Pl. Cub., 220. 1866; *Acoelorraphe arborescens* (Sargent) Beccari; *A. pinetorum* Bartlett; *Paurotis androsana* O. F. Cook; *P. arborescens* (Sargent) O. F. Cook; *P. wrightii* (Grisebach & H. Wendland) Britton

Stems multiple, brown, to 7 m, covered in tardily deciduous leaf bases. **Leaves:** petiole strongly (rarely weakly) armed; hastula present on adaxial surface. **Inflorescences** exceeding leaves, appearing secund because of pendulous rachillae, 15–22 cm, orange in fruit. **Fruits** ripening from green through orange to black, globose, 7.5–8.5 mm diam. $2n = 36$.

Flowering spring–summer. Thin, rocky soil over limestone in hydric hammocks, wet savannas, and swamps of Everglades; of conservation concern; 0–10 m; Fla.; se Mexico; West Indies (Bahamas, Cuba); Central America; n South America (Colombia–Isla de Providencia).

Little is known about this palm in its natural habitats. In Florida, it occurs in the Everglades National Park and is categorized as threatened by the state of Florida. Like *Serenoa repens*, *Acoelorraphe* is tolerant of occasional burning. It is probably bee-pollinated, and the seeds are dispersed by birds (G. Galeano-Garcés 1986).

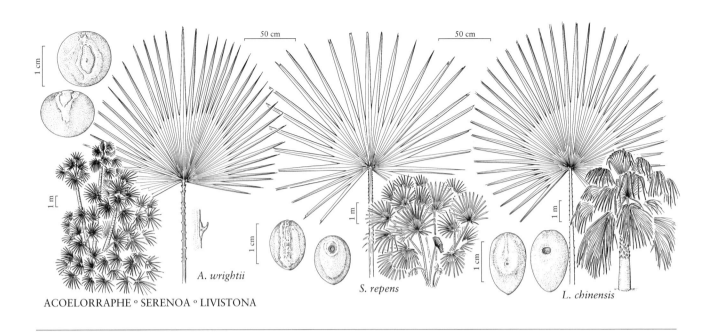

ACOELORRAPHE ∘ SERENOA ∘ LIVISTONA

A. *wrightii*

S. *repens*

L. *chinensis*

6. SERENOA Hooker f. in G. Bentham and J. D. Hooker, Gen. Pl. 3: 879, 926, 1228. 1883 • Saw palmetto [for Sereno Watson, 1826–1892, botanist]   E

Plants moderate, clustered, shrubby. **Stems** branched or unbranched, procumbent or ascending, covered with leafbases or becoming striate or smooth with age. **Leaves:** sheath fibers soft; petiole base not split; petiole armed with fine teeth, sometimes only at base of petiole; abaxial hastula not well developed, obtuse; adaxial hastula usually well developed, obtuse; costa absent; blade palmate; plication induplicate; segments lanceolate, basally connate, apices acute or 2-cleft; cross veins conspicuous. **Inflorescences** axillary within crown of leaves, paniculate, ascending, about as long as leaves, with 2–3 orders of branching; main axis bearing 2 peduncular bracts above prophyll; rachillae pubescent. **Flowers** bisexual, borne singly or in pairs along rachillae, sessile; perianth 2-seriate; calyx cupulate, 3-lobed; petals 3, imbricate, elliptic, reflexed, alternate with outer whorl of stamens, basally adnate to filaments; stamens 6 in 2 whorls; filaments narrowly triangular, basally connate; anthers dorsifixed, versatile; pistils 3, distinct basally, glabrous; ovules 3, only 1 developing into fruit; styles connate, elongate, glabrous; nectaries 3, septal; stigma minutely 3-lobed, dry. **Fruits** drupes, ellipsoid; exocarp black, smooth; mesocarp blackish; endocarp brown, bony. **Seeds** brown, ellipsoid, with conspicuous longitudinal raphe; endosperm bony, homogeneous; embryo nearly basal; eophyll undivided, lanceolate. $x = 18$.

Species 1: North America.

Serenoa, with a single polymorphic species, is endemic to the southeastern United States. It grows in a variety of habitats and communities including pine flatwoods, sand-pine scrub, and coastal sand dunes. It sometimes occurs in vast stands nearly excluding all other understory shrubs.

The relationships of *Serenoa* are with *Acoelorraphe*, of the Caribbean Basin (N. W. Uhl et al. 1995), and perhaps *Brahea*, of Mexico and Central America (N. W. Uhl and J. Dransfield 1987).

SELECTED REFERENCES Bennett, B. C. and J. R. Hicklin. 1998. Uses of saw palmetto (*Serenoa repens*, Arecaceae) in Florida. Econ. Bot. 52: 381–393. Hawkes, A. D. 1950. Notes on the palms 2. Saw palmetto *Serenoa repens* Small. Natl. Hort. Mag. 29: 93–95. Hilmon, J. B. 1968. Autecology of Saw Palmetto (*Serenoa repens* (Bartr.) Small). Ph.D. dissertation. Duke University. Smith, D. 1972. Fruiting in the saw palmetto. Principes 16: 30–33.

1. Serenoa repens (W. Bartram) Small, J. New York Bot. Gard. 27: 197. 1926 E F

Corypha repens W. Bartram, Travels Carolina, 61. 1791; *Brahea serrulata* (Michaux) H. Wendland; *Chamaerops serrulata* Michaux; *Corypha obliqua* W. Bartram; *Sabal serrulata* (Michaux) Nuttall ex Schultes & Schultes f.; *Serenoa serrulata* (Michaux) G. Nicholson

Stems usually creeping, branched, sometimes ascending, 1–3 m. **Leaves** yellow-green, green, or silvery green, stiff; petioles finely to strongly serrate; hastula present abaxially and adaxially. **Flowers** creamy white, fragrant, 4–5 mm. **Fruits** ripening from green through orange to black, length ca. 2 cm, diam. 1 cm. $2n = 36$.

Flowering spring. Pinelands, dunes, sand pine scrub, mesic hammocks and woodlands, plants colonial, often forming dense stands in the understory; 0–50 m; Ala., Fla., Ga., La., Miss., S.C.

The range of this species extends from Beaufort, Jasper, Colletin, and Charleston counties, South Carolina to St. Tammany Parish, Louisiana. Reports of it from Arkansas, North Carolina, and Texas are in error.

Two or three leaf-color morphs are found in *Serenoa* (A. D. Hawkes 1950). The green type is more widespread, and the glaucous or blue-gray type seems to be more abundant in coastal sites of southeastern Florida. *Serenoa repens* forma *glauca* Moldenke was described as differing from the typical form of the species in having glaucous leaf blades (H. N. Moldenke 1967). Since a type specimen for *S. repens* is lacking, it is impossible to know if Moldenke's form differs from the type. The genetic basis for these color differences is not understood.

Serenoa repens apparently is pollinated by bees. Flowering and fruiting are not necessarily annual events, and some years see more abundant flowering than others (J. B. Hilmon 1968). Even when flowering is abundant, fruit production is erratic (D. Smith 1972); the causes are quite unknown. When fruits are present, they are eagerly sought by black bears (D. S. Maehr 1984; D. S. Maehr and J. R. Brady 1984), white-tailed deer (R. F. Harlow 1961), raccoons, foxes, opossums, and various birds (J. B. Hilmon 1968).

Serenoa fruits are the source of a steroidal drug that inhibits the conversion of testosterone to dihydrotestosterone, which binds to receptors in the prostate gland and in hair follicles (B. C. Bennett and J. R. Hicklin 1998). This inhibition is the biochemical basis for the use of *Serenoa* extracts in treating benign prostrate swelling and baldness. Bennett and Hicklin provided a complete review of the uses of *Serenoa* in traditional and modern medicine, as well as its use as a fiber and thatch plant.

7. WASHINGTONIA H. Wendland, Bot. Zeitung (Berlin) 37: 68, 148. 1879, name conserved • Fan palm, palmier évantail de Californie [for George Washington, 1732–1799, American patriot and first president of the United States]

Neowashingtonia Sudworth

Plants tree palms. **Stems** solitary, erect, tall, massive in one species (100–150 cm diam.), partly or completely covered with old leaf bases and marcescent dry leaves forming conspicuous skirt around trunk. **Leaves:** sheath fibers soft; petiole split at base, conspicuously armed with teeth along margins, sometimes unarmed in tall plants; abaxial hastula absent; adaxial hastula irregularly shaped, margin becoming tattered, fibrous; costa prominent; blade costapalmate, plication induplicate; segments lanceolate, basally connate, bearing fibers between segments; apices 2-cleft or irregularly tattered into fibers. **Inflorescences** axillary within crown of leaves, paniculate, arching well beyond leaves, with 2 orders of branching; prophyll leathery; rachis bracts very conspicuous, tubular at base, distally flattened and leathery, long; rachillae glabrous. **Flowers** bisexual, borne singly along rachillae, short pedicellate; perianth 2-seriate; calyx cupulate, 3-lobed, apices and margins irregular; petals 3, long, chaffy, basally connate into tube; stamens 6, adnate briefly to petals; pistils 3, distinct basally, glabrous; styles connate, slender, long; stigma inconspicuous. **Fruits** drupes, blackish, ellipsoid; exocarp smooth; mesocarp thin, fleshy; endocarp thin. **Seeds** ellipsoid; endosperm homogeneous; embryo basal; eophyll undivided, lanceolate. $x = 18$.

Species 2 (2 in the flora): w North America, nw Mexico.

The two species are easy to distinguish when cultivated side by side, but they are sometimes difficult to identify from herbarium specimens. Analysis of flavonoids (S. Zona and R. Scogin 1988) can reliably distinguish the two species.

SELECTED REFERENCES Bailey, L. H. 1936. *Washingtonia*. Gentes Herb. 4: 51–82.   Zona, S. and R. Scogin. 1988. Flavonoid aglycones and C-glycosides of the palm genus *Washingtonia* (Arecaceae: Coryphoideae). SouthW. Naturalist 33: 498.

1. Stems 100 cm diam. or greater. . . . . 1. *Washingtonia filifera*
1. Stems 80 cm diam. or less. . . . . 2. *Washingtonia robusta*

1. Washingtonia filifera (Linden ex André) H. Wendland ex de Bary, Bot. Zeitung (Berlin) 37: LXI. 1879
• California fan palm, palmier évantail de Californie C
F

Pritchardia filifera Linden ex André, Ill. Hort. 21: 28. 1874; *Neowashingtonia filifera* (Linden) Sudworth; *Pritchardia filamentosa* H. Wendland ex Fenzi; *Washingtonia filamentosa* (H. Wendland ex Fenzi) Kuntze

Stems massive, 100–150 cm diam. $2n = 36$.

Flowering spring–summer. Native to desert washes, seeps, and springs where underground water is continuously available; of conservation concern; 100–1200 m; Ariz., Calif., Nev.; Mexico (Baja California).

L. H. Bailey (1936) cited S. Watson, not H. Wendland, as the combining authority of this species, believing that Wendland did not explicitly make the new combination. The combination was made, however, by the journal editor, Anton de Bary, in the index to the volume, appearing on page (column) LXI. Hence, de Bary, not Watson, is the combining author (J. L. Strother, pers. comm.).

Naturalized populations of this species were reported at four sites in Nevada (J. W. Cornett 1987) and in Death Valley National Monument, California (J. W. Cornett 1988). L. R. McClenaghan and A. C. Beauchamp (1986) found low genetic variation among populations of *Washingtonia filifera*. V. J. Miller (1983) discussed the history and setting of *W. filifera* in Arizona.

SELECTED REFERENCES Cornett, J. W. 1987. The occurrence of the desert fan palm, *Washingtonia filifera*, in southern Nevada. Desert Pl. 8: 169–171.   Cornett, J. W. 1988. Naturalized populations of the desert fan palm, *Washingtonia filifera*, in Death Valley National Monument. In: C. A. Hall Jr. and V. Doyle-Jones, eds. 1988. Plant Biology of Eastern California. Los Angeles. Pp. 167–174.   McClenaghan,

L. R. Jr. and A. C. Beauchamp. 1986. Low genetic differentiation among isolated populations of the California fan palm (*Washingtonia filifera*). Evolution 40: 315–322.   Miller, V. J. 1983. Arizona's own palm: *Washingtonia filifera*. Desert Pl. 5: 99–104.   Small, J. K. 1931. The fanleaf-palm—*Washingtonia filifera*. J. New York Bot. Gard. 32: 33–43.

2. Washingtonia robusta H. Wendland, Gart.-Zeitung (Berlin) 2: 198. 1883 • Mexican fan palm, palmier évantail du Mexique I

Neowashingtonia robusta (H. Wendland) A. Heller; *N. sonorae* (S. Watson) Rose; *Washingtonia filifera* var. *robusta* (H. Wendland) Parish; *W. filifera* var. *sonorae* (S. Watson) M. E. Jones; *W. gracilis* Parish; *W. robusta* var. *gracilis* (Parish) Beccari; *W. sonorae* S. Watson

Stems taller, more slender than preceding species, 50–80 cm diam. $2n = 36$.

Flowering spring. Native to desert washes where underground water is continuously available, naturalized in disturbed areas with moderate rainfall; introduced; Calif., Fla.; Mexico (native, Baja California Sur and Sonora).

The name *Washingtonia filifera* var. *robusta* (H. Wendland) Parish is nomenclaturally a synonym of *W. robusta*, although it was applied by S. B. Parish to palms now known as *W. filifera*. Hence, Parish's variety has been included erroneously by L. H. Bailey (1936) and others in synonymy under *W. filifera*.

Washingtonia robusta is widely cultivated in warm areas in the United States and has occasionally naturalized in Florida (S. Zona 1997) and southern California (J. W. Cornett 1986).

SELECTED REFERENCE Cornett, J. W. 1986. *Washingtonia robusta* naturalized in southeastern California. Bull. S. Calif. Acad. Sci. 85: 56–57.

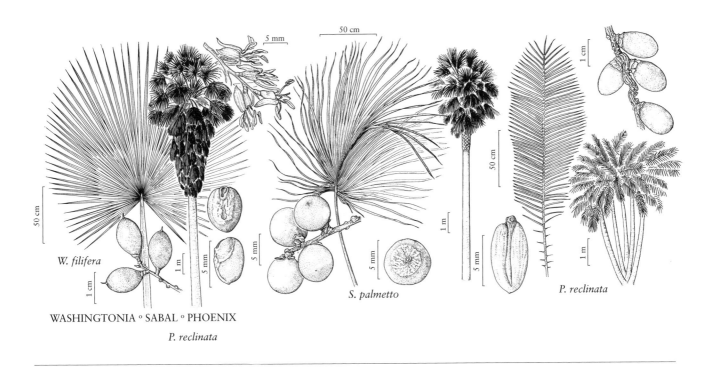

W. filifera

WASHINGTONIA ∘ SABAL ∘ PHOENIX

P. reclinata

S. palmetto

P. reclinata

201a.1c. ARECACEAE Schultz-Schultzenstein (tribe CORYPHEAE) subtribe **SABALINAE** Martius in S. L. Endlicher, Gen. Pl. 4: 252. 1837

8. SABAL Adanson ex Guersent, Bull. Sci. Soc. Philom. Paris 87: 205. 1804 • Palmetto [derivation of name unknown]

Inodes O. F. Cook

Plants dwarf, moderate, or tall, usually robust. **Stems** solitary, aerial or subterranean, covered with leaf bases or clean, obscurely [strongly] ringed, becoming striate or smooth with age. **Leaves** few to many; sheath fibers soft; petiole split at base, completely unarmed; adaxial hastula well developed, obtuse to acuminate-triangular; costa present; blade weakly to strongly costapalmate; plication induplicate; segments lanceolate, basally connate to connate for ½ length [or in groups of 2 or 3 segments connate for nearly entire length], often bearing thread-like fibers between segments, apices acute or 2-cleft, stiff or lax. **Inflorescences** axillary within crown of leaves, paniculate, erect or arching beyond leaves [shorter than leaves], with 2 or 3 [–4] orders of branching; peduncular bracts 2–5, tightly clasping, inconspicuous; rachillae glabrous. **Flowers** bisexual, borne singly along rachillae, sessile, creamy white, fragrant; perianth 2-seriate; calyx cupulate; 3-lobed; petals 3, imbricate, elliptic, obovate or spatulate, alternate with outer whorl of stamens [basally connate], basally adnate to filaments; stamens 6 in 2 whorls; filaments narrowly triangular, basally connate; anthers dorsifixed, versatile; pistils 1, 1-carpellate, glabrous; nectaries 3, septal; ovules 3, but usually only one develops into seed; stigma minutely 3-lobed, papillose. **Fruits** drupes, berrylike, spheroid [oblate or pyriform] or lobed when more than 1 seed develops; exocarp black; mesocarp blackish, dry to fleshy; endocarp brown, membranaceous. **Seeds** 1–3, oblate, glossy; endosperm bony, homogeneous; embryo nearly apical, lateral or nearly lateral; eophyll undivided, linear-lanceolate. $x = 18$.

Species 16 (5 in the flora): North America, Mexico, West Indies, Central America, South America.

Sabal flowers are bisexual and are pollinated mostly by native bees, especially of the families Halictidae and Megachilidae (S. Zona 1987, 1990), European honeybees, and wasps (Vespidae; P. F. Ramp 1989). Fruits are eagerly sought by both mammals (bears, deer, raccoons) and birds (S. Zona and A. Henderson 1989). *Sabal minor*, perhaps more than other species, is also dispersed by water (P. F. Ramp 1989; S. Zona 1990).

SELECTED REFERENCES   Lockett, L. 1991. Native Texas palms north of the lower Rio Grande Valley: Recent discoveries. Principes 35: 64–71.   Zona, S. 1990. A monograph of *Sabal* (Arecaceae: Coryphoideae). Aliso 12: 583–666.

1. Inflorescences with 2 (rarely 3 in basal branches) orders of branching (not counting main axis).
    2. Leaf strongly costapalmate and curved, bearing fibers between segments; inflorescence bushy and compact, ± as long as leaves. . . . . 1. *Sabal etonia*
    2. Leaf weakly costapalmate and little if at all curved, not bearing fibers between segments; inflorescence sparsely branched, much longer than leaves. . . . . 4. *Sabal minor*
1. Inflorescence with 3 orders of branching.
    3. Stems subterranean; leaves 6 or fewer. . . . . 3. *Sabal miamiensis*
    3. Stems usually aerial; leaves more than 10.
        4. Fruit 8.1–13.9 mm diam. . . . . 5. *Sabal palmetto*
        4. Fruit 14.8–19.3 mm diam. . . . . 2. *Sabal mexicana*

1. **Sabal etonia** Swingle ex Nash, Bull. Torrey Bot. Club 23: 99. 1896 • Scrub palmetto, dwarf palmetto [E]

Sabal adansonii Guersent var. *megacarpa* Chapman; *S. megacarpa* (Chapman) Small

Stems usually subterranean. **Leaves** 4–7, yellow-green, strongly costapalmate, curved, bearing fibers between segments; hastula narrowly triangular, 1.6–3.3 cm; segments 35–64 × 1.3–3.1 cm; apices 2-cleft. **Inflorescences** bushy and compact, densely branched with 2 orders of branching (not counting main inflorescence axis), ascending, ± as long as leaves. **Flowers** 4.9–6.1 mm. **Fruits** brownish black, oblate-spheroid, 8.5–13.1 mm, diam. 9–15.4 mm; mesocarp thick, fleshy. **Seeds** 5.4–6.7 mm, diam. 6.4–9.9 mm.

Flowering spring–summer. Deep white sand in sand pine scrub; 10–50 m; Fla.

Sabal etonia is found in the deep white sand of sand pine (*Pinus clausa*) scrub communities of the Central Florida Ridge and the Atlantic Coastal Ridge (S. Zona and W. S. Judd 1986). The habitat of *S. etonia* is under pressure from both agriculture and urbanization, but fortunately vast tracts are preserved in the Ocala National Forest.

SELECTED REFERENCES   Zona, S. 1987. Phenology and pollination biology of *Sabal etonia* (Palmae) in southeastern Florida. Principes 31: 177–182.   Zona, S. and W. S. Judd. 1986. *Sabal etonia* (Palmae): Systematics, distribution, ecology, and comparisons to other Florida scrub endemics. Sida 11: 417–427.

2. **Sabal mexicana** Martius in C. F. P. von Martius et al., Hist. Nat. Palm. 3: 246, plate 8. 1838 • Mexican palmetto, sabal du Mexique

Inodes exul O. F. Cook; *I. mexicana* (Martius) Standley; *I. texana* O. F. Cook ; *Sabal exul* (O. F. Cook) L. H. Bailey; *S. texana* (O. F. Cook) Beccari

Stems aerial, 20–35 cm diam. **Leaves** 10–30, strongly costapalmate; hastula acute to acuminate, 9.5–15.5 cm; segments filiferous, 80–145 × 3.2–5.3 cm; apices 2-cleft. **Inflorescences** with 3 orders of branching (not counting main inflorescence axis), arching, ± as long as leaves. **Flowers** 3.7–6.5 mm. **Fruits** black, oblate-spheroid, 13.8–17 mm, diam. 14.8–19.3 mm. **Seeds** 5.4–7.4 mm, diam. 8.6–13.3 mm. $2n = 36$.

Flowering spring–summer (all year in southern part of range). Mesic hammocks, floodplains, levees, river banks, swamps; 0–50 m; Tex.; Mexico; Central America (Guatemala, Honduras, El Salvador, Nicaragua).

Hybridization between *Sabal mexicana* and *S. minor* is possibly evidenced by a small population of caulescent palms in Brazoria County, Texas (L. Lockett 1991). Further research is needed to test this hypothesis.

3. **Sabal miamiensis** Zona, Brittonia 37: 366, figs. 1–2. 1985 • Miami palmetto C E

Stems subterranean. **Leaves** 3–6, yellow-green, strongly costapalmate; hastula narrowly triangular, 2.4–7.7 cm; segments filiferous, 50–85 × 2.8–3.0 cm; apices 2-cleft. **Inflorescences** paniculate, loosely branched with 3 orders of branching (not counting main inflorescence axis), horizontal-arching, about as long as leaves. **Flowers** 5–5.5 mm. **Fruits** black, shiny, oblate-spheroid, 14.3–16.9 mm, diam. 15.7–19 mm; pericarp thick, fleshy. **Seeds** 6.2–6.7 mm, diam. 10.2–11 mm.

Flowering spring–summer. Rocky calcareous soil of Miami pinelands; of conservation concern; 0–10 m; Fla.

This species is restricted to the oölitic limestone of the pine rocklands of Dade County (S. Zona 1985). The natural habitat of *Sabal miamiensis* has been urbanized, so this species is likely extinct. Despite differences in habitat, this species may not be distinct from *S. etonia*.

SELECTED REFERENCE Zona, S. 1985. A new species of *Sabal* (Palmae) from Florida. Brittonia 37: 366–368.

4. **Sabal minor** (Jacquin) Persoon, Syn. Pl. 1: 399. 1805 • Dwarf palmetto, little blue stem, latanier

Corypha minor Jacquin, Hort. Bot. Vindob. 3: 8, plate 8. 1776; *Chamaerops acaulis* Michaux; *C. louisiana* W. Darby; *Corypha pumila* Walter; *Sabal adansonii* Guersent; *S. adiantinum* Rafinesque; *S. deeringiana* Small; *S. louisiana* (W. Darby) Bomhard; *S. pumila* (Walter) Elliott

Stems usually subterranean. **Leaves** 4–10, dark green, weakly costapalmate, little if at all curved, not bearing fibers between segments; hastula obtuse, 0.8–4.7 cm; segments not filiferous, 34–84 × 1.4–3.7 cm; apices weakly if at all 2-cleft. **Inflorescences** sparsely branched with 2 orders of branching (not counting main inflorescence axis), erect, much longer than leaves. **Flowers** 3.5–5.2 mm. **Fruits** brownish black, spheroid, 6.2–8.5 mm, diam. 6.4–9.7 mm; pericarp thin. **Seeds** 3.5–5.1 mm, diam. 4.4–6.9 mm. 2n = 36.

Flowering spring–summer. Mesic hammocks, flood-plains, levees, riverbanks, swamps, but occurring on much drier sites in wc Tex.; 10–600 m; Ala., Ark., Fla., Ga., La., Miss., N.C., Okla., S.C., Tex.; Mexico.

This species has recently been found in Tamaulipas, Mexico (D. H. Goldman 1999).

Sabal minor is usually a small palm with a subterranean trunk; however, one can find individuals with larger features and well-developed aerial stems. In Louisiana, these individuals were recognized as separate species (J. K. Small 1929; M. L. Bomhard 1935), but more recently they have been treated as merely ecological variants of a single widespread species (A. Henderson et al. 1995; P. F. Ramp and L. B. Thien 1995; S. Zona 1990). Large emergent forms of *S. minor* were even thought by B. J. Simpson (1988) to be hybrids of that species with *S. palmetto*, but his claim is undocumented and unsubstantiated.

An unusual habitat for this species, a dry hillside in central Texas, was illustrated by L. Lockett (1991).

SELECTED REFERENCES Bomhard, M. L. 1935. *Sabal louisiana*, the correct name for the polymorphic palmetto of Louisiana. J. Wash. Acad. Sci. 25: 35–44. Goldman, D. H. 1999. Distribution update: *Sabal minor* in Mexico. Palms 43: 40–44. Ramp, P. F. 1989. Natural History of *Sabal minor*: Demography, Population Genetics and Reproductive Ecology. Ph.D. dissertation. Tulane University. Ramp, P. F. and L. B. Thien. 1995. A taxonomic history and reexamination of *Sabal minor* in the Mississippi Valley. Principes 39: 77–83. Small, J. K. 1929. Palmetto-with-a-stem–*Sabal deeringiana*. J. New York Bot. Gard. 30: 278–284.

5. **Sabal palmetto** (Walter) Loddiges ex Schultes & Schultes f. in J. J. Roemer et al., Syst. Veg. 7(2): 1487. 1830 • Cabbage-palm, chou palmiste F

Corypha palmetto Walter, Fl. Carol., 119. 1788; *Chamaerops palmetto* (Walter) Michaux; *Corypha palma* W. Bartram; *Inodes palmetto* (Walter) O. F. Cook; *Sabal jamesiana* Small

Stems usually aerial, 20–35 cm diam. **Leaves** 15–30, strongly costapalmate, bearing threadlike fibers between segments; hastula acute to acuminate, 5.3–18 cm; segments 55–120 × 2.5–4.2 cm; apices 2-cleft. **Inflorescences** with 3 orders of branching (not counting main inflorescence axis), arching, equaling or exceeding leaves in length. **Flowers** 4.1–6.7 mm. **Fruits** black, spheroid, 8–13.8 mm, diam. 8.1–13.9 mm. **Seeds** 4–7 mm, diam. 5.4–9.7 mm. 2n = 36.

Flowering spring–summer (n part of range) or year around (s part of range). Hammocks, pinelands, riverbanks, dunes, tidal flats; 0–40 m; Fla., Ga., N.C., S.C.; West Indies (Bahamas, Cuba).

Sabal palmetto grows in a variety of habitats, from pine and oak associations to coastal dunes and to coastal marshes (K. E. Brown 1976; S. Zona 1990). Like *S. minor*, it is polymorphic at the extremes of its range; however, differences in stature, size, and trunk characteristics are not of a magnitude to warrant taxonomic rank. In the pine rocklands of Dade County, Florida, *S. palmetto* may flower and fruit with little or no aboveground trunk.

Although *Sabal palmetto* is a moderately important honey plant, its greatest economic use is as an ornamental.

SELECTED REFERENCE Brown, K. E. 1976. Ecological studies of the cabbage palm, *Sabal palmetto*. Principes 20: 3–10, 49–56, 98–115, 148–157.

201a.2. ARECACEAE Schultz-Schultzenstein (subfam. CORYPHOIDEAE) tribe PHOENICEAE Drude in C. F. P. von Martius et al., Fl. Bras. 3(2): 279. 1881

9. PHOENIX Linnaeus, Sp. Pl. 2: 1188. 1753; Gen. Pl. ed. 5, 496. 1754 • Date palm, palmier dattier [derivation uncertain, perhaps for the Phoenicians, known for a dye that was similar in color to ripening dates; name used by Theophrastus for the date palm] [I]

Dachel Adanson; *Elate* Linnaeus; *Palma* Miller

Stems solitary or clustered, erect or ascending [subterranean], slender to massive, often clothed in old leaf bases. **Leaves:** sheath fibers soft; petiole not split at base, armed, not forming crownshaft; blade pinnate; plication induplicate; segments lanceolate, in 1 or more planes, apices acute; basal segments modified into stout spines. **Inflorescences** axillary within crown of leaves, paniculate, ascending, much shorter than leaves, with 1 order of branching, alike in staminate and pistillate plants; prophyll often caducous, conspicuous, becoming boat-shaped, short; peduncular bracts absent; rachillae glabrous. **Staminate flowers** borne singly along rachillae; calyx cupulate, 3-lobed; petals 3, free, valvate; stamens 6, free; pistillode inconspicuous or absent. **Pistillate flowers** borne singly on rachillae; calyx cupulate, 3-lobed; petals 3, imbricate, free; staminodial ring cupulate or deeply 6-lobed; pistils 3 (only 1 developing), distinct; stigmas small. **Fruits** drupes, berrylike, fleshy; exocarp blackish brown, smooth; mesocarp fleshy or fibrous; endocarp papery. **Seeds** 1, elongate; endosperm homogeneous; embryo lateral [basal]; eophyll undivided, lanceolate. $x = 18$.

Species 13 (2 in the flora): introduced; widespread, native to Eastern Hemisphere, including the Canary and Cape Verde islands, s Europe, Africa (including Madagascar), s Asia, and Philippines.

Several species of *Phoenix* are cultivated as ornamentals in Florida and California, although identification can be difficult because the species are dioecious and apparently hybridize with great ease. *Phoenix dactylifera* Linnaeus, the date palm, is grown as a commercial crop in southern California and Arizona and as an ornamental palm in Florida, but it seems noninvasive. It can be recognized by its massive trunk (eventually bearing basal offshoots) and its stiff, ascending, glaucous leaves. In Florida, *P. roebelenii* O'Brien (pygmy date palm), with its solitary trunk less than 15 cm diam., is also cultivated as an ornamental although it does not seem to escape. Other species of *Phoenix* are occasionally cultivated in warm parts of the United States. Elements of cultivated species of *Phoenix* entering the flora may be of uncertain parentage.

Two species, *Phoenix canariensis* and *P. reclinata*, have escaped and are sporadically naturalized in southern Florida and, to a much lesser extent, in California. *Phoenix dactylifera* is reportedly naturalized in California (E. McClintock 1993), but I have seen no specimens.

SELECTED REFERENCES Austin, D. F. 1978. Exotic plants and their effects in southeastern Florida. Environm. Conservation 5: 25–34.   Barrow, S. C. 1998. A monograph of *Phoenix* L. (Palmae: Coryphoideae). Kew Bull. 53: 513–575.

1. Trunk solitary, 55–70 cm diam. . . . . 1. *Phoenix canariensis*
1. Trunks multiple, 10–15 cm diam. . . . . 2. *Phoenix reclinata*

1. Phoenix canariensis Hort ex Chabaud, Prov. Agric. Hort. Ill. 19: 293. 1882 • Canary Island date palm, dattier des Canaries [I]

Stems solitary, erect, to 15 m, diam. 55–70 cm. **Fruits** ripening from green through orange to reddish purple, ellipsoid, 24–27 mm, diam. 10–12 mm.

Flowering throughout the year. Volunteer in waste places and disturbed areas; 0–1000 m; introduced; Calif., Fla.; Europe; Atlantic Islands (native, Canary Islands).

Phoenix canariensis is widely cultivated, especially in Florida and California. This species is naturalized in the San Francisco Bay area, as well as in southern California (E. McClintock 1993).

2. Phoenix reclinata Jacquin, Fragm. Bot., 27, plate 24. 1801 • Senegal date palm, dattier du Sénégal [F] [I]

Stems multiple, usually ascending, to 8 m, diam. 10–15 cm. **Fruits** ripening from green through orange to reddish brown, ellipsoid, 12–18 mm, diam. 7–8 mm. $2n = 36$.

Flowering throughout year. Volunteer in waste places and disturbed areas; 0–10 m; introduced; Fla.; native, Africa (including Madagascar).

Phoenix reclinata is commonly cultivated in central and southern Florida. Its seeds are dispersed by birds and raccoons.

201b. ARECACEAE Schultz-Schultzenstein subfam. CEROXYLOIDEAE Drude in C. F. P. von Martius et al., Fl. Bras. 3(2): 271. 1881

201b.1. ARECACEAE Schultz-Schultzenstein (subfam. CEROXYLOIDEAE) tribe CYCLOSPATHEAE O. F. Cook, Mem. Torrey Bot. Club 12: 24. 1902

10. PSEUDOPHOENIX H. Wendland ex Sargent, Bot. Gaz. 11: 314. 1886 • Cherry palm [from Greek *pseudos*, false, and *Phoenix*, the date palm]

Cyclospathe O. F. Cook; *Sargentia* H. Wendland & Drude ex Salomon

Stems solitary, erect, robust, unarmed, smooth. **Leaves:** leaf bases not completely tubular, forming partial crownshaft, unarmed; petiole unarmed; blade pinnate; plication reduplicate; segments irregularly spaced along unarmed rachis, in multiple planes, linear-lanceolate, apices acute. **Inflorescences** axillary within leafy crown, ascending, becoming pendulous in fruit, with 4 orders of branching; prophyll small, tubular, leathery; peduncular bracts 2, tubular, enclosing most of peduncle, leathery. **Flowers** bisexual (distal flowers with reduced gynoecia), borne singly along rachillae on pedicel-like stalks, subtended by minute bract; sepals 3, connate, forming shallow triangular calyx; petals 3, ovate, basally slightly connate, persistent in fruit;

stamens 6, basally adnate to corolla; filaments linear; anthers dorsifixed; pistils 1, 3-loculate; ovules 1 per locule; styles indistinct; stigmas minute. **Fruits** drupes, globose, or 2- or 3-lobed if more than 1 ovule matures; stigmatic scar basal or in 3-lobed fruits, apical; exocarp red, smooth; mesocarp fleshy; endocarp bony. **Seeds** 1–3, globose; endosperm homogeneous; embryo basal; eophyll undivided, lanceolate. $x = 17$.

Species 4 (1 in the flora): North America, Mexico, West Indies, Central America (Belize).

Pseudophoenix sargentii occupies a most precarious position in the North American flora: only small populations remain on Elliott Key in southern Florida (B. R. Ledin et al. 1959; C. Lippincott 1992, 1995). Since its discovery some 100 years ago, botanists have witnessed a precipitous decline in *Pseudophoenix* populations in the Florida Keys as these attractive palms were removed for use in landscaping (F. C. Craighead Sr. and D. B. Ward n.d.). Recently naturalists have replanted palms on Long Key, more than 30 years after the last wild palms had been seen there (C. Lippincott 1992, 1995).

Pseudophoenix is probably insect pollinated, since its greenish yellow flowers produce nectar and attract numerous bees. The fruits of *P. sargentii*, which float when dry, are thought to be dispersed by water (R. W. Read 1968); their attractive red coloration and fleshy mesocarp, however, suggest animal dispersal (S. Zona and A. Henderson 1989).

SELECTED REFERENCES   Craighead, F. C. Sr. and D. B. Ward. N.d. Endangered: Buccaneer palm. In: D. B. Ward, ed. N.d. Rare and Endangered Biota of Florida. Vol. 5. Plants. Gainesville. Pp. 54–55.   Ledin, R. B., S. C. Kiem, and R. W. Read. 1959. *Pseudophoenix* in Florida. I. The native *Pseudophoenix sargentii*. Principes 3: 23–133.   Lippincott, C. 1992. Restoring Sargent's cherry palm on the Florida Keys. Fairchild Trop. Gard. Bull. 47: 12–21.   Lippincott, C. 1995. Reintroduction of *Pseudophoenix sargentii* in the Florida Keys. Principes 39: 5–13.   Read, R. W. 1968. A study of *Pseudophoenix* (Palmae). Gentes Herb. 10: 169–213.

1. Pseudophoenix sargentii H. Wendland ex Sargent, Bot. Gaz. 11: 314. 1886   [F]

Subspecies 2 (1 in the flora): North America; Mexico; West Indies; Central America.

1a. Pseudophoenix sargentii H. Wendland ex Sargent subsp. **sargentii** · Sargent's cherry palm, buccaneer palm   [F]

Stems 4–8 m, smooth. **Leaves** 1.5–3 m; segments inserted on rachis at divergent angles, usually in groups of 3–5, waxy graygreen abaxially. **Inflorescences** arching, 7.5–12 dm. **Fruits** 1–2 cm diam. $2n = 34$.

Flowering summer. Rocky, calcareous soil of coastal hammocks and scrub; 0–10 m; Fla.; Mexico; Central America (Belize).

Pseudophoenix sargentii subsp. *sargentii* also occurs on the coast of the Yucatan Peninsula. The other subspecies, *P. sargentii* subsp. *saonae* (O. F. Cook) Read, is thought to comprise two varieties distributed in Cuba, the Bahamas, and Hispaniola (R. W. Read 1968). Read now thinks that perhaps none of these subspecies or varieties should be maintained (C. Lippincott 1992).

For an excellent account of the history of this palm in the Florida Keys and modern efforts to restore this species to habitats that it originally occupied, see C. Lippincott (1992).

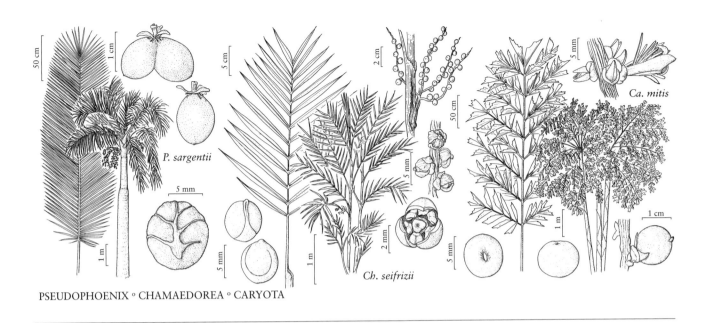

P. sargentii

Ch. seifrizii

Ca. mitis

PSEUDOPHOENIX ∘ CHAMAEDOREA ∘ CARYOTA

201b.2. ARECACEAE Schultz-Schultzenstein (subfam. CEROXYLOIDEAE) tribe HYOPHORBEAE Drude in C. F. P. von Martius et al., Fl. Bras. 3(2): 275. 1881

11. CHAMAEDOREA Willdenow, Sp. Pl. 4(2): 638, 800. 1806, name conserved
• Bamboo palm, parlor palm [Greek *chamai*, on the ground, and *dorea*, gift, in reference to small, low-growing palms of great beauty]   ⊡

Plants small, usually low-growing, unarmed. **Stems** clustered [solitary], erect [creeping, lianoid], slender, unarmed. **Leaves:** sheaths tubular, unarmed, forming crownshaft; blade pinnate [undivided], with leaf segments regularly spaced along unarmed rachis, in 1 plane [many planes]; plication reduplicate; segments linear-lanceolate, apical pair of segments sometimes wider than others. **Inflorescences** axillary below crown of leaves, ascending, with 1 order of branching [spicate or 2 orders]; prophyll small; peduncular bracts 5–6, tubular, papery; rachillae green at anthesis, turning orange in fruit. **Flowers** unisexual, sessile, staminate and pistillate flowers on different plants. **Staminate flowers** borne singly, partially sunken into fleshy rachillae; sepals 3, briefly connate at base [distinct]; petals 3, ovate, basally briefly connate [connate by tips]; stamens 6, distinct; anthers dorsifixed; pistillode minute. **Pistillate flowers** borne singly, slightly sunken into fleshy rachillae; sepals 3, free; petals 3, free, ovate; staminodes 6, minute; pistil 1, 3-loculate; ovules 1 per locule; style indistinct; stigmas minute. **Fruits** drupes, globose; stigmatic scar basal, exocarp black, smooth; mesocarp thin; endocarp bony. **Seeds** globose; endosperm homogeneous; embryo subapical; eophyll 2-cleft [pinnate], segments linear. $x = 13$.

Species 77–100 or more (1 in the flora): introduced; Fla., Mexico, Central America, and n South America.

Chamaedorea species are understory palms in Mexico, Central America, and northern South America. Many species are cultivated in North America both as houseplants and, in warm areas, as garden plants.

1. Chamaedorea seifrizii Burret, Notizbl. Bot. Gart. Berlin-Dahlem 14: 268. 1938 [F][I]

Chamaedorea donnell-smithii Dammer, name rejected; *C. erumpens* H. E. Moore

Stems clustered. **Leaves** pinnately divided; segments ca. 20 cm, uniform in width or terminal segments wider than other segments. **Fruits** black, globose, 5–8 mm diam. $2n = 26$.

Flowering summer. Moist organic soil over limestone in mesic hammocks and disturbed wooded areas; 0–10 m; introduced; Fla.; native, Mexico, Central America.

The fruits of *Chamaedorea seifrizii*, ripening from green through orange to black, are taken by birds, and seedlings thrive in shaded, moist habitats. The segments may be narrow or broad, terminal segments may be equal or wider than other segments, and segments may be upright or flat.

201c. ARECACEAE Schultz-Schultzenstein subfam. ARECOIDEAE

201c.1. ARECACEAE Schultz-Schultzenstein tribe CARYOTEAE Drude in C. F. P. von Martius et al., Fl. Bras. 3(2): 278. 1881

12. CARYOTA Linnaeus, Sp. Pl. 2: 1189. 1753; Gen. Pl. ed. 5, 497. 1754 • Fishtail palm [Greek *caryon*, nut] [I]

Stems solitary or clustered, slender to massive, smooth, with conspicuous nodal rings. **Leaves:** blade 2-pinnate (1-pinnate in juvenile plants); plication induplicate; segments cuneate, in 1 plane, apices jagged and irregular; basal segments not modified into spines. **Inflorescences** initiated basipetally, first one appearing terminal, successive one borne axillary among leaves, and later ones below leaves, pendulous, paniculate, with 1 order of branching [spicate]; prophyll small; peduncular bracts numerous, tubular. **Flowers** unisexual, sessile, borne in triads of 1 pistillate flower flanked by 2 staminate flowers. **Staminate flowers:** sepals 3, imbricate, free; petals 3, connate basally, valvate; stamens numerous [6], free; pistillode absent. **Pistillate flowers:** sepals 3, imbricate, free; petals 3, connate for nearly ½ length, valvate; staminodial lobes present or absent; pistils 1, 3-loculate; ovules 1 per locule; styles indistinct; stigmas 3-lobed. **Fruits** berries, globular; exocarp purple, smooth; mesocarp fleshy, containing irritating raphides; endocarp absent. **Seeds** globular; endosperm ruminate [homogeneous]; embryo lateral; eophyll 2-cleft, segments fan-shaped. $n = 17$.

Species 12 (2 in the flora): native to Asia, Pacific Islands, Australia.

Caryota, readily distinguished by its 2-pinnate leaves, is native to India, southeastern Asia, Malesia, the Philippines, the Solomon Islands, Vanuatu, and Australia. Two species of *Caryota* are commonly cultivated in southern Florida: *C. urens* and *C. mitis*. Both species have naturalized in Dade County, Florida.

1. Stems solitary, 15 cm diam. or more. . . . . 1. *Caryota urens*
1. Stems clustered, 15 cm diam. or less. . . . . 2. *Caryota mitis*

1. Caryota urens Linnaeus, Sp. Pl. 2: 1189. 1753
• Solitary fishtail palm I

Stems solitary, greater than 15 cm diam. **Leaves** 5–7 m. $2n = 32$.

Flowering in summer. Moist organic soil over limestone in mesic hammocks and disturbed wooded areas; 0–10 m; introduced; Fla.; native, India.

2. Caryota mitis Loureiro, Fl. Cochinch. 2: 569. 1790
• Clustering fishtail palm, caryote doux F I

Stems clustered, 15 cm or less diam. **Leaves** 2–3 m. $2n = 28,$ 32, 34.

Flowering summer. Moist organic soil over limestone in mesic hammocks and disturbed wooded areas; 0–10 m; introduced; Fla.; native, India to the Philippine Islands.

201c.2. ARECACEAE Schultz-Schultzenstein (subfam. ARECOIDEAE) tribe ARECEAE

201c.2a. ARECACEAE Schultz-Schultzenstein (tribe ARECEAE) subtribe DYPSIDINAE Beccari, Palme Madagascar, 2. 1912

13. DYPSIS Noroña ex Martius in C. F. P. von Martius et al., Hist. Nat. Palm. 3: 180. 1838 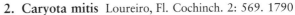 I

Trunks erect, clustered, less than 15 cm diam., with conspicuous nodal rings, unarmed, occasionally branching near base. **Leaves:** leaf bases unarmed, yellowish, forming distinct crownshaft; blade pinnate [undivided], unarmed; plication reduplicate; segments lanceolate, evenly spaced, strongly ascending, apices acuminate. **Inflorescences** axillary below crown of leaves, paniculate, with 3 orders of branching; prophyll small; peduncular bracts caducous, tubular. **Flowers** unisexual, sessile, in triads of 1 pistillate flower flanked by 2 staminate flowers. **Staminate flowers:** sepals 3, imbricate, free; petals 3, briefly connate basally, valvate; stamens 6, free; anthers dorsifixed; pistillode present. **Pistillate flowers:** sepals 3, imbricate, free; petals 3, imbricate, free; staminodes 6, minute; pistil 1; ovules 1; stigmas 3. **Fruits** drupes, ellipsoid; exocarp yellow, smooth; mesocarp fleshy; endocarp thin, fibrous, terete in cross section. **Seeds** ovoid; endosperm homogeneous; embryo subbasal; eophyll 2-cleft; segments lanceolate.

Species 140 (1 in the flora): native to Africa (Madagascar and adjacent islands).

Dypsis is a morphologically diverse genus restricted to Madagascar and adjacent islands. The genus has recently been revised and substantially expanded by J. Dransfield and H. J. Beentje (1995). The most familiar species, *D. lutescens* (H. Wendland) Beentje & J. Dransfield, is hardly representative of the astonishing diversity in growth form, leaf shape, floral details, and fruit morphology found in the genus.

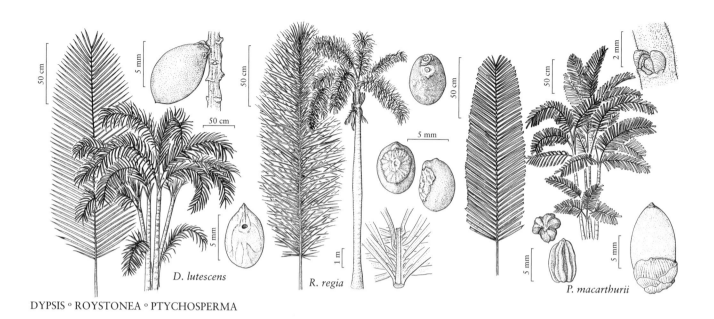

D. lutescens

R. regia

P. macarthurii

DYPSIS ∘ ROYSTONEA ∘ PTYCHOSPERMA

1. Dypsis lutescens (H. Wendland) Beentje & J. Dransfield, Palms Madagascar, 212. 1995 • Areca palm, butterfly palm F I

Chrysalidocarpus lutescens H. Wendland, Bot. Zeitung (Berlin) 36: 117. 1878

Leaves 2–2.5 m; segments 60–70 cm, strongly ascending. **Fruits** yellow, ellipsoid, 2.0–2.5 cm; apex acute; stigmatic scar basal. $2n = 32$.

Flowering   spring–summer.

Moist organic soil over limestone in mesic hammocks and disturbed wooded areas; 0–10 m; introduced; Fla.; native, Africa (Madagascar).

This species is a commonly cultivated ornamental palm in Florida, where it has escaped and sporadically naturalized in Dade County.

201c.2b. ARECACEAE Schultz-Schultzenstein (tribe ARECEAE) subtribe **ROYSTONEINAE** J. Dransfield & N. W. Uhl, Principes 30: 7. 1986

14. ROYSTONEA O. F. Cook, Science, ser. 2, 12: 479. 1900 • Royal palm, palmier royal [for Roy Stone, 1836–1905]

Stems solitary, erect, greater than 20 cm in diam., smooth, unarmed. **Leaves:** leaf bases unarmed, forming crownshaft, crownshaft prominent, green, smooth; petiole unarmed; blade pinnate; plication reduplicate; segments linear-lanceolate, in more than 1 plane. **Inflorescences** axillary below crown of leaves, paniculate, with 2 or 3 orders of branching; prophyll tubular; peduncular bract greatly exceeding prophyll, leathery, splitting longitudinally on abaxial side and circumscissilly at base; rachillae covered with copious caducous dendritic trichomes, becoming glabrous. **Flowers** unisexual, sessile, in triads of 1 pistillate flower flanked by 2 staminate flowers. **Staminate flowers:** sepals 3, imbricate, membranaceous; petals 3, valvate; stamens

6(–10, rarely); anthers dorsifixed, often twisting upon drying; pistillode minute, obscurely 3-cleft. **Pistillate flowers** globose to conic; sepals 3, imbricate; petals 3, basally connate, distally valvate; staminodes 6, basally connate, adnate to corolla basally; pistil 1; ovules 1; style indistinct; stigmas 3. **Fruits** drupes, fibrous; stigmatic scar basal; exocarp ripening from green to red to purplish black at maturity, thin, leathery; mesocarp fleshy, oily; endocarp hard. **Seeds** 1, nearly globose [obovoid], dorsiventrally compressed, abaxially attached to endocarp; endosperm homogeneous; embryo basal; eophyll undivided, linear-lanceolate. $x = 18$.

Species 11 (1 in the flora): Florida, Mexico, West Indies, Central America, South America.

Most species of *Roystonea* are widely known as royal palms. They are cultivated worldwide and are especially favored as avenue trees. Long rows of gray-white columnar trunks are unmatched for their magnificence and stateliness. In some parts of the Caribbean, especially Cuba, *Roystonea* is a significant resource for thatch (leafbases only), timber, livestock feed, *palmito* (palm cabbage or heart-of-palm), and edible oil (F. A. Reynoso 1976; C. Ruebens 1968; T. A. Zanoni 1991; S. Zona 1991, 1996).

SELECTED REFERENCES Cook, O. F. 1936. Royal palms in upper Florida. Science, ser. 2, 84: 60–61.   Reynoso, F. A. 1976. Importancia económica de la palma real dominicana (*Roystonea hispaniolana*). Agroconocimiento 1: 8–9.   Ruebens, C. 1968. Industrialización del palmiche en Cuba. Industr. Alimenticia 1: 8–25.   Small, J. K. 1937. Facts and fancies about our royal palm. J. New York Bot. Gard. 38: 49–58.   Zanoni, T. A. 1991. The royal palm on the island of Hispaniola. Principes 35: 49–54.   Zona, S. 1991. Notes on *Roystonea* in Cuba. Principes 35: 225–233.   Zona, S. 1996. *Roystonea* (Arecaceae: Arecoideae). In: Organization for Flora Neotropica. 1968+. Flora Neotropica. 75+ nos. New York. No. 71, pp.1–36.

1. Roystonea regia (Kunth) O. F. Cook, Science, ser. 2, 12: 479. 1900 • Cuban royal palm, Florida royal palm F

Oreodoxa regia Kunth in A. von Humboldt et al., Nov. Gen. Sp. 1: 305. 1816, name conserved; *Roystonea elata* (W. Bartram) F. Harper; *R. floridana* O. F. Cook

Stems gray-white, 30 m, diam. 35–41 cm, smooth. **Leaves:** segments inserted on rachis in several ranks at divergent angles, leaf appearing almost plumose. **Inflorescences** to 1 m; rachillae 11–31 cm, stiff. **Flowers** white; anthers pinkish. **Fruits** ripening from green through red to purplish black at maturity, dorsiventrally compressed obovoid, 9.5–10.5 mm, diam. 7.4–7.8 mm. $2n = 36$.

Flowering winter–summer (Jan–Jul). Tropical hardwood hammocks and mixed swamp vegetation in peat soils over limestone; 0–10 m; Fla.; se Mexico; West Indies (Cuba, Bahamas, Cayman Islands).

The Florida populations, once known as *Roystonea elata*, are conspecific with the Cuban *R. regia* (S. Zona 1996, 1997). A proposal to conserve *Oreodoxa regia* over *Palma elata* W. Bartram, the older basionym, has been accepted, because the name *R. regia* is so widely used (R. K. Brummitt 1996; S. Zona 1994).

Roystonea attracted the attention of William Bartram (1791), who described the palm growing near Lake Dexter, along the St. Johns River in what are now Lake and Volusia counties. That this palm naturally grew so far north of its present range in historical times is an intriguing puzzle. Possibly the severe freeze of 1835 may have extirpated the northern populations (J. G. Cooper 1861). Perhaps fires used by early settlers to clear land reduced the number of *Roystonea* palms in northern and central Florida (O. F. Cook 1936). Or, in addition to freezes in 1835 and 1894–95, overexploitation by humans may have extirpated the palms (J. K. Small 1937); Small reported that a factory in the vicinity of Bartram's population turned out walking sticks made of palm wood. Whatever the historical reason for their decline, indigenous populations of *R. regia* are now found only in Collier and Dade counties.

Flowers of *Roystonea* attract numerous bees and are probably insect pollinated. The fruits, available April through October, are eaten by birds and bats, which are the likely dispersers of the seeds (S. Zona and A. Henderson 1989).

201c.2c. ARECACEAE Schultz-Schultzenstein (tribe ARECEAE) subtribe PTYCHOSPERMATINAE Hooker f. in G. Bentham and J. D. Hooker, Gen. Pl. 3: 874. 1883 (as Ptychospermeae)

15. PTYCHOSPERMA Labillardière, Mém. Cl. Sci. Math. Inst. Natl. France 1808(2): 252. 1809 • [Greek *ptyx*, folded, and *sperma*, seed, in reference to the ridged endocarp] I

Stems solitary or clustered, less than 15 cm diam., unarmed. **Leaves:** sheath green, unarmed, forming crownshaft; blade pinnate, unarmed; plication reduplicate; segments in 1 plane, usually cuneate, apices jagged and irregular, cleft. **Inflorescences** axillary below crown of leaves, paniculate, with 2 or 3 orders of branching, pendulous; prophyll short; peduncular bract tubular; secondary peduncular bracts often present, incomplete. **Flowers** unisexual, sessile, in triads of 1 pistillate flower flanked by 2 staminate flowers. **Staminate flowers:** sepals 3, imbricate, free; petals 3, valvate, free; stamens [9–]25–34 [or more], in whorls of 3 or more; pistillode present, with slender style. **Pistillate flowers:** sepals 3, imbricate, free; petals 3, imbricate, free; staminode lobes or scales present; pistils 1; ovules 1; styles indistinct; stigmas 3. **Fruits** drupes, small; exocarp red [black], smooth; mesocarp fleshy; endocarp bony, 5-lobed in cross section [obscurely lobed]. **Seeds** ovoid; endosperm homogeneous or ruminate; embryo basal; eophyll 2-cleft, segments linear, apices jagged and irregular.

Species 28 (2 in the flora): introduced; Florida; Pacific Islands, Australia.

This genus is popular in cultivation. Two species have naturalized sporadically in Dade County, Florida. The fruits are red and fleshy, and birds undoubtedly disperse the seeds.

SELECTED REFERENCE Zona, S. 1999. New perspectives on generic limits and relationships in the Ptychospermatinae (Palmae: Arecoideae). Mem. New York Bot. Gard. 83: 255–263.

1. Stems solitary. . . . . 1. *Ptychosperma elegans*
1. Stems clustered. . . . . 2. *Ptychosperma macarthurii*

1. Ptychosperma elegans (R. Brown) Blume, Rumphia 2: 118. 1843 (as eleganti) • Solitaire palm I

Seaforthia elegans R. Brown, Prodr., 267. 1810; *Archontophoenix elegans* (R. Brown) H. Wendland & Drude

Stems solitary. $2n = 32$.

Flowering spring–summer. Moist organic soil over limestone in mesic hammocks and disturbed wooded areas; 0–10 m; introduced; Fla.; native to Australia (Queensland).

2. Ptychosperma macarthurii (H. Wendland ex anonymous) G. Nicholson, Ill. Dict. Gard. 3: 248. 1886 • MacArthur palm F I

Kentia macarthurii H. Wendland ex anonymous in James Veitch and Sons, Catalogue 1879, 26, 15. 1879; *Actinophloeus macarthurii* (H. Wendland ex anonymous) O. Beccari ex Wigman

Stems clustered. $2n = 32$.

Flowering spring–summer. Moist organic soil over limestone in mesic hammocks and disturbed wooded areas; 0–10 m; introduced; Fla.; n Australia; Pacific Islands; native, Pacific Islands (s New Guinea); native, n Australia.

201c.3. A R E C A C E A E Schultz-Schultzenstein (subfam. ARECOIDEAE) tribe COCOEAE Martius in S. L. Endlicher, Gen. Pl. 4: 254. 1837

201c.3a. A R E C A C E A E Schultz-Schultzenstein (tribe COCOEAE) subtribe BUTIINAE Saakov, Palmy S.S.S.R., 193. 1954

16. C O C O S Linnaeus, Sp. Pl. 2: 1188. 1753; Gen. Pl. ed. 5, 495. 1754 • Coconut palm, cocotier [derivation of name uncertain] ⊡

Stems solitary, erect or ascending, often leaning, robust, unarmed, trunks with conspicuous nodal rings. **Leaves:** sheath coarse, clothlike, not forming crownshaft; petiole unarmed; petiole base entire; blade pinnate, unarmed; plication reduplicate; segments regularly arranged, apices acute. **Inflorescences** axillary within crown of leaves, paniculate, with 1 or 2 orders of branching, stiffly ascending; prophyll short; peduncular bract woody, beaked, splitting abaxially, becoming boat-shaped; rachillae glabrous. **Flowers** unisexual, sessile, in triads of 1 pistillate flower flanked by 2 staminate flowers, staminate flowers borne singly along distal portions of rachillae. **Staminate flowers:** sepals 3, imbricate, free; petals 3, valvate; stamens 6, free; anthers sagittate; pistillode with 3 slender lobes. **Pistillate flowers** borne basally on rachillae, massive; sepals 3, imbricate, free, ± reniform; petals 3, imbricate, free, large; staminode a thin ring at base of pistil; pistils 1, large; ovules 3, usually only 1 ovule fertile; styles indistinct; stigmas 3. **Fruits** drupes, greater than 10 cm diam., strongly to obscurely 3-angled, dry; exocarp brown, thin, smooth; mesocarp very fibrous, dry; endocarp thick, bony, with 3 basal germination pores. **Seeds** very large, hollow and fluid-filled; endosperm lining cavity of seed, homogeneous, oily; embryo basal; eophyll 2-cleft, segments lanceolate. $x = 16$.

Species 1: pantropical.

Originally from Melanesia (H. C. Harries 1978, 1992), the coconut is now almost universally cultivated in tropical and subtropical areas of the world. The earliest record of its introduction into Florida is from the early nineteenth century (D. F. Austin 1978b). The plant persists after cultivation, and fruits with germinating seeds may be found in abandoned or disturbed sites along coastal southern Florida. The species is included here because it regenerates in the managed and disturbed strand vegetation in southern Florida.

SELECTED REFERENCES Austin, D. F. 1978b. The coconut in Florida. Principes 22: 83–87. Harries, H. C. 1978. The evolution, dissemination, and classification of *Cocos nucifera* L. Bot. Rev. (Lancaster) 44: 265–320. Harries, H. C. 1992. Biogeography of the coconut *Cocos nucifera* L. Principes 36: 155–162.

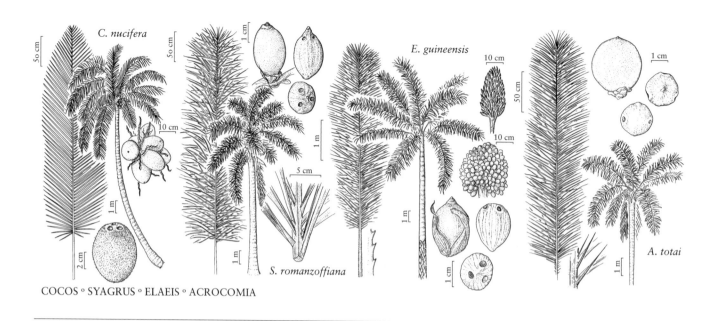

C. nucifera

E. guineensis

S. romanzoffiana

A. totai

COCOS ∘ SYAGRUS ∘ ELAEIS ∘ ACROCOMIA

1. Cocos nucifera Linnaeus, Sp. Pl. 2: 1188. 1753

[F] [I]

Stems erect or leaning, smooth. **Leaves:** segments inserted on rachis in 2 ranks; bract persistent, peduncular, to 1 m, woody. **Staminate flowers** creamy yellow, 11–13 mm. **Fruits** green, yellow, or bronzy red when immature, brown when mature; mesocarp dry, fibrous; endocarp brown, bearing 3 germination pores. $2n = 32$.

Flowering throughout the year. Coastal dune vegetation in sandy soils; ca. 0–10 m; introduced; Fla.; native, Pacific Islands (Melanesia).

This is the coconut of commerce, although it is cultivated in the U.S. solely for its ornamental value. Although not native, the coconut persists long after cultivation and is essentially naturalized in coastal southern Florida. Lethal yellowing disease eliminated a large number of susceptible coconuts from the landscape. Presently, most cultivated individuals are resistant cultivars.

17. SYAGRUS Martius, Palm. Fam., 18. 1824; in C. F. P. von Martius et al., Hist. Nat. Palm. 2: 129, plates 89, 90, 101, figs. 4, 5. 1826 • Queen palm [classical name, derivation unknown, but used by Pliny for a kind of palm] [I]

Stems solitary, erect, robust, unarmed, bearing conspicuous nodal rings. **Leaves:** petiole unarmed; sheath not forming crownshaft; blade pinnate, unarmed; plication reduplicate; segments regularly arranged in multiple planes, apices acute to 2-cleft. **Inflorescences** axillary within crown of leaves, paniculate, 1 order of branching, ascending, becoming pendulous in fruit; prophyll short; peduncular bract woody, beaked, splitting abaxially, becoming boat-shaped. **Flowers** unisexual, sessile, borne in triads of 1 pistillate flower flanked by 2 staminate, staminate flowers borne singly along distal portions of rachillae. **Staminate flowers:** sepals 3, connate; petals 3, free, valvate, leathery; stamens 6, free; anthers linear; pistillode with 3 minute lobes. **Pistillate flowers** borne basally on rachillae, massive; sepals 3, imbricate, free; petals 3, imbricate, free; staminode a low annular ring at base of pistil; pistils 1, large; ovules 1; styles indistinct; stigmas 3. **Fruits** drupes, ovoid, less than 4 cm diam.; exocarp orange, thin, smooth; mesocarp fleshy, oily; endocarp thick, bony, irregularly folded into seed, with 3 basal germination pores. **Seeds** irregular with hollow cavity; endosperm homogeneous; embryo basal; eophyll undivided, lanceolate.

Species 32 (1 in the flora): North America, West Indies (Lesser Antilles), South America.

1. Syagrus romanzoffiana (Chamisso) Glassman, Fieldiana, Bot. 31: 382. 1968 [F] [I]

Cocos romanzoffiana Chamisso in L. Choris, Voy. Pittor. 6: 5, plates 5, 6. 1822; *Arecastrum romanzoffianum* (Chamisso) Beccari; *Cocos plumosa* Loddiges ex Hooker

Stems solitary, smooth, with conspicuous nodal rings. **Leaves** 5 m. **Fruits** 3–3.5 cm, ovoid, yellowish orange; endocarp ovoid, brown, with 3 germination pores. $2n = 32$.

Flowering throughout the year. Disturbed hammocks and woodlands; 0–30 m; introduced; Fla.; native, South America.

Syagrus is tenuously represented in the flora by the cultivated ornamental *S. romanzoffiana*, still known in the nursery trade as *Cocos plumosa*. This South American species is widely planted throughout much of southern and central Florida. Although it is not yet widely established in the flora, seedlings volunteer in natural areas, and mature plants persist after cultivation. A closely related ornamental palm from South America, *Butia capitata* (Martius) Beccari, jelly palm, is widely grown in the southeastern United States and crosses with *Syagrus romanzoffiana*, producing ×*Butyagrus nabonnandii* (A. R. Proschowsky) Vorster, largely sterile hybrids. *Butia* shows little inclination for escaping.

201c.3b. ARECACEAE Schultz-Schultzenstein (tribe COCOEAE) subtribe ELAEIDINAE Hooker f. in G. Bentham and J. D. Hooker, Gen. Pl. 3: 882. 1883 (as Elaeideae)

18. ELAEIS Jacquin, Select. Stirp. Amer. Hist., 280. 1763 • Oil palm [Greek *elaia*, olive, in reference to the oily fruits] [I]

Alfonsia Kunth; *Corozo* Jacquin ex Giseke

Stems solitary, erect, robust, covered with persistent leaf bases or bare, unarmed. **Leaves:** crownshaft absent; petiole margins armed with lignified, indurate bases of midveins persisting as spines after blade erodes; blade pinnate, unarmed; plication reduplicate; segments regularly arranged in multiple planes [in 1 plane], apices acute to 2-cleft. **Inflorescences** within crown of leaves, densely paniculate, with 1 order of branching, either staminate or pistillate, partially obscured by leaf bases; peduncle short; prophyll short; peduncular bract woody, splitting abaxially; rachillae thick, apices stiff, sharp. **Flowers** unisexual, sessile, borne singly along rachillae. **Staminate flowers** in pits in rachillae; sepals 3, free; petals 3, free, valvate, leathery; stamens 6, filaments briefly connate; anthers rectangular; pistillode with 3 minute lobes. **Pistillate flowers:** sepals 3, imbricate, free; petals 3, imbricate, free; staminodial ring bearing 6 short points; pistils 1; ovules 3; styles indistinct; stigmas 3. **Fruits** drupes, ovoid; exocarp orange-yellow and black, thin; mesocarp fleshy, oily; endocarp thick, bony, with 3 apical germination pores. **Seeds** spheroid; endosperm homogeneous; embryo subapical; eophyll undivided, lanceolate.

Species 2 (1 in the flora): tropics; Central America, South America, Africa.

1. **Elaeis guineensis** Jacquin, Select. Stirp. Amer. Hist., 280, plate 172. 1763 • African oil palm, palmier à huile d'Afrique [F] [I]

Leaves to 8 m; segments 120 cm. **Fruits** ripening from green to yellow to red or black, 4 cm; endocarp black, 3-sided, bearing 3 germination pores. $2n = 32$.

Flowering spring–summer. Moist organic soil over limestone in mesic hammocks and disturbed wooded areas; 0–10 m; introduced; Fla.; native, Africa.

Elaeis guineensis is widespread in wet tropical Africa. It is now cultivated throughout the tropics, where it is the most important perennial oil crop, the source of both palm oil and palm kernel oil. *E. guineensis* has escaped in the vicinity of Miami, Florida, and may be counted as an element of the flora. Birds and small mammals eat the oil fruits and disperse the seeds. Juveniles are easily recognized by the very sharp, spinescent remains of leaf segment midribs at the base of each leaf.

201c.3c. ARECACEAE Schultz-Schultzenstein (tribe COCOEAE) subtribe BACTRIDINAE Hooker f. in G. Bentham and J. D. Hooker, Gen. Pl. 3: 873, 881. 1883 (as Bactrideae)

19. ACROCOMIA Martius, Palm. Fam., 22. 1824; in C. F. P. von Martius et al., Hist. Nat. Palm. 2: 66. 1824 • Coyol, gru-gru [Greek *akron*, summit, and *kome*, hairs of the head, in reference to the high crown of leaves; *akrokomos*, with leaves at the top, said especially of palms] [I]

Acanthococos Barbosa Rodrigues

Stems solitary, robust, armed, covered with persistent leaf bases or bare. **Leaves:** petioles armed with needlelike prickles; petiole margins unarmed; blade pinnate, armed with prickles; plication reduplicate; segments regularly arranged, apices acute. **Inflorescences** axillary within crown of leaves, paniculate, arching, becoming pendulous in fruit, with 1 order of branching; prophyll short; peduncular bract woody, prickly, splitting abaxially, curling downward; rachis armed with prickles. **Flowers** unisexual, sessile, borne in triads of 1 pistillate flower flanked by 2 staminate flowers, staminate flowers borne singly along distal portions of rachillae. **Staminate flowers:** sepals 3, free; petals 3, valvate, basally connate, leathery; stamens 6, free; anthers rectangular; pistillode with 3 minute lobes. **Pistillate flowers:** sepals 3, imbricate, free; petals 3, imbricate, basally connate or nearly free; staminodes well developed, bearing short, sterile anthers; pistils 1, large, tomentose; ovules 3; styles indistinct; stigmas 3. **Fruits** drupes, globose; exocarp brownish green, thin, pubescent near fruit apex [bristly]; mesocarp fleshy, oily; endocarp thick, bony, with 3 equatorial germination pores. **Seeds** irregular; endosperm homogeneous; embryo lateral; eophyll undivided [2-cleft], linear-lanceolate. $x = 15$.

Species 2–30 (1 in the flora): extreme se United States, Mexico, West Indies, Central America, and South America.

Although as many as 30 species of *Acrocomia* have been described, a recent study (A. Henderson et al. 1995) recognized only two, *A. aculeata* (Jacquin) Loddiges (including *A. totai* Martius) and *A. hassleri* (Barbosa Rodrigues) W. J. Hahn. I maintain *A. aculeata* and *A. totai* as two separate species, both of which are cultivated in Florida, where the latter species is naturalized. A general comparison of these two species was made by B. Peterson (1991) in which he noted that the eophyll of *A. aculeata* is 2-cleft and that of *A. totai* is undivided. He also found several subtle differences between these species when mature. For example, the trunk spines of *A. totai* are ca. 12–13 cm and those of *A. aculeata* are ca. 6–10 cm. Clearly, additional study is warranted to resolve the prickly systematic problems in this genus.

SELECTED REFERENCES   Peterson, B. 1991. A comparison of some central Florida acrocomias. Centr. Florida Palm Bull. 11(1): 11–12.   Peterson, B. 1991b. *Acrocomia* naturalized in central Florida. Principes 35: 110–111.

1. Acrocomia totai Martius in A. D. d'Orbigny, Voy. Amér. Mér. 7(3): 78. 1844; 8(1): plate 9, fig. 1. 1842

[I]

Stems armed with spines 12–13 cm. **Leaves** ca. 4 m; segments glabrous abaxially. **Fruits** brownish green, globose, 2–3 cm diam.; endocarp globose, brown, bearing 3 germination pores. **Seeds:** eophyll undivided.

Flowering summer. Woodlands and hammocks; 0–30 m; introduced; Fla.; native, South America.

Acrocomia totai is sparingly naturalized in central Florida in Brevard County (B. Peterson 1991b), as well as in Dade County in the vicinity of Miami.

202. ACORACEAE C. Agardh

• Sweet-flag Family

Sue A. Thompson

Herbs, perennial, wetland, usually with aromatic oil, especially in rhizomes. **Rhizomes** horizontal, creeping at or near surface, branched. **Stems** repent, branched rhizomes. **Cataphylls** absent. **Leaves** not differentiated into petiole and blade, equitant, sword-shaped, larger than 1.5 cm; venation parallel along length of leaf. **Inflorescences** spadices, from 3-angled axis (peduncle fused with proximal portion of sympodial leaf, i.e., leaf encircling terminal inflorescence), distal sympodial leaf extending beyond spadix; true spathe absent; spadix nearly cylindric, tapering, apex obtuse. **Flowers** bisexual; tepals 6; stamens 6, distinct; ovaries 1, (1–)3-locular, sessile; stigmas sessile (styles essentially absent), minute. **Fruits** berries; pericarp thin, leathery. **Seeds** 1–6(–14), from apex of locule.

Genera 1, species 3–6 (2 species in the flora): temperate Northern Hemisphere, tropical Asia at higher elevations, and sporadically introduced into Southern Hemisphere.

Acorus historically was recognized as an aberrant genus within Araceae, but much evidence supports its treatment as a separate family and the removal of this family from Arales (M. H. Grayum 1987). Other than the absence of a close association with Arales, the phylogenetic affinities of Acoraceae remain unclear. Evidence based on DNA sequences fails to show any close relationships between *Acorus* and other genera, and instead supports *Acorus* as the oldest extant lineage of monocotyledons (M. R. Duvall et al. 1993).

The removal of *Acorus* from Araceae is supported by the absence of a spathe and the unique vasculature of the structure traditionally interpreted as a spathe (T. S. Ray 1987). The structure that has been called a spathe in *Acorus* is not morphologically equivalent to the spathe of Araceae; instead it is interpreted as the distal part of the sympodial leaf. The proximal part of the sympodial leaf is adnate to the peduncle, forming a 3-angled axis that bears the inflorescence.

SELECTED REFERENCES Grayum, M. H. 1987. A summary of evidence and arguments supporting the removal of *Acorus* from the Araceae. Taxon 36: 723–729. Thompson, S. A. 1995. Systematics and Biology of the Araceae and Acoraceae of Temperate North America. Ph.D. dissertation. University of Illinois.

1. A C O R U S Linnaeus, Sp. Pl. 1: 324. 1753; Gen. Pl. ed. 5, 151. 1754 • Sweet-flag [Latin form of Greek *akoron*, presumably an ancient plant name]

Herbs, wetlands or rocky stream banks, when bruised or broken producing pleasant and distinctive aromatic odor. **Leaves** bright green; sheathing base 2-facial (proximal part of leaf); distal part of leaf 1-facial, flattened in median rather than transverse plane; prominent veins 1–6, parallel along length of leaf. **Inflorescences** solitary. **Flowers:** tepals light brown; anthers yellow, introrse; ovaries green. **Fruits** light brown to reddish with darker streaks. **Seeds** embedded in mucilage. $x = 12$.

Species 3–6 (2 in the flora): temperate Northern Hemisphere, tropical Asia at higher elevations, sporadically introduced in Southern Hemisphere.

Considerable confusion exists in the taxonomic literature over the status of *Acorus* in North America. Whether *Acorus* is native or introduced, as well as the number of taxa in North America, has been debated for years. Evidence based on studies of morphology, essential oil chemistry, cytology, isozymes, and ethnobotany supports the existence of two species in North America—*A. calamus*, an introduced sterile triploid, and *A. americanus*, a native fertile diploid (J. G. Packer and G. S. Ringius 1984; S. A. Thompson 1995). *Acorus americanus* is not only morphologically distinct from triploid *A. calamus* but also from diploid and tetraploid *A. calamus* populations occurring in Asia (L. C. M. Röst 1979; S. A. Thompson 1995). Although the existence of two species of *Acorus* in North America was discussed by J. G. Packer and G. S. Ringius (1984) and *A. americanus* has been listed in several North American floras, this is the first flora that formally treats two species of *Acorus* in North America and provides a key to distinguish these two species.

Traditionally, the name *Acorus calamus* has been applied to all populations of *Acorus* in North America without regard for the biological species involved. Other authors (e.g., E. T. Browne and R. Athey 1992; K. A. Wilson 1960) have adopted *A. americanus* as the "correct" name for all *Acorus* in North America, including populations in regions where only *A. calamus* occurs or is the predominant species. The use of either name in the literature requires further study to determine which species is being cited.

The rhizomes of *Acorus calamus* contain an aromatic oil that has been used medicinally since ancient times and has been harvested commercially. Native Americans exploited *Acorus* as a medicine and for ceremonial uses. Although this plant is cited in the ethnographic and ethnobotanical literature as *A. calamus*, the distribution of the tribes reported to use *Acorus* corresponds to the range of the native species (S. A. Thompson 1995).

The combination of equitant, sword-shaped leaves plus an elongate inflorescence borne about midway on a sympodial leaf is not found in any other North American plant. Fresh material of *Acorus* is easily distinguished from other plants by the unique pleasant fragrance given off by rhizomes and leaves when broken. The bright green color of the leaves is also distinctive.

SELECTED REFERENCE Packer, J. G. and G. S. Ringius. 1984. The distribution and status of *Acorus* (Araceae) in Canada. Canad. J. Bot. 62: 2248–2252.

1. Midvein prominently raised above leaf surface, other veins barely or not raised; mature fruits not produced, sterile triploid. . . . . 1. *Acorus calamus*
1. Midvein plus 1–5 additional veins equally raised above leaf surface; mature fruits produced, fertile diploid. . . . . 2. *Acorus americanus*

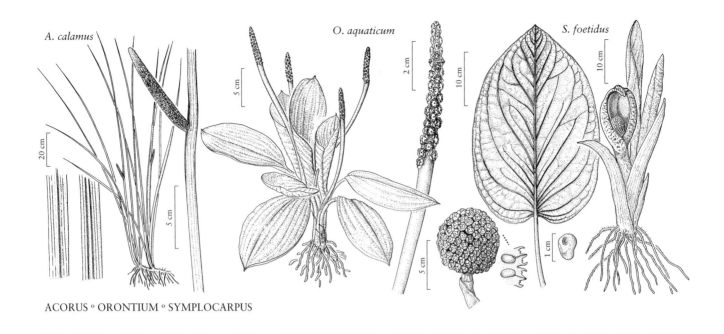

A. calamus   *O. aquaticum*   *S. foetidus*

ACORUS ∘ ORONTIUM ∘ SYMPLOCARPUS

1. Acorus calamus Linnaeus, Sp. Pl. 1: 324. 1753 • Calamus [F] [I]

Leaves basally white with pink or red, otherwise bright green; single midvein (secondary midrib) prominently raised above leaf surface, usually somewhat off-center, other veins barely or not raised; cross section rhomboid. **Vegetative leaves** to 1.75 m; sheathing base (proximal part of leaf) 22.1–66.5(–73.3) cm; distal part of leaf 31.9–95.8(–117.6) × 0.5–2 cm, 1.4–1.8 times longer than proximal part of leaf, margins sometimes undulate or crisped. **Sympodial leaf** (29.9–)34.7–159.1(–183.9) cm, usually shorter than to nearly equal to vegetative leaves; sheathing base 16.1–76.4(–100.1) cm; distal part of leaf 13.5–86.2(–101.2) × 0.4–1.9 cm. **Spadix** (3.8–)4.9–8.9 cm × 5.3–10.8 mm at anthesis, post-anthesis spadix 5.5–8.7 cm × 6–12.6 mm. **Flowers** 3–4 mm; pollen grains not staining in aniline blue. **Fruits** not produced in North America. $2n = 36$.

Flowering early spring–early summer. Wet open areas, marshes, swales, and along edges of quiet water; 0–900 m; introduced; N.B., N.S., Ont., Que.; Ala., Ark., Calif., Colo., Conn., Del., D.C., Ga., Ill., Ind., Iowa, Kans., Ky., La., Maine, Md., Mass., Mich., Minn., Miss., Mo., Nebr., N.H., N.J., N.Y., N.C., Ohio, Okla., Oreg., Pa., R.I., S.C., S.Dak., Tenn., Tex., Vt., Va., W.Va., Wis.; Europe; Asia; Africa; Indian Ocean Islands; Pacific Islands.

Acorus calamus, a sterile triploid, was introduced to North America by early European settlers, who grew it for medicinal uses. Rhizomes propagate easily, and the species has spread throughout northeast and central United States. Scattered populations occur elsewhere. Although leaf and spadix size of *A. calamus* and *A. americanus* overlap, those measurements differ significantly, with *A. calamus* in general having longer and wider leaves and longer spadices.

SELECTED REFERENCE   Motley, T. J. 1994. The ethnobotany of sweet flag, *Acorus calamus* (Araceae). Econ. Bot. 48: 397–412.

2. Acorus americanus (Rafinesque) Rafinesque, New Fl. 1: 57. 1836 • Belle-angélique [E] [F]

Acorus calamus Linnaeus var. *americanus* Rafinesque, Med. Fl. 1: 25. 1828; *A. calamus* var. *americanus* (Rafinesque) H. Wulff

Leaves basally white with pink or red, otherwise bright green; major veins 2–6, equally raised above leaf surface; cross section swollen in center, gradually tapering to ends. **Vegetative leaves** to 1.45 m; sheathing base (proximal part of leaf) 18.1–51.8(–58.8) cm; distal part of leaf 31.2–88.6(–100.4) × 0.3–1.2 cm, usually slightly longer to more than 2 times length of distal leaf, margins usually entire. **Sympodial leaf** (46–)56.8–148 (–166.7) cm, usually equal to or slightly longer than vegetative leaves; sheathing base (20.9–)25.3–74.1 (–100.2) cm; distal part of leaf (20.9–)27.9–77.9(–92.6) × 0.3–1.3 cm. **Spadix** 3.3–7.4(–8.7) cm × 4.7–10 (–13.3) mm at anthesis; fruiting spadix 3.5–7.8(–8.8) cm × 6.9–18.2 mm. **Flowers** 2–3 mm; pollen grains usually deeply staining in aniline blue. **Fruits** obpyramidal, 4–6 mm. **Seeds** (1–)6(–14), tan, narrowly oblong to obovate, (2–)3–4 mm. $2n = 24$.

Flowering late spring–mid summer. Wet open areas, marshes, swales, and along edges of quiet water; 0–900 m; Alta., B.C., Man., N.B., Nfld. and Labr. (Nfld.), N.W.T., N.S., Ont., P.E.I., Que., Sask.; Alaska, Conn., D.C., Idaho, Ill., Ind., Iowa, Maine, Mass., Mich., Minn., Mont., Nebr., N.H., N.J., N.Y., N.Dak., Ohio, Pa., R.I., S.Dak., Vt., Va., Wash., Wis.

Acorus americanus, a fertile diploid, occurs from northeastern United States across Canada and the northern plains. Specimens from central Siberia with similar leaf venation were examined, and the species is perhaps holarctic in distribution. Examination of additional material is necessary to determine if northern Asian diploid plants are conspecific with *A. americanus*. In North America, Native Americans probably played a significant role in the present-day distribution of *A. americanus* because sweet-flag rhizomes and plants were valued by many groups and were objects of trade. Disjunct populations occur in localities that are often near old Native American village sites or camping areas (M. R. Gilmore 1931).

Acorus americanus is susceptible to infection by *Uromyces sparganii* (Uredinales).

203. ARACEAE Jussieu

• Arum Family

Sue A. Thompson

Herbs, perennial, wetland or terrestrial, occasionally emergent or floating, [often epiphytic or climbing], usually with milky or watery latex, rarely colored. **Rhizomes, corms, or stolons** present; rhizomes vertical or horizontal, creeping at or near surface, sometimes branched; corms underground, starchy; stolons at or near surface. **Stems** absent [sometimes aboveground or aerial]. **Cataphylls** usually present. **Leaves** rarely solitary, alternate or clustered; petiole rarely absent, with sheathing base; blade simple or compound [occasionally perforate], elliptic to obovate or spatulate, occasionally sagittate-cordate, larger than 1.5 cm; venation parallel or pinnate- or palmate-netted. **Inflorescences** spadices, each with 3–900 usually tightly grouped, sessile flowers, subtended by spathe; spathe rarely absent, persistent (sometimes only proximally) or deciduous, variously colored; spadix cylindric or ovoid, various parts occasionally naked or with sterile flowers. **Flowers** bisexual or unisexual, staminate and pistillate usually on same plants or functionally on different plants, staminate flowers distal to pistillate when unisexual; perianth absent or present; stamens 2–12, distinct or connate in synandria; ovaries 1, 1–3(–many)-locular, sessile or embedded in spadix; styles 1; stigmas hemispheric, capitate, or discoid [sometimes strongly lobed]. **Fruits** berries, distinct or connate at maturity. **Seeds** 1–40(–many) per berry.

Genera 105, species more than 3300 (8 genera, 10 species in the flora; species in 10 additional genera may persist locally within flora area, see table 203.1): nearly worldwide, primarily tropical regions.

Araceae are best characterized by the inflorescence, a fleshy cylindric or ovoid, unbranched spadix subtended or surrounded by a spathe. True spathes are absent in the Nearctic genus *Orontium* and in the Australian genus *Gymnostachys*. Other plant families with a compressed spadix-like inflorescence, such as Piperaceae and Cyclanthaceae, either do not have a structure equivalent to a spathe (Piperaceae) or have early-deciduous bracts (Cyclanthaceae). Plants are usually glabrous, rarely pubescent or spiny (pubescent in *Pistia*). Many Araceae exhibit typical monocotyledonous parallel leaf venation, but some genera have net leaf venation more typical of dicotyledons.

Table 203.1. *Cultivated Araceae that Occasionally Naturalize or Persist from Cultivation.*

Taxon	Will Key To	Distinguishing Characteristics	Comments
Aglaonema commutatum Schott	*Peltandra*	Leaves oblanceolate to oblong with thick midveins and parallel venation; creeping or erect aboveground stem; spathe not differentiated into tube and blade	Dade Co., Florida; persisting after cultivation, rarely found
Alocasia macrorrhizos (Linnaeus) Schott	*Peltandra*	Leaves net-veined between primary lateral veins; aboveground stem; spadix with sterile appendage	Dade Co., Florida; persisting for short periods, rarely found
Arum italicum Linnaeus	*Peltandra*	Leaves net-veined between primary lateral veins; long sterile appendage	California, Louisiana, North Carolina, Oregon; escaping locally and forming colonies
Caladium bicolor (Aiton) Ventenat	*Colocasia*	Leaves usually variegated with pink, red, or white; spadix without sterile appendage	Florida, Louisiana; rarely found
Dracunculus vulgaris Schott	*Arisaema dracontium*	Spadix about as long as spathe; spathe with large distinct blade	California, Oregon; rarely found
Epipremnum pinnatum (Linnaeus) Engler cv. 'Aureum'	will not key well to any genus	Stem scandent; juvenile leaves entire, adult leaves irregularly pinnatifid, leaves variegated with yellow; rarely flowering	Florida; escaping and persisting in shady hammocks and on roadside trees
Pinellia ternata (Thunberg) Makino	*Arisaema triphyllum*	Petiole with bulbils basally and at apex; base of spadix adnate to spathe on one side	California, District of Columbia, Maryland, New Jersey, New York, Ohio, Pennsylvania, West Virginia; very local, forming colonies usually in gardens and nurseries
Syngonium podophyllum Schott	*Arisaema*	Stems scandent; milky sap; juvenile leaves entire, adult leaves pedatisect	Florida; established on roadside trees and hammocks, local; other species of *Syngonium* may persist, but are rarely found
Xanthosoma spp.	*Peltandra*	Leaves not peltate, net-veined between primary lateral veins; may have aboveground stem; spadix without sterile appendage	S Florida, Louisiana, and Texas; persisting from cultivation, rarely spreading
Zantedeschia aethiopica (Linnaeus) Sprengel	*Peltandra*	Spathe funnel-shaped, not constricted, white	California; uncommon

Infrafamilial classification of the Araceae is under active study. The only classification of the family to date to utilize modern phylogenetic techniques (S. J. Mayo et al. 1997) recognizes seven subfamilies, of which three are represented in native temperate North American aroid flora: Orontioideae (*Orontium, Symplocarpus, Lysichiton*); Calloideae (*Calla*); and Aroideae (*Peltandra, Arisaema,* and *Pistia*). *Acorus*, a genus historically included in Araceae, is treated as a separate family in the flora based on extensive morphologic and chemical evidence that supports its removal from Arales (M. H. Grayum 1987).

The number of genera of Araceae occurring in temperate North America is low in comparison with other continents, and primitive taxa are disproportionately represented. Orontioideae and Calloideae, which include four of the seven native genera found in the flora area, are the basal clades within Araceae. Plants in these subfamilies possess the primitive states for many characters in Araceae and share few derived characters with other aroid genera (M. H. Grayum

1990). The more advanced genera native to the flora area include one genus endemic to eastern North America (*Peltandra*), a pantropical genus with an uncertain native distribution (*Pistia*), and a genus clearly Eurasian in origin (*Arisaema*).

Araceae contain crystals of calcium oxalate, which are often cited as causing the intense irritation experienced when handling or consuming the raw plant tissue of many genera in the family. This supposition is contradicted by the fact that although irritation generally is not produced by properly cooked plants, the crystals remain after heating. Other compounds must therefore be involved with causing this reaction. Studies of *Dieffenbachia* demonstrated that a proteolytic enzyme, as well as other compounds, are responsible for the severe irritation caused by this plant and that raphides of calcium oxalate do not play a major role (J. Arditti and E. Rodriguez 1982). Whether irritation is caused by enzymes or crystals, that aspect of Araceae has resulted in aroid genera being included in many lists of poisonous plants (e.g., K. F. Lampe and M. A. McCann 1985; G. A. Mulligan and D. B. Munro 1990; K. D. Perkins and W. W. Payne 1978).

Despite the toxic effects of Araceae, species of several genera are cultivated as food plants, mainly as subsistence crops in tropical areas. The major edible Araceae are *Colocasia esculenta* and several species of *Xanthosoma*, grown primarily for their corms and sometimes for their leaves. Most North American species of Araceae were historically used by Native Americans, as both food and medicine (T. Plowman 1969). The family, currently more valued for its many ornamental species, is the most important family in North America for indoor foliage plants (T. B. Croat 1994). Araceae commonly grown as ornamentals in American homes include species of *Aglaonema* (Chinese-evergreen), *Anthurium*, *Caladium*, *Dieffenbachia* (dumbcane), *Epipremnum* (golden pothos), *Philodendron*, *Spathiphyllum*, *Syngonium*, and *Zantedeschia* (calla-lily).

Plants of some cultivated species of Araceae escape and may persist or naturalize, especially in warmer climates. One of these species, *Colocasia esculenta*, is widespread enough to warrant full inclusion in the flora, but other introduced species of Araceae are very local in occurrence. Uncommon species represented by herbarium specimens or literature reports as escaped or persisting from cultivation are listed (table 203.1) with distinguishing characters and areas of occurrence.

SELECTED REFERENCES Bown, D. 1988. Aroids: Plants of the Arum Family. Portland. Grayum, M. H. 1990. Evolution and phylogeny of the Araceae. Ann. Missouri Bot. Gard. 77: 628–697. Mayo, S. J., J. Bogner, and P. C. Boyce. 1997. The Genera of Araceae. 1 vol. + laser disc. [London.] Plowman, T. 1969. Folk uses of New World aroids. Econ. Bot. 23: 97–122. Thompson, S. A. 1995. Systematics and Biology of the Araceae and Acoraceae of Temperate North America. Ph.D. dissertation. University of Illinois. Wilson, K. A. 1960. The genera of the Arales in the southeastern United States. J. Arnold Arbor. 41: 47–72.

1. Leaves palmately or pedately divided; flowers unisexual. . . . . .7. *Arisaema*, p. 139
1. Leaves simple; flowers unisexual or bisexual.
　2. All flowers unisexual.
　　3. Leaves sessile to nearly sessile; plant floating. . . . . .8. *Pistia*, p. 141
　　3. Leaves petiolate; plant rooted.
　　　4. Leaves peltate. . . . . .6. *Colocasia*, p. 137
　　　4. Leaves not peltate. . . . . .5. *Peltandra*, p. 135
　2. All or most flowers bisexual.
　　5. Spathe absent. . . . . .1. *Orontium*, p. 131
　　5. Spathe present, different in appearance from foliage leaves.
　　　6. Flowers without perianth; spathe white, often green or partially green abaxially. . . . .4. *Calla*, p. 134
　　　6. Flowers with perianth; spathe not white.

7. Spathe yellowish green to dark red-purple, usually spotted or striped with both, open only apically at maturity, enclosing spadix; spadix ovoid to globose. . . . . 2. *Symplocarpus*, p. 132

7. Spathe bright yellow, open fully at maturity, not enclosing spadix; spadix nearly cylindric. . . . . 3. *Lysichiton*, p. 133

1. ORONTIUM Linnaeus, Sp. Pl. 1: 324. 1753; Gen. Pl. ed. 5, 151. 1754 • Golden-club [ancient Greek name for plant that grew on River Orontes] E

Herbs, wetland. **Rhizomes** vertical. **Leaves** appearing before flowers, several, emergent or sometimes floating, clustered; petiole longer than blade; blade abaxially paler green, adaxially dark bluish green with distinctive velvety sheen, simple, not peltate, oblong-elliptic, base acute to obtuse or sometimes oblique, apex rounded-apiculate or short-acuminate; venation parallel to midvein except near base. **Inflorescences:** peduncle reclining in fruit, usually equal to or longer than leaves, apex swollen; spathe absent; spadix long-conic. **Flowers** mostly bisexual, the distal staminate; perianth present. **Fruits** partially embedded in spadix, green to blue-green. **Seeds** 1, embedded in mucilage. $x = 13$.

Species 1: e United States, primarily on coastal plain.

The leaf sheathing the peduncle, which has been interpreted as the spathe in *Orontium*, is a sympodial leaf and is not equivalent to the spathe of other Araceae (T. S. Ray 1988). *Orontium* also differs in several other characteristics from the presumably related genera *Lysichiton* and *Symplocarpus* and has been segregated as a monogeneric tribe (M. H. Grayum 1990).

SELECTED REFERENCES Grear, J. W. Jr. 1966. Cytogeography of *Orontium aquaticum* (Araceae). Rhodora 68: 25–34. Klotz, L. H. 1992. On the biology of *Orontium aquaticum* L. (Araceae), golden club or floating arum. Aroideana 15: 25–33.

1. Orontium aquaticum Linnaeus, Sp. Pl. 1: 324. 1753 E F

Roots contractile. **Rhizomes** 1.5–3 cm diam. **Leaves:** petiole dark green to red-green, 10–40(–60) cm; blade (6–)10–30(–45) cm, about ⅓ as wide (greater than 3 cm wide), main veins from base, curving apically parallel to midvein. **Inflorescences:** spadix bright yellow, 2–10 cm × 5–10 mm, apex rounded. **Flowers** covering spadix; tepals (2–)6, arched over ovaries, yellow; stamens (2–)6, sometimes with 1 or 2 staminodes; ovaries 1-locular, ovules 1. **Fruits** 1–2 cm diam.; pericarp thin. **Seeds** 1–1.5 cm diam.; endosperm absent; germinating soon after falling from parent plant. $2n = 26$.

Flowering late winter (s range)–spring. Shallow water of bogs, marshes, swamps, and streams; 0–900 m; Ala., Conn., Del., D.C., Fla., Ga., Ky., La., Md., Mass., Miss., N.J., N.Y., N.C., Pa., R.I., S.C., Tenn., Tex., Va., W.Va.

Orontium aquaticum is probably the most distinctive species of Araceae growing in North America. It is the only species that does not have a spathe, and the characteristic blue-green velvety leaves are not easily confused with those of any other plants. The unique appearance of the leaf blade of *O. aquaticum* is partly because of the waxy epidermal layer, which readily sheds water. The leaves also have large intercellular air spaces that contribute to their buoyancy.

Specimens of *Orontium aquaticum* supposedly collected in Iowa are almost certainly a labeling error. Populations occur primarily on the Atlantic and Gulf coastal plains and less frequently in the Appalachian region.

Orontium aquaticum and *Peltandra virginica* may have been used somewhat interchangeably by Native Americans and may be confused at times in the literature. Seeds of *O. aquaticum* were either dried or boiled repeatedly in water before being eaten by Native Americans; Swedish settlers also used the seeds in a similar manner (P. Kalm 1770–1771). Fresh seeds microwaved for 5 minutes in tap water have a "firm texture and pleasant, nutty flavor" and produce no irritation (L. H. Klotz 1992). *Orontium aquaticum* is sometimes grown for its attractive foliage and bright yellow spadices in aquatic gardens. It is available through catalog sources.

2. SYMPLOCARPUS Salisbury ex Nuttall in W. P. C. Barton, Veg. Mater. Med. U.S. 1: 124. 1817, name conserved · Skunk-cabbage, tabac-du-diable, chou puant [Greek *symplokos*, connected, and *karpos*, fruit, in reference to the infructescence]

Herbs, wetland or subterrestrial. **Rhizomes** vertical. **Leaves** appearing after flowers, several, erect, clustered; petiole equal to or shorter than blade; blade green to dark green, simple, not peltate, oblong to ovate, base truncate or cordate, apex acute to obtuse; primary veins lateral, branching apically. **Inflorescences:** peduncle partly underground, much shorter than leaves, apex not swollen; spathe yellowish green to dark red-purple, usually spotted or striped with both, open only apically at maturity, enclosing spadix; spadix ovoid to globose. **Flowers** bisexual; perianth present. **Fruits** embedded in enlarged spongy spadix, dark purple-green to dark red-brown. **Seeds** 1, mucilage absent. $x = 15$.

Species 1 or 2 (1 in the flora): ne North America and ne Asia.

Symplocarpus is one of the earliest plants to flower in spring in northeastern North America, sometimes with spathes emerging through snow. Because inflorescences are developed during the previous summer, flowering can occur during any warmer than normal weather throughout winter. The spadices of both Asian and American plants produce heat during flowering and can reach temperatures to 25°C above ambient air temperature (R. M. Knutson 1972; S. Uemura et al. 1993). These elevated temperatures probably play a role in pollination and in facilitating floral development at cold temperatures.

SELECTED REFERENCES Barabé, D. 1982. Vascularisation de la fleur de *Symplocarpus foetidus* (Araceae). Canad. J. Bot. 60: 1536–1544.   Shull, J. M. 1925. *Spathyema foetida*. Bot. Gaz. 79: 45–59.   Williams, K. A. 1919. A botanical study of skunk cabbage, *Symplocarpus foetidus*. Torreya 19: 21–29.

1. Symplocarpus foetidus (Linnaeus) Salisbury ex W. P. C. Barton, Veg. Mater. Med. U.S. 1: 123. 1817   [E] [F]

Dracontium foetidum Linnaeus, Sp. Pl. 2: 967. 1753; *Spathyema foetida* (Linnaeus) Rafinesque

Roots fleshy, contractile. **Rhizomes** thick, to 30 cm or more. **Leaves:** petiole sheathed basally, 5–57 cm; blade thick, 10–60 × 7–40 cm; primary lateral veins parallel, branching apically, interprimary veins anastomosing. **Inflorescences** at ground level; spathe hoodlike, 6–13(–18) cm, fleshy, apex acuminate, twisted or incurved, not persisting in fruit; spadix short-stipitate, somewhat flattened dorsiventrally, 2–3 × 1.5–3 cm. **Flowers** covering spadix; tepals 4, yellowish to dark red-purple; stamens 4, dehiscing longitudinally; ovaries 1-locular; ovules 1. **Infructescences** dark purple-green to dark red-brown, globose to oblong or ovoid, 4–7(–10) cm. **Seeds** brown, 7–15 mm diam. $2n = 60$.

Flowering late winter–spring. Swamps, wet woods, along streams, and other wet low areas; 0–1100 m; N.B., N.S., Ont., Que.; Conn., Del., D.C., Ill., Ind., Iowa, Maine, Md., Mass., Mich., Minn., N.H., N.J., N.Y., N.C., Ohio, Pa., R.I., Tenn., Vt., Va., W.Va., Wis.

Disagreement exists regarding correct author citation of the combination *Symplocarpus foetidus*. According to J. T. Kartesz and K. N. Gandhi, the correct citation is *S. foetidus* (Linnaeus) Nuttall (1992). The citation I have adopted is *S. foetidus* (Linnaeus) Salisbury ex W. Barton because Barton, in his 1817 publication, did not specifically attribute this combination to Nuttall; he did cite Salisbury in his first mention of the name *S. foetidus*.

Symplocarpus foetidus was included as occurring in Manitoba (H. J. Scoggan 1957) based on a misidentified specimen (B. Boivin 1967–1979, part 4). Although the species has been listed for Georgia (W. H. Duncan and J. T. Kartesz 1981), I have seen no specimens from that state. A specimen collected by S. B. Buckley labeled "Hab. Florida" is probably not from that state.

When bruised or broken, all parts of *Symplocarpus foetidus* give off an unpleasant odor. Various species of insects, including those from the orders Diptera, Coleoptera, Hymenoptera, and Hemiptera, many of which are attracted by the odor, have been collected from the inflorescences (W. W. Judd 1961), but the specific mechanism of pollination remains unknown. Although insects are likely pollen vectors, wind tunnel observations of the inflorescence suggest a capacity also for wind pollination (S. Camazine and K. J. Niklas 1984). Whatever the pollination mechanism, fertilization is limited, and few inflorescences develop into infructescences (J. A. Small 1959; personal observation).

Symplocarpus foetidus, in various forms and often combined with other plants, was used medicinally by Native Americans for a variety of ailments, including swellings, coughs, consumption, rheumatism, wounds,

convulsions, cramps, hemorrhages, toothaches, and headaches (D. E. Moerman 1986). Skunk cabbage was officially listed as the drug "dracontium" in the *U. S. Pharmacopoeia* from 1820 to 1880 for treating diseases of respiratory organs, nervous disorders, rheumatism, and dropsy (A. Henkel 1907). Plants are sparingly cultivated as a curiosity in North American gardens and are reported to be highly prized in aquatic gardens in European estates and public parks (F. W. Case 1992).

3. LYSICHITON Schott, Oesterr. Bot. Wochenbl. 7: 62. 1857 (as Lysichitum) • Skunk-cabbage [Greek *lysis*, dissolve, and *chiton*, a tunic, referring to the spathe, which withers soon after flowering]

Herbs, wetland. **Rhizomes** vertical. **Leaves** appearing at or soon after flowering, several, clustered, erect; petiole short; blade shiny medium green, simple, not peltate, elliptic to oblong-ovate or oblanceolate, base cuneate to subtruncate, apex obtuse to acute; primary veins pinnate. **Inflorescences:** peduncle absent; spathe bright yellow [white], boat-shaped, open fully at maturity, not enclosing spadix; spadix nearly cylindric. **Flowers** bisexual; perianth present. **Fruits** embedded in white pulpy axis of spadix, green. **Seeds** 2(–4), mucilage probably present. $x = 14$.

Species 2 (1 in the flora): nw North America and ne Asia.

Prior to recognition of the North American *Lysichiton* as a distinct species, the genus contained a single taxon, *L. camtschatcensis*. This name (in several spelling variations) was previously applied to both Asian and American *Lysichiton* and is now the valid name for Asian populations. Asian and American plants differ primarily in spathe color, with *L. camtschatcensis* having white spathes and *L. americanus*, yellow spathes.

SELECTED REFERENCES Barabé, D. and M. Labrecque. 1984. Vascularisation de la fleur de *Lysichitum camtschatcense* (Araceae). Canad. J. Bot. 62: 1971–1983. Hultén, E. and H. St. John. 1931. The American species of *Lysichitum*. Svensk Bot. Tidskr. 25: 453–464. Pellmyr, O. and J. M. Patt. 1986. Function of olfactory and visual stimuli in pollination of *Lysichiton americanum* (Araceae) by a staphylinid beetle. Madroño 33: 47–54.

1. Lysichiton americanus Hultén & H. St. John, Svensk Bot. Tidskr. 25: 455. 1931 (as Lysichitum americanum)

E

Roots white, contractile. **Rhizomes** to 30 cm or more, 2.5–5 cm diam. **Leaves** large; petiole stout, 5–40 cm; blade to 135 × 70 cm, base cuneate to subtruncate, apex obtuse to acute; midvein thick, grading into petiole. **Inflorescences** erect, malodorous, distinctive indoloid odor remaining even in old herbarium specimens; spathe wilting shortly after anthesis, basal portion 8–40 cm; blade basally contracted into long sheath enclosing stipe, 8–25 cm, stipe elongating in fruit; spadix stipitate, 4–12(–14) cm, initially shorter than spathe, eventually long-exserted because of elongation of stipe. **Flowers** yellow-

ish green; tepals 4; stamens 4, dehiscing longitudinally; ovaries (1–)2-locular; ovules 1–2. **Infructescences** oblong-ovate, 4–15 × 1.5–4 cm. **Seeds** gray-brown to red-brown, (3–)5–11 mm. $2n = 28$.

Flowering late winter–spring. Swamps, wet woods, along streams, and other low wet areas; 0–1400 m; B.C.; Alaska, Calif., Idaho, Mont., Oreg., Wash.; naturalized in Europe.

Reports of *Lysichiton americanus* in Wyoming (R. D. Dorn 1977; E. Hultén and H. St. John 1931) are based on a single specimen labeled "Yellowstone National Park." Despite recent attempts to locate *Lysichiton* at this locality, it has not been found (R. Hartman, pers. comm.). In addition, Wyoming is extralimital to the range of this species, which follows a biogeographic pattern typical of many plant and animal species in western North America.

Lysichiton americanus is pollinated by adults of *Pe-*

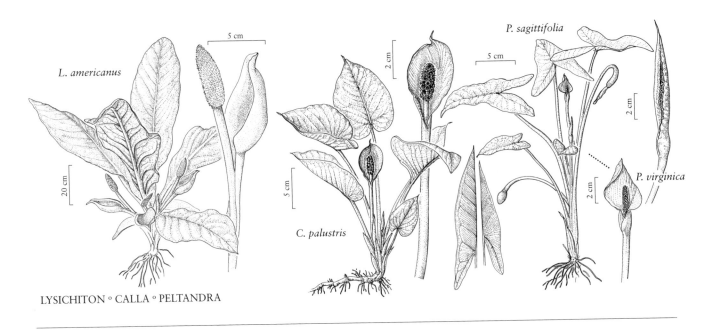

LYSICHITON ∘ CALLA ∘ PELTANDRA

lecomalius testaceum (Coleoptera: Staphylinidae), which feed on the pollen and use the inflorescences as a mating site. The distinctive odor produced by the inflorescences acts as an initial attractant for the beetles, which respond by initiating search behavior for yellow spathes (O. Pellmyr and J. M. Patt 1986).

Plants of this species were used as food, as medicine, and also in the material culture of Native Americans of northwestern North America (S. A. Thompson 1995). Although considered to be a famine food and rarely part of the diet under normal conditions, almost all parts were eaten. Perhaps the most important and widely used parts of *Lysichiton americanus* were the large, waxy leaves, which served the same functions as waxed paper does today. Medicinal use of the leaves, especially as a poultice for burns and injuries, was widespread among northwestern Native Americans. Like *Symplocarpus foetidus*, this species is widely planted in European gardens (F. W. Case 1992).

4. CALLA Linnaeus, Sp. Pl. 2: 968. 1753; Gen. Pl. ed. 5, 414. 1754 • Water arum, wild calla [a plant name used by Pliny, perhaps from Greek *kallos*, beauty]

Herbs, wetland. **Rhizomes** horizontal. **Leaves** appearing before flowers, several, emergent, arising along rhizome, also clustered terminally; petiole 1.5–2 or more times as long as blade; blade bright green, simple, not peltate, ovate to nearly round, base cordate, apex short-acuminate to apiculate; lateral veins parallel. **Inflorescences:** peduncle erect, as long as or longer than petiole, apex not swollen; spathe white, often green or partially green abaxially, not enclosing spadix; spadix cylindric. **Flowers** all bisexual or distal ones staminate; perianth absent. **Fruits** not embedded in spadix, red. **Seeds** 4–9(–11), embedded in mucilage. $x = 18$.

Species 1: circumboreal.

Numerous cytogenetic studies have been conducted on *Calla* with both diploid ($2n = 36$) and apparently tetraploid ($2n = 72$) populations reported (see G. Petersen 1989). All North American populations counted have a somatic number of 36 chromosomes.

SELECTED REFERENCES Dudley, M. G. 1937. Morphological and cytological studies of *Calla palustris*. Bot. Gaz. 98: 556–571. Lehmann, N. L. and R. Sattler. 1992. Irregular floral development in *Calla palustris* (Araceae) and the concept of homeosis. Amer. J. Bot. 79: 1145–1157. Scribailo, R. W. and P. B. Tomlinson. 1992. Shoot and floral development in *Calla palustris*. Int. J. Pl. Sci. 153: 1–13.

1. Calla palustris Linnaeus, Sp. Pl. 2: 968. 1753 [F]

Roots adventitious, arising from nodes. **Rhizomes** creeping at or near surface, elongate, 1–3 cm diam. **Leaves:** petiole 6–30(–40) cm; blade 4–14 cm wide; lateral veins curved-ascending, parallel. **Inflorescences:** spathe ovate to elliptic, 3–6(–8) cm, apex long-apiculate, 4–10 mm; spadix on thick short stipe, cylindric, shorter than spathe, apex rounded. **Flowers** covering spadix; stamens (6–)9–12, of 2 types, outer with broad filaments and inner with narrow filaments; ovaries 1-locular; ovules 6–9, anatropous. **Infructescences** 2–5 × 1.5–3.5 cm. **Fruits** pear-shaped, 6–12 × 5–10 mm. **Seeds** brown with dark spots at chalazal end, cylindric, 3–5 mm. 2*n* = 36.

Flowering late spring–summer. Bogs, marshes, wooded swamps, and marshy shores of rivers, ponds, and lakes; 0–900 m; Alta., B.C, Man., N.B., Nfld. and Labr., N.W.T., N.S., Nunavut, Ont., P.E.I., Que., Sask., Yukon; Alaska, Conn., Ill., Ind., Maine, Md., Mass., Mich., Minn., N.H., N.J., N.Y., N.Dak., Ohio, Pa., R.I., Vt., Wis.; Eurasia.

Plants with two or three spathes per inflorescence occur; this anomaly is recognized as *C. palustris* forma *polyspathacea* Victorin & Rousseau. *Calla palustris* has been reported from Iowa (T. G. Lammers and A. G. van der Valk 1979), but the only specimen is an unreliable record. The species was also reported from Mendocino County, California (A. Eastwood 1900), but the specimen on which this record is based was destroyed (H. L. Mason 1957). Perhaps this collection was an incorrect identification of *Zantedeschia aethiopica*, the cultivated calla-lily, which has escaped and naturalized in areas along the California coast (see table 203.1).

Preliminary field studies indicate that species of syrphid flies (*Sphegina* spp., Diptera: Syrphidae: Milesiinae) are especially common on inflorescences, with occasional visits from the widespread flower fly, *Toxomerus geminatus* (Diptera: Syrphidae: Syrphinae) (S. A. Thompson 1995).

Flour can be made from the seeds (M. L. Fernald et al. 1958), and plants are sometimes sold in aquatic garden catalogs for ornamental plantings in bog gardens.

5. PELTANDRA Rafinesque, J. Phys. Chim. Hist. Nat. Arts 89: 103. 1819, name conserved • Arrow arum [Greek *pelte*, small shield, and *andros*, male, referring to the shield-shaped tops of the staminate flowers] [E]

Herbs, wetland. **Rhizomes** vertical. **Leaves** appearing before flowers, several, clustered apically, erect; petiole equal to or longer than blade; blade green or glaucous light green, simple, not peltate, lanceolate to widely ovate, base hastate to sagittate, rarely cordate, apex acuminate to rounded or mucronate; lateral veins parallel. **Inflorescences:** peduncle recurving in fruit, ½ as long as to slightly longer than petiole, apex not swollen; spathe tube green, enclosing base of spadix; spathe blade green to white, opening slightly to fully at anthesis; spadix cylindric. **Flowers** unisexual, staminate and pistillate on same plant, pistillate flowers covering basal portion of spadix, staminate flowers apical, consisting of 4–5 connate stamens forming flat-topped synandrium; sterile flowers proximal to and usually distal to staminate flowers; perianth absent. **Fruits** not embedded in spadix, red or green to dark purple-green. **Seeds** 1–2(–4), mucilage present. *x* = 14.

Species 2 (2 in the flora): eastern North America.

Only one species of *Peltandra*, *P. virginica*, with two subspecies, was recognized by W. H. Blackwell and K. P. Blackwell (1974), a treatment followed by some botanists. They synonymized *P. luteospadix* with *P. sagittifolia*, based primarily on their contention that no *Peltandra* species have red fruits, and called that taxon *P. virginica* subsp. *luteospadix* (Fernald) Blackwell & Blackwell. All of the specimens cited in their treatment of the genus are *P. virginica*.

The two species of *Peltandra* can be distinguished not only on reproductive characters, but also on features of leaf venation. Fossil leaves congeneric with modern *Peltandra* are known from very late Paleocene through early Eocene deposits of North America. *Peltandra* is one of two genera of Araceae endemic to the flora area (the other is *Orontium*).

SELECTED REFERENCE   Blackwell, W. H. Jr. and K. P. Blackwell. 1974. The taxonomy of *Peltandra* (Araceae). J. Elisha Mitchell Sci. Soc. 90: 137–140.

1. Lateral leaf veins of ± same thickness; spadix about 1/2 as long as spathe; spathe blade white; fruits red. . . . . .1. *Peltandra sagittifolia*
1. Lateral leaf veins of 2 different thicknesses; spadix more than 1/2 to almost as long as spathe; spathe blade green to green with white or yellow-green along margin; fruits pea green to mottled green or dark purple-green. . . . . .2. *Peltandra virginica*

1. **Peltandra sagittifolia** (Michaux) Morong, Mem. Torrey Bot. Club 5: 102. 1894 (as sagittaefolia) · Spoonflower, white arrow arum E F

Calla sagittifolia Michaux, Fl. Bor.-Amer. 2: 187. 1803; *Peltandra glauca* (Elliott) Feay ex A. W. Wood

Leaves: petiole pink with green spots basally to green mottled with dark green to purple-green spots apically, 31–59 cm; blade light green, glaucous, 8–31 × (2.6–)4–11 cm, smaller on average and less variable in shape than in *Peltandra virginica*; lateral veins all ± same thickness. **Inflorescences** 5–12 cm; peduncle 30–58 cm; spathe tube light green inside and out, closed, 1.3–3.4 × 0.6–1.7 cm; spathe blade white, widely open, 3.8–9.4 × 1.9–4 cm, margins not undulate, spadix nearly cylindric, ± ½ as long as spathe. **Flowers:** pistillate flowers pale green, ovaries 1-locular, ovules 1; staminate portion of spadix yellow; sterile flowers between pistillate and staminate flowers and also frequently a few at apex. **Infructescences** enclosed by spathe tube, opening as fruits mature. **Fruits** red, 8.6–12.6 × 5.4–7 mm. **Seeds** 1, mucilage present, 7–10 mm. Chromosome number not available.

Flowering spring–mid summer. Acidic bogs and swampy woodlands; Atlantic and Gulf Coastal Plains; of conservation interest; 0–60 m; Ala., Fla., Ga., Miss., N.C., S.C.

Peltandra sagittifolia, the least common species of Araceae in North America, is monitored by conservation agencies in all states in which it occurs. Although its range is completely within the range of *P. virginica*, the two species occur in different habitats. Plants of *P. sagittifolia* grow only on acidic substrates with a mean pH of 4.8 (range 4.0 to 6.0).

2. **Peltandra virginica** (Linnaeus) Schott in H. W. Schott and S. L. Endlicher, Melet. Bot., 19. 1832 · Arrow arum, tuckahoe, peltandre E

Arum virginicum Linnaeus, Sp. Pl. 2: 966. 1753; *Peltandra luteospadix* Fernald; *P. tharpii* F. A. Barkley

Leaves: petiole green to purple-green, 38–98 cm; blade medium green, not glaucous or slightly glaucous abaxially, 9–57 × (2.5–)5–15(–31) cm, larger on average and more variable in shape than in *Peltandra sagittifolia*; lateral veins of 2 thicknesses. **Inflorescences** 7–25 cm; peduncle 20–56 cm; spathe tube green outside, paler green within, closed; 1.5–3.5(–5.2) × 0.7–1.9 cm; spathe blade green to green with white or yellow-green along margins, loosening only to slightly open to fully open at anthesis, (5.9–)8.5–21.4 × 0.5–2.3 cm, margins undulate; spadix tapering apically, more than ½ to almost as long as spathe. **Flowers:** pistillate flowers pale green to greenish white, ovaries 1-locular; ovules 1–4; staminate portion of spadix white, cream white, or pale yellow; sterile flowers between pistillate and staminate flowers; sterile tip 0.5–2 cm. **Infructescences** enclosed by spathe tube, rotting away to release fruits. **Fruits** pea green to mottled green or very dark purple-green, 10–18 × 6–16 mm. **Seeds** 1–2(–4), embedded in mucilage, 8–17 mm. $2n = 112$.

Flowering spring–late summer, also fall and winter in extreme s areas of range. Wetland habitats, including bogs, swamps, freshwater to low-salinity tidal marshes, and ditches, as well as along edges of ponds, lakes, and rivers; 0–1200 m; Ont., Que.; Ala., Ark., Conn., Del., D.C., Fla., Ga., Ill., Ind., Iowa, Kans., Ky., La., Maine, Md., Mass., Mich., Minn., Miss., Mo., N.H., N.J., N.Y., N.C., Ohio, Okla., Pa., R.I., S.C., Tenn., Tex., Vt., Va., W.Va., Wis.

Leaf shape is highly variable in *Peltandra virginica*, and different forms have been recognized taxonomically, both at the specific and infraspecific levels. Because leaf shape varies within populations and even within an individual clump of plants, *P. virginica* is treated here as a single taxon.

Populations of *Peltandra virginica* are most common along the Atlantic Coastal Plain, but its range appears to be actively expanding. Since 1978, the species was reported as new to the floras of Iowa, Kansas, Minnesota, West Virginia, and Wisconsin, and introduced populations may persist in Oregon and California. Fruits and seeds of *P. virginica* are a food for wildlife, especially waterfowl, and their use by migratory birds is an important factor in the spread of this species.

The flowers of *Peltandra virginica* are pollinated by a chloropid fly, *Elachiptera formosa* (Diptera: Chloropidae), which uses the inflorescence as a mating site and a larval food source. Eggs are deposited within the inflorescence, and the emerging larvae feed on the rotting male portion of the spadix. The fruits are primarily dispersed by water, although animals also play a role.

Peltandra virginica may have been an important food plant for eastern Native Americans, especially in the mid-Atlantic coastal region from Pennsylvania to Virginia, where the plants are now common and grow in large, dense populations. Historical accounts mention use of the rhizomes as well as the leaves, fruits, and seeds as food.

6. COLOCASIA Schott in H. W. Schott and S. L. Endlicher, Melet. Bot., 18. 1832, name conserved • Taro, dasheen [classical Greek name derived from an old Middle Eastern name *colcas* or *culcas*]   [1]

Herbs, wetland [or terrestrial]. **Stolons** with nodes produced at or near surface; corms underground [aboveground], tuberous. **Leaves** appearing before flowers, several, clustered apically, erect; petiole usually longer than blade; blade green to dark green or glaucous blue-green adaxially, simple, peltate, ovate- or sagittate-cordate, basal lobes rounded, apex mucronate; primary lateral veins parallel, secondary lateral veins netted. **Inflorescences:** peduncle erect, shorter than leaves, apex not swollen; spathe tube green; spathe blade orange, opening basally and reflexing apically at anthesis to expose spadix; spadix slender, tapering, usually terminated by sterile appendage. **Flowers** unisexual, staminate and pistillate on same plant; pistillate flowers covering base of spadix, staminate flowers apical, sterile flowers between pistillate and staminate flowers; perianth absent. **Fruits** greenish to whitish or red. **Seeds** 0–5(–35), mucilage probably present. $x = 7$.

Species 7 (1 in the flora): southeastern Asia, 1 species cultivated and escaping in the tropics and subtropics worldwide.

Species in the genus *Colocasia* have received little attention except for *C. esculenta*, commonly called taro, which is cultivated throughout the tropics and subtropics for its starchy, edible corms. The origin of the cultivated species is uncertain, but all other species in the genus occur in northeastern India and southeastern Asia. Prior to modernization, taro was especially important on the Pacific Islands. Hawaii was a main center of taro cultivation, and the crop played an important role in native culture. About 150 varieties of *C. esculenta* were developed on Hawaii, including those specifically grown for *poi*, a fermented paste made from crushed, cooked corms (A. B. Greenwell 1947).

Colocasia esculenta was probably brought to the Caribbean and North America from Africa as part of the slave trade. In the southeastern United States, taro was commonly cultivated in the kitchen gardens of slaves and their free descendants (W. Bartram 1791). In the early 1900s, the U.S. Department of Agriculture attempted a campaign to introduce taro as a new root crop in frost-free zones in the southern United States (O. W. Barrett and O. F. Cook 1910). Promotional literature, which included cultivation techniques and recipes for use, was distributed to encourage farmers to try this new crop, but taro was never accepted as a substitute for potatoes (R. A. Young 1936).

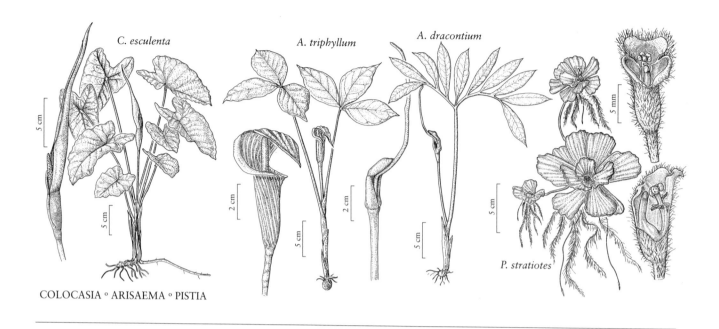

C. esculenta   *A. triphyllum*   *A. dracontium*

P. stratiotes

COLOCASIA ∘ ARISAEMA ∘ PISTIA

Plants of *Colocasia esculenta* are known by many common names, including taro, cocoyam, dasheen, eddo, malanga, tannia, and others. Many of these same names are also applied to species of *Xanthosoma*, a New World aroid also cultivated for its starchy corms, which is often confused with *Colocasia* (see S. K. O'Hair and M. P. Asokan 1986 for a review of edible aroids). Edible taxa in the two genera can be readily distinguished by the peltate leaves in *Colocasia* and absence of a sterile tip on the spadix in *Xanthosoma*.

SELECTED REFERENCES   Matthews, P. 1991. A possible tropical wildtype taro: *Colocasia esculenta* var. *aquatilis*. Bull. Indo-Pacific Prehist. Assoc. 11: 69–81.   Wang J.-K. and S. Higa, eds. 1983. Taro, a Review of *Colocasia esculenta* and Its Potentials. Honolulu.

1. Colocasia esculenta (Linnaeus) Schott in H. W. Schott and S. L. Endlicher, Melet. Bot.,18. 1832 • Elephant's-ear, wild taro [F] [I]

Arum esculentum Linnaeus, Sp. Pl. 2: 965. 1753; *Colocasia antiquorum* Schott

Corms underground, starchy; stolons elongate, with nodes produced at or near surface, spreading horizontally. **Leaves:** petiole green, often purple apically, 30–80(–180) cm, spongy and filled with air spaces; blade green to dark green or glaucous blue-green on adaxial surface, usually with red or purple spot at point of petiole attachment, peltate for 2.5–7 cm, 17–70 × 10–40 cm; primary lateral veins parallel, secondary lateral veins netted, forming collective vein between primary lateral veins; apex mucronate. **Inflorescences:** spathe 20–35 cm; tube green; blade orange outside and in, opening basally and reflexing apically at anthesis to expose spadix, more than 3 times longer

than tube; spadix 9–15 cm. **Flowers:** pistillate flowers pea green, interspersed with white pistillodes; ovaries 1-locular; ovules 36–67; sterile flowers white to pale yellow; staminate flowers and sterile tip pale orange, stamens 3–6, connate. **Fruits** orange. **Seeds** 1–1.5 mm, not observed in flora area. $2n = 28, 42$ (Old World).

Flowering late spring–late fall. Usually forming large colonies along streams, ponds, ditches, canals, and other wet areas; 0–100 m; introduced; Ala., Fla., Ga., La., Miss., Tex.; Mexico; West Indies; Central America; South America; Asia; Africa; Atlantic Islands (Bermuda); Indian Ocean Islands; Pacific Islands; Australia.

Weedy plants of *Colocasia esculenta* in the United States are essentially all one morphologic form (usually with long stolons and with a red to purple spot adaxially on the leaf opposite the junction of petiole and blade). This taxon has been called *C. esculenta* var. *aquatilis* Hasskarl in some treatments (K. A. Wilson 1960). Other forms of *C. esculenta* are cultivated in the flora area both for food and as ornamentals. The species is extremely variable and many varieties have been recognized taxonomically with lack of agreement on the application of names. Because of their weedy status and their infrequent flowering, specimens of *C. esculenta* are not frequently collected, and the distribution indicated here reflects this deficiency. Plants may occur beyond the boundary outlined on the map, but the species does not become established in areas subjected to cold temperatures.

7. ARISAEMA Martius, Flora 14: 459. 1831 • Jack-in-the-pulpit [Greek *aris*, plant name used by Pliny, and *haima*, blood, in reference to the red-spotted leaves of some species]

Muricauda Small

Herbs, terrestrial or wetland. **Corms** [rhizomes] nearly globose. **Leaves** usually appearing with flowers, 1–2(–3), erect; petiole longer than blade; blade medium to dark green, sometimes glaucous adaxially, palmately or pedately [radiately] divided, not peltate, leaflet elliptic to broadly ovate or oblanceolate, base rounded to obtuse or attenuate, apex obtuse or acute to acuminate; primary lateral veins of each leaflet pinnate. **Inflorescences:** peduncle erect, nearly equal to leaves [to very short], apex not swollen; spathe variously colored or striped, distal part open at maturity, exposing tip to 1/2 or more of spadix appendage; spadix ± cylindric, surmounted by sterile appendage of variable shape. **Flowers** unisexual, staminate and pistillate on same or different spadix; pistillate flowers congested; staminate flowers usually scattered, distal to pistillate flowers when both are present; perianth absent. **Fruits** not embedded in spadix, glossy orange to bright red. **Seeds** 1–6, mucilage sometimes present (not present in *Arisaema triphyllum*). $x = 13, 14$.

Species ca. 170 (2 in the flora): mostly temperate Asia, also North America, Mexico, and Africa.

The phenomenon of sex changing in *Arisaema* has been investigated by many authors (e.g., P. Bierzychudek 1982; K. Clay 1993; E. Kinoshita 1986). Smaller plants produce only staminate flowers, and larger plants produce either staminate and pistillate flowers simultaneously or pistillate flowers only. Changes in gender expression are directly correlated with size and are also influenced by the environment in which the plants are growing. Reversions in phenotypic gender have been experimentally induced by such factors as removing leaf area or changing soil nutrient levels.

Although some species are cultivated as ornamentals, the genus is not of great economic importance.

SELECTED REFERENCES Huttleston, D. G. 1953. A Taxonomic Study of the Temperate North American Araceae. Ph.D. dissertation. Cornell University. Murata, J. 1990. Present status of *Arisaema* systematics. Bot. Mag. 103: 371–382. Treiber, M. 1980. Biosystematics of the *Arisaema triphyllum* Complex. Ph.D. dissertation. University of North Carolina.

1. Leaflets 3(–5); spadix blunt apically, shorter than spathe. . . . . 1. *Arisaema triphyllum*
1. Leaflets (5–)7–13(–21); spadix tapering apically, longer than spathe. . . . . 2. *Arisaema dracontium*

1. Arisaema triphyllum (Linnaeus) Schott in H. W. Schott and S. L. Endlicher, Melet. Bot., 17. 1832 • Jack-in-the-pulpit, Indian-turnip, petit-prêcheur E F

Arum triphyllum Linnaeus, Sp. Pl. 2: 965. 1753; *Arisaema acuminatum* Small; *A. atrorubens* (Aiton) Blume; *A. polymorphum* (Buckley) Chapman; *A. pusillum* (Peck) Nash; *A. quinatum* (Nuttall) Schott; *A. stewardsonii* Britton; *A. triphyllum* subsp. *pusillum* (Peck) Huttleston; *A. triphyllum* subsp. *quinatum* (Nuttall) Huttleston; *A. triphyllum* subsp. *stewardsonii* (Britton) Huttleston; *A. triphyllum* var. *pusillum* Peck; *A. triphyllum* var. *stewardsonii* (Britton) Stevens ex Wiegand & Eames

Plants 1.5–9 dm. **Roots** radiating from apex of corm; corm to 8 cm diam. **Leaves** 1–2(–3); petiole green, purple-marked, or purple; blade palmately divided; leaflets 3(–5), sessile or short-petiolulate, elliptic to broadly ovate, to 30 × 20 cm, apex obtuse or acute to acuminate; lateral leaflets often asymmetric, sometimes lobed or divided. **Inflorescences:** spathe convolute basally, 6–16 cm; tube green, sometimes striped or variously marked with purple or white; blade outside green, often with purple or white, inside varying from uniformly green to purple, or variously striped, forward-curving, forming hood over tube, broadly ovate to lanceolate, apex acute to long-acuminate; spadix 3–9 cm, shorter than spathe, apex blunt. **Staminate flowers** with 3–5 stamens. **Fruits** widely obovoid to obconic, 6–15 mm. **Seeds** 1–3(–6), 3–5 mm diam. 2n = 28, 42, 56.

Flowering early–late spring. Moist to dry deciduous woods and thickets, bottoms, swamps, and bogs; 0–2000 m; Man., N.B., N.S., Ont., P.E.I., Que.; Ala., Ark., Conn., Del., D.C., Fla., Ga., Ill., Ind., Iowa, Kans., Ky., La., Maine, Md., Mass., Mich., Minn., Miss., Mo., Nebr., N.H., N.J., N.Y., N.C., N.Dak., Ohio, Okla., Pa., R.I., S.C., S.Dak., Tenn., Tex., Vt., Va., W.Va., Wis.

Considerable diversity of opinion exists as to the delimitation of taxa and application of names in the *Arisaema triphyllum* complex. Three subspecies are recognized in the most recent study of this group (M. Treiber 1980): subsp. *triphyllum*, subsp. *pusillum*, and subsp. *stewardsonii*. This treatment of infraspecific taxa is essentially in agreement with that of the previous revision of the complex (D. G. Huttleston 1949, 1953, 1981), except for *A. triphyllum* subsp. *quinatum* (Nuttall) Huttleston, which is regarded by Huttleston as a distinct subspecies and is included in subsp. *pusillum* by Treiber.

As defined by M. Treiber (1980), *Arisaema triphyllum* subsp. *triphyllum* is a widespread tetraploid (2n =

56) ranging from Gaspé Peninsula to Manitoba and North Dakota south to central Florida and eastern Texas. The distribution of *A. triphyllum* subsp. *pusillum*, a diploid (2n = 28), approaches that of subsp. *triphyllum* except in the northeast and midwest, where it is uncommon. *Arisaema triphyllum* subsp. *stewardsonii*, also a diploid, has the most restricted range of the three subspecies and is found primarily in the northeastern United States ranging south along the mountains to North Carolina and Tennessee. In addition to having a restricted range, subsp. *stewardsonii* is also restricted in habitat and grows primarily in swamps, marshes, floodplains, and other moist areas. The other two subspecies, especially subsp. *triphyllum*, grow in a broad range of habitats. Sympatric populations of the different subspecies occur.

Distinguishing characteristics of *Arisaema triphyllum* subsp. *triphyllum* include leaves that are abaxially glaucous, a club-shaped (rarely cylindric) spadix appendage, and 4.5–9.5 mm wide spathe flanges. The other two subspecies have leaves that are not abaxially glaucous, spathes with narrower flanges (1–3 mm wide), and a cylindric (rarely club-shaped) spadix appendage. *Arisaema triphyllum* subsp. *stewardsonii* is distinguished primarily by its strongly fluted or ribbed spathe tube and a spathe hood that is usually green with white or purple stripes. The spathe tube of *A. triphyllum* subsp. *pusillum* is at most weakly fluted, and the spathe hood is usually uniformly green or purple.

Although these subspecies are not recognized here, the following key (M. Treiber 1980) is provided for those wishing to determine plants at the infraspecific level.

1. Leaves glaucous beneath at maturity; spathe flange 4.5–9 mm wide; spathe hood green or green with purple stripes; sterile spadix appendage club-shaped, rarely cylindric; stomatal guard cell length mostly greater than 40 μm. . . . . *Arisaema triphyllum* subsp. *triphyllum*
1. Leaves polished or lustrous beneath, not glaucous; spathe flange 1–3 mm wide; sterile spadix appendage cylindric, rarely club-shaped; stomatal guard cell length mostly less than 40 μm.
  2. Spathe tube strongly fluted; spathe hood green with white or purple stripe. . . . . *Arisaema triphyllum* subsp. *stewardsonii*
  2. Spathe tube not fluted or rarely weakly fluted; spathe hood green or purple. . . . . *Arisaema triphyllum* subsp. *pusillum*

Although these morphologic forms may be recognizable in the field, distinguishing these differences in herbarium specimens is often difficult, and much overlap occurs in expression of the characteristics supposedly defining infraspecific taxa. Numerous intermediate forms exist, including putative hybrid populations between the subspecies with $2n = 42$ (D. G. Huttleston 1949, 1953). Given these problems and the sympatric ranges of the "subspecies" recognized by previous workers, *A. triphyllum* is treated here as one highly variable species.

In addition to the above variability within the *Arisaema triphyllum* complex, putative hybrid populations between *A. triphyllum* and *A. dracontium* also occur naturally (L. L. Sanders and C. J. Burk 1992). These plants do not produce mature fruits but do reproduce vegetatively.

2. **Arisaema dracontium** (Linnaeus) Schott in H. W. Schott and S. L. Endlicher, Melet. Bot., 17. 1832 • Green-dragon, dragon-root F

Arum dracontium Linnaeus, Sp. Pl. 2: 964. 1753; *Muricauda dracontium* (Linnaeus) Small

Plants 1.5–9 dm. **Roots** radiating from apex of corm; corm to 8 cm diam. **Leaves** usually solitary; petiole medium green or purple-marked; blade pedately divided, leaflets (5–)7–13(–21), sessile or petiolulate, elliptic to oblanceolate, to 28 × 10 cm, apex acute to acuminate; central leaflet usually shorter than neighboring ones, these leaflets longest, outer progressively smaller. **Inflorescences:** spathe light green, sometimes marked with purple, convolute, 3–6(–12) cm; blade usually scarcely distinguished from tube; spadix 6–20 cm (or longer), longer than spathe, apex tapering in long slender appendage to 15 cm. **Staminate flowers** with 2–4 stamens. **Fruits** oblong or pear-shaped, 7–13 mm. **Seeds** 1–2(–6), 3–5 mm diam. $2n = 28, 56$.

Flowering late winter (s part of range)–late spring. Mesic to wet deciduous woods, thickets, and bottoms; 30–1200 m; Ont., Que.; Ala., Ark., Conn., Del., D.C., Fla., Ga., Ill., Ind., Iowa, Kans., Ky., La., Md., Mass., Mich., Minn., Miss., Mo., Nebr., N.J., N.Y., N.C., Ohio, Okla., Pa., S.C., Tenn., Tex., Vt., Va., W.Va., Wis.; e Mexico.

Reports of *Arisaema dracontium* occurring in New Hampshire and Rhode Island have not been substantiated by specimens. The species has also been reported from Nuevo León and Veracruz, Mexico (E. Matuda 1954); more study is needed to determine if these plants are conspecific. Specimens with a wider spathe blade than is typical in *A. dracontium* have been collected in Florida and Georgia, and these forms may represent intermediates between *A. dracontium* and the Mexican species *A. macrospathum* Bentham, which has an expanded spathe blade. D. G. Huttleston (1953) treated *A. macrospathum* as a subspecies of *A. dracontium* in his dissertation, but this change in taxonomic status was never formally published.

8. PISTIA Linnaeus, Sp. Pl. 2: 963. 1753; Gen. Pl. ed. 5, 411. 1754 • Water-lettuce

[Greek *pistra*, watertrough, in reference to the aquatic habitat]

Herbs, aquatic, floating on still or slow-moving water. **Stolons** frequently produced, terminating in young plants. **Leaves** appearing before flowers, several, floating to ascending or nearly erect, in dense rosette, sessile to subsessile; blade light green to grayish green, simple, not peltate, obovate to spatulate, base obtuse or rounded, apex subtruncate, rounded, or notched, pubescent; primary veins nearly parallel. **Inflorescences:** peduncle erect, very short, pubescent; spathe greenish, closed basally, open apically, exposing tip of female flower and male spadix; spadix basally adnate to spathe. **Flowers** unisexual; staminate and pistillate on same plant; staminate flowers distal to single pistillate flower; perianth absent. **Fruits** not embedded in spadix, green, turning brown at maturity. **Seeds** (1–)4–10(–20), mucilage probably present. $x = 14$.

Species 1: tropical and subtropical regions worldwide.

Although many names have been published in *Pistia*, it is currently considered to contain a single, highly variable species. Morphological variation is strongly influenced by environmental conditions and population density. Some botanists consider the genus to have been introduced into the United States and many regional floras state that fruits and seeds are not produced in the flora area. Seeds with high rates of germination have been reported from many sites in Florida, however (F. A. Dray Jr. and T. D. Center 1989). The status of *Pistia* as native to the United States has not been resolved; available evidence suggests that it is indigenous.

Pistia is economically and biologically important as a breeding site for insects, especially species of mosquitoes in the genus *Mansonia*, and it is often an aquatic weed. Although a serious pest in some areas of the world, *Pistia* has not caused major problems in the United States (M. C. Bruner 1982). At least 23 species of insects worldwide have been reported to feed on *P. stratiotes*; those shown to be effective in biological control include *Neohydronomus affinis* (Coleoptera: Curculionidae: Erirhininae) and *Samea multiplicalis* (Lepidoptera: Pyralidae: Pyraustinae) (M. C. Bruner 1982; K. L. Harley et al. 1990). Plants are cultivated as aquatic ornamentals in greenhouses and gardens.

SELECTED REFERENCES   Mitra, E. 1966. Contributions to our knowledge of Indian freshwater plants 5. On the morphology, reproduction and autecology of *Pistia stratiotes* Linn. J. Asiat. Soc. 8: 115–135.   Stoddard, A. A. 1989. The phytogeography and paleoflo[r]istics of *Pistia stratiotes* L. Aquatics 11(3): 21–24.

1. **Pistia stratiotes** Linnaeus, Sp. Pl. 2: 963. 1753 • Water-lettuce F

Pistia spathulata Michaux

Roots to 50 cm, with short branches. **Leaves** light green to grayish green, 2–15(–20) cm, spongy, pubescence dense, white; major veins 5–13(–15), nearly parallel, abaxially prominent. **Inflorescences:** spathe convolute basally, slightly constricted above middle, spreading apically, white to pale green, pubescent outside, glabrous inside; spadix adnate to spathe more than ½ its length, shorter than spathe; axis naked at base of staminate part and sometimes extending beyond staminate flowers. **Flowers:** staminate flowers (2–)6–8, in single whorl around central stalk, stamens 2, connate; pistillate flower solitary, ovaries 1-locular, 4–5 mm, ovules 4–15(–20), orthotropous, styles 3 mm, stigmas obtuse, with small hairs. **Fruits** with thin pericarp. **Seeds** light brown, cylindric, 2 × 1 mm. $2n = 28$ (India, Borneo).

Flowering mainly late summer–early winter (although plants have been collected in flower in almost all months). Slow-moving streams, canals, drainage ditches, ponds, lakes, and springs; 0–10 m; Fla., La., Tex.; Mexico; West Indies; Central America; South America; Asia; Africa; Pacific Islands; Australia.

Pistia stratiotes has been collected north and west of its U.S. range in Arizona, California, Georgia, Missouri, New Jersey, Ohio, and South Carolina. These are probably populations that do not persist. Although *P. stratiotes* is often reported as occurring from Florida to Texas, I saw no verifying herbarium specimens from Alabama or Mississippi, nor were any populations located during my fieldwork in those states.

204. LEMNACEAE Gray

• Duckweed Family

Elias Landolt

Herbs, mostly perennial (*Lemna aequinoctialis* and *L. perpusilla* also annual), aquatic, floating or submersed, reduced to small green bodies called fronds corresponding partly to leaf and partly to stem. **Roots** 0 or 1–21; root hairs absent. **Stems:** distinct stems absent. **Cataphylls** absent. **Fronds** 1 or 2–20 or more, not differentiated into petiole and blade (thin white stipe or green stalk attaching new fronds to mother frond), coherent at base, flattened or globular, fronds smaller than 1.5 cm, venation from node, outer veins sometimes branching distally from inner ones, or veins absent; new fronds (daughter fronds) arising successively in 1–2 pouches or in cavity at base of mother frond; turions present in some species. **Inflorescences** usually solitary (mostly 2 per frond for *L. perpusilla*). **Flowers** mostly bisexual, 1(–2) per frond (rare in many species); sepals absent; petals absent; stamens 1–2; ovaries 1, bottle-shaped, 1-locular, tapering into short styles; stigmas funnel-shaped. **Fruits** follicles; pericarp membranous, opening by bursting. **Seeds** 1–5, nearly as long as fruit.

Genera 4, species 37 (4 genera, 19 species in the flora): nearly worldwide, very rare in regions with high or very low precipitation; not in Greenland or Aleutian Islands.

In this treatment the terms upper/lower and above/below are always used in relation to the position of the frond in the water. The positions toward the base and the apex of the frond are called proximal and distal, respectively.

The "leaf" in the Lemnaceae does not correspond to the leaf of higher plants. It is supposed to consist of a stem in the proximal part (from base to node) and a leaf in the distal part (from node to apex). In Lemnaceae literature it is called a frond. Some authors hold that the flower of the Lemnaceae corresponds to an inflorescence consisting of 1 to 2 male flowers (anthers) and 1 female flower (ovary).

With respect to chromosome number, hundreds of Lemnaceae clones were counted, and three levels of cytological variations were identified: (1) intra-individual variation (aneusomaty and/or mixoploidy, (2) intra-populational variation (aneuploidy or polyploidy), and (3) "racial" differentiation (K. Urbanska-Worytkiewicz 1980). Therefore, many chromosome numbers within a species were counted even if only North American plants are listed.

Lemnaceae are the smallest and most reduced flowering plants. Therefore, closely related species have but few distinguishing characteristics that are easy to recognize. In addition, most of these features are strongly modifiable and overlap considerably. Occasionally Lemnaceae have small, 1–5-celled papillae on the upper surface of the frond. Some species have turions (compact fronds reduced in size and structure, filled with starch grains, forming under unfavorable conditions). Turions remain for several days with the mother frond before they sink to the bottom.

Because of the rarity of flowering and fruiting, only vegetative characteristics are available in most cases. *Lemna gibba*, *L. aequinoctialis*, and *L. perpusilla* are the only American species fruiting rather frequently. The smallness of the plants requires good magnification and some technical preparations for identification. To analyze anatomic structures (e.g., number of veins or extension of air spaces), transparent slides are necessary. Fronds preserved in 70% ethanol become transparent. Dried fronds must be boiled first in 70% ethanol and treated afterwards with 10% NaOCl to clear them. Preserving them in ethanol, however, removes pigments that are sometimes important for determination. Under optimal growth conditions the typical anthocyanin pattern of some species does not develop. Therefore, one should observe the species in another season or cultivate it under different conditions. Another difficulty of determination occurs because several (up to 10) Lemnaceae species very often grow together in nature. Therefore, many collection samples contain more than one species, some of which may not be recognized as different at first glance.

Lemnaceae are easily distributed by birds over short distances. In many places, they live only so long as conditions are favorable. Afterwards they disappear. A species of a southern area might suddenly occur farther north and remain there for one or several years. If the species is exposed to water fed by a warm spring, it might persist far beyond expectations. The distribution maps of the Lemnaceae show the area where the species once was collected and do not represent the actual distribution area, which might be considerably smaller and change within a few years. Generally, most Lemnaceae species have expanded during the last years because of the warming of the climate and eutrophication of the waters.

Lemnaceae have a high productivity (some species can double in number within 24 hours) and a very high percentage of amino acids (up to 45% of dry weight). They are used in many regions as food for poultry, pigs, and cows. *Wolffia* fronds are eaten as a vegetable in southeastern Asia; *Lemna gibba* is cultivated in Israel for use as a vegetable and salad. Lemnaceae are also used for waste-water purification and as test and indicator plants.

SELECTED REFERENCES Daubs, E. H. 1965. A monograph of Lemnaceae. Illinois Biol. Monogr. 34. Hegelmaier, C. F. 1868. Die Lemnaceen. Eine monographische Untersuchung.... Leipzig. Landolt, E. 1986. The family of Lemnaceae—A monographic study, vol. 1. Veröff. Geobot. Inst. E. T. H. Stiftung Rübel Zürich 71. Landolt, E. and R. Kandeler. 1987. The family of Lemnaceae—A monographic study, vol. 2. Veröff. Geobot. Inst. E. T. H. Stiftung Rübel Zürich 95. Urbanska-Worytkiewicz, K. 1980. Cytological variation within the family of Lemnaceae. Veröff. Geobot. Inst. E. T. H. Stiftung Rübel Zürich 70: 30–101.

1. Roots 1–21 per frond; fronds with 1–21 veins; daughter fronds and flowers from 2 lateral pouches at frond base; flowers surrounded by small utricular, membranous scale; stamens 2, 4-locular; seeds longitudinally ribbed.
    2. Roots (1–)2–21 per frond; fronds with (3–)5–16(–21) veins, surrounded at base by small scale covering point of attachment of roots; pigment cells present (visible in dead fronds as brown dots). . . . . .1. *Spirodela*, p. 145
    2. Roots 1 per frond; fronds with 1–5(–7) veins, without scale at base; pigment cells absent (red pigmentation present in some species). . . . . .2. *Lemna*, p. 146
1. Roots absent; fronds without veins; daughter fronds from single terminal pouch or cavity

at frond base; flower(s) in cavity of upper frond surface; flower(s) not surrounded by utricular scale; stamens 1, 2-locular; seeds nearly smooth.

3. Fronds flat (linear, ribbon-, sabre-, or tongue-shaped, or ovate), with air spaces; daughter fronds from terminal flat pouch at mother-frond base; flower(s) in cavity at side of median line of upper frond surface. . . . . 3. *Wolffiella*, p. 150
3. Fronds 3-dimensional (globular, ovoid, or boat-shaped), without air spaces; daughter fronds from terminal conic pouch or cavity at mother-frond base; flower in cavity on median line of upper frond surface. . . . . 4. *Wolffia*, p. 151

1. SPIRODELA Schleiden, Linnaea 13: 391. 1839 • Duck-meal [Greek *speira*, winding, and *delos*, distinct] [W]

Roots (1–)2–21 per frond, abaxial surface at node. **Fronds** floating (only turions sink to bottom), 2–10, coherent in groups, each obovate-circular, flat or gibbous, 1.5–10 mm, margins entire; air spaces in tissue; reproductive pouches 2, lateral, at base from which daughter fronds and flowers originate, triangular; veins (3–)5–16(–21), from point in proximal part of frond (node); small 2-cleft membranous scale (interpreted as prophyllum) enveloping base of frond, covering point of attachment of roots; anthocyanins present (especially on abaxial surface); pigment cells present (visible in dead fronds as brown dots); turions present or absent, brownish olive, circular-reniform, smaller than growing frond. **Flowers** 1(–2) per frond, surrounded by small utricular, membranous scale; stamens 2, 4-locular. **Seeds** 1–3, longitudinally ribbed. $x = 10, 18, 23$.

Species 3 (2 in the flora): nearly worldwide.

1. Fronds 1–1.5 times as long as wide, with 7–16(–21) veins and 7–21 roots, 1 or 2 perforating scale. . . . . 1. *Spirodela polyrrhiza*
1. Fronds 1.5–2 times as long as wide, with (3–)5–7 veins and (1–)2–7(–12) roots, all perforating scale. . . . . 2. *Spirodela punctata*

1. Spirodela polyrrhiza (Linnaeus) Schleiden, Linnaea 13: 392. 1839 • Spirodèle polyrhize [F] [W]

Lemna polyrrhiza Linnaeus, Sp. Pl. 2: 970. 1753 (as *polyrhiza*)

Roots 7–21, to 3 cm, 1 or 2 perforating scale. **Fronds** obovate to circular, flat or rarely gibbous, 2–10 mm, 1–1.5 times as long as wide, apex rounded or pointed, upper surface sometimes with red spot in center; veins 7–16(–21); turions sometimes present, rootless, brownish olive, circular-reniform, 1–2 mm diam. **Flowers:** ovaries 1–2-ovulate. **Fruits** 1–1.5 mm, laterally winged to apex. **Seeds** with 12–20 distinct ribs. $2n = 30, 38, 40, 50, 80$.

Flowering (very rare) early summer–early fall. Eutrophic, quiet waters, in temperate to tropical regions; 0–2500 m; Alta., B.C., Man., N.B., N.S., Ont., P.E.I., Que., Sask.; Ala., Ariz., Ark., Calif., Colo., Conn., Del., D.C., Fla., Ga., Idaho, Ill., Ind., Iowa, Kans., Ky., La., Maine, Md., Mass., Mich., Minn., Miss., Mo., Mont., Nebr., Nev., N.H., N.J., N.Mex., N.Y., N.C., N.Dak., Ohio, Okla., Oreg., Pa., R.I., S.C., S.Dak., Tenn., Tex., Utah, Vt., Va., Wash., W.Va., Wis., Wyo.; nearly worldwide.

2. Spirodela punctata (G. Meyer) C. H. Thompson, Rep. (Annual) Missouri Bot. Gard. 9: 28. 1898 [F] [I] [W]

Lemna punctata G. Meyer, Prim. Fl. Esseq., 262. 1818; *Spirodela oligorrhiza* (Kurz) Hegelmaier

Roots (1–)2–7(–12), to 7 cm, all perforating scale. **Fronds** obovate, flat or gibbous, 1.5–8 mm, 1.5–2 times as long as wide, apex mostly pointed, upper surface without red spot; veins (3–)5–7; distinct turions absent. **Flowers:** ovaries 1–2-ovulate. **Fruits** 0.8–1 mm, laterally winged to apex. **Seeds** with 10–15 distinct ribs. $2n = 40, 46, 50$.

Flowering (rare) summer–early fall. Eutrophic, quiet

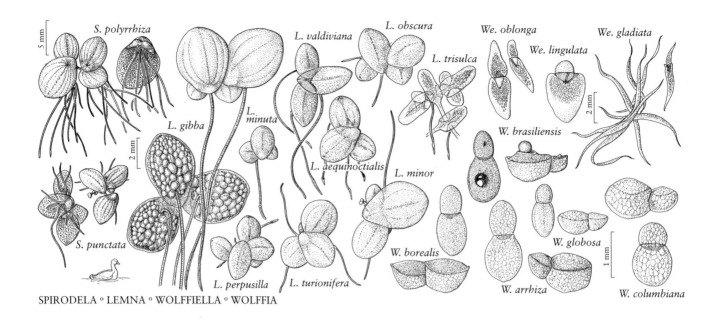

waters, in warm-temperate to subtropical regions with mild winters; 0–900 m; introduced; Ala., Ariz., Ark., Calif., D.C., Fla., Ga., Ill., Ky., La., Miss., Mo., N.C., Okla., Oreg., Pa., S.C., Tenn., Tex., Va., Wash.; South America; Asia; Africa; Atlantic Islands; Pacific Islands; Australia.

2. LEMNA Linnaeus, Sp. Pl. 2: 970. 1753; Gen. Pl. ed. 5, 417. 1754 • Duckweed, lenticules [Greek name of a water plant] [W]

Roots 1 per frond. **Fronds** floating or submersed, 1 or 2–20 or more, coherent in groups or forming chains, lanceolate-ovate, flat or gibbous, 1–15 mm, margins entire or denticulate, upper surfaces sometimes with small conic papillae along veins (especially at node and near apex); air spaces in tissue; reproductive pouches 2, lateral, at base from which daughter fronds and flowers originate, triangular; veins 1–5(–7), originating from point in proximal part of frond (node) or if more than 3 veins present, outer ones sometimes branching distally from inner ones; scale at base of frond absent; anthocyanins sometimes present; pigment cells absent (red pigmentation present in some species); turions absent (sometimes present in *L. turionifera*). **Flowers** 1(–2) per frond, surrounded by small utricular, membranous scale; stamens 2, 4-locular. **Seeds** 1–5, longitudinally ribbed. $x = 10, 21, 22$.

Species 13 (9 in the flora): worldwide except arctic and antarctic regions.

In *Lemna* the connection to the mother frond is formed by a thin white stipe at the base that falls off or decays after the frond is grown (frond of *L. trisulca* is narrowed at base into a green stalk persisting on frond).

1. Fronds submersed (except when flowering or fruiting), margins denticulate distally, 2–3.5 times as long as wide, base suddenly narrowed into green stalk, 2–20 mm.. . . . . 5. *Lemna trisulca*
1. Fronds floating, margins entire, 1–3 times as long as wide.
   2. Fronds with 1 vein.

3. Vein mostly prominent, longer than extension of air spaces, or running through at
   least 3/4 distance between node and apex. . . . . 8. *Lemna valdiviana*
3. Vein sometimes indistinct, very rarely longer than extension of air spaces, not longer
   than 2/3 distance between node and apex. . . . . 9. *Lemna minuta*
2. Fronds with 3–5(–7) veins.
   4. Root sheath winged at base; root tip usually sharp pointed; roots to 3(–3.5) cm;
      fronds without reddish color or spots of anthocyanin, mostly with 1 very distinct
      papilla near apex on upper surface.
      5. Seeds with 35–70 indistinct ribs (to count ribs remove membranous pericarp),
         staying within fruit wall after ripening; root sheath wing 2–3 times as long as
         wide. . . . . .6. *Lemna perpusilla*
      5. Seeds with 8–26 distinct ribs, falling out of fruit wall after ripening; root sheath
         wing 1–2.5 times as long as wide. . . . . 7. *Lemna aequinoctialis*
   4. Root sheath not winged; root tip mostly rounded; roots often longer than 3 cm;
      fronds often with reddish tinge or spots of anthocyanin, with or without distinct
      papilla near apex on upper surface.
      6. Plants forming small, olive to brown, rootless turions, which sink to bottom. . .
         . . . . 4. *Lemna turionifera* (in part)
      6. Plants usually without distinct turions.
         7. Largest air spaces longer than 0.3 mm; if red-colored on lower surface, color-
            ing beginning from margin; ovary (1–)2–7-ovulate. . . . . 1. *Lemna gibba*
         7. Largest air spaces 0.3 mm or shorter; if pigmented on lower surface, red
            coloring beginning from attachment point of root; ovary 1-ovulate.
            8. Fronds not reddish on lower surface (or at least much less so than
               on upper); greatest distance between lateral veins near or proximal to
               middle. . . . . 3. *Lemna minor*
            8. Fronds often reddish on lower surface (more intensely so than on upper);
               greatest distance between lateral veins near or distal to middle.
               9. Fronds flat, with mostly distinct papillae on midline of upper surface;
                  seeds with 30–60 indistinct ribs. . . . . 4. *Lemna turionifera* (in part)
               9. Fronds often gibbous, with very distinct papillae above node and near
                  apex on upper surface but not between node and apex; seeds with 10–
                  16 distinct ribs. . . . . 2. *Lemna obscura*

1. **Lemna gibba** Linnaeus, Sp. Pl. 2: 970. 1753
   F W

Roots to 15 cm, tip mostly rounded; sheath not winged. **Stipes** white, thin, often decaying. **Fronds** floating, 1 or 2–5 or more, coherent in groups, obovate, often gibbous, 1–8 mm, 1–1.5 times as long as wide, margins entire; veins (3–)4–5(–7) (all originating from node), greatest distance between lateral veins near or distal to middle; papillae often indistinct; lower surface sometimes red colored, coloring beginning from margin, upper surface occasionally with distinct red spots beginning from margins near apex; largest air spaces longer than 0.3 mm; distinct turions absent. **Flowers:** ovaries (1–)2–7-ovulate, utricular scale with narrow opening at apex. **Fruits** 0.6–1 mm, laterally winged. **Seeds** with 8–16 dis-

tinct ribs, falling out of fruit wall after ripening. $2n = 40, 42, 44, 50$.

Flowering (rather frequent) spring–fall. Eutrophic, quiet waters in temperate regions with mild winters; 0–1900 m; Ariz., Calif., Ill., Nebr., Nev., N.Mex., Tex., Wyo.; n Mexico; South America; Eurasia; Africa; Atlantic Islands.

2. **Lemna obscura** (Austin) Daubs, Illinois Biol. Monogr. 34: 20. 1965   F W

Lemna minor Linnaeus var. *obscura* Austin in A. Gray, Manual ed. 5, 479. 1867

Roots to 15 cm, tip mostly rounded; sheath not winged. **Stipes** white, small, often decaying. **Fronds** floating, 1 or 2–5 or more, coherent in groups, obovate, flat or gibbous, 1–3.5 mm,

1–1.5 times as long as wide, margins entire; veins 3, greatest distance between lateral veins near middle; very distinct papillae near apex, some smaller indistinct ones on midline of upper surface; lower surface very often red colored (more intensely than on upper), coloring beginning at attachment point of root, upper surface sometimes with red spots; air spaces shorter than 0.3 mm; distinct turions absent. **Flowers:** ovaries 1-ovulate, utricular scale with narrow opening at apex. **Fruits** 0.5–0.7 mm, not winged. **Seeds** with 10–16 distinct ribs, staying within fruit wall after ripening. $2n = 40, 42, 50$.

Flowering (occasional) spring–fall. Mesotrophic to eutrophic, quiet waters, in temperate to subtropical regions with mild winters; 0–800 m; Ala., Ark., Del., D.C., Fla., Ga., Ill., Ind., Iowa, Kans., Ky., La., Md., Minn., Miss., Mo., Nebr., N.J., N.Y., N.C., Ohio, Okla., Pa., S.C., S.Dak., Tenn., Tex., Va., Wis.; c Mexico; South America (Colombia, Ecuador); Hawaii.

3. **Lemna minor** Linnaeus, Sp. Pl. 2: 970. 1753 · Lenticule mineure F W

Roots to 15 cm, tip mostly rounded; sheath not winged. **Stipes** white, small, often decaying. **Fronds** floating, 1 or 2–5 or more, coherent in groups, ovate, scarcely gibbous, flat, 1–8 mm, 1.3–2 times as long as wide, margins entire; veins 3(–5) (if more than 3, outer ones branching from inner ones), greatest distance between lateral veins near or proximal to middle; papillae not always distinct (one near apex usually larger); lower surface very seldom slightly reddish (much less than on upper), coloring beginning from attachment point of root, upper surface occasionally diffusely reddish; air spaces 0.3 mm or shorter; distinct turions absent. **Flowers:** ovaries 1-ovulate, utricular scale with narrow opening at apex. **Fruits** 0.8–1 mm, laterally winged toward apex. **Seeds** with 8–15 distinct ribs, staying within fruit wall after ripening. $2n = 40, 42, 50, 63, 126$.

Flowering (rare) late spring–early fall. Mesotrophic to eutrophic, quiet waters, in suboceanic, cool-temperate regions with relatively mild winters; 0–2000 m; B.C., Ont., Que., Sask.; Ala., Ariz., Ark., Calif., Conn., Del., D.C., Fla., Idaho, Ill., Ind., Iowa, Kans., Ky., La., Maine, Md., Mass., Mich., Minn., Mo., Mont., Nebr., N.H., N.J., N.Mex., N.Y., N.C., N.Dak., Ohio, Okla., Oreg., Pa., R.I., S.Dak., Tenn., Utah, Vt., Va., Wash., W.Va., Wis.; w Eurasia; Africa; Atlantic Islands; introduced, Australia; introduced, New Zealand.

Indication of this species in Newfoundland (H. J.

Scoggan 1978–1979) probably refers to *Lemna turionifera*.

A specimen in the Gray Herbarium from St. Pierre and Miquelon may represent *Lemna minor* or *L. turionifera*; its determination is questionable.

4. **Lemna turionifera** Landolt, Aquatic. Bot. 1: 355, fig. 4g–h. 1975 · Lenticulare turionifère F W

Roots shorter than 15 cm, tip mostly rounded; sheath not winged. **Stipes** white, small, often decaying. **Fronds** floating, 1 or 2–few, coherent in groups, obovate, scarcely gibbous, flat, 1–4 mm, 1–1.5 times as long as wide, margins entire; veins 3, greatest distance between lateral veins near or distal to middle; papillae distinct on midline of upper surface (apical papilla scarcely larger than others); lower surface often red (more intensely so than on upper), coloring beginning at attachment point of root, upper surface (especially near apex) sometimes with red spots; air spaces to 0.3 mm; turions sometimes present, rootless, olive to brown, 0.8–1.6 mm diam., sinking to bottom. **Flowers:** ovaries 1-ovulate, utricular scale with narrow opening at apex. **Fruits** 0.5–0.6 mm, not winged. **Seeds** with 30–60 indistinct ribs, staying within fruit wall after ripening. $2n = 40, 42, 50, 80$.

Flowering (occasional) summer. Mesotrophic to eutrophic, quiet waters, in continental, temperate regions; 0–3700 m; Alta., B.C., Man., N.B., N.W.T., N.S., Ont., P.E.I., Que., Sask., Yukon; Ala., Alaska, Ariz., Calif., Colo., Conn., Idaho, Ill., Ind., Iowa, Kans., Mass., Mich., Minn., Mo., Mont., Nebr., Nev., N.Mex., N.Y., N.Dak., Ohio, Okla., Oreg., Pa., R.I., S.Dak., Tex., Utah, Vt., Wash., W.Va., Wis., Wyo.; Mexico (Baja California); Eurasia.

I know of no specimens of *Lemna turionifera* from St. Pierre and Miquelon or from New Jersey, but the species is to be expected there.

5. **Lemna trisulca** Linnaeus, Sp. Pl. 2: 970. 1753 · Lenticule trisulquée F W

Roots to 2.5 cm (sometimes not developed), tip pointed; sheath not winged. **Green stalks** 2–20 mm. **Fronds** submersed (except when flowering or fruiting), 3–50, coherent and very often forming branched chains, narrowly ovate, flat, thin, 3–15 mm (excluding stalk), 2–3.5 times as long as wide, base suddenly narrowed into green stalk, margins denticulate distally; veins (1 or) 3, lateral veins only in proximal part of frond; papillae absent; anthocyanin often present; air spaces shorter than 0.3 mm; turions absent. **Flowers:** ovaries 1-ovulate, utricular scale with narrow opening at apex. **Fruits** 0.6–0.9 mm, laterally winged toward apex. **Seeds** with 12–18 distinct ribs, staying within fruit wall after ripening. $2n = 40$, 42, 44, 60, 63, 80.

Flowering (rare) late spring–summer. Mesotrophic, quiet waters rich in calcium, in cool-temperate regions; 0–3000 m; Alta., B.C., Man., N.B., N.W.T., N.S., Nunavut, Ont., P.E.I., Que., Sask., Yukon; Alaska, Ariz., Ark., Calif., Colo., Conn., Idaho, Ill., Ind., Iowa, Kans., Maine, Md., Mass., Mich., Minn., Mo., Mont., Nebr., Nev., N.H., N.J., N.Mex., N.Y., N.Dak., Ohio, Oreg., Pa., R.I., S.Dak., Tenn., Utah, Vt., Va., Wash., W.Va., Wis., Wyo.; nearly worldwide, except arctic and antarctic regions and South America; in warm regions only in mountains.

The report of *Lemna trisulca* in Florida is dubious because the climate is too warm. The species may be temporarily introduced there by birds.

6. **Lemna perpusilla** Torrey, Fl. New York 2: 245. 1843
• Lenticule très petite E F W

Roots to 3.5 cm, tip usually sharp pointed; sheath narrowly winged at base (wing 2–3 times as long as wide). **Stipes** white, small, often decaying. **Fronds** floating, 1 or 2–few, coherent in groups, ovate-obovate, flat, 1–4 mm, 1–1.7 times as long as wide, margins entire; veins 3, greatest distance between lateral veins near or distal to middle; 1 distinct papilla near apex on upper surface, 2–3 very distinct papillae above node; anthocyanin absent, no reddish color; air spaces much shorter than 0.3 mm; distinct turions absent. **Flowers:** ovaries 1-ovulate, utricular scale open on 1 side. **Fruits** 0.7–1 mm, not winged. **Seeds** with 35–70 indistinct ribs, staying within fruit wall after ripening. $2n = 40, 42$.

Flowering (frequent) late spring–fall. Mesotrophic to eutrophic, quiet waters in temperate regions with relatively mild winters; 0–600 m; Que.; Ark., Del., D.C., Ill., Ind., Iowa, Kans., Ky., Maine, Md., Mass., Minn., Mo., Nebr., N.J., N.Y., N.C., Ohio, Okla., Pa., R.I., Tenn., Tex., Vt., Va., W.Va., Wis.

I know of no specimens of *Lemna perpusilla* from Connecticut.

7. **Lemna aequinoctialis** Welwitsch, Bol. Ann. Cons. Ultramar. (Portugal) 55: 578. 1859 F W

Lemna paucicostata Hegelmaier

Roots to 3 cm; tip usually sharp pointed; sheath winged at base (wing 1–2.5 times as long as wide). **Stipes** small, white, often decaying. **Fronds** floating, 1 or 2–few, coherent in groups, ovate-lanceolate, flat, 1–6 mm, 1–3 times as long as wide, margins entire; veins 3, greatest distance between lateral veins near or proximal to middle; 1 often very distinct papilla near apex on upper surface and 1 above node; anthocyanin absent, no reddish color; largest air spaces much shorter than 0.3 mm; distinct turions absent. **Flowers:** ovaries 1-ovulate, utricular scale open on 1 side. **Fruits** 0.5–0.8 mm, not winged. **Seeds** with 8–26 distinct ribs, falling out of fruit wall after ripening. $2n = 40, 42, 50, 60, 80, 84$.

Flowering (frequent) spring–fall. Mesotrophic to eutrophic, quiet waters in warm-temperate to tropical regions; 0–1300 mm; Ala., Ariz., Ark., Calif., Fla., Ga., Ill., Ind., Iowa, Kans., Ky., La., Miss., Mo., Nebr., N.Mex., N.C., Okla., S.C., Tenn., Tex., Va., Wis.; West Indies; Central America; South America; s Eurasia; Africa; Atlantic Islands; Indian Ocean Islands; Pacific Islands; Australia.

Some authors did not distinguish between *Lemna aequinoctialis* and *L. perpusilla* and used the latter name for both species.

8. **Lemna valdiviana** Philippi, Linnaea 33: 239. 1864 F W

Lemna cyclostasa (Elliott) C. H. Thompson; *L. torreyi* Austin

Roots to 1.5 cm, tip rounded to pointed; sheath not winged. **Stipes** white, small, often decaying. **Fronds** floating or (rarely) submersed, 1 or 2–few, coherent in groups, ovate to lanceolate, flat, thin, 1–5 mm, 1.3–3 times as long as wide, margins entire; veins 1, mostly prominent, longer than extension of air spaces, or running through at least 3/4 of distance between node and apex; with or without small papillae along midline of upper surface; anthocyanin absent; largest air spaces much shorter than 0.3 mm; turions absent. **Flowers:** ovaries 1-ovulate, utricular scale open on 1 side. **Fruits** 1–1.35 mm, not winged. **Seeds** with 15–29 distinct ribs. $2n = 40, 42$.

Flowering (very rare) spring–fall. Mesotrophic, quiet waters in temperate to tropical regions; 0–2000 m; Ala.,

Ariz., Ark., Calif., Conn., D.C., Fla., Ga., Ill., Ind., Kans., Ky., La., Md., Mass., Mich., Miss., Mo., Mont., Nebr., N.H., N.J., N.Mex., N.Y., N.C., Ohio, Okla., Pa., R.I., S.C., Tenn., Tex., Va., W.Va., Wyo.; Mexico; West Indies (Bermuda); Central America; South America.

I know of no specimens of *Lemna valdiviana* from Delaware, but the species is to be expected there.

9. **Lemna minuta** Kunth in A. von Humboldt et al., Nov. Gen. Sp. 1: 372. 1816 [F] [W]

Lemna minima Philippi; *L. minuscula* Herter

Roots to 1.5 cm, tip rounded to pointed; sheath not winged. **Stipes** white, small, often decaying. **Fronds** floating, 1 or 2–few, coherent in groups, obovate, flat to thickish (but not gibbous), 0.8–4 mm, 1–2 times as long as wide, margins entire; veins 1, sometimes indistinct, very rarely longer than extension of air spaces, not longer than 2/3 of distance between node and apex; with or without small papillae along midline; anthocyanin absent; largest air spaces much shorter than 0.3 mm; turions absent. **Flowers:** ovaries 1-ovulate, utricular scale open on 1 side. **Fruits** 0.6–1 mm, not winged. **Seeds** with 12–15 distinct ribs. $2n = 36, 40, 42$.

Flowering (very rare) late spring–early fall. Mesotrophic to eutrophic, quiet waters in temperate to subtropical regions with relatively mild winters; 0–2600 m; Ala., Ariz., Ark., Calif., Colo., Fla., Ga., Ill., Ind., Kans., Ky., La., Mich., Mo., Nebr., Nev., N.Mex., Ohio, Okla., Oreg., Tenn., Tex., Utah, Wash., W.Va., Wyo.; Mexico; West Indies; Central America; South America; introduced, Eurasia.

I know of no specimens of *Lemna minuta* from Mississippi, but the species is to be expected there.

3. **WOLFFIELLA** (Hegelmaier) Hegelmaier, Bot. Jahrb. Syst. 21: 303. 1895 • [For Johann Friedrich Wolff, 1778–1806, German physician, and Latin *-ella*, diminutive] [W]

Wolffia Horkel ex Schleiden subg. *Wolffiella* Hegelmaier, Lemnac., 131. 1868

Roots absent. **Fronds** submersed (except when flowering or fruiting), proximal part near surface, 1 or 2–20 or more, coherent, linear, ribbon-, sabre- or tongue-shaped, or ovate, flat, longer than 2 mm, margins entire; air spaces in tissue; pouch 1, terminal, at base from which daughter fronds (no flowers) originate, triangular, lower wall of pouch with tract of elongated cells forming connection between node and attachment to mother frond; veins 0; scale at base of frond absent; anthocyanins absent; pigment cells present (visible in dead fronds as brown dots); turions absent. **Flowers** 1(–2) per frond, originating in cavity at side of median line of upper frond surface, not surrounded by utricular scale; stamen 1, 2-locular. **Seeds** 1, nearly smooth. $x = 10, 20, 21$.

Species 10 (3 in the flora): North America, West Indies, Central America, South America, Africa.

1. Fronds (4–)6–15(–20) times as long as wide; angle of pouch 25°–50°. . . . . 3. *Wolffiella gladiata*
1. Fronds 1.5–8 times as long as wide; angle of pouch 45°–120°
   2. Angle of pouch 70°–120°; tract of elongated cells running between median line and edge of lower wall of pouch; area of air spaces within frond rarely longer than wide. . . . . 1. *Wolffiella lingulata*
   2. Angle of pouch 45°–90°; tract of elongated cells running along or close to edge of lower wall of pouch; area of air spaces within frond mostly longer than wide. . . . . 2. *Wolffiella oblonga*

1. Wolffiella lingulata (Hegelmaier) Hegelmaier, Bot. Jahrb. Syst. 21: 303. 1895 [F][W]

Wolffia lingulata Hegelmaier, Lemnac., 132. 1868

Fronds 1 or 2–4 coherent in small groups, wide tongue-shaped or ovate, 3–9 mm, 1.5–4 times as long as wide, rounded at tip; tract of elongated cells running between median line and edge of lower wall of pouch; area of air spaces within frond rarely longer than wide; angle of pouch 70°–120°; flowering fronds narrower than most vegetative ones. **Fruits** 0.41–0.44 mm. $2n = 40, 42$.

Flowering (very rare) throughout year. Mesotrophic, quiet waters in subtropical to tropical regions with mild winters; 0–200 m; Calif., Fla., La., Tex.; Mexico; West Indies; Central America; South America.

2. Wolffiella oblonga (Philippi) Hegelmaier, Bot. Jahrb. Syst. 21: 303. 1895 [F][W]

Lemna oblonga Philippi, Linnaea 29: 45. 1857

Fronds 1 or 2–8 coherent together in often starlike groups, narrowly tongue-shaped to ribbonlike, 1.2–7.5 mm, 3–8 times as long as wide, rounded at tip or pointed; tract of elongated cells running along or near edge of lower wall of pouch; area of air spaces within frond much longer than wide, often spreading throughout most of frond; angle of pouch 45°–90°; flowering fronds similar to vegetative ones. **Fruits** 0.3–0.4 mm. $2n = 40, 42$.

Flowering (very rare) throughout year. Mesotrophic, quiet waters in warm-temperate to subtropical regions with mild winters; 0–400 m; Calif., Fla., La., Miss., Tex.; Mexico; West Indies; Central America; South America.

3. Wolffiella gladiata (Hegelmaier) Hegelmaier, Bot. Jahrb. Syst. 21: 304. 1895 [F][W]

Wolffia gladiata Hegelmaier, Lemnac., 133. 1868; *W. gladiata* var. *floridana* Donnell Smith; *Wolffiella floridana* (Donnell Smith) C. H. Thompson

Fronds 1 or 2–20 or more, coherent in starlike groups, narrowly sabre-shaped, 3–9 mm, (4–)6–15(–20) times as long as wide, pointed at tip; tract of elongated cells running along edge of lower wall of pouch; area of air spaces much longer than wide, spreading throughout most of frond; angle of pouch 25°–50°; flowering fronds much wider at base than vegetative ones. **Fruits** 0.3–0.4 mm. $2n = 40, 42$.

Flowering (very rare) spring and fall. Mesotrophic, quiet waters in warm-temperate regions with relatively mild winters; 0–400 m; Ala., Ark., Del., D.C., Fla., Ga., Ill., Ind., Ky., La., Md., Mass., Miss., Mo., N.J., N.C., Ohio, Okla., Pa., S.C., Tenn., Tex., Va., Wash.; Mexico (Federal District).

The type collection of *Wolffiella gladiata*, from Mexico City, is taxonomically identical with plants from Florida called *W. gladiata* var. *floridana* by J. D. Smith. Some authors included narrow specimens of *W. oblonga* in *W. gladiata*, causing confusion.

4. WOLFFIA Horkel ex Schleiden, Beitr. Bot., 233. 1844 • Water-meal [for Johann Friedrich Wolff, 1778–1806, German physician] [W]

Roots absent. **Fronds** floating or submersed (only turions sink to bottom), 1 or 2, coherent, each frond globular, ovoid, or boat-shaped, 3-dimensional, smaller than 1.6 mm, margins entire; air spaces not in tissue; reproductive cavity terminal, at base from which daughter fronds (but no flowers) originate, conic, lower side of cavity with short tract of elongated cells along median line forming connection between node and attachment to mother frond; veins 0; scale at base of frond absent; anthocyanins absent; in some species pigment cells present (visible in dead fronds as brown dots); turions light green, globular, smaller than growing fronds. **Flowers** 1 per frond, originating in cavity on median line of upper frond surface, not surrounded by utricular scale; stamen 1, 2-locular. **Seeds** 1, nearly smooth. $x = 10, 20, 21, 22, 23$.

Species 11 (5 in the flora): nearly worldwide (except arctic and antarctic regions).

1. Fronds boat-shaped, 0.3–1 times as deep as wide, with pigment cells in vegetative tissue.
  2. Fronds 1–1.5 times as long as wide, rounded at apex, papilla usually prominent in center
    of upper surface. . . . . .1. *Wolffia brasiliensis*
  2. Fronds 1.3–2 times as long as wide, with point at apex bent upwards, papilla absent. . .2. *Wolffia borealis*
1. Fronds globular to ovoid, 1–1.5 times as deep as wide, without pigment cells in the
  vegetative tissue.
    3. Fronds with 10–100 stomates, slightly pointed at apex, the upper surface intensely green
    (not transparent). . . . . .3. *Wolffia arrhiza*
    3. Fronds with 1–10(–30) stomates, rounded or slightly pointed at apex, upper surface
    transparently green.
      4. Fronds 1.3–2 times as long as wide, 0.3–0.5 mm wide. . . . . .4. *Wolffia globosa*
      4. Fronds 1–1.3 times as long as wide, 0.4–1.2 mm wide. . . . . .5. *Wolffia columbiana*

1. Wolffia brasiliensis Weddell, Ann. Sci. Nat., Bot., sér. 3, 12: 170. 1849　F W

Wolffia papulifera C. H. Thompson; *W. punctata* Grisebach

Fronds boat-shaped, 0.5–1.6 mm, 1–1.5 times as long as wide, 0.3–0.7 times as deep as wide, rounded at apex, papilla usually prominent in center of upper surface (tent-shaped); upper surface intensely green, with 50–100 stomates; pigment cells present in vegetative tissue (visible in dead fronds as brown dots). 2*n* = 20, 40, 42, 50, 60, 80.

Flowering (rare) late spring–early fall. Mesotrophic to eutrophic, quiet waters in temperate to subtropical regions; 0–1000 m; Ala., Ark., Calif., Conn., Del., D.C., Fla., Ga., Ill., Ind., Kans., Ky., La., Md., Mass., Mich., Miss., Mo., N.J., N.Y., N.C., Ohio, Okla., Oreg., Pa., S.C., Tenn., Tex., Va., Wash., W.Va., Wis.; Mexico; West Indies; Central America; South America.

2. Wolffia borealis (Engelmann) Landolt, Ber. Geobot. Inst. E. T. H. Stiftung Rübel 44: 137. 1977 • Wolffie boréale　E F W

Wolffia brasiliensis Weddell var. *borealis* Engelmann in C. F. Hegelmaier, Lemnac., 127. 1868

Fronds boat-shaped, 0.7–1.5 mm, 1.3–2 times as long as wide, 0.7–1 times as deep as wide, with point at apex bent upward; papilla absent; upper surface intensely green, with 50–100 stomates; pigment cells present in vegetative tissue (visible in dead fronds as brown dots). 2*n* = 20, 22, 30, 40.

Flowering (very rare) summer–early fall. Mesotrophic to eutrophic, quiet waters in temperate regions; 0–1400 m; Alta., B.C., Ont., Que.; Calif., Colo., Idaho, Ill., Ind., Iowa, Kans., Ky., Mass., Mich., Minn., Mo.,

Mont., Nebr., N.H., N.Y., Ohio, Okla., Oreg., Pa., S.Dak., Tenn., Utah, Vt., Wash., Wis.

The name *Wolffia punctata* has been applied to this species in error.

3. Wolffia arrhiza (Linnaeus) Horkel ex Wimmer, Fl. Schles. ed. 3, 140. 1857　F I W

Lemna arrhiza Linnaeus, Mant. Pl. 2: 294. 1771

Fronds globular to ovoid, 0.5–1.5 mm, 1–1.3 times as long as wide, 1.2–1.5 times as deep as wide, slightly pointed at apex, papilla absent; upper surface intensely green (not transparent), with 10–100 stomates; pigment cells absent in vegetative tissue. 2*n* = 30, 40, 42, 44–46, 50, 60, 62 (Europe, Africa).

Flowering (very rare) late spring–summer. Mesotrophic to eutrophic, quiet waters in temperate to subtropical regions with relatively mild winters; 0–800 m; introduced; Calif.; South America (Brazil); w Eurasia; Africa.

Wolffia arrhiza was reported from Alberta because of the size of epidermal cells. The plants are within the variation shown by *W. columbiana*, however.

4. Wolffia globosa (Roxburgh) Hartog & Plas, Blumea 18: 367. 1970　F I W

Lemna globosa Roxburgh, Fl. Ind. ed. 1832, 3: 565. 1832

Fronds ovoid, 0.4–0.8 mm (× 0.3–0.5 mm), 1.3–2 times as long as wide, 1–1.5 times as deep as wide, rounded or slightly pointed at apex, papilla absent; adaxial surface transparently green, with 1–10(–30) stomates; pigment cells absent in vegetative tissue. 2*n* = 30, 60.

Flowering (very rare) late spring–fall. Mesotrophic to

eutrophic, quiet waters in warm-temperate to tropical regions with mild winters; 0–600 m; probably introduced; Calif., Fla.; South America (Colombia, Ecuador); e Asia; Pacific Islands (Hawaii).

5. **Wolffia columbiana** H. Karsten, Bot. Untersuch. (Berlin) 1: 103, figs. 2g, 3g. 1865 • Wolffie de Colombie F W

Fronds nearly globular, 0.5–1.4 mm (× 0.4–1.2 mm), 1–1.3 times as long as wide, 1–1.3 times as deep as wide, rounded at apex, papilla absent; upper surface transparently green, with 1–10(–30) stomates; pigment cells absent in vegetative tissue. $2n = 30$, 40, 42, 50, 70, 80.

Flowering (very rare) summer–early fall. Mesotrophic to eutrophic, quiet waters in temperate to subtropical regions; 0–1100 m; Alta., Man., Ont., Que., Sask.; Ala., Ark., Calif., Colo., Conn., Del., D.C., Fla., Ga., Ill., Ind., Iowa, Kans., Ky., La., Maine, Md., Mass., Mich., Minn., Miss., Mo., Mont., Nebr., N.H., N.J., N.Y., N.C., N.Dak., Ohio, Okla., Oreg., Pa., S.C., S.Dak., Tenn., Tex., Vt., Va., W.Va., Wis.; Mexico; West Indies (Curaçao); Central America; South America.

I know of no specimens of *Wolffia columbiana* from Rhode Island.

[Note added in press: Recent research shows that *Spirodela punctata* should be treated in a distinctive genus as *Landoltia punctata* (G. Meyer) D. H. Les & D. J. Crawford (Novon 9: 530–533. 1999).—Ed.]

205. XYRIDACEAE C. A. Agardh

• Yellow-eyed-grass Family

Robert Kral

Herbs, perennial or annual, mostly heliophytes of acidic wetlands, rarely aquatic. **Leaves** alternate, 2-ranked (occasionally many-ranked); base equitant, sheathing or open, sometimes with ligule or auricles at junction; blade mostly linear or filiform, flat to variously thickened and/or sulcate [lingulate], margins entire to papillate or scabrous; venation parallel. **Inflorescences** terminal; scape sheaths proximally tubular, distally open, bladed; scapes erect, linear to filiform, not bracteate [bracteate]; spikes single [cluster, rarely panicle of spikes], conelike; bracts chaffy. **Flowers** bisexual, radially [bilaterally] symmetric; sepals [2–]3, unequal [equal], outer 2 (lateral sepals) nearly opposite, connivent to variously connate, clasping mature capsule, inner sepal membranous (occasionally similar to others or absent), abcissing as flower opens; petals 3, equal [unequal], distinct [connate]; stamens 3; staminodes (0–)3, 2-brachiate, free [adnate to perianth or absent]; anthers 4-sporangiate, 2-locular at anthesis; pollen monosulcate or inaperturate; ovaries superior, 1[–3]-locular; ovules anatropous; styles 1, distally 3-branched [simple]; stigmas 3 [1, 3-lobed, capitate, or funnelform]. **Fruits** capsular, mostly loculicidal. **Seeds** [1–6–]15–90 or more, mostly under 2[–4] mm, mostly ridged or lined.

Genera 5, species ca. 300 (1 genus, 21 species in the flora): tropical and subtropical regions.

All North American *Xyris* increase by axillary buds, and thus the solitary habit is rare, often a result of poor or droughty habitat. A few can be termed annual or short-lived perennials (i.e., *X. brevifolia, X. flabelliformis, X. jupicai*). Of these only *S. jupicai* tends to die to the base at the end of a growing season.

In North American *Xyris*, leaf arrangement is distichous (2-ranked) and equitant (much as it is in *Iris*), the sheath margins converging at the junction of sheath and blade to form one edge of the leaf blade, with the midzone (or keel) of the sheath making up the other edge (a "unifacial" blade). A few species show some trends toward a spiral arrangement, typically displayed by a bulbous habit (*X. caroliniana, X. platylepis, X. scabrifolia, X. tennesseensis, X. torta*). In these species, the leaf sheath is abruptly dilated at the base, imparting a swollen, bulblike aspect to the plant base.

All North American *Xyris* have unifacial leaf blades; some neotropical species do not. In the latter the sheath margins do not completely converge when passing into the blade, and the blade may have a deep, narrow or broad groove or sulcus adaxially. A tendency toward this is shown in *X. baldwiniana* and *X. isoetifolia*.

The staminode in nearly all *Xyris* is much like a petal, being long-clawed at the base. The staminodial blade is typically 2-lobed, each lobe with a margin of long, moniliform hairs, each hair a chain of tiny cells connected much like a string of beads. Species-by-species taxonomic evaluations of those hairs, particularly SEM studies, have yet to be done. Of the species treated herein, only *X. baldwiniana* lacks bearded staminodes.

The keys and descriptions below are constructed from study of mature, preferably fruiting specimens. Bracts and sepals tend to grow substantially after flowering, and the measurements of lateral sepals are based on sepals that subtend ripe fruit.

North American *Xyris* have no nectaries, and the pollen reward is relatively scanty. A few bees and flies are occasional visitors. Usually pollination of these species appears to be by wind or by pollen-feeding insects.

SELECTED REFERENCE Kral, R. 1983d. The Xyridaceae in the southeastern United States. J. Arnold Arbor. 64: 421–429.

1. XYRIS Linnaeus, Sp. Pl. 1: 42. 1753; Gen. Pl. ed. 5, 25. 1754 • Yellow-eyed-grass [Greek *xyron*, razor, in reference to a plant with two-edged leaves]

Jupica Rafinesque; *Kotsjiletti* Adanson; *Ramotha* Rafinesque; *Schizmaxon* Steudel; *Xuris* Adanson; *Xyroides* Thouars

Herbs, perennial, occasionally annual, rosulate. **Stems** simple, erect, sometimes caudiciform, short to branching, elongate. **Leaves** alternate, 2-ranked, equitant; blade mostly linear to filiform, flattened to nearly terete, margins smooth to variously papillate or scabrous. **Inflorescences:** scapes variously elongate; spike bracts imbricate, the proximal sterile, decussate, distal ones fertile, in spiral [decussate], bearing medially, usually subapically, a distinct and mostly differently textured and colored "dorsal area" or this indistinct or absent. **Flowers** solitary in bract axils; sepals 3, distinct, unequal, lateral (outer) sepals distinct [connate], boat-shaped, chaffy, inner one membranous, capping corolla; keel papery, or scarious, variously winged, variously margined; petals 3, distinct [connate], equal, strongly clawed; blade ephemeral, spreading, yellow to white; staminodes 3, distinct, strongly clawed, apically yoked, bearded with moniliform hairs (rarely reduced, beardless); stamens 3, epipetalous; anthers 2–4-locular; ovaries thin-walled, placentation marginal to parietal [basal, free-central, axile]; styles elongate, tubular, 3-branched; stigmas 3, U-shaped or funnelform. **Capsules** 3-valved, usually thin, loculicidal. **Seeds** [1–]15–90 or more, mealy or translucent, ovoid to cylindric, 0.3–1(–4) mm, variously lined and cross-lined; endosperm starchy and proteinaceous; embryo basal-lateral, minute. $x = 9$.

Species more than 250 (21 in the flora): North America, West Indies, Central America, South America, Asia, Africa, South Pacific Islands.

Most species of *Xyris* occur in the Guiana Highlands, Amazonia, and Brazilia in South America, with other smaller centers of endemism in Africa and Australasia.

The key below is constructed to work with healthy plants that are in flower and/or fruit. Identifications based on sterile material are doubtful at best, then attempted only after long experience with such plants.

In the descriptions, scapes are described as linear if, 1 cm below the spike, they are 1 mm or more wide, or as filiform if, at that level, they are less than 1 mm wide. Petal color is assumed to be yellow, although *X. caroliniana* and *X. platylepis* have white-petaled forms. Leaf widths or thicknesses are measured at widest or thickest part of blade. Spike measurements are taken from mature spikes, those having at least some ripe fruit.

SELECTED REFERENCES   Kral, R. 1966b. The genus *Xyris* (Xyridaceae) in the southeastern United States and Canada. Sida 2: 177–260.   Kral, R. 1988. The genus *Xyris* (Xyridaceae) in Venezuela and contiguous northern South America. Ann. Missouri Bot. Gard. 75: 522–722.

1. Keel of lateral sepals usually firm, papillate, ciliate, ciliolate, or fimbriate, or in various combinations of these.
  2. Plants mostly under 30 cm, annual or perennial; scapes narrowly linear; scape sheaths mostly equaling or exceeding principal leaves; keels of lateral sepals papillate or ciliate; seeds 0.3–0.5 mm.
    3. Margins of fertile bracts scarious, lacerate, often squarrose with red inner band; spikes mostly as broad as long; lateral sepal keel reddish apically; plants annual. . . . 1. *Xyris brevifolia*
    3. Margins of fertile bracts entire or erose; spikes mostly longer than broad; lateral sepal keel not reddish apically, concolorous; plants annual or perennial.
      4. Base of leaf sheaths with distinct chestnut brown patch; leaf blades olive green; plants perennial. . . . . .2. *Xyris drummondii*
      4. Base of leaf sheaths without chestnut brown patch; leaf blades strongly maroon-tinged; plants annual. . . . . .3. *Xyris flabelliformis*
  2. Plants usually over 30 cm, perennial; scapes linear; scape sheaths exceeded by leaves; keels of lateral sepals ciliate, long-fimbriate, or lacerate; seeds 0.5 mm or longer.
    5. Lateral sepal tips exserted even on closed spikes; lateral sepal keels fimbriate or lacero-fimbriate.
      6. Lateral sepal keel apex lacero-fimbriate; fertile bracts with scarious, lacerate, often squarrose borders; seeds mostly 0.5–0.6 mm. . . . . .4. *Xyris elliottii* (in part)
      6. Lateral sepal keel apex distinctly long-fimbriate; fertile bracts entire to erose; seeds 0.7–1 mm.
        7. Plant base deeply set; leaf base chestnut brown; spikes ellipsoid to lanceoloid or cylindric, acute; leaves not in fans, narrowly linear, twisted, fleshy; sepal fimbriae red; scape ribs smooth or somewhat scabrous. . . . . .5. *Xyris caroliniana*
        7. Plant base shallowly set; leaf base dull, straw-colored, green, or pinkish; spikes ovoid to ellipsoid, blunt; leaves in fans, broadly linear, slightly twisted, flat; sepal fimbriae pale; scape ribs harsh. . . . . .6. *Xyris fimbriata*
    5. Lateral sepal tips included; lateral sepal keels ciliate.
      8. Leaves in fans; blade flat or slightly twisted.
        9. Seeds mealy; leaf base maroon; leaf blade deep green with strong maroon tints; petal blade obtriangular, 3–4 mm. . . . . .7. *Xyris stricta*
        9. Seeds translucent; leaf base tan, straw-colored, or pale pink, but not maroon; leaf blade usually pale or olive green; petal blade obovate, 10 mm. . . . . .8. *Xyris ambigua*
      8. Leaves erect or ascending; blade twisted.
        10. Plant base abruptly bulbous; leaf blade elongate-linear, 2–5 mm wide. . . . . .9. *Xyris torta*
        10. Plant base not abruptly bulbous; leaf blade narrowly linear to filiform, to 1 mm wide. . . . . .10. *Xyris isoetifolia*
1. Keel of lateral sepals scarious, lacerate to (rarely) nearly entire.
  11. Leaf sheaths or sheath base light green, straw-colored, or dull brown.
    12. Plants annual; leaves smooth; seeds translucent; scape ribs papillate. . . . . .20. *Xyris jupicai*
    12. Plants perennial; leaves papillate to rugose, minutely scabrous; seeds mealy; scape ribs minutely scabrous. . . . . .21. *Xyris serotina*
  11. Leaf sheaths or sheath base with red, pink, or purple tints, or glossy brown or red-brown.
    13. Leaf sheaths firm, glossy red-brown to tan.

14. Staminodes beardless; seeds fusiform to cylindric, (0.7–)0.8–1 mm; leaf blade rarely to 1 mm wide. . . . . 11. *Xyris baldwiniana*
14. Staminodes bearded; seeds ellipsoid, 0.5–0.6 mm; leaf blade rarely less than 1 mm wide. . . . . 4. *Xyris elliottii* (in part)
13. Leaf sheaths soft, pink, purple, or red.
  15. Tips of lateral sepals exserted.
    16. Apex of lateral sepals firm, red; seeds narrowly ellipsoid, 0.7–0.9 mm; leaf sheaths papillate. . . . . 12. *Xyris montana*
    16. Apex of lateral sepals thin, not red; seeds ellipsoid, 0.4–0.8 mm; leaf sheaths smooth.
      17. Leaf blade 1–2(–3) mm wide; spikes 10–16 mm; fertile bracts 4–6 mm; seeds 0.4–0.6 mm. . . . . 13. *Xyris longisepala*
      17. Leaf blade 5–15 mm wide; spikes 10–20(–25) mm; fertile bracts 5–8 mm; seeds (0.6–)0.7(–0.8) mm. . . . . 14. *Xyris smalliana*
  15. Tips of lateral sepals included.
    18. Seeds opaque or mealy.
      19. Base of mature plant bulbous. . . . . 15. *Xyris tennesseensis*
      19. Base of mature plant not bulbous.
        20. Leaf blade smooth, at least 10 mm wide; scape distally with 1–2 smooth ribs. . . . . 16a. *Xyris laxifolia* var. *iridifolia*
        20. Leaf blade papillate or somewhat scabrous, under 7 mm wide; scape distally with 3 or more somewhat scabrous ribs. . . . . 17. *Xyris difformis* (in part)
    18. Seeds translucent or mealy.
      21. Base of plant not bulbous; leaves in fans; plants flowering morning. . . . . 17. *Xyris difformis* (in part)
      21. Base of plant bulbous or nearly bulbous; principal leaves erect or ascending; plants flowering midday to afternoon.
        22. Seeds 0.5–0.6 mm. . . . . 18. *Xyris platylepis*
        22. Seeds 0.6–1 mm. . . . . 19. *Xyris scabrifolia*

1. **Xyris brevifolia** Michaux, Fl. Bor.-Amer. 1: 23. 1803

F

Xyris intermedia Malme

Herbs, annual, cespitose, rarely solitary, (4–)10–30(–60) cm. **Stems** compact. **Leaves** in fans, 2–10(–15) cm; sheath base greenish to pink; blade green or red-tinged, filiform to linear, 1–4 mm wide, smooth, margins smooth to papillate. **Inflorescences:** scape sheaths mostly exceeding leaves, blade prominent (in filiform-leaved extremes, overtopped by leaves); scapes wiry, nearly terete, 0.5–1 mm wide, low-ribbed apically; spikes oblate to globose or ovoid, mostly as broad as long, 5–7(–10) mm, apex blunt; fertile bracts 3–6 mm, margins often squarrose, lacerate, scarious, with red inner band, apex broadly rounded. **Flowers:** lateral sepals included, slightly curved, 3–5 mm, keel firm, nearly entire to papillate or ciliolate, apex reddish; petals unfolding in morning, blade obovate, 2.5–3 mm; staminodes bearded. **Seeds** amber, broadly ellipsoid, 0.5 mm, finely lined. $2n = 18$.

Flowering spring–summer(–winter in south). Acid, sandy, moist savanna and cleared areas; 0–200 m; Ala., Fla., Ga., N.C., S.C.; West Indies; South America (Brazil).

Plants in south Florida and Cuba often have longer and narrower leaf blades than is typical, and G. O. K. Malme (1925) named Cuban material *Xyris intermedia* on the basis of longer, narrower leaf blades and entire sepal keels. In Florida, however, all of these characters intergrade. The red sepal tips, almost a homology with the bract dorsal area, are a constant (yet neglected) character. This species is often weedy.

SELECTED REFERENCE Malme, G. O. K. 1925. Xyridaceen der Insel Cuba. Ark. Bot. 19(19): 1–6.

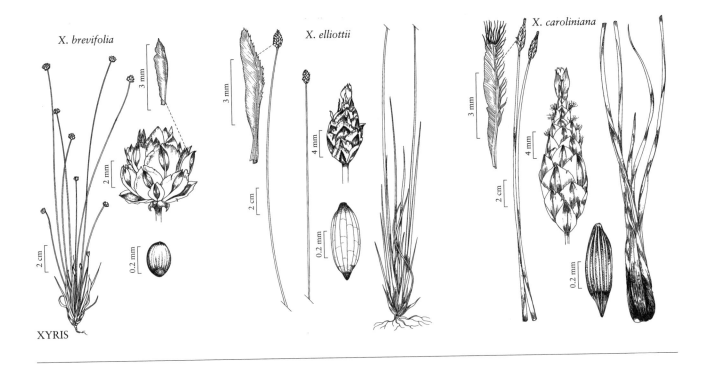

X. *brevifolia*

X. *elliottii*

X. *caroliniana*

XYRIS

2. Xyris drummondii Malme, Ark. Bot. 25A(12): 14. 1933 E

Herbs, perennial, cespitose, rarely solitary, 4–25 cm. Stems compact. Leaves in fans, 3–8(–10) cm; sheath base with distinct chestnut brown patch; blade olive green, narrowly linear, flat, 1.5–5 mm wide, smooth, margins smooth. Inflorescences: scape sheaths equaling principal leaves; scapes narrowly linear, 0.5–1 mm wide, distally 2-ribbed; spikes ovoid, mostly longer than broad, 3–8 mm, apex acute; fertile bracts 4–5 mm, margins entire, apex keeled. Flowers: lateral sepals included, strongly curved, 2.5–3.5 mm, keel concolorous, firm, ciliate; petals unfolding in morning, blade obovate, 3 mm; staminodes bearded. Seeds translucent, broadly ellipsoid, 0.3 mm, finely ridged longitudinally, with fainter cross-ridges. $2n = 18$.

Flowering summer–fall. Acid, sandy-peaty pine flatwoods, seeps, bogs, coastal plain; of conservation concern; 0–200 m; Ala., Fla., Ga., La., Miss., Tex.

3. Xyris flabelliformis Chapman, Fl. South. U.S., 499. 1860 E

Herbs, annual, cespitose, rarely solitary, 7–30 cm. Stems short. Leaves in fans, (1–)2–3(–10) cm; sheath base without chestnut brown patch; blade strongly maroon-tinged, lance-linear, 1–3(–4) mm wide, margins papillate or finely tuberculate-scabrous. Inflorescences: scape sheaths much exceeding leaves; scapes filiform, nearly terete, 0.5 mm wide, many ribbed; spikes mostly ovoid, mostly longer than broad, 4–8(–10) mm, apex acute; fertile bracts 3–5 mm, margins entire or erose, apex keeled. Flowers: lateral sepals included, strongly curved, 2–3 mm, keel concolorous, firm, ciliate; petals unfolding in morning, blade obovate, 2.5–3 mm; staminodes bearded. Seeds translucent, broadly ellipsoid to round, 0.3 mm, finely lined longitudinally. $2n = 18$.

Flowering spring–early summer (all year south). Acid, sandy, peaty flatwoods, clearings, disturbed moist sands, coastal plain; 0–200 m; Ala., Fla., Ga., Miss., N.C., S.C.

4. Xyris elliottii Chapman, Fl. South. U.S., 500. 1860

Xyris elliottii var. *stenotera* Malme

Herbs, perennial, densely cespitose, 40–60(–70) cm. Stems compact. Leaves in narrow fans to erect, 10–30(–40) cm; sheath base tan to brown, firm; blade mostly green or tinged with maroon, narrowly linear, flattened, plane or slightly twisted, 1–2(–2.5) mm wide, smooth, margins pale, narrow, incrassate, smooth or papillate. Inflorescences: scape sheaths exceeded by leaves; scapes linear, nearly terete, 0.7–1 mm wide, apically 2-ribbed, ribs smooth or papillate; spikes mostly ovoid or ellipsoid, 6–15 mm, apex acute; fertile bracts 5–6 mm, margins pale, strongly scarious, lacerate, often squarrose, submarginally often reddish, apex low-keeled. Flowers: lateral sepals included or slightly exsert, slightly curved, (5.5–)6–7 mm, keel concolorous, firm, finely lacerate or apically lacero-fimbriate, not papillate or ciliate; petals unfolding in morning, blade obovate, 5 mm; staminodes bearded. Seeds translucent, ellipsoid, 0.5–0.6 mm, prominently longitudinally lined. 2n = 18.

Flowering late spring–summer (all year south). Acid, sandy flatwoods, sandy shores, swales in pinelands, bog edges, coastal plain; Ala., Fla., Ga., S.C.; West Indies; Central America.

Xyris elliottii, with its densely cespitose habit, its glossy brown or red-brown, chaffy leaf sheath bases, and narrow leaves, is a part of a complex including *Xyris baldwiniana* and *X. isoetifolia*. Usually it is readily distinguished by its taller habit, thicker scapes, and larger spikes, but particularly by its strongly contrasting pale, incrassate leaf blade borders. In peninsular Florida, however, this leaf border is not consistently present, particularly in the narrower-bladed populations (in these, leaf blades may be less than 1 mm wide). Such plants can be distinguished from *X. baldwiniana* by the staminodial brush, absent in *X. baldwiniana*, and from *X. isoetifolia* by the different spike shape, the ragged (rather than entire) bracts, and by a different seed sculpture. Hybrids between *X. elliottii* and *X. brevifolia* occur in southern Florida.

5. Xyris caroliniana Walter, Fl. Carol., 69. 1788

Xyris arenicola Small 1903, not Miquel 1844; *X. conocephala* Wright; *X. flexuosa* Muhlenberg ex Elliott; *X. flexuosa* var. *pallescens* (Mohr) Barnhart; *X. pallescens* Small; *X. torta* Kunth 1843, not Smith 1818; *X. torta* var. *pallescens* C. Mohr

Herbs, perennial, usually cespitose, 30–80(–100) cm, base deeply set. Stems compact. Leaves erect or ascending, 20–50 cm; sheath base chestnut brown; blade green, narrowly linear, twisted, 2–5 mm wide, compressed but fleshy, margins minutely tuberculate. Inflorescences: scape sheaths much exceeded by leaves; scapes linear, wiry, flexuous, nearly terete, 1.2–1.5 mm wide, distally 1-ribbed, rib smooth or somewhat scabrous; spikes ellipsoid to lanceoloid or cylindric, 10–30 mm, apex acute; fertile bracts 5–10(–13) mm, margins entire or erose, apex rounded. Flowers: lateral sepals conspicuously exserted, tan to reddish brown, nearly straight, 13–15 mm (including long hairs), keel firm, apex long-fimbriate, fimbriae red; petals unfolding in afternoon, blade obovate, 8–10 mm; staminodes bearded. Seeds translucent, cylindro-fusiform, 0.8–1 mm, with flat, longitudinal ribs. 2n = 18.

Flowering summer–fall. Moist sands of pine savannas, bog edges, upper shores, flatwoods, sandy coastal dune swales, coastal plain; 0–300 m; Ala., Fla., Ga., La., Miss., N.J., N.C., S.C., Tex., Va.; West Indies (Cuba).

In Florida and the lower Gulf Coastal Plain west to Mississippi are paler-leaved, white-petaled examples that have been treated as *Xyris flexuosa* var. *pallescens* (C. Mohr) Barnhart. These indeed are always in uniform populations but differ in no other significant way from typical plants. Because some other *Xyris* have white-flowered morphs (particularly *X. platylepis*), these are regarded as too close to be distinguished as varieties. Populational studies are in order, however.

6. Xyris fimbriata Elliott, Sketch Bot. S. Carolina 1: 52. 1816

Herbs, perennial, usually cespitose, 60–150 cm, base shallowly set. Stems compact. Leaves in fans, 40–70 cm; base dull, straw-colored, green, or pinkish; blade lustrous green, broadly linear, flat or slightly twisted, 5–25 mm, smooth, margins papillate. Inflorescences: scape sheaths much exceeded by leaves; scapes linear, 1–1.5 mm wide, distally scabrous, 2-ribbed, ribs harsh; spikes ovoid to ellipsoid, 12–25 mm, apex blunt; fertile bracts 5–8 mm, margins nearly entire, apex rounded. **Flowers:** lateral sepals conspicuously exserted, yellow-brown, nearly straight, (5.5–)6–8(–9) mm, keel firm, long-fimbriate with pale hairs; petals unfolding in late morning, blade obovate, 5–8 mm; staminodes bearded. **Seeds** translucent, cylindro-fusiform, 0.8–1 mm, with broad, flat, pale longitudinal ribs. $2n = 18$.

Flowering summer–fall (all year south). Peats and mucks of pond shallows, sluggish shallow streams, bogs, seeps, low pine savanna; 0–300 m; Ala., Del., Fla., Ga., La., Md., Miss., N.J., N.C., Pa., S.C., Tenn., Tex., Va.

7. **Xyris stricta** Chapman, Fl. South. U.S., 500. 1860
[E]

Herbs, perennial, cespitose, 20–90(–100) cm. **Stems** compact. **Leaves** in narrow fans, 10–60 cm; sheaths smooth to papillate, base maroon; blade deep green with red or strong maroon tints, linear-tapering, flat, (2–)4–10 mm wide, smooth or papillate, margins smooth or minutely scabrous. **Inflorescences:** scape sheaths much exceeded by leaves; scapes linear, distally oval to 2-edged, 1.5–3 mm wide, 2–several-ribbed, ribs papillate or minutely scabrous; spikes ellipsoid to ovoid or cylindric, (10–)15–30 mm, apex blunt to acute; fertile bracts 5–7 mm, margins entire, apex slightly keeled, convex. **Flowers:** lateral sepals included, reddish brown, curved, 5–7 mm, keel firm, ciliate; petals unfolding in morning, blade obtriangular, 3–4 mm; staminodes bearded. **Seeds** mealy, cylindro-fusiform, 0.6–0.8 mm, lined longitudinally with papillae, less distinctly cross-lined. $2n = 18$.

Varieties 2 (2 in the flora): North America.

Xyris stricta forms a complex with *X. ambigua* and possibly arose as an independently breeding hybrid of *X. ambigua* and *X. laxifolia* within whose ranges it grows. I have labored since 1966 to determine its taxonomic status and at first believed it to be an extreme of *X. ambigua* with farinous seeds and smaller petals; then later I determined it to be a distinct species (*X. obscura* Kral, nom. nud.). Since then I have continued to make field observations and collections of the plant and have arrived at a different conclusion, allowing a more accurate and consistent taxonomy. On the one hand, the plant recently described as *X. louisianica* ranges to drier habitat than is usual for *X. stricta* and there commonly mingles with *X. ambigua*. But in character the plant overlaps most strongly with *X. stricta*, certainly in habit, pigmentation, leaf shape, bract and sepal charac-

ter, corolla, and seed. The morphometrics are an indication that these plants have evolved in the Gulf Coastal Plain from *X. stricta* and are backcrossing with it.

1. Plants 50–90(–104) cm high; leaves 20–60 cm; blade 3–8 mm wide, edges smooth to ciliolate or papillate; scapes distally sharply 2-ribbed, these making edges, but with no or few additional low ribs; ribs smooth, scabrous, or ciliolate; spikes lance-cylindric, or cylindric, 2–3 cm; fertile bracts 6–7 mm; sepals averaging 5–6.5 mm. . . . . .7a. *Xyris stricta* var. *stricta*
1. Plants mostly 40–80 cm high; leaves 20–40 cm; blade (2–)2.5–3(–4) mm wide, edges scabrous or scabro-ciliolate; scapes distally strongly 2-ribbed, these making edges, but with several additional ribs between; all ribs minutely scabrous or papillate; spikes narrowly ovoid, lanceoloid, or ellipsoid, under 2 cm; fertile bracts 5–6.5 mm; sepals averaging 5–6 mm. . . . . .7b. *Xyris stricta* var. *obscura*

7a. **Xyris stricta** Chapman var. **stricta**   [E]

Herbs, 50–90(–104) cm. **Leaves** 20–60 cm; blade mostly 3–8 mm wide, margins smooth to ciliolate or papillate. **Inflorescences:** scapes often compressed distally, nearly 2-edged, (1.7–)2–3(–4) mm wide, ribs 2, strong, sharp, making edges, edges smooth to somewhat scabrous or ciliolate, rarely with few additional intervening, low, smooth ribs; spikes lance-cylindric or cylindric, (1.5–)2–3(–3.5) cm; fertile bracts nearly orbiculate, (5.5–)6–7(–7.5) mm, margins erose, narrow, scarious. **Flowers:** lateral sepals 5–6.5 mm. **Seeds** (5.5–)6–8 mm.

Flowering summer–fall. Wet sands, peats, and peat-muck of shallows of acidic pineland ponds, bayheads, ditches, and seeps; 0–200 m; Ala., Fla., Ga., La., Miss., S.C., Tex.

Xyris stricta var. *stricta* is the taller, wider leaved, smoother, broader scaped, larger bracted of the two varieties and prefers a wetter habitat.

7b. Xyris stricta Chapman var. **obscura** Kral, Novon 9: 218, fig. 8. 1999 E

Xyris louisianica E. L. Bridges & Orzell

Herbs, 40–80(–85) cm. **Leaves** (15–)20–40(–44) cm; blade mostly (2–)2.5–3(–5) mm wide, margins papillate to ciliolate. **Inflorescences:** scapes terete, ribs 2, strong, distal, with additional narrower, but still strong ribs; all ribs scabrous or scabro-ciliolate; spikes ellipsoid to ovoid or lanceoloid, 1–2(–2.7) cm; fertile bracts nearly orbiculate, 5–6.5(–7) mm, margins entire. **Flowers:** lateral sepals (4.5–)5–6(–7) mm. **Seeds** 6–7(–8) mm.

Flowering summer–fall. Moist sands and sandy peats of pine savanna, bogs, clearings, ditchbanks, coastal plain; 0–200 m; Ala., Fla., Ga., La., Miss., Tex.

In Louisiana, Alabama, northwest Florida, and Georgia are what appear to be intergrades with *Xyris stricta* var. *stricta*.

8. Xyris ambigua Beyrich ex Kunth, Enum. Pl. 4: 13. 1843 F

Herbs, perennial, cespitose or solitary, 30–100 cm. **Leaves** in broad or narrow fans, (5–)10–40(–50) cm; sheath base tan, straw-colored, or pale pink; blade pale or olive green, linear-triangular, (2–)3–7(–10) mm wide, smooth or slightly papillate, margins smooth to minutely scabrous or papillate. **Inflorescences:** scape sheaths much exceeded by leaves; scapes linear, rarely flexuous, terete or nearly terete, 1–1.5 mm wide, distally usually 2-ribbed, sometimes with additional ribs; spikes ovoid to ellipsoid, lanceoloid, or cylindric, 10–20(–30) mm; fertile bracts 5–8 mm, margins entire, apex rounded. **Flowers:** lateral sepals included, dark brown, strongly curved, 4–7 mm, keel firm, ciliate, petals unfolding in morning, blade obovate, 10 mm; staminodes bearded. **Seeds** translucent, ovoid to ellipsoid, 0.5–0.6 mm, finely multiribbed and cross ribbed. $2n = 18$.

Flowering late spring–fall (all year south). Acid, sandy, moist pine or oak savanna, pine flatwoods, pond shores, ditches, bogs; 0–300 m; Ala., Ark., Fla., Ga., La., Miss., N.C., S.C., Tenn., Tex., Va.; Mexico; West Indies (Cuba); Central America (Belize, Honduras, Nicaragua).

Xyris ambigua is one of the more widespread and weedy of xyrids, frequently invading disturbed moist, sandy areas. It is also one of the most variable in habit and apparently forms intermediates with *X. stricta* and *X. torta*.

9. Xyris torta Smith in A. Rees, Cycl. 39: Xyris no. 11. 1819 E F

Kotsjelottia flexuosa Nieuwland; *Xyris bulbosa* Kunth

Herbs, perennial, cespitose or solitary, 15–80(–100) cm, base abruptly bulbous. **Stems** compact. **Leaves** ascending to erect, 20–50 cm; sheath base often reddened or pink; blade green, elongate-linear, twisted, 2–5 mm wide, smooth to papillate, with strongly raised veins, margins smooth or papillate. **Inflorescences:** scape sheaths exceeded by leaves; scapes linear, flexuous, 1–1.5(–2) mm wide, distally 5–6-ribbed, ribs smooth or papillate; spikes globose to ovoid, ellipsoid, lanceoloid, or cylindric, 8–25 mm, apex acute or blunt; fertile bracts 5–7 mm, margins entire except for red fimbriolation at rounded apex. **Flowers:** lateral sepals included, strongly curved, 4.5–5.5 mm; keel firm, ciliate except for red-fimbriolate tip; petals unfolding in morning, blade obovate, 4 mm; staminodes bearded. **Seeds** translucent, ellipsoid, 0.5 mm, strongly ridged longitudinally with finer cross-lines. $2n = 18$.

Flowering late spring–fall. Sphagnous bogs, streambanks, pond shores, wet sandy swales, moist disturbed sites, various physiographic provinces; 100–1200 m; Ala., Ark., Conn., Del., Ga., Ill., Ind., Iowa, Ky., La., Md., Mass., Mich., Minn., Miss., Mo., N.H., N.J., N.Y., N.C., Ohio, Okla., Pa., R.I., S.C., Tenn., Tex., Va., W.Va., Wis.

Although I have seen no records from Maine, *Xyris torta* is to be expected there.

Xyris torta is the widest-ranging of all North American xyrids and the most expressive ecologically; thus it is not surprising that it varies so much morphologically. It is often confused in older literature with *X. caroliniana*. *Xyris torta* is the type species for the genus.

10. Xyris isoetifolia Kral, Sida 2: 227, plate [p. 252], fig 2. 1966 C E F

Herbs, perennial, densely cespitose, (15–)20–30(–40) cm, base not abruptly bulbous. **Stems** compact. **Leaves** erect or ascending, 4–15 cm; sheaths glossy brown or red-brown, chaffy; blade green, filiform or narrowly linear, twisted, to 1 mm wide, smooth. **Inflorescences:** scape

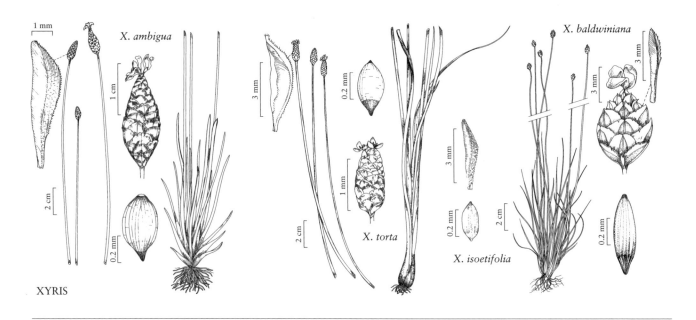

XYRIS

sheaths exceeded by leaves; scapes linear, nearly terete, 0.5(–0.7) mm wide, smooth, not ribbed; spikes ellipsoid to obovoid, 5–7(–10) mm; fertile bracts ca. 4.5 mm, margins nearly entire, apex rounded. **Flowers:** lateral sepals included, reddish brown, linear-curvate, 4 mm, keel firm, ciliate; petals unfolding in morning, blade obovate, 4 mm; staminodes bearded. **Seeds** translucent, ellipsoid, 0.5 mm, distinctly longitudinally ribbed, with fainter transverse lines. $2n = 18$.

Flowering summer–fall. Sphagnous bogs, low pine savanna, shores of dolines, coastal plain; 0–100 m; Ala., Fla.

Xyris isoetifolia, locally abundant only in northwest Florida, is most often mistaken for *X. baldwiniana* but is distinguishable by its bearded staminodes and more distinctly ribbed, shorter seeds.

11. Xyris baldwiniana Schultes in J. A. Schultes and J. H. Schultes, Mant. 1: 351. 1822 [F]

Xyris baldwiniana var. *tenuifolia* (Chapman) Malme; *X. juncea* Baldwin ex Elliott 1816, not R. Brown 1810; *X. setacea* Chapman

Herbs, perennial, densely cespitose, 15–40(–50) cm. **Leaves** erect or ascending, 10–30 cm; sheaths glossy light brown or red-brown, firm; blade green, linear to filiform, often angularly terete, or sulcate, rarely to 1 mm wide. **Inflorescences:** scape sheaths exceeded by leaves; scapes linear, straight or flexuous, terete, 1 mm wide, rarely 1-ribbed; spikes ovoid to ellipsoid, 4–7 mm, apex acute; fertile bracts 4–5 mm, margins entire

or erose, apex rounded. **Flowers:** lateral sepals included, reddish brown, slightly curved, less than 5 mm, keel scarious, lacerate from middle to tip; petals unfolding in morning, blade obovate, to 5 mm; staminodes beardless. **Seeds** translucent, fusiform to cylindric, (0.7–)0.8–1 mm, finely lined longitudinally. $2n = 18$.

Flowering late spring–fall. Moist to wet sands, sandy peats of bogs, pine savanna, ditches and low cleared areas, coastal plain; 0–200 m; Ala., Ark., Fla., Ga., La., Miss., N.C., S.C., Tex.; Mexico (Chiapas); Central America (Belize, Honduras, Nicaragua).

The beardless staminodes and the long, translucent seeds distinguish *Xyris baldwiniana*. Its leaf blades vary from terete to flat, and in eastern Texas and North Carolina the flat-leaved ones have been mistaken for *X. elliottii*. This same problem exists in Floridian narrow-leaved *X. elliottii*, which bears a strong resemblance to *X. baldwiniana* but has bearded staminodes and larger spikes.

12. Xyris montana Ries, Bull. Torrey Bot. Club 19: 38. 1892 • Xyris de montagne [E]

Xyris flexuosa Muhlenberg. var. *pusilla* A. Gray, Manual ed. 5, 548. 1867; *X. papillosa* Fassett

Herbs, perennial, cespitose, 5–30 cm. **Stems** compact. **Leaves** in narrow fans, 4–15 cm; sheaths reddish, soft, papillate; blade deep green, narrowly linear, 0.8–2(–3) mm wide, smooth, margins smooth to papillate. **Inflorescences:** scape sheaths exceeded by leaves; scapes linear, wiry, terete, (0.25–)0.5–

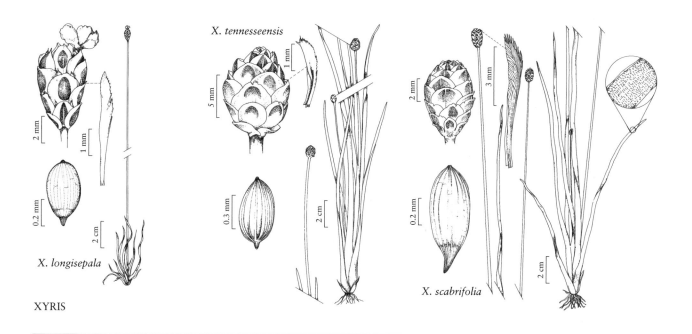

X. tennesseensis

X. longisepala

XYRIS

X. scabrifolia

0.8(–1) mm wide, distally with 2–4 ribs, ribs papillate; spikes broadly to narrowly ellipsoid or ovoid, 4–8 mm; fertile bracts 3–4(–4.5) mm, margins erose or minutely fimbriolate, sometimes with narrow reddish border, apex very slightly to slightly keeled. **Flowers:** lateral sepals slightly exserted, straight, 4.2–4.7 mm, keel scarious, entire or apically lacerate, apex red, narrow, firm; petals unfolding in morning, blade obovate, 3–4 mm; staminodes bearded. **Seeds** translucent, narrowly ellipsoid, (0.6–)0.7–0.9(–1) mm, finely lined.

Flowering summer–fall. Sphagnous bogs, poor fens, acid seeps, shores of glacial lakes, streams, muskegs, or floating bog mats; 0–500 m; N.B., Nfld. and Labr. (Nfld.), N.S., Ont., Que.; Conn., Maine, Mass., Mich., Minn., N.H., N.J., N.Y., Pa., R.I., Vt., Wis.

Most or all known populations of this species fall within the boundaries of Wisconsin glaciation. The long stems (a trait not known for other North American species) are a response to the burial of the clump bases in deep sphagnum.

13. Xyris longisepala Kral, Sida 2: 245. 1966 ⊂ E F

Herbs, perennial, cespitose, 30–90 cm. **Stems** compact. **Leaves** ascending in narrow fans, 6–30 cm; sheaths pinkish, soft, smooth; blade pale green to pale red-brown, narrowly linear, 1–2 (–3) mm wide, smooth, margins smooth. **Inflorescences:** scape sheaths exceeded by leaves; scapes linear, slightly twisted, nearly terete, 0.7–1 mm,

smooth, distally striate or 1-ribbed; spikes ellipsoid to narrowly obovoid, 10–16 mm; fertile bracts 4–6 mm, margins entire, aging to erose, apex rounded. **Flowers:** lateral sepals exserted, linear-curvate, 0.5–1 mm longer than subtending bract, keel scarious, lacero-ciliate, apex not red, thin, lacero-fimbriate; petals unfolding around noon or in afternoon, blade obovate, 3.5 mm; staminodes bearded. **Seeds** translucent, ellipsoid, 0.4–0.6 mm, distinctly longitudinally ribbed with few, indistinct cross ribs.

Flowering summer–fall. Moist to wet sandy borders of dolines, sandy swales in longleaf pine hills; of conservation concern; 0–200 m; Ala., Fla.

Taxonomically, *Xyris longisepala* is nearest to *X. smalliana*, a frequent associate around the dolines. *Xyris smalliana* is taller, with broader leaves, broader scapes, longer spikes, and longer seeds. The flowering time of day overlaps, but *X. longisepala* blooms first.

14. Xyris smalliana Nash, Bull. Torrey Bot. Club 22: 159. 1895

Xyris caroliniana Walter var. *olneyi* A. W. Wood; *X. congdonii* Small; *X. smalliana* Nash var. *congdonii* (Small) Malme; *X. smalliana* var. *olneyi* (A. W. Wood) Gleason ex Malme

Herbs, perennial, cespitose, 50–150 cm. **Stems** compact. **Leaves** ascending in narrow fans, (19–) 30–50(–60) cm; sheaths pinkish or pale red, soft, smooth; blade pale green proximally, distally deep lustrous green, linear, flattened, plane, 5–15 mm wide, margins smooth. **Inflorescences:** scape sheaths exceeded by leaves; scapes linear, straight, distally slightly compressed, 1–2 mm wide, smooth, 2-ribbed; spikes ovoid to ellipsoid or cylindric, 10–20(–25) mm, apex usually blunt; fertile bracts 5–8 mm, margins entire, apex rounded. **Flowers:** lateral sepals slightly to conspicuously exserted, slightly curved, 6–8 mm, keel scarious, apex not red, broad, thin, lacerate; petals unfolding in afternoon, blade obovate, 5–6 mm; staminodes bearded. **Seeds** translucent, narrowly ellipsoid to ovoid, (0.6–)0.7(–0.8 mm), irregularly ridged and cross-ridged. $2n = 18$.

Flowering summer–fall. Sandy or peaty shallows and shores of ponds and sluggish acidic streams; 0–100 m; Ala., Conn., Del., Fla., Ga., La., Maine, Md., Mass., Miss., N.H., N.J., N.Y., N.C., R.I., S.C., Tex.; Mexico (Tabasco); West Indies (Cuba); Central America (Belize, Honduras, Nicaragua).

Although no specimens were seen for Virginia, *Xyris smalliana* is to be expected there.

15. Xyris tennesseensis Kral, Rhodora 80: 444, figs. a–e. 1978 [C][E][F]

Herbs, perennial, cespitose, 30–70 cm, base bulbous in maturity. **Stems** compact. **Leaves** ascending, 15–45 cm; sheaths soft, base pink or red; blade deep green, linear, flattened, somewhat twisted, 5–10 mm wide, smooth, margins somewhat scabrous. **Inflorescences:** scape sheaths exceeded by leaves; scapes linear, terete, distally somewhat compressed, 1 mm wide, 2–5-ribbed, ribs papillate; spikes ellipsoid to broadly ovoid, 10–15 mm; fertile bracts (5.5–)6–6.5(–7) mm wide, margins entire, apex rounded. **Flowers:** lateral sepals included, slightly curved, (4.5–)5 mm, keel scarious, lacerate; petals unfolding midday, blade obovate, 4.5 mm; staminodes bearded. **Seeds** opaque, ellipsoid, 0.5-0.6 mm, finely and irregularly ridged. $2n = 18$.

Flowering summer–fall. Fens and seeps over calcareous rock, streambanks in calcareous districts; of conservation concern; 0–250 m; Ala., Ga., Tenn.

Xyris tennesseensis is unusual in habitat because its sites are usually basic and its associates are fen or basic soil plants.

16. Xyris laxifolia Martius, Flora 24(2,Beibl.): 53. 1841

Varieties 2 (1 in the flora): North America, Mexico, Central America, South America.

16a. Xyris laxifolia Martius var. iridifolia (Chapman) Kral, Novon 9: 214. 1999

Xyris iridifolia Chapman, Fl. South. U.S., 501. 1860

Herbs, perennial, cespitose, 60–90 cm, base not bulbous. **Stems** compact. **Leaves** ascending or in narrow fans, 40–70 cm × 10–25 mm; sheaths strongly tinged with red, soft; blade deep shining green, broadly linear, flattened, plane, 10–25 mm wide, smooth. **Inflorescences:** scape sheaths exceeded by leaves; scapes terete, 2-edged, compressed distally to 3–4 mm wide, 1–2-ribbed distally, ribs smooth; spikes broadly ovoid to lanceoloid or cylindric, 20–35 mm; fertile bracts 6–7 mm, margins entire, apex rounded. **Flowers:** lateral sepals included, curved, 6–7 mm, keel scarious, lacerate or lacero-ciliate; petals unfolding in morning, blade obtriangular, 3 mm; staminodes bearded. **Seeds** mealy, oblong-fusiform, 0.7–0.9(–1) mm. $2n = 18$.

Flowering summer–fall. Swamps, pond and lake shallows, shores, ditch banks, wet savanna, wet disturbed areas; 0–300 m; Ala., Ark., Fla., Ga., La., Miss., N.C., S.C., Tenn., Tex., Va.; Mexico.

Xyris laxifolia is the tallest of the weedier species of *Xyris*, often sharing habitat with *X. jupicai* (with which it has often been confused taxonomically). It differs in having taller, wider foliage and scapes, in the leaf sheaths being reddened (rather than greenish or straw-colored), in the larger, longer, thicker, darker spikes, and in the longer, mealy (rather than translucent) seeds. Scape ribs in *X. jupicai* are papillate distally; those of *X. iridifolia* are smooth.

The type variety, *Xyris laxifolia* Martius var. *laxifolia*, is a common wetland weed in large areas of Central America and South America, and it is distinguished primarily by its comparably large, acute (rather than blunt) spikes. The earliest name for the species is *X. macrocephala* Vahl (Enum. Pl. 2: 204. 1805), but the type material at Copenhagen is a mix of two other species.

17. Xyris difformis Chapman, Fl. South. U.S., 500. 1860

Herbs, perennial, cespitose, (10–)15–70(–90) cm, base not bulbous. **Stems** compact. **Leaves** in narrow to broad fans, 5–50 cm; sheaths soft, base reddish; blade deep green, linear to linear–sword-shaped, flat, not twisted, under 7 mm wide, papillate or smooth, margins smooth or papillate, somewhat scabrous. **Inflorescences:** scape sheaths exceeded by leaves; scapes linear to filiform, terete prox-imally, terete or 2-edged–winged distally, 0.5–3 mm wide, variably ribbed, distally with 3 or more somewhat scabrous ribs, ribs papillate to scabrous; spikes preva-lently ovoid, 5–20 mm, apex usually acute; fertile bracts (4–)5–7(–9) mm, margins entire, apex rounded. **Flow-ers:** lateral sepals included, slightly curved, 5–7 mm, keel brown, scarious, lacerate; petals unfolding in morning, blade obtriangular, 4 mm; staminodes bearded. **Seeds** translucent to mealy, ovoid to ellipsoid, 0.5 mm, finely lined longitudinally with small papillae.

Varieties 3 (3 in the flora): North America, Central America.

1. Leaves and scapes (except for edges and ribs) smooth; scapes somewhat to much widened dis-tally; 2 ribs comparably wider, making wings, smooth or papillate; seeds translucent. . . . . . .
. . . . 17a. *Xyris difformis* var. *difformis*
1. Leaves and scapes variously papillate or minutely scabrous; scapes not much widened distally; ribs all equally prominent, somewhat scabrous; seeds translucent or mealy.
    2. Seeds translucent; leaves spreading-ascending in broad fans, rarely as long as 10 cm, linear–sword-shaped; spikes rarely longer than 5 mm. . . . . 17b. *Xyris difformis* var. *curtissii*
    2. Seeds mealy; leaves ascending in broad to nar-row fans, rarely shorter than (10–)15 cm, lin-ear; spikes 6–10(–15) mm. . . . .
. . . . 17c. *Xyris difformis* var. *floridana*

17a. Xyris difformis Chapman var. **difformis**

Xyris difforme [E]

Xyris elata Chapman

Herbs, 20–70(–90) cm. **Leaves** in broad to narrow fans,10–30(–50); sheath blade deep green, lin-ear–sword-shaped, 0.2–1.5 cm wide, smooth, margins papillate. **Inflorescences:** scape sheaths ba-sally pinkish; scapes widened dis-tally, smooth or very finely pap-illate, 2–several-ribbed, 2 ribs wider, making wings, other ribs low or absent, all ribs smooth or papillate; spikes (in s part of range) mostly 10 mm or more. **Seeds** translucent. *n* = 9.

Flowering summer–fall. Sands and peats or gravels of edges of acidic swamps, pond shores, bogs, poor fens, seeps, open sphagnous areas; 0–500 m; N.S., Ont.; Ala., Ark., Conn., Del., D.C., Fla., Ga., Ind., Ky., La., Maine, Md., Mass., Mich., Miss., N.H., N.J., N.Y., N.C., Ohio, Okla., Pa., R.I., S.C., Tenn., Tex., Vt., Va., Wis.

The tallest, broadest-leaved, broadest-scaped, and largest-spiked plants are mostly in the southern Atlantic and Gulf coastal plains. Those of the Appalachians and

from Delaware north become progressively lower and smaller in dimension. In the maritime provinces, in New England, and along the Great Lakes, the plants become more similar to the often associated *Xyris montana* Ries, and some morphologic intermediates appear. *Xy-ris montana*, however, has generally smaller spikes with darker bracts and with less distinct dorsal areas; leaves generally of narrower outline, with papillate sheaths; and lateral sepals that sometimes have slightly exserted tips, usually with firmer keels, and especially with thicker and redder apices. Seeds of *X. montana* are distinctly longer and narrower in outline.

17b. Xyris difformis Chapman var. **curtissii** (Malme) Kral, Sida 2: 255. 1966

Xyris curtissii Malme, Ark. Bot. 13(8): 24. 1913; *Xyris bayardii* Fernald; *X. neglecta* Small 1894, not Alb. Nilsson 1892; *X. serotina* Chapman var. *curtissii* (Malme) Kral

Herbs, 15–20 cm. **Leaves** spreading-ascending in broad fans, mostly under 10 cm; sheaths reddish or purplish, rugulose-papillate; blade dull green, linear–sword-shaped, 1–4 mm wide, often papillate, margins somewhat scabrous. **Inflorescences:** scape sheath base rich, deep red-brown, lustrous; scapes linear-filiform, angulately terete, not much widened dis-tally, several-ribbed, ribs equally prominent, somewhat scabrous; spikes mostly 5 mm or less. **Seeds** translucent. 2*n* = 18.

Flowering summer–fall. Moist to wet sands or sandy peats of bog slicks, sphagnous areas, pine savanna, pond shores, and acid seeps; 0–200 m; Ala., Ark., Fla., Ga., La., Miss., N.C., S.C., Tex., Va.; Central America (Belize).

Xyris difformis var. *curtissii* remains distinct even when mixed with the other two varieties, a common event in the southern Atlantic and Gulf coastal plains.

17c. Xyris difformis Chapman var. **floridana** Kral, Sida 2: 256. 1966

Herbs, mostly 20–50(–60) cm. **Leaves** ascending in broad to narrow fans, (10–)15–25(–28) cm; sheaths reddish or purplish, rugulose-papillate; blade dull green, linear, 1.5–6 mm wide, papillate, margins somewhat scabrous. **Inflorescences:** scape sheaths proximally lustrous brown, shading distally to red tones; scapes linear, not widened distally, papillate or minutely scabrous, sharply multiribbed distally, ribs all equally prominent, somewhat scabrous; spikes 6–10(–15) mm. **Seeds** mealy.

Flowering summer–fall. Moist to wet sands or sandy peats of bogs, pine savanna, shores, and seeps; 0–200 m; Ala., Fla., Ga., La., Miss., N.C., S.C.; Central America (Belize, Honduras, Nicaragua).

18. **Xyris platylepis** Chapman, Fl. South. U.S., 501. 1860 E

Herbs, perennial, cespitose, occasionally solitary, 2–8(–10) cm, base bulbous. **Stems** compact. **Leaves** erect or ascending, 15–30 (–50) cm; sheaths pinkish to red, soft; blade green, linear, flattened, twisted, 5–10 mm wide, smooth, margins smooth. **Inflorescences:** scape sheaths exceeded by leaves; scapes linear, often flexuous, terete, to (1.5–)2(–3) mm wide, distally 2–4(–6) ribbed, ribs smooth or papillate; spikes ovoid to cylindric, 8–30 mm; fertile bracts 5–7 mm, margins entire, apex rounded. **Flowers:** lateral sepals included, light brown, slightly curved, 5–7 mm, keel scarious, lacerate; petals, unfolding at midday, blade broadly obovate, 5 mm; staminodes bearded. **Seeds** translucent, ellipsoid, 0.5–0.6 mm, longitudinally irregularly ribbed, with fainter cross lines. $2n = 18$.

Flowering summer–fall (all year south). Moist to wet acid, sandy seeps, bogs, low pine flatwoods, savannas, and ditch banks; 0–300 m; Ala., Fla., Ga., La., Miss., N.C., S.C., Tex., Va.

Xyris platylepis, which may be associated with other bulbous-based species such as *X. torta* and *X. caroliniana*, appears very similar to larger extremes of the former but differs in its plane (rather than prominently ribbed) leaf surfaces and its lacerate (rather than ciliate) sepal keels, and from the latter in its more shallowly set and pinkish or red (rather than chestnut brown) bases, as well as in its sepal keels that are lacerate rather than fimbriate.

19. **Xyris scabrifolia** R. M. Harper, Bull. Torrey Bot. Club 30: 325. 1903 E F

Xyris chapmanii E. L. Bridges & Orzell

Herbs, perennial, cespitose, occasionally solitary, 2–10(–11) dm, base bulbous to nearly bulbous. **Stems** compact. **Leaves** erect to ascending, 10–50 cm; sheaths pinkish, rugulose, papillate, or scabrous to nearly smooth, soft; blade dull green, linear, flattened, slightly to very

twisted, 2.5–10 mm wide, smooth to papillate or scabrous, margins smooth to scabrous. **Inflorescences:** scape sheaths exceeded by leaves; scapes linear, terete, to 2.5 mm wide, smooth to minutely scabrous, 2–4-ribbed distally, ribs papillate to minutely scabrous; spikes prevalently ovoid-ellipsoid, (7–)10–17(–20) mm; fertile bracts 6–8 mm, margins entire, apex rounded. **Flowers:** lateral sepals included, slightly curved, 6–8 mm, keel scarious, lacerate to lacero-fimbriate; petals unfolding midday or afternoon, blade broadly obovate to nearly orbiculate, 3–5 mm; staminodes bearded. **Seeds** translucent, ellipsoid-cylindric, 0.6–1 mm, longitudinally multiribbed with fainter cross ribs. $2n = 18$.

Flowering summer–fall. Sandy peats of deep pineland bogs and seeps, bog edges; 0–200 m; Ala., Fla., Ga., La., Miss., N.C., Tex.

Although I have seen no records from South Carolina, *Xyris scabrifolia* is to be expected there.

Several examples of what Bridges and Orzell have named *Xyris chapmanii*, together with a series of my own of this morph, show such intergradation that it is impossible to break the two out even as varieties.

20. **Xyris jupicai** Richard, Actes Soc. Hist. Nat. Paris 1: 106. 1792 F

Xyris anceps Persoon 1805, not Lamarck 1791; *X. arenicola* Miquel; *X. communis* Kunth; *X. gymnoptera* Grisebach; *X. jupicae* Michaux; *X. jupicai* var. *brachylepis* Malme

Herbs, annual, rarely biennial, cespitose or solitary, 10–100 cm. **Stems** compact. **Leaves** erect or ascending in narrow fans, 5–60 cm; sheaths straw-colored, light green, or brown, smooth; blade green, linear, flattened, 2–5(–15) mm wide, smooth, margins smooth or papillate. **Inflorescences:** scape sheaths exceeded by principal leaves; scapes linear, terete, distally oval, (0.5–)1–1.5(–2) mm wide, smooth, 1–2-ribbed, ribs papillate; spikes ovoid to ellipsoid or cylindro-lanceoloid, 7–15(–25) mm, apex acute; fertile bracts 5–7 mm, margins entire, apex rounded. **Flowers:** lateral sepals included, slightly curved, 5–7 mm, keel scarious, lacerate, thin; petals unfolding in morning, blade obtriangular, 3 mm; staminodes bearded. **Seeds** translucent, ellipsoid, 0.4–0.5 mm, faintly ribbed. $2n = 18$.

Flowering summer–fall (all year south). Moist sands, sandy peats of savannas, flatwoods, swales, shores, ditches, and roadsides, particularly in disturbed situations; 0–350 m; Ala., Ark., Del., Fla., Ga., La., Md., Miss., N.J., N.C., Okla., S.C., Tenn., Tex., Va.; Mexico; West Indies; Central America; South America.

The widest ranging of all New World xyrids and the

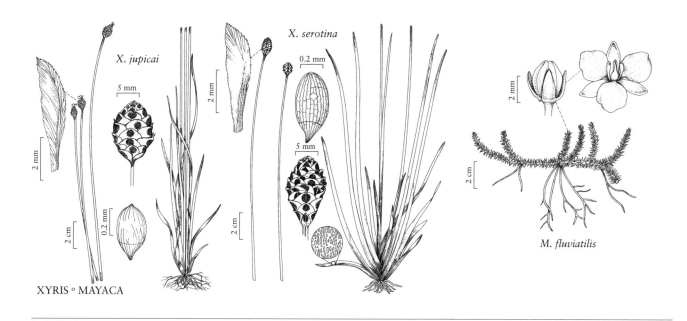

X. jupicai

X. serotina

XYRIS ° MAYACA

M. fluviatilis

most ample ecologically, *Xyris jupicai* is a frequent invader of disturbed or fallow open wetlands within its extensive range. In the southeastern United States it frequently shares habitat with two other species of its complex, namely *X. difformis* var. *difformis* and *X. laxifolia*. It differs from both in its lack of red pigmentation, from *X. difformis* by its more erect leaves and narrower, less prominently ribbed scapes, and from *X. laxifolia* by its narrower leaves and scapes, shorter, narrower, paler spikes, and translucent (rather than mealy), shorter seeds. Nonetheless, some difficult "calls" arise because all three flower at the same time, and occasional chance hybrids do form.

21. **Xyris serotina** Chapman, Fl. South. U.S., 500. 1860
[E] [F]

Herbs, perennial, densely cespitose, 20–60 cm, base thickened. **Stems** compact. **Leaves** in fans, 15–20 cm; sheaths straw-colored to dark brown, dull, papillate to rugose, minutely scabrous; blade dull green, flattened, plane, 3–12 mm wide, papillate to minutely scabrous, margins minutely scabrous. **Inflorescences:** scape sheaths exceeded slightly by leaves; scapes straight, linear, terete, distally slightly compressed or broadly oval, to 2 mm, rugulose, minutely scabrous, strongly 2-ribbed, ribs minutely scabrous; spikes prevalently ovoid to ellipsoid, 8–20 mm, apex acute; fertile bracts 5–7 mm, margins entire, apex shape rounded. **Flowers:** lateral sepals included, dark brown, slightly curved, 5–7 mm, keel scarious, lacerate; petals unfolding in morning, blade obtriangular, 3–3.5 mm; staminodes bearded. **Seeds** mealy, ellipsoid, 6 mm, lined longitudinally with papillae, less conspicuously cross-lined. $2n = 18$.

Flowering summer–fall (all year south). Sands and peats of wet pine savanna, cypress pond edges, bogs, ditches; 0–150 m; Ala., Fla., Ga., Miss., N.C., S.C.

Xyris serotina is perhaps the most distinctive species of xyrid over its range. Certainly, it is overall the most scabrous in foliage, and its leaves are unusually long in relation to its scapes.

206. MAYACACEAE Kunth

• Mayaca or Bog-moss Family

Robert B. Faden

Herbs, perennial. **Leaves** cauline, alternate; sheath absent; blade simple, linear to filiform, margins entire, 1-veined. **Inflorescences:** solitary flowers that appear lateral. **Flowers** bisexual, radially symmetric; sepals 3, sepaloid, distinct, equal; petals 3, petaloid, distinct, equal; stamens 3; staminodes absent; anthers dehiscing by apical pores or porelike slits; ovary superior, 3-carpellate, 1-locular; ovules 2-seriate; styles 1, simple; stigmas 1, simple or slightly 3-fid. **Fruits** loculicidal capsules. **Seeds:** embryotega present.

Genera 1, species ca. 4 (1 species in the flora): 3 in warm temperate to tropical America and 1 in West Africa.

1. MAYACA Aublet, Hist. Pl. Guiane 1: 42, plate 15. 1775

Herbs, aquatic or terrestrial. **Leaves** spirally arranged, many ranked; blade linear to filiform, apex commonly 2-fid. **Inflorescences:** solitary flowers, morphologically terminal, soon appearing lateral because of sympodial growth of shoots. **Flowers:** sepals subvalvate; petals imbricate, shortly clawed; nectar and nectaries absent; stamens antisepalous; filaments slender, glabrous; anthers basifixed, sometimes at end of tubular, apical appendage; styles terminal; stigmas terminal. **Fruits** 3-valved. **Seeds** ovoid to globose; seed coat reticulate.

Species ca. 4 (1 in the flora): 3 in warm temperate to tropical America and 1 in w Africa.

SELECTED REFERENCES Lourteig, A. 1952. Mayacaceae. Notul. Syst. (Paris) 14: 30–33. 1952. Lourteig, A. 1960. Distribution géographique des Mayacacées. Compt. Rend. Sommaire Séances Soc. Biogéogr. 36[323]: 31–35. Thieret, J. W. 1975. The Mayacaceae in the southeastern United States. J. Arnold Arbor. 56: 248–255.

1. Mayaca fluviatilis Aublet, Hist. Pl. Guiane 1: 42, plate 15. 1775 F

Mayaca aubletii Michaux; *M. caroliniana* Gandoger; *M. longipes* Gandoger

Herbs, submerged or terrestrial. **Stems** sometimes much branched and matted, decumbent to erect, to 60 cm, usually much less. **Leaves:** blade narrowly lanceolate to linear, 2–20(–30) × 0.5 mm, apex entire or 2-fid. **Flowers** to 1 cm wide; pedicels 2–12 mm, elongating in fruit to 20 mm; sepals persistent, ovate to lanceolate-elliptic, 2–4.5 × 0.7–1.5 mm; petals rose to maroon or lilac, sometimes whitish basally, broadly ovate, 3.5–5 × 3–4.5 mm; stamens 1.5–3 mm; filaments 1–2 mm; anthers 0.5–1 mm, dehiscing by means of apical, porelike slits; pistils 2–2.3 mm, stigmas 3-lobed to 3-fid. **Capsules** nearly globose to ellipsoid, often irregular because of abortion, to 4 × 3.4 mm. **Seeds** 2–25 per capsule, nearly globose, to 1.3 × 0.9 mm; seed coat scrobiculate or ridged and pitted.

Flowering spring–fall ([Mar–]May–Oct[–Nov]). Marshes, seepage areas, springy places, and in or at margins of ponds, pools, streams, and on coastal plain; Ala., Fla., Ga., La., Miss., N.C., S.C., Tex.; Mexico; Central America; South America.

Some floras have recognized two species in our area, *Mayaca fluviatilis*, with leaves mostly 4–20 mm, pedicels shorter than the leaves, and capsules oblong-ellipsoid, nearly twice as long as wide; and *M. aubletii*, with leaves mostly 3–5 mm, pedicels longer than the leaves, and capsules ovoid to nearly globose, nearly as broad as long. *Mayaca fluviatilis* appears to be the more aquatic form and *M. aubletii* the more terrestrial one of a single species, and it is quite possible that all or most of the morphologic variation is because of differences in ecology. Experimental work is needed to test the variability in this species as a function of habitat.

The literature report for Virginia could not be confirmed.

207. COMMELINACEAE R. Brown

• Spiderwort Family

Robert B. Faden

Herbs, perennial or annual. **Leaves** basal or cauline, alternate; sheaths closed; blade simple, often succulent, margins entire, venation parallel. **Inflorescences** terminal or terminal and axillary [sometimes all axillary], sometimes becoming leaf-opposed, cymose (cymes scorpioid), thyrsiform or variously reduced, sometimes umbel-like, sometimes enclosed in spathaceous bracts. **Flowers** bisexual or bisexual and staminate on same plants, rarely bisexual and pistillate on same plants [bisexual and unisexual (staminate and pistillate), all on same plants], bilaterally or radially symmetric; sepals 3, sepaloid [occasionally petaloid], distinct or occasionally connate, usually subequal; petals 3, deliquescent, petaloid, distinct or connate, equal or unequal; stamens 6, all fertile or some staminodial or absent (rarely all stamens absent); anthers with longitudinal [rarely poricidal] dehiscence; ovaries superior, 2–3-locular; ovules 1-seriate [2-seriate]; styles 1, simple, usually slender; stigmas 1, simple [rarely slightly 3-lobed], enlarged or not. **Fruits** loculicidal capsules [rarely indehiscent or berries]. **Seeds** 1–several [rarely many] per locule; hilum dotlike or linear; lidlike embryotega covering embryo.

Genera 40, species ca. 630 (6 genera, 51 species in the flora): pantropical and nearly pantemperate, primarily tropical.

The flowers lack nectar and are ephemeral, lasting only a few hours. Their structure is seldom preserved in dried specimens. In the absence of well-pressed flowers, mature buds can be readily dissected *in situ*, and the arrangement and degree of development of the androecium and gynoecium easily determined.

Some familiar genera, such as *Setcreasea*, *Zebrina*, *Rhoeo*, and *Cuthbertia*, have been reduced into synonymy under either *Tradescantia* or *Callisia* (D. R. Hunt 1975, 1986, 1986b). Further research is needed to corroborate this treatment, especially for the segregate genera of *Callisia*, such as *Cuthbertia*. The same generic delimitation has been followed by R. B. Faden (1998), R. B. Faden and D. R. Hunt (1991), and G. C. Tucker (1989).

SELECTED REFERENCES Faden, R. B. 1998. Commelinaceae. In: K. Kubitzki et al., eds. 1990+. The Families and Genera of Vascular Plants. 4+ vols. Berlin etc. Vol. 4, pp. 109–128. Faden, R. B. and D. R. Hunt. 1991. The classification of the Commelinaceae. Taxon 40: 19–31. Hunt, D. R. 1975. The reunion of *Setcreasea* and *Separotheca* with *Tradescantia*. American

Commelinaceae: I. Kew Bull. 30: 443–458. Hunt, D. R. 1986. *Campelia, Rhoeo* and *Zebrina* united with *Tradescantia*. American Commelinaceae: XIII. Kew Bull. 41: 401–405. Hunt, D. R. 1986b. Amplification of *Callisia* Loefl. American Commelinaceae: XV. Kew Bull. 41: 407–412. Tucker, G. C. 1989. The genera of Commelinaceae in the southeastern United States. J. Arnold Arbor. 70: 97–130.

1. Inflorescences composed of pairs of contracted, sessile, umbel-like cymes; flowers radially symmetric; stamens 6 (rarely fewer), all fertile.
  2. Cyme pairs enclosed in or subtended by pairs of large, conspicuous spathaceous or foliaceous bracts. . . . . .3. *Tradescantia*, p. 173
  2. Cyme pairs subtended by small, inconspicuous bracts. . . . . .4. *Callisia* (in part), p. 187
1. Inflorescences composed of 1–several elongate cymes, not umbel-like; flowers radially or bilaterally symmetric; stamens 6 or fewer, usually some staminodial (rarely all fertile).
  3. Inflorescences enclosed in or closely subtended by leafy bracts (spathes); flowers strongly bilaterally symmetric, usually blue (rarely lilac, lavender, peach, apricot, or white).
    4. Fertile stamens 6, all but proximalmost stamen with filaments densely bearded; capsules 3-valved; leaves glaucous. . . . . .1. *Tinantia*, p. 171
    4. Fertile stamens 3, filaments glabrous; capsules 2–3-valved; leaves not glaucous. . . .6. *Commelina*, p. 192
  3. Inflorescences not enclosed in or closely subtended by leafy bracts (although sometimes axillary); flowers radially or bilaterally symmetric, variously colored.
    5. Flowers sessile or subsessile; filaments glabrous or bearded; ovaries and capsules 2–3-locular. . . . . .4. *Callisia* (in part), p. 187
    5. Flowers distinctly pedicellate; some or all filaments bearded; ovaries and capsules 3-locular.
      6. Flowers white to purple or violet; fertile stamens 2–3; staminodes 3–4. . . . . . .5. *Murdannia*, p. 190
      6. Flowers white; fertile stamens 6; staminodes 0. .2. *Gibasis*, p. 172

1. TINANTIA Scheidweiler, Allg. Gartenzeitung 7: 365. 1839 • [For François Tinant, Luxemburger forester]

Commelinantia Tharp

Herbs, annual (or perennial). **Roots** thin. **Leaves** spirally arranged; blade occasionally sessile [usually petiolate], glaucous. **Inflorescences** terminal or terminal and axillary, 1[–several] elongate cyme[s], [occasionally axillary and perforating sheaths], subtended by spathaceous or foliaceous bracts; bracts leaflike; bracteoles persistent. **Flowers** bisexual [bisexual and staminate], bilaterally symmetric; pedicels well developed; sepals distinct, subequal; petals distinct, unequal, not clawed, proximal petal small [large], distal 2 blue or blue-violet [white to pink], equal, large; stamens 6, all fertile, polymorphic; proximal 3 stamens long, lateral filaments bearded, medial glabrous, anthers large; distal 3 stamens short, filaments densely bearded, anthers small; filaments connate basally; ovary 3-locular; ovules 2–several per locule, 1-seriate. **Capsules** 3-valved, 3-locular. **Seeds** 2–several per locule; hilum linear; embryotega lateral. $x = 13, 14, 16, 17$.

Species ca. 14 (1 in the flora): Texas; tropical America, especially Mexico to Nicaragua.

The reproductive features of B. C. Tharp's two species of *Commelinantia* (1922) clearly link them with *Tinantia* (O. Rohweder 1962). My research indicates that although some attributes, including pollen and chromosome number, can still be used to separate *Commelinantia* from *Tinantia*, these characters are not of sufficient import to merit separate generic status.

SELECTED REFERENCES Anderson, E. and K. Sax. 1936. A cytological monograph of the American species of *Tradescantia*. Bot. Gaz. 97: 433–476. Jetton Castro, K. 1978. The Biology of *Commelinantia anomala*. M.S. thesis. Angelo State University. Rohweder, O. 1962. Zur systematischen Stellung der Commelinaceen-Gattung *Commelinantia* Tharp. Ber. Deutsch. Bot. Ges.

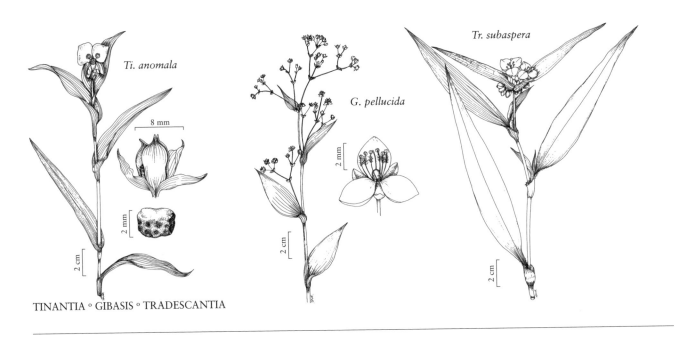

TINANTIA ∘ GIBASIS ∘ TRADESCANTIA

75: 51–56. Simpson, B. B., J. L. Neff, and G. Dieringer. 1986. Reproductive biology of *Tinantia anomala* (Commelinaceae). Bull. Torrey Bot. Club 113: 149–158. Tharp, B. C. 1922. *Commelinantia*, a new genus of the Commelinaceae. Bull. Torrey Bot. Club 49: 269–275.

1. **Tinantia anomala** (Torrey) C. B. Clarke in A. L. P. de Candolle and C. de Candolle, Monogr. Phan. 3: 287. 1881 • False dayflower, widow's-tears [F]

Tradescantia anomala Torrey in W. H. Emory, Rep. U.S. Mex. Bound. 2(1): 225. 1859; *Commelinantia anomala* (Torrey) Tharp

Herbs, annual, cespitose, to 80 cm. **Leaves** dimorphic, glaucous; basal leaves tapered into long petiole, linear-spatulate; distal cauline leaves sessile, broadly to narrowly lanceolate, 6– 20 cm, base commonly cordate-clasping, apex acute to acuminate. **Inflorescences** terminal, cymes solitary; bracteoles 3–5 mm. **Flowers** 2–2.5 cm wide; sepals boat-shaped, 9–12 mm, glabrous; proximal petal white, rhombic, 3–4 mm; distal petals blue to lavender, obovate, 15–18 mm; proximal lateral stamens with purple hairs; distal stamens bearded with yellow-tipped hairs; ovules 2 per locule. **Capsules** oblong, 6–8 mm. **Seeds** gray-brown. $2n = 26$.

Flowering spring–early summer. Limestone talus slopes, granitic slopes, edges of woods and ravines, prefers some shade; Tex.; Mexico (Durango)

Only one record of *Tinantia anomala* is known from Durango.

2. **GIBASIS** Rafinesque, Fl. Tellur 2: 16. 1837 • [Latin *gibbus*, swollen, and *basis*, base] [I]

Herbs, perennial or annual. **Roots** fibrous [tuberous]. **Leaves** 2-ranked [spirally arranged]; blade sessile. **Inflorescences** terminal and sometimes axillary, pairs or umbels of cymes, cymes pedunculate, axis sharply angled at junction with peduncle; spathaceous bract absent; bracteoles persistent. **Flowers** bisexual, radially symmetric; pedicels well developed; sepals distinct, subequal; petals distinct, white [to pink or blue], equal; stamens 6, all fertile, equal; filaments bearded; ovaries 3-locular, ovules 2 per locule, 1-seriate. **Capsules** 3-valved, 3-locular. **Seeds** 2 per locule; hilum elongate-punctiform to linear; embryotega abaxial. $x = 4, 5, 6$.

Species 11 (1 in the flora): introduced; neotropical, centered in Mexico.

SELECTED REFERENCES Hunt, D. R. 1986c. A revision of *Gibasis* Rafin. American Commelinaceae: XII. Kew Bull. 41: 107–129. Rohweder, O. 1956. Commelinaceae in die Farinosae in der Vegetation von El Salvador. Abh. Auslandsk., Reihe C, Naturwiss. 18: 98–178.

1. Gibasis pellucida (M. Martens & Galeoti) D. R. Hunt, Kew Bull. 38: 132. 1983 • Tahitian bridal-veil F

Tradescantia pellucida M. Martens & Galleoti, Bull. Acad. Roy. Sci. Bruxelles 2: 376. 1842; *Gibasis schiedeana* (Kunth) D. R. Hunt

Herbs, perennial, decumbent. **Leaves** 2-ranked, blade lanceolate-oblong, 4.5–7 cm × 1.4–2.3 cm. **Inflorescences** terminal and distally axillary, pairs of pedunculate cymes in compound inflorescences; peduncles of cymes 1.7–2.5 cm; bracteoles overlapping. **Flowers:** sepals keeled, 2.5–3 mm, glabrous; petals white, 4 mm; filaments distal to middle, bearded at base. **Capsules** ovoid, 2.5 mm. **Seeds** light brown, 1 mm, rugose. $2n = 10, 16$ (Mexico).

Fruiting fall. Weedy in citrus groves and waste places; introduced; Fla.; native, Mexico.

Gibasis pellucida has often been confused with *G. geniculata*, the only other commonly cultivated species; the leaves of *G. geniculata* are more pubescent, however, and its filaments are bearded only at the base. For about 10 years *G. pellucida* was known as *G. schiedeana*.

Aneilema linearis (Bentham) Woodson, a species of *Gibasis*, was reported as occurring in Mexico just south of the Big Bend region of Texas (L. H. Shinners 1962). That species is likely to be *G. venustula* (Kunth) D. R. Hunt (D. R. Hunt 1986c). No specimens have been seen from the United States.

3. TRADESCANTIA Linnaeus, Sp. Pl. 1: 288. 1753; Gen. Pl. ed. 5, 139. 1754

• Spiderwort, wandering-Jew, spider-lily, éphémères [for John Tradescant, gardener to Charles I of England]

Rhoeo Hance; *Setcreasea* Schumann & Sydow; *Zebrina* Schnizlein

Herbs, perennial. **Roots** thin or tuberous. **Leaves** spirally arranged or 2-ranked; blade sessile or rarely petiolate [petiolate]. **Inflorescences** terminal or terminal and axillary, pairs of cymes, cymes sessile, umbel-like, contracted, subtended by spathaceous bract; bract similar to leaves or differentiated, margins distinct; bracteoles persistent. **Flowers** bisexual, radially symmetric; pedicels very short or well developed; sepals distinct (basally connate in *T. zebrina*), subequal; petals distinct (rarely connate basally), white to pink, blue, or violet, equal, rarely clawed; stamens 6, all fertile, equal; filaments bearded or glabrous; ovaries 3-locular, ovules (1–)2 per locule, 1-seriate. **Capsules** 3-valved, 3-locular. **Seeds** 2 per locule (1 in *T. spathacea*); hilum oblong to linear; embryotega abaxial to lateral. $x = 6–8$, probably others.

Species ca. 70 (30 in the flora): neotemperate and neotropical.

The species described by E. Anderson and R. E. Woodson Jr. (1935) are narrowly defined and typological. Nevertheless, they are recognizable entities even if some of them may prove eventually unworthy of specific rank. Where specific problems have been recognized, they are noted in the discussions at the end of the species.

Tradescantia species hybridize freely when growing together (E. Anderson and R. E. Woodson Jr. 1935). My observations in the field and garden tend to confirm this. The definite or probable hybrids are listed after the species. The list is almost certainly incomplete. The questionable records are based on uncertain determinations. The record of a possible hybrid between *T. ohiensis* and *Callisia rosea* (as *Cuthbertia rosea*), cited by Anderson and Woodson, is omitted: the specimens appear to be merely gracile plants of *T. ohiensis*. Some native species are occasionally cultivated, although most garden plants seem to be hybrids of *T. virginiana* and other species (E. Anderson 1952). They are usually sold as *Tradescantia* ×*andersoniana* (an invalid name) followed by a cultivar epithet.

SELECTED REFERENCES Anderson, E. 1954. A field survey of chromosome numbers in the species of *Tradescantia* closely related to *Tradescantia virginiana*. Ann. Missouri Bot. Gard. 41: 305–327. Anderson, E. and K. Sax. 1936. A cytological monograph of the American species of *Tradescantia*. Bot. Gaz. 97: 433–476. Anderson, E. and R. E. Woodson Jr. 1935. The species of *Tradescantia* indigenous to the United States. Contr. Arnold Arbor. 9: 1–132. Hunt, D. R. 1980. Sections and series in *Tradescantia*. American Commelinaceae: IX. Kew Bull. 35: 437–442. MacRoberts, D. T. 1980. Notes on *Tradescantia* (Commelinaceae) V. *Tradescantia* of Louisiana. Bull. Mus. Life Sci. Louisiana State Univ. 4: 1–15. Sinclair, C. 1967. Studies on the Erect Tradescantias. Ph.D. dissertation. University of Missouri.

1. Flowers subsessile; petals clawed, claws connate at least basally; stamens epipetalous.
  2. Leaves 2-ranked, bases oblique, cuneate; blade usually variegated; sepals connate basally. . . . . .30. *Tradescantia zebrina*
  2. Leaves spirally arranged, bases symmetric, rounded to broadly cuneate; blade not variegated; sepals distinct.
    3. Leaves narrowly lanceolate, apex acuminate; stamen filaments glabrous. . . . .29. *Tradescantia leiandra*
    3. Leaves lanceolate-elliptic to oblong-elliptic or ovate, apex obtuse to abruptly acute-apiculate; stamen filaments glabrous or bearded.
      4. Leaves oblong-elliptic to lanceolate-elliptic, mostly 7–15 cm; peduncle (3.5–)4–13 cm; leaves usually purplish violet. . . . . .28. *Tradescantia pallida*
      4. Leaves oblong-elliptic to ovate, mostly 3–7 cm; peduncle 1–5(–6) cm; leaves green.
        5. Filaments and ovary glabrous; flowering Jun–Oct. . . . . .26. *Tradescantia brevifolia*
        5. Filaments bearded, ovary densely bearded; flowering Feb–May. . . . . . .27. *Tradescantia buckleyi*
1. Flowers distinctly pedicellate; petals neither clawed nor connate; stamens free.
  6. Sprawling to decumbent plants rooting at nodes; leaves lanceolate to lanceolate-elliptic or lanceolate-oblong to ovate-lanceolate or ovate-elliptic.
    7. Leaves lanceolate-elliptic to ovate-lanceolate, to 5 × 2 cm; cyme pairs usually 1–2 per shoot; bracts all or mostly foliaceous, occasionally reduced. . . . . .24. *Tradescantia fluminensis*
    7. Leaves lanceolate-oblong to ovate-elliptic, to 10 × 3.5 cm; cyme pairs 2–4 per shoot; bracts, especially those of axillary inflorescences, usually reduced. . . . . .25. *Tradescantia crassula*
  6. Erect or ascending plants, rarely rooting at nodes; leaves mostly linear-lanceolate to lanceolate-oblong.
    8. Inflorescences all or chiefly axillary.
      9. Inflorescences pedunculate in axils well proximal to shoot apex, enclosed in boat-shaped spathes; leaves glabrous; flowers white. . . . . .23. *Tradescantia spathacea*
      9. Inflorescences mostly sessile in axils of distal leaves; boat-shaped spathes absent; leaves usually arachnoid-villous; flowers blue to purple. . . . . .22. *Tradescantia crassifolia*
    8. Inflorescences terminal, commonly terminal and axillary.
      10. Distal leaf blades wider than opened, flattened sheaths.
        11. Pedicels 1–1.7 cm; proximal leaves petiolate; stems frequently flexuous; plants flowering mainly May–Sep. . . . . .1. *Tradescantia subaspera*
        11. Pedicels (1.5–)2–3.2 cm; proximal leaves narrowed directly into sheath; stems not flexuous; plants flowering mainly Feb–May.
          12. Sepals 9–16 mm, ± inflated, eglandular-pilose; flowers usually deep blue, purple, or rose-red. . . . . .4. *Tradescantia ernestiana*
          12. Sepals 6–12 mm, not inflated, glandular-pilose or mixed glandular- and eglandular-pubescent; flowers usually white or pale pink to pale lavender.
            13. Leaves not glaucous; capsules 8–10 mm. . . . . .2. *Tradescantia edwardsiana*
            13. Leaves ± glaucous; capsules 6–8 mm. . . . . .3. *Tradescantia ozarkana*
      10. Distal leaf blades equal to or narrower than opened, flattened sheaths.

14. Sepals glabrous or with eglandular hairs only (very rarely a few minute glandular hairs at base).
    15. Sepals glabrous (or with apical tuft of eglandular hairs or a few minute glandular hairs at base).
        16. Stems 5–18 cm; pedicels, sepal bases often with minute glandular hairs; petals 10 mm. . . . . 20. *Tradescantia wrightii*
        16. Stems 15–115 cm; pedicels glabrous; sepals glabrous or with apical tuft of eglandular hairs; petals usually 0.8–20 mm.
            17. Plants distinctly glaucous; leaves 5–45 cm, arcuate, forming acute angle with stem (also see *Tradescantia occidentalis* var. *scopulorum*). . . . . 5. *Tradescantia ohiensis*
            17. Plants not at all to slightly glaucous; leaves 4–11 cm, straight, forming nearly right angle with stem. . . . . 6. *Tradescantia paludosa*
    15. Sepals covered with eglandular hairs.
        18. Bracts saccate at base, blades reduced, densely, minutely velvety . . . . 7. *Tradescantia gigantea*
        18. Bracts not saccate at base, blades well developed, sparsely to densely pilose.
            19. Flowering stems 2–7 cm (elongating to 20 cm in fruit), pilose to villous; sepals purple or rose-colored (rarely pale green), not inflated; rocky prairies. . . . . 8. *Tradescantia tharpii*
            19. Flowering stems 5–50 cm, glabrous to pilose or hirsute; sepals various; habitat various but rarely rocky prairies.
                20. Roots (1.5–)2–4 mm thick; stems commonly glabrous proximal to inflorescence; sepals usually ± inflated; ne United States, Appalachia. . . . . 9. *Tradescantia virginiana*
                20. Roots 1–1.5(–2) mm thick; stems usually pilose to hirsute throughout; sepals not inflated; se United States. . . . . 10. *Tradescantia hirsutiflora* (in part)
14. Sepals pubescent with glandular and often eglandular hairs.
    21. Pedicels 0.8–1 cm, glandular-puberulent; sepals 4–6 mm; petals 9–12 mm; hilum much shorter than seed. . . . . 21. *Tradescantia pinetorum*
    21. Pedicels (0.8–)1–6 cm, glandular- or eglandular-pubescent; sepals (4–)6–16 mm; petals (6–)10–19 mm; hilum as long as seed.
        22. Sepals with mostly glandular pubescence.
            23. Internodes and leaves glabrous. . . . . 11. *Tradescantia occidentalis* (in part)
            23. Internodes pubescent (rarely glabrous in *Tradescantia roseolens*)
                24. Pedicels 1–2.8 cm; roots all thin and fibrous; South Carolina to Florida and Alabama. . . . . 12. *Tradescantia roseolens* (in part)
                24. Pedicels 2.5–4.5 cm; at least some roots thick and tuberous; Texas. . . . . 13. *Tradescantia pedicellata*
        22. Sepals with mixture of glandular, eglandular pubescence.
            25. Stems, leaves completely glabrous; plants glaucous; sepal hairs mainly glandular (also see *Tradescantia roseolens*). . . . . 11. *Tradescantia occidentalis* (in part)
            25. Stems, leaves usually sparsely to densely pubescent, if glabrous then sepal hairs mainly eglandular; plants usually not glaucous (somewhat glaucous in *T. roseolens*); sepal hairs various.
                26. Stems densely arachnoid-pubescent; roots thick, brownish tomentose.
                    27. Stems erect or ascending, unbranched or sparsely branched, 30–105 cm. . . . . 14. *Tradescantia reverchonii*

27. Stems spreading, diffusely branched, 10–30 cm. . . . . 15. *Tradescantia subacaulis*
26. Stems variously pubescent, not arachnoid-pubescent; roots various, not brownish tomentose.
  28. Plants diffuse, spreading; stems much branched. . . . . 16. *Tradescantia humilis*
  28. Plants erect or ascending; stems unbranched or sparsely branched.
    29. Sepal hairs mainly eglandular, glandular hairs few, inconspicuous. . . . . 10. *Tradescantia hirsutiflora* (in part)
    29. Sepal hairs mainly glandular or eglandular, glandular hairs numerous, conspicuous.
      30. Sepals puberulent, hairs all less than 1 mm; roots relatively thin, 0.5–1(–2) mm thick . . . . 12. *Tradescantia roseolens* (in part)
      30. Sepals pilose-puberulent, longer hairs 1.5–6 mm; roots relatively stout, 1–3 mm thick.
        31. Plants bright green; stems, leaves usually glabrous. . . . . 17. *Tradescantia bracteata*
        31. Plants dull green; stems, leaves usually pubescent (rarely glabrescent).
          32. Stems (2–)15–40 cm; pedicels 1.5–3.5 cm; leaves, bracts puberulent, usually sparsely to densely pilose, margins ± densely ciliolate. . . . . 18. *Tradescantia hirsuticaulis*
          32. Stems 2–10 cm; pedicels (2–)4–6 cm; leaves, bracts pilose, not puberulent, margins sparsely ciliate. . . . . 19. *Tradescantia longipes*

1. Tradescantia subaspera Ker Gawler, Bot. Mag. 39: plate 1597. 1814 E F

Tradescantia pilosa Lehmann

Herbs, erect or ascending, rarely rooting at nodes. **Stems** often flexuous, 30–100 cm; internodes pilose to glabrescent. **Leaves** spirally arranged, at least proximal ones distinctly petiolate; blade dark green, lanceolate-oblong to lanceolate-elliptic or lanceolate, 6–30 × (0.4–)1–6.5 cm (distal leaf blades wider than sheaths when sheaths opened, flattened), apex acuminate, glabrous to puberulent. **Inflorescences** terminal, usually also axillary at distal nodes, axillary inflorescences sessile or variously pedunculate; bracts foliaceous. **Flowers** distinctly pedicillate; pedicels 1–1.7 cm, pilose to glabrous; sepals 4–10 mm, puberulent with glandular, eglandular, or mixture of glandular, eglandular hairs, occasionally glabrous; petals distinct, light to dark blue, rarely white, broadly ovate, not clawed, 1–1.5 cm; stamens free; filaments bearded. **Capsules** 4–6 mm. **Seeds** 2–3 mm. $2n = 12, 24$.

Flowering spring–fall (May–Sep). Rich woods along streams and on slopes and bluffs, less commonly dry woods, roadsides, fields, or along railroads; Ala., Ark., D.C., Fla., Ga., Ill., Ind., Ky., La., Miss., Mo., N.C., Ohio., S.C., Tenn., Va., W.Va.

Two varieties were recognized by E. Anderson and R. E. Woodson Jr. (1935): *Tradescantia subaspera* var. *subaspera*, with the stems more or less conspicuously flexuous distally and the distal lateral inflorescences sessile (western extensions of Appalachian Plateau from western West Virginia, central Kentucky, and Tennessee to Illinois and Missouri); and *T. subaspera* var. *montana* (Britton) Anderson & Woodson, with the stems straight or only slightly flexuous distally and all the lateral inflorescences pedunculate (southern Appalachians from southwestern Virginia to northern Alabama and Georgia, also the coastal plain from northern Florida to Louisiana). Many specimens can only be determined by their locale, so I do not find the separation of the two varieties very meaningful. The distribution record for the District of Columbia is based on a specimen believed to be a garden escape; that from southern Florida on a specimen cited by C. Sinclair (1967).

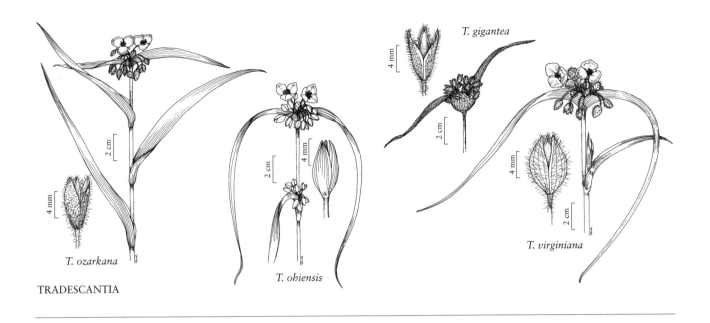

T. gigantea

T. ozarkana

TRADESCANTIA

T. ohiensis

T. virginiana

2. Tradescantia edwardsiana Tharp, Rhodora 34: 57, fig. 1. 1932 [E]

Herbs, erect or ascending, rarely rooting at nodes. **Stems** not flexuous, 25–70 cm, densely puberulent to glabrescent. **Leaves** gradually narrowed into sheath; blade light green, elliptic-lanceolate, 7–39 × 1.5–4.5 cm (distal leaf blades wider than sheaths when sheaths opened, flattened), minutely puberulent to glabrescent. **Inflorescences** terminal, usually also axillary from distal nodes; bracts foliaceous. **Flowers** distinctly pedicillate; pedicels 1.5–3 cm, densely puberulent; sepals green, 6–9 mm, glandular-puberulent or mixed glandular- and eglandular-puberulent; petals distinct, white to pale mauve, rarely bright pink, broadly ovate, not clawed, 1–1.2 cm; stamens free; filaments bearded. **Capsules** 8–10 mm. **Seeds** 3–4 mm. $2n = 12$.

Flowering late winter–spring (Feb–May). Rich woods, and along moist terraces and ravines; Tex.

3. Tradescantia ozarkana E. S. Anderson & Woodson, Contr. Arnold Arbor. 9: 56, plate 12, map 3. 1935 [E] [F]

Herbs, erect or ascending, rarely rooting at nodes. **Stems** not flexuous, 10–50 cm; internodes glabrous to pilose. **Leaves** spirally arranged, sessile; blade silvery or gray-green, lanceolate to linear-lanceolate or linear-oblong, 8–28 × 1–6 cm (distal leaf blades wider than sheaths when sheaths opened, flattened), base ± rounded to cuneate, apex acuminate, ± glaucous, usually glabrous. **Inflorescences** all or mostly terminal; bracts foliaceous. **Flowers** distinctly pedicillate; pedicels 2–3.2 cm, glandular-pilosulose; sepals 6–12 mm, sparsely to densely glandular-pilosulose; petals distinct, white or pale pink to pale lavender, broadly ovate, not clawed, 1.2–1.6 cm; stamens free. **Capsules** 6–8 mm. **Seeds** 3–4 mm. $2n = 12, 24$.

Flowering spring (Apr–May). Rich woods, mainly on rocky slopes and along cliffs, occasionally in bottomlands; Ark., Mo., Okla.

Tradescantia ozarkana is endemic to the Ozarks.

4. Tradescantia ernestiana E. S. Anderson & Woodson, Contr. Arnold Arbor. 9: 58, plate 8, map 4. 1935 [E]

Herbs, erect or ascending, rarely rooting at nodes. **Roots** (1–)1.5–5 mm thick, fleshy. **Stems** not flexuous, 5–40 cm; internodes usually glabrous. **Leaves** spirally arranged, sessile; blade dull green, linear-lanceolate to lanceolate-oblong, 9–27 × 1–4 cm (distal leaf blades wider than sheaths when sheaths opened, flattened), base cuneate to rounded, apex acuminate, not glaucous, glabrous or sparsely puberulent. **Inflorescences** terminal; bracts foliaceous. **Flowers** distinctly pedicillate; pedicels 2–3.2 cm, minutely pilose; sepals somewhat inflated, 9–16 mm, uniformly eglandular-pilose; petals distinct, deep blue, purple, or rose-red, broadly ovate, not clawed, 1.2–1.5 cm; stamens free; filaments bearded. **Capsules** 5–7 mm. **Seeds** 2–3 mm. $2n = 12$.

Flowering spring (Mar–May). Wooded hillsides, ledges and bluffs, occasionally along streams or in pastures; Ala., Ark., Ga., Mo., Okla., Tex.

Tradescantia ernestiana is sympatric with, and easily confused with, *T. virginiana* in northern Alabama and perhaps northern Georgia [reported from Georgia by C. Sinclair (1967, p. 87), but no specimens were cited and I have seen none]. At present, the two species can be separated only by the relative width of the blade and sheath of the distal leaves. They are obviously closely related and should be studied in the field in the southern Appalachians where their ranges overlap. The Texas record is taken from C. Sinclair (1967).

The hybrid *Tradescantia ernestinana* × *T. ozarkana* is known from Arkansas and Missouri.

5. Tradescantia ohiensis Rafinesque, Précis Découv. Somiol., 45. 1814 [E] [F]

Tradescantia canaliculata Rafinesque; *T. foliosa* Small; *T. incarnata* Small; *T. reflexa* Rafinesque

Herbs, erect or ascending, rarely rooting at nodes. **Stems** 15–115 cm; internodes glabrous or occasionally pilose, glaucous. **Leaves** spirally arranged, sessile, forming acute angle with stem, arcuate; blade linear to linear-lanceolate, 5–45 × 0.4–4.5 cm (distal leaf blades equal to or narrower than sheaths when sheaths opened, flattened), apex acuminate, glaucous, usually glabrous, sometimes pilose near sheath. **Inflorescences** terminal and often axillary; bracts foliaceous. **Flowers** distinctly pedicillate; pedicels 0.7–3 cm, glabrous; sepals glaucous, 4–15 mm, glabrous or with

apical tuft of eglandular hairs; petals distinct, deep blue to rose, rarely white, broadly ovate, not clawed, 0.8–2 cm; stamens free; filaments bearded. **Capsules** 4–6 mm. **Seeds** 2–3 mm. $2n = 12, 24$.

Flowering late winter–fall (Feb, Fla.; Mar–Sep). Roadsides, railroad rights-of-way, fields, thickets, less commonly in woods, occasionally along streams; Ont.; Ala., Ark., Conn., Del., D.C., Fla., Ga., Ill., Ind., Iowa, Kans., Ky., La., Md., Mass., Mich., Minn., Miss., Mo., Nebr., N.J., N.Y., N.C., Ohio, Okla., Pa., R.I., S.C., Tenn., Tex., Va., W.Va., Wis.

Tradescantia ohiensis is the most common and widespread tradescantia species in the United States. It hybridizes with many of the other species.

Tradescantia ohiensis var. *foliosa* (Small) MacRoberts has been recognized for the forms with pilose leaves and sheaths (D. T. MacRoberts 1977). I have found such plants scattered among populations of glabrous plants, and I do not consider them worthy of formal taxonomic status.

The following hybrids are known: *Tradescantia ohiensis* × *T. gigantea*, in Louisiana and Texas; *T. ohiensis* × *T. hirsuticaulis*, Arkansas; *T. ohiensis* × *T. occidentalis*, Arkansas, Louisiana; *T. ohiensis* × *T. ozarkana*, Arkansas; *T. ohiensis* × *T. paludosa*, Louisiana (reported by MacRoberts, 1980); *T. ohiensis* × *T. roseolens*, Alabama, Florida; *T. ohiensis* × *T. subaspera*, Alabama, Georgia, Mississippi, North Carolina, South Carolina, Tennessee, Virginia, West Virginia; and *T. ohiensis* × *T. virginiana*, Georgia, Kentucky, Maryland, Mississippi, North Carolina, South Carolina, and Virginia.

6. Tradescantia paludosa E. S. Anderson & Woodson, Contr. Arnold Arbor. 9: 83; plate 2, fig. 4; plate 4, fig. 6; plate 11; map 10. 1935 [E]

Tradescantia ohiensis Rafinesque var. *paludosa* (E. S. Anderson & Woodson) MacRoberts

Herbs, erect, ascending, or occasionally decumbent, rarely rooting at nodes. **Stems** often much branched distally, 15–60 cm; internodes not at all to slightly glaucous, glabrous. **Leaves** spirally arranged, sessile, forming nearly right angle with stem, straight; blade narrowly oblong-elliptic to linear-lanceolate, 4–11(–20) × 0.4–1.2 cm (distal leaf blades equal to or narrower than sheaths when sheaths opened, flattened), base often constricted, apex acuminate, not at all to slightly glaucous, glabrous. **Inflorescences** terminal, often axillary; bracts foliaceous. **Flowers** distinctly pedicillate; pedicels 0.8–1.5 cm, glabrous; sepals 0.6–0.8 mm, glabrous or with apical tuft of eglandular hairs; petals distinct, pale blue, ovate, not clawed, 1.3–1.5 cm; stamens free; filaments bearded. **Capsules** 2–5 mm. **Seeds** 2–3 mm. $2n = 12$.

Flowering spring (Mar–May), sporadically to early fall. Alluvial bottoms and swamps, forests, roadsides, railroad rights-of-way, fields, ditches, and lawns; Ala., Ark., Fla., La., Miss., Tex.

Tradescantia paludosa is clearly Anderson and Woodson's weakest species, and D. T. MacRoberts (1979) may be correct in treating it as a variety of *Trandescantia ohiensis*. In view of its importance as a research tool, however, I prefer to maintain *T. paludosa* as a species until a more rigorous analysis of its variation is published. Plants of this species do not seem to require a winter dormancy, hence they can be cultivated in greenhouses year-round.

7. Tradescantia gigantea Rose, Contr. U. S. Natl. Herb. 5: 205. 1899 E F

Herbs, erect or ascending, rarely rooting at nodes. **Stems** 16–100 cm; proximal internodes glabrous, distal glabrous to densely eglandular-puberulent. **Leaves** spirally arranged, sessile (with sheaths ± saccate); blade linear-lanceolate, 10–40 × 0.5–2.5 cm (distal leaf blades equal to or narrower than sheaths when sheaths opened, flattened), glaucous, glabrous or adaxially densely and minutely eglandular-velvety. **Inflorescences** terminal, axillary; bracts reduced, bases saccate, minutely velvety. **Flowers** distinctly pedicillate; pedicels 0.9–2.8 cm, densely eglandular-puberulent; sepals 5–13 mm, densely, minutely eglandular-puberulent; petals distinct, magenta to blue or violet, broadly obovate, not clawed, 1.5–1.8 cm; stamens free; filaments bearded. **Capsules** 6–7 mm. **Seeds** 2–3 mm. $2n = 12$.

Flowering spring (Mar–May). Rocky limestone areas, pasturelands, weedy lots, roadsides, and along railroad tracks; La., Tex.

Plants of *Tradescantia gigantea* growing around Ruston, Louisiana, may have originated from cultivated plants. They hybridize with *T. ohiensis* there.

8. Tradescantia tharpii E. S. Anderson & Woodson, Contr. Arnold Arbor. 9: 70; plate 4, fig. 9; plate 9; map 7. 1935 E

Herbs, erect or ascending, rarely rooting at nodes. **Stems** absent or 2–7 cm in flower, to 20 cm in fruit, shaggy pilose to villous. **Leaves:** blade green, linear-lanceolate, 4–30 × 0.9–2.5 cm (distal leaf blades equal to or narrower than sheaths when sheaths opened, flattened), margins frequently clear or edged with rose, laxly and irregularly pilose or villous. **Inflorescences** terminal, solitary; bracts foliaceous, well developed, not saccate, sparsely to densely pilose. **Flowers** distinctly pedicillate; pedicels 4–6 cm, laxly pilose; sepals usually purple or rose-colored (rarely pale green), not inflated, 1.2–1.6 cm, uniformly eglandular-pilose; petals distinct, deep rose or purple, or frequently blue, broadly ovate, not clawed, 1.8–2.2 cm; stamens free; filaments bearded. **Capsules** 5–7 mm. **Seeds** 2–3 mm. $2n = 24$.

Flowering spring. Clay soils of rocky prairies and open woodlands; Kans., Mo., Okla., Tex.

9. Tradescantia virginiana Linnaeus, Sp. Pl. 1: 288. 1753 • Éphémère de Virginie E F

Tradescantia brevicaulis Rafinesque

Herbs, erect or ascending, rarely rooting at nodes. **Roots** (1.5–)2–4 mm thick, fleshy. **Stems** 5–35 cm; internodes glabrous or occasionally distal internodes sparsely puberulent. **Leaves** spirally arranged, sessile; blade linear-lanceolate, 13–37 × 0.4–2.5 cm (distal leaf blades equal to or narrower than sheaths when sheaths opened, flattened), apex acuminate, glabrous or occasionally puberulent. **Inflorescences** terminal and (rarely) axillary; bracts foliaceous, well developed, not saccate, sparsely to densely pilose. **Flowers** distinctly pedicillate; pedicels 1.2–3.5 cm, eglandular-pilose or puberulent; sepals ± inflated, 7–16 mm, uniformly eglandular-pilose; petals distinct, blue to purple, occasionally rose or white, broadly ovate, not clawed, 1.2–2 cm; stamens free; filaments bearded. **Capsules** 4–7 mm. **Seeds** 2–3 mm. $2n = 12, 24$.

Flowering spring–summer (Mar–Jul). Woods, thickets, fields, roadsides and railroad rights-of-way; Ont.; Ala., Ark.(?), Conn., Del., D.C., Ga., Ill., Ind., Iowa, Ky., Maine, Md., Mass., Mich., Miss.(?), Mo., N.H., N.J., N.Y., N.C., Ohio, Okla., Pa., R.I., S.C., Tenn., Vt., Va., W.Va., Wis.

The records from the northern parts of the range of

Tradescantia virginiana may all represent garden escapes (E. Anderson 1954). The uncertainty about the records from Arkansas and Mississippi reflects the difficulty in identifying some specimens. The specimens in question come from areas in which *T. hirsutiflora* (but not *T. virginiana*) has been recorded (E. Anderson and R. E. Woodson Jr. 1935). The exact geographic boundaries between these putatively allopatric species are uncertain. D. T. MacRoberts (1980b) has made a useful contribution toward our knowledge of these species.

10. Tradescantia hirsutiflora Bush, Trans. Acad Sci. St. Louis 14: 184. 1904 [E]

Herbs, erect or ascending, rarely rooting at nodes. Roots 1–1.5(–2) mm thick, scarcely fleshy. Stems unbranched or sparsely branched, 5–50 cm; internodes densely spreading, pilose or hirsute to glabrous. Leaves spirally arranged, sessile; blade linear-lanceolate, 10–32 × 0.6–2 cm (distal leaf blades equal to or narrower than sheaths when sheaths opened, flattened), apex acuminate, usually pilose, occasionally glabrous or glabrescent. Inflorescences terminal, sometimes axillary; bracts foliaceous, well developed, not saccate, sparsely to densely pilose. Flowers distinctly pedicillate; pedicels 1–3 cm, usually pilose; sepals not inflated, 7–16 mm, usually uniformly eglandular-pilose, rarely a few inconspicuous glandular hairs present; petals distinct, bright blue to rose, rarely white, broadly ovate, not clawed, 1.2–1.9 cm; stamens free; filaments bearded. Capsules 5–7 mm. Seeds 2–3 mm. 2*n* = 12, 24.

Flowering spring–summer (Mar–Aug). Roadsides, fields, clearings, railroad rights-of-way, scrub, bottomlands, and pine or pine-mixed hardwood woods, usually in sandy soil; Ala., Ark., Fla., Ga., La., Miss., Mo., Okla., Tex.

Tradescantia hirsutiflora was considered their most ill-defined species by E. Anderson and R. E. Woodson Jr. (1935). The difficulties in separating it from *T. virginiana* have been mentioned under that species. A specimen from Beaufort County, South Carolina, appears to be a hybrid between *T. hirsutiflora* and *T. ohiensis*, but there is no record of *T. hirsutiflora* from the state. Some specimens from Highlands County, Florida, will key to, and probably are, *T. hirsutiflora*. They represent a range disjunction from the Florida panhandle. Their relationships with the co-occurring *T. roseolens* are being investigated.

This species commonly has been confused with *Tradescantia hirsuticaulis* (J. K. Small 1933; R. P. Wunder-

lin 1982), perhaps because of the similar name. They are not closely related.

Specimens of *Tradescantia hirsutiflora* with glandular hairs on the sepals were not found by D. T. MacRoberts (1980b). In Texas, plants with glandular hairs are frequent, and the glandular hairs may be numerous and conspicuous. These plants, which have been referred to *T. bracteata* by MacRoberts, need to be investigated further. I have also seen three sheets of *T. hirsutiflora* from Louisiana and one from Mississippi that have a few inconspicuous glandular hairs among the numerous longer, eglandular ones.

The following hybrids are known: *Tradescantia hirsutiflora* × *T. occidentalis*, from Alabama, Louisiana; *T. hirsutiflora* × *T. ohiensis*, Alabama, Arkansas, Florida, Georgia, Louisiana, Mississippi, South Carolina; *T. hirsutiflora* × *T. paludosa*, Arkansas, Louisiana; and *T. hirsutiflora* × *T. roseolens*, Florida.

11. Tradescantia occidentalis (Britton) Smyth, Trans. Kansas Acad. Sci. 16: 163. 1899 [E]

Tradescantia virginiana Linnaeus var. *occidentalis* Britton in N. L. Britton and A. Brown, Ill. Fl. N. U.S. 1: 377. 1896

Herbs, erect or ascending, rarely rooting at nodes. Stems 5–90 cm; internodes glaucous, glabrous. Leaves spirally arranged, sessile; blade linear-lanceolate, 5–50 × 0.2–3 cm (distal leaf blades equal to or narrower than sheaths when sheaths opened, flattened), apex acuminate, glaucous, glabrous. Inflorescences terminal, often axillary; bracts foliaceous. Flowers distinctly pedicillate; pedicels 0.8–3 cm, glandular-puberulent, rarely glabrous or glabrescent; sepals 4–11 mm, glandular-puberulent, usually with apical tuft of eglandular hairs, occasionally with scattered eglandular hairs among glandular, rarely glabrous or glabrescent; petals distinct, bright blue to rose or magenta, broadly ovate, not clawed, 0.6–1.6 cm; stamens free; filaments bearded. Capsules 4–7 mm. Seeds 2–4 mm.

Varieties 2 (2 in the flora): North America.

All of the chromosome counts cited by E. Anderson (1954) for this species are attributable to *Tradescantia occidentalis* var. *occidentalis*.

1. Sepals and pedicels ± uniformly glandular-puberulent, rarely nearly glabrous. . . . .
. . . . . . 11a. *Tradescantia occidentalis* var. *occidentalis*
1. Sepals and pedicels completely glabrous. . . . . .
. . . . . . 11b. *Tradescantia occidentalis* var. *scopulorum*

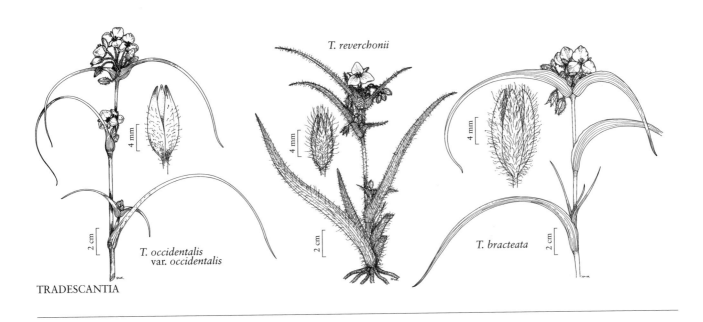

T. reverchonii

T. occidentalis
var. *occidentalis*

T. bracteata

TRADESCANTIA

11a. Tradescantia occidentalis (Britton) Smyth var. **occidentalis** E F

Stems 5–90 cm. **Leaves:** blade 5–50 × 0.2–3 cm. **Flowers:** pedicels 0.8–3 cm, glandular-puberulent, rarely nearly glabrous; sepals 4–11 mm, glandular-puberulent, rarely nearly glabrous; petals bright blue to rose or magenta, 1.2–1.6 cm. $2n = 12, 24$.

Flowering spring–summer (Mar–Jul). Prairies, plains, fields, thickets, woods, roadsides, and along railroads, mostly in sandy or rocky soils; Man.; Ariz., Ark., Colo., Iowa, Kans., La., Minn., Mont., Nebr., N.Mex., N.Dak., Okla., S.Dak., Tex., Utah, Wis., Wyo.

11b. Tradescantia occidentalis (Britton) Smyth var. **scopulorum** (Rose) E. S. Anderson & Woodson, Contr. Arnold Arbor. 9: 106. 1935 E

Tradescantia scopulorum Rose, Contr. U. S. Natl. Herb. 5: 205. 1899

Stems 14–35 cm. **Leaves:** blade 5–45 × 0.4–0.9 cm. **Flowers:** pedicels 1–2 cm, glabrous; sepals 5–8 mm, glabrous; petals bright blue, 0.7–1 cm.

Flowering spring–summer, fall (Apr–Aug, Oct). Moist canyons and stream banks; Ariz., N.Mex.

The two varieties intergrade; the only absolute difference between them is sepal pubescence. *Tradescantia*

occidentalis var. *scopulorum* evidently was defined more broadly by W. C. Martin and C. R. Hutchins (1980), so the distribution I have shown for New Mexico could be too wide. A broader circumscription of this variety has also been used by others (C. Sinclair 1967; M. Bolick 1986), but I have not accepted these records.

Most specimens of *Tradescantia occidentalis* from Arkansas and Louisiana have an apical tuft of eglandular hairs on the sepals, a character acquired through introgression with *T. ohiensis* (E. Anderson and R. E. Woodson Jr. 1935).

Tradescantia occidentalis var. *melanthera* has been described for plants from Arkansas, Louisiana, eastern Texas, and eastern Oklahoma that have dark anther connectives (D. T. MacRoberts 1977). I have not been able to recognize this character in dried specimens with any consistency.

12. Tradescantia roseolens Small, Bull. Torrey Bot. Club 51: 379. 1924 E

Herbs, erect or ascending, rarely rooting at nodes. **Roots** thin, fibrous, 0.5–1(–2) mm thick. **Stems** unbranched or sparsely branched, 19–60 cm; internodes puberulent with glandular or eglandular hairs, rarely pilose or glabrous. **Leaves** spirally arranged, sessile; blade linear-lanceolate, 10–42 × 0.5–1.6 cm (distal leaf blades equal to or narrower than sheaths when sheaths opened, flattened), apex acuminate, somewhat glaucous, puberulent to pilosulose, rarely glabrescent. **Inflorescences** terminal, often axillary; bracts foliaceous. **Flowers** rose-scented, distinctly pedicillate; pedicels 1–2.8 cm, glandular-puberulent; sepals 6–12 mm, glandular-puberulent, glandular hairs numerous and conspicuous, often mixed with eglandular hairs, usually with apical tuft of eglandular hairs, all hairs less than 1 mm; petals distinct, broadly deep blue to magenta, ovate, not clawed, 1–1.4 cm; stamens free; filaments bearded. **Capsules** 5–7 mm. **Seeds** 3–4 mm. $2n = 24$.

Flowering late winter–summer (Feb–Aug). Oak and oak-palmetto scrub, oak woods, pine woods, hammocks, sandhills, roadsides, and open areas, sandy soil; Ala., Fla., Ga., S.C.

Although E. Anderson and R. E. Woodson Jr. (1935) did not report eglandular hairs on the sepals in *Tradescantia roseolens*, they are present in nearly all specimens. While some eglandular hairs might be the result of hybridization, in general they seem to be part of the normal variation in this species.

The illegitimate name *Tradescantia longifolia* Small (a later homonym of *T. longifolia* Sessé and Mociño 1894) has been used for this species.

13. Tradescantia pedicellata Celarier, Field & Lab. 24: 6. 1956 E

Herbs, erect or ascending, rarely rooting at nodes. **Roots:** some tuberous, thick. **Stems** spreading and diffuse, much branched, mainly from base, pubescent with glandular hairs, eglandular hairs, or mixture. **Leaves** recurved and somewhat lax; blade dark green to yellowish green, linear-lanceolate, mostly 20–30 × 0.5–1.0 cm (distal leaf blades equal to or narrower than sheaths when sheaths opened, flattened), sparsely to densely pubescent. **Inflorescences** terminal, solitary, or more commonly also with lateral, pedunculate inflorescences; bracts foliaceous, similar to leaves in form, sparsely to densely pubescent. **Flowers** distinctly pedicillate; pedicels 2.5–4.5 cm, densely pubescent with medium to long, glandular hairs; sepals 6–11 mm, densely pubescent with glandular hairs like those of pedicels, occasionally with a few eglandular hairs; petals distinct, pink to dark blue, broadly ovate, not clawed; stamens free; filaments bearded. **Seeds** 2–4 mm; hilum as long as seed. $2n = 12$.

Flowering spring. Rocky soil; Tex.

Tradescantia pedicellata is a most unsatisfactory species. The species may have arisen as a hybrid between *Tradescantia humilis* and *T. occidentalis* and been recognized as a species because of its constant morphology

and high pollen fertility (R. P. Celarier 1956). C. Sinclair (1967) concluded, however, that there was no evidence for the species' existence, and I have found it very difficult to recognize specimens that agree with the original description (no type has been located).

Tradescantia diffusa Bush, a name overlooked by E. Anderson and R. E. Woodson Jr. (1935), has been considered the correct name for this plant (D. T. MacRoberts 1978). After examining the type of *T. diffusa*, I concluded that it was conspecific with the type of *T. humilis*.

14. Tradescantia reverchonii Bush, Trans. Acad. Sci. St. Louis 14: 190. 1904 E F

Herbs, erect or ascending, rarely rooting at nodes. **Roots** thick, fleshy, densely brownish tomentose. **Stems** erect or ascending, unbranched or sparsely branched, 30–105 cm; internodes arachnoid-pubescent. **Leaves** spirally arranged, sessile; blade linear-lanceolate, 10–35 × 0.7–2.8 cm (distal leaf blades equal to or narrower than sheaths when sheaths opened, flattened), apex acuminate, arachnoid-pubescent, especially on sheaths. **Inflorescences** terminal, often axillary; bracts foliaceous. **Flowers** distinctly pedicillate; pedicels 1–2.3 cm, pilose or villous with eglandular or mixed eglandular, glandular hairs; sepals 5–14 mm, pubescent with mixture of glandular, eglandular hairs; petals distinct, bright blue-violet, rarely rose or white, broadly ovate, not clawed, 1.5–1.8 cm; stamens free; filaments bearded. **Capsules** 6–8 mm. **Seeds** 3–4 mm; hilum as long as seed. $2n = 12, 24$.

Flowering spring–summer (Mar–Jul). Sandhills with oaks, pine woods, rocky open woods, rarely seepage areas, and roadsides; Ark., La., Tex.

15. Tradescantia subacaulis Bush, Trans. Acad. Sci. St. Louis 14: 185. 1904 E

Herbs, erect or ascending, rarely rooting at nodes. **Roots** thick, brownish tomentose. **Stems** spreading, diffusely branched, particularly at base, 10–30 cm, arachnoid-villous or rarely nearly glabrescent. **Leaves** relatively lax and flaccid; blade deep green, linear-lanceolate, 10–18 × 0.5–1.5 cm (distal leaf blades equal to or narrower than sheaths when sheaths opened, flattened), ± arachnoid-villous as on stems. **Inflorescences** terminal, solitary, often with axillary, pedunculate inflorescences from distal nodes; bracts foliaceous, similar to leaves in form, ± arachnoid-villous. **Flowers** violet-scented; distinctly pedicillate; pedicels 2–3 cm, puberulent or pilose with mixed glandular, eglandular hairs; sepals green or suffused with rose or purple, 7–8 mm, pubescent with mixture of glandular, eglandular hairs; petals distinct, bright blue, broadly ovate, not clawed, 1.3–1.4 cm; stamens free; filaments bearded. **Capsules** 5–6 mm. **Seeds** 2–3 mm; hilum as long as seed. $2n = 12$.

Flowering spring (Mar–Jun). Sandy soil; Tex.

16. Tradescantia humilis Rose, Contr. U.S. Natl. Herb. 5: 204. 1899   E

Tradescantia diffusa Bush

Herbs, erect or ascending, rarely rooting at nodes. **Roots** tuberous in part, not brownish tomentose. **Stems** spreading, diffusely branched, particularly at base, 0.5–20(–45) cm, densely pubescent to glabrescent. **Leaves** somewhat recurved or falcate; blade deep green, or paler and somewhat glaucous, linear-lanceolate, 11–20 × 1–2 cm (distal leaf blades equal to or narrower than sheaths when sheaths opened, flattened), margins usually tinged with purple, crisped, puberulent to glabrescent. **Inflorescences** terminal, solitary, or more frequently also axillary and pedunculate from distal nodes; bracts foliaceous, similar to leaves in form, puberulent to glabrescent. **Flowers** distinctly pedicillate; pedicels 1.5–2.5 cm, puberulent or pilose with mixed glandular, eglandular hairs; sepals dull green or occasionally edged or suffused with purple, 9–11 mm, pubescent with mixed glandular, eglandular hairs; petals distinct, bright blue or occasionally pink, broadly ovate, not clawed, 1.1–1.9 cm; stamens free; filaments bearded. **Capsules** 6–7 mm. **Seeds** 2–3 mm; hilum as long as seed. $2n = 12$.

Flowering spring (Mar–Jun). Sandy and rocky soil, formerly also in rich black soil at the edge of the coastal plain, now more commonly in disturbed sites, such as roadsides, fencerows, and railroad rights-of-way; Tex.

17. Tradescantia bracteata Small in N. L. Britton and A. Brown, Ill. Fl. N. U.S. 3: 510. 1898   E F W

Herbs, erect or ascending, rarely rooting at nodes. **Roots** not brownish tomentose. **Stems** sparsely branched, 5–45 cm, glabrous, or puberulent distally. **Leaves** stiff; blade bright green, linear-lanceolate, 15–29 × 0.9–2 cm (distal leaf blades equal to or narrower than sheaths when sheaths opened, flattened), apex long-acuminate, glabrous. **Inflorescences** terminal, solitary, sometimes also lateral and pedunculate from distal nodes; bracts foliaceous, glabrous, or rarely sheath puberulent. **Flowers** distinctly pedicillate; pedicels 1.8–3.3 cm, pubescent with mixture of glandular, eglandular hairs; sepals, 10–13 mm, densely pubescent with mixture of glandular, eglandular hairs, glandular hairs numerous, conspicuous, longer hairs 1.5–6 mm; petals distinct, usually bright rose, less commonly blue, ovate, not clawed, 1.8–1.9 cm; stamens free; filaments bearded. **Capsules** 5–6 mm. **Seeds** 2–3 mm; hilum as long as seed. $2n = 12$, 24.

Flowering spring (Apr–Jun). Prairies, spreading to thickets, roadsides, and railroad rights-of-way; Colo., Ill., Ind., Iowa, Kans., Mich., Minn., Mo., Mont., Nebr., N.Dak., Okla., S.Dak., Wis., Wyo.

The record of the species from Indiana (E. Anderson and R. E. Woodson Jr. 1935) was based on an unnumbered specimen collected by Mason, deposited at the Field Museum in Chicago, that Anderson later treated as a depauperate specimen of *Tradescantia virginiana* (E. Anderson 1954). The internodes are puberulent with glandular and eglandular hairs, however, a character that I have seen in an occasional specimen of *T. bracteata* but never in *T. virginiana*.

Tradescantia bracteata was distinguished from *T. occidentalis* partly by the former's unbranched stems versus the freely branched in *T. occidentalis* (M. Bolick 1981). By using this feature, branching specimens from Minnesota would be identified as *T. occidentalis*, although their sepal pubescence and lax, green, pubescent-margined bracts and leaves clearly place them in *T. bracteata*.

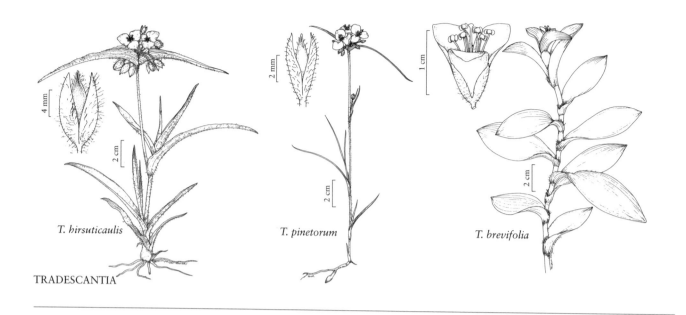

T. *hirsuticaulis*

T. *pinetorum*

T. *brevifolia*

TRADESCANTIA

18. Tradescantia hirsuticaulis Small, Bull. Torrey Bot. Club 24: 233. 1897 [E] [F]

Herbs, erect or ascending, rarely rooting at nodes. **Roots** not brownish tomentose. **Stems** unbranched or sparsely branched, (2–)15–40 cm; internodes densely pilose to glabrescent. **Leaves** spirally arranged, sessile; blade dull green, linear-lanceolate, 10–30 × 0.6–1.6 cm (distal leaf blades equal to or narrower than sheaths when sheaths opened, flattened), margins ± densely ciliolate, sometimes also sparsely ciliate, apex acuminate, puberulent and sparsely to densely pilose (rarely glabrescent). **Inflorescences** terminal, rarely axillary; bracts foliaceous, usually spreading, puberulent, usually sparsely to densely pilose, margins ± densely ciliolate. **Flowers** distinctly pedicillate; pedicels 1.5–3.5 cm, densely pubescent with mixture of glandular, eglandular hairs; sepals suffused with rose or purple, 6–13 mm, pilose with mixture of glandular, eglandular hairs, glandular hairs numerous, conspicuous, longer hairs 1.5–6 mm; petals distinct, rose to bright blue, ovate, not clawed, 1.1–1.6 cm; stamens free; filaments bearded. **Capsules** 5–6 mm. **Seeds** 2–3 mm; hilum as long as seed. $2n = 12$.

Flowering spring (Mar–May). Rocky woods on hillsides, also sandy woods, thickets, ledges, rock outcrops, stream banks and roadsides; Ala., Ark., Ga., N.C., Okla., S.C.

The separation of specimens nearly without evident stems from *Tradescantia longipes* is discussed by S. L. Timme and R. B. Faden (1984). The record of *T. hir-*

suticaulis from Texas (C. Sinclair 1967) is not considered credible. The hybrid *T. hirsuticaulis* × *T. virginiana* was reported from Alabama (E. Anderson and R. E. Woodson Jr. 1935).

19. Tradescantia longipes E. S. Anderson & Woodson, Contr. Arnold Arbor. 9: 91, plate 5, fig. 10; plate 6, fig. 9; plate 10; map 13. 1935 [E]

Herbs, usually ± rosette. **Roots** not brownish tomentose. **Stems** unbranched or sparsely branched, 1–10 cm; internodes pilose or villous. **Leaves** spirally arranged, sessile; blade dull green, linear-lanceolate, 5–33 × 0.3–1.2 cm (distal leaf blades equal to or narrower than sheaths when sheaths opened, flattened), margins sparsely ciliate, apex acute to acuminate, pilose. **Inflorescences** terminal; bracts foliaceous, ascending, pilose, margins sparsely ciliate. **Flowers** distinctly pedicillate; pedicels (2–)4–6 cm, pilose with glandular or glandular and eglandular hairs; sepals suffused with pink to purple, 5–11 mm, pilose with glandular, eglandular hairs, glandular hairs numerous, conspicuous, longer hairs 1.5–6 mm; petals distinct, rose to blue or purplish, broadly ovate, not clawed, 0.8–1.6 cm; stamens free; filaments bearded. **Capsules** 4–6 mm. **Seeds** 2–3 mm; hilum as long as seed. $2n = 24$.

Flowering spring (Apr–May). Wooded slopes on rocky hillsides; Ark., Mo.

20. **Tradescantia wrightii** Rose & Bush, Trans. Acad. Sci. St. Louis 14: 188. 1904

Herbs, erect or ascending, rarely rooting at nodes. **Stems** unbranched, 5–18 cm. **Leaves:** blade linear-lanceolate, 4–10 × 0.2–0.5 cm, firmly membranaceous to subsucculent, glaucous or glaucescent, glabrous. **Inflorescences** terminal, solitary; bracts foliaceous. **Flowers** distinctly pedicillate; pedicels 1.2–1.7 cm, with few to many minute glandular hairs (or glabrous); sepals glaucous or glaucescent, 0.5–0.6 cm, glabrous or with a few minute glandular hairs at base; petals distinct, rose to magenta or purple, broadly ovate, not clawed, 1 cm; stamens free; filaments bearded. **Capsules** 3–4 mm. **Seeds** 2–3 mm. $n = 6$.

Flowering spring–fall (May–Sep). Moist canyon stream banks; N.Mex., Tex.; Mexico (Chihuahua, Coahuila).

Tradescantia wrightii var. *glandulopubescens* was described for Mexican plants with glandular-pubescent pedicels and sepals (B. L. Turner 1983), and this variety was listed for Texas (S. L. Hatch et al. 1990). All U.S. collections that I have examined, however, including the holotype of *T. wrightii*, have at least some glandula hairs on these parts. Marshall Johnston believes that this is a valid variety, and I may not have examined typical specimens, but the diagnosis is not differential from the typical variety.

Tradescantia wrightii and *T. pinetorum* belong to sect. *Tradescantia* ser. *Tuberosae* D. R. Hunt and differ from the species of ser. *Virginianae* D. R. Hunt (species 1–19) by being geophytes (instead of hemicryptophytes) and in having the hilum much shorter than the seed (instead of ± equal to the seed). *Tradescantia wrightii* differs from *T. pinetorum* by its lack of root tubers, its glabrous leaves and internodes, and the absence of lateral inflorescences. The glandular hairs on the pedicels and sepal bases, much shorter than those of *T. pinetorum*, are easily overlooked.

21. **Tradescantia pinetorum** Greene, Erythea 1: 247. 1893 [F]

Herbs, erect to ascending, rarely rooting at nodes. **Roots** sometimes tuberous. **Stems** sparsely branched, 8–39 cm, scabridulous or rarely glabrescent. **Leaves:** blade linear-lanceolate, 1–10 × 0.15–0.8 cm, firmly membranaceous, glaucous, glabrous. **Inflorescences** terminal, solitary, or frequently with 1–3 axillary inflorescences from distal nodes; bracts foliaceous. **Flowers** distinctly pedicillate; pedicels 0.8–1 cm, glandular-puberulent; sepals frequently suffused with red, glaucous, 4–6 mm, glandular-puberulent; petals distinct, bright blue to rose and purple, not clawed, 0.9–1.2 cm; stamens free; filaments bearded. **Capsules** 3–4 mm. **Seeds** 1.5–2 mm; hilum much shorter than seed.

Flowering summer–fall (Jul–Sep). Moist canyons and stream banks; 1700–3000 m; Ariz., N.Mex.; Mexico (Chihuahua, Durango, Sonora).

22. **Tradescantia crassifolia** Cavanilles, Icon. 1: 54, plate 75. 1791

Herbs, erect, rarely rooting at nodes. **Roots** tuberous, thick. **Stems** unbranched or sparsely branched, arachnoid-villous. **Leaves** 2-ranked, narrowing toward shoot apex; blade lanceolate-elliptic to ovate, 4–15 × 2–3.5 cm, base cordate-clasping, sparsely pilose to densely arachnoid-villous. **Inflorescences** terminal, 1–4 axillary, sessile in axils of distal reduced leaves, boat-shaped spathes absent. **Flowers** distinctly pedicillate; pedicels 0.6–1.4 cm, densely arachnoid-villous; sepals 6.5–8 mm, pilose to densely arachnoid-pubescent; petals distinct, rose-purple to bluish, not clawed, 1–1.5 cm; stamens free; filaments bearded. **Capsules** 3.5–4 mm. **Seeds** 2 mm. $2n = 12, 24, 24+$ (Mexico).

Flowering summer–early fall. Specific habitat unknown; Tex.; Mexico; Central America (Guatemala).

Although not recorded in any U.S. flora or checklist, *Tradescantia crassifolia* has been added to the flora based on an old collection from the Chenati Mountains in Texas. Another specimen labeled New Mexico is considered less credible.

SELECTED REFERENCE Faden, R. B. 1993. *Tradescantia crassifolia* (Commelinaceae), an overlooked species in the southwestern United States. Ann. Missouri Bot. Gard. 80: 219–222.

23. **Tradescantia spathacea** Swartz, Prodr., 57. 1788
• Moses-in-the-cradle, oyster-plant, boat-lily [I] [W]

Rhoeo discolor (L'Héritier) Hance ex Walpers; *R. spathacea* (Swartz) Stearn

Herbs, erect or ascending, rarely rooting at nodes. **Stems** unbranched, short. **Leaves** spirally arranged; blade usually abaxially purple, adaxially green, strapshaped, to 35 × 5 cm (distal leaf blades wider or narrower than sheaths when sheaths opened, flattened), leathery, succulent, glabrous. **Inflorescences** axillary, sessile, or pedunculate in axils well below shoot apex, cymes enclosed in pairs of boat-shaped spathes. **Flowers** distinctly pedicillate; pedicels glabrous; sepals distinct, white, 3–6 mm, glabrous; petals distinct, white, ovate, not clawed; stamens free; filaments bearded. **Capsules** 3- or (by abortion) 2-locular, 3–4 mm. **Seeds** 1 per locule, 3–4 mm. $2n = 12$ (Belize).

Flowering winter (Jan). Occasionally escaped to hammocks and weedy places; introduced; Fla.; Mexico; West Indies; Central America.

Tradescantia spathacea is native to southern Mexico, Central America, and the West Indies.

24. Tradescantia fluminensis Vellozo, Fl. Flumin., 140; plate vol. 3, 152. 1829 ⬜I

Herbs, decumbent, rooting at nodes. **Leaves** 2-ranked; blade lanceolate-elliptic to ovate-lanceolate, 2.5–5 × 1–2 cm (distal leaf blades wider or narrower than sheaths when sheaths opened, flattened), margins ciliolate, apex acute, glabrous. **Inflorescences** terminal, becoming leaf-opposed, sometimes axillary from distalmost leaf axil, 1–2 cyme pairs per stem; bracts mostly foliaceous, occasionally reduced. **Flowers** distinctly pedicillate; pedicels 1–1.5 cm, glandular-pilose; sepals 5–7 mm, midrib pilose with eglandular hairs; petals distinct, white, not clawed, 8–9 mm; stamens free; filaments white, densely bearded with white hairs.

Flowering spring–fall. Woods, roadsides, and open areas, sometimes as weed; introduced; Ala., Calif., Fla., La.; introduced; native, South America (Brazil–Argentina); Africa (South Africa); Australia.

This species was recorded north to North Carolina (J. K. Small 1933), but I have not seen any supporting records from Georgia or North Carolina.

25. Tradescantia crassula Link & Otto, Icon. Pl. Rar. 2: 13, plate 7. 1828 ⬜I

Herbs, decumbent, rooting at nodes. **Leaves** 2-ranked; blade lanceolate-oblong to ovate-elliptic, 5–10 × 2–3.5 cm, margins ciliolate, apex acute, glabrous. **Inflorescences** terminal and axillary from distal leaf axils, 2–4 cyme pairs per stem; bracts foliaceous or more commonly reduced (especially in axillary inflorescences). **Flowers** distinctly pedicillate; pedicels 1–2 cm, glandular-pilose especially toward apex; sepals 5–7 mm, midrib pilose

with glandular and eglandular hairs; petals distinct, white, not clawed, 0.9 cm; stamens free; filaments white, densely bearded with white hairs.

Flowering spring (May). Under live oaks; introduced; Fla.; South America.

26. Tradescantia brevifolia (Torrey) Rose, Contr. U. S. Natl. Herb. 3: 323. 1895 ⬜F

Tradescantia leiandra Torrey var. *brevifolia* Torrey in W. H. Emory, Rep. U.S. Mex. Bound. 2(1): 225. 1859; *Setcreasea brevifolia* (Torrey) Schumann & Sydow

Herbs, prostrate, coarsely rhizomatous. **Stems** unbranched or sparsely branched, to 30 cm. **Leaves** spirally arranged; blade green, not variegated, broadly ovate to elliptic or lanceolate-elliptic, to 8 × 3 cm, succulent, base symmetric, rounded to broadly cuneate, ± glaucous. **Inflorescences** terminal, pedunculate, peduncles 1–5 cm; bracts similar to leaves. **Flowers** subsessile; pedicels densely pilose; sepals distinct, glabrous; petals deep pink to reddish purple, obovate, clawed, claws basally connate forming tube; stamens epipetalous; filaments glabrous; ovary glabrous. **Capsules** obovoid to subglobose, 4 mm diam., glabrous. $2n = 18$ (Mexico).

Flowering summer–fall (Jun–Oct). Rocks or crevices, grassy slopes, or sheltered cliffs on north-facing slopes; Tex.; Mexico (Coahuila, Nuevo Leon).

Setcreasea, in which *Tradescantia brevifolia, T. buckleyi, T. pallida,* and *T. leiandra* were formerly placed, is synonymous with *Tradescantia* (D. R. Hunt 1975).

27. Tradescantia buckleyi (I. M. Johnston) D. R. Hunt, Kew Bull. 30: 451. 1975

Setcreasea buckleyi I. M. Johnston, J. Arnold Arbor. 25: 54. 1944

Roots tuberous, tufted. **Stems** loosely branched, to 50 cm, glabrous. **Leaves** spirally arranged; blade not variegated, elliptic, to 12 × 3.5 cm, succulent, base symmetric, rounded to broadly cuneate, margins ciliolate, apex acute, glabrous. **Inflorescences** terminal, pedunculate; peduncles 3–5.5 cm; bracts similar to leaves but somewhat smaller. **Flowers** subsessile; pedicels glabrous except for long silky hairs at summit; sepals distinct, base densely pilose, other surfaces sparsely pilose; petals ovate, clawed, 1 cm, claws basally connate forming tube, limb pale pink or whitish; stamens epipetalous; filaments bearded, connective orange, broad; ovaries densely bearded. **Capsules**

subglobose, 3.5 mm diam., pubescent. **Seeds** 2–3 mm.
Flowering mainly late winter–spring (Feb–May). Clay mounds in chaparral; Tex.; Mexico (Tamaulipas).

28. Tradescantia pallida (Rose) D. R. Hunt, Kew. Bull. 30: 452. 1975 [I]

Setcreasea pallida Rose, Contr. U.S. Natl. Herb. 13: 294. 1911; *S. purpurea* Boom

Herbs, perennial, succulent. **Stems** suffused with purplish violet. **Leaves** spirally arranged; blade not variegated, suffused with purplish violet, lanceolate-oblong to oblong-elliptic, (4–)7–15 × 1.5–3 cm, base symmetric, rounded to broadly cuneate, margins ciliate or ciliolate, apex acute, glabrous or glabrescent. **Inflorescences** terminal, often becoming leaf-opposed, pedunculate; peduncles (3.5–)4–13 cm; bracts similar to leaves but usually greatly reduced. **Flowers** subsessile; pedicels densely white-pilose at summit; sepals distinct, 7–10 mm, pilose basally; petals ± connate at base, pink, clawed, 1.5–2 cm; stamens epipetalous; filaments very sparsely bearded. **Capsules** 3.5 mm, glabrous. **Seeds** 2.5–3 mm. 2*n* = 24 (Mexico).

Flowering summer–fall. Landfill and old home sites; introduced; Fla., La.; native, Mexico.

29. Tradescantia leiandra Torrey in W. H. Emory, Rep. U.S. Mex. Bound. 2(1): 224. 1859

Setcreasea leiandra (Torrey) Pilger

Roots clustered, fibrous-thickened. **Stems** sparsely branched, 30–50 cm, tufted, glabrous. **Leaves** spirally arranged, distant; blade not variegated, narrowly lanceolate, 7.5–16 × 1–2.5 cm, base symmetric, rounded to broadly cuneate, margins

smooth or ciliate-scabrous, apex acuminate, glabrous to sparsely pubescent. **Inflorescences** terminal; bracts very unlike proximal stem leaves, connate, ovate-lanceolate, 2.4–4.5 cm, base cordate, somewhat dilated. **Flowers** subsessile; pedicels densely covered with white, long, eglandular hairs or occasionally only with colorless, short, glandular hairs; sepals distinct, usually glandular-pubescent as well as villous; petals purplish red, clawed, claws connate basally forming tube; stamens epipetalous; filaments glabrous. **Capsules** 3.5 mm, glabrous. **Seeds** 1.5 mm.

Flowering summer–fall (Jul–Oct). Moist, rocky places, on ledges, among shrubs and in canyons; Tex.; Mexico (Coahuila, Nuevo León).

The plants from the Capote Falls region of Presidio County, Texas have pedicels with short, colorless, glandular hairs instead of long, white, eglandular hairs and were separated as *Tradescantia leiandra* var. *glandulosa* Correll (D. S. Correll 1968).

30. Tradescantia zebrina Hort ex Bosse, Vollst. Handb. Bl.-Gärtn. 4: 655. 1849 • Wandering-Jew [I]

Zebrina pendula Schnizlein

Herbs, decumbent. **Leaves** 2-ranked; blade variegated, abaxially reddish purple, adaxially striped green and white, lanceolate-elliptic to ovate-elliptic, 3–9 × 1.5–3 cm, base oblique, cuneate, apex acute to acuminate. **Inflorescences** terminal, consisting of pairs of sessile cymes enclosed in sheaths of spathaceous bracts, pedunculate; spathaceous bracts foliaceous, reduced. **Flowers** subsessile; sepals basally connate, 4–5 mm; petals pink, clawed, claws basally connate forming tube; stamens epipetalous; filaments bearded. **Capsules** 3-locular; locules 2-seeded.

Flowering fall–winter (Sep–Feb). Hammocks and weedy places; introduced; Fla.; native, tropical America.

4. CALLISIA Loefling, Iter Hispan., 305. 1758 • [Greek *kallos*, beauty, referring to the attractive leaves]

Cuthbertia Small; *Leiandra* Rafinesque; *Phyodina* Rafinesque; *Tradescantella* Small

Herbs, perennial or rarely annual. **Roots** thin, rarely tuberous. **Leaves** spirally arranged or 2-ranked; blade sessile. **Inflorescences** terminal and/or axillary, cyme pairs (often aggregated into larger spikelike or paniclelike units), cymes sessile, umbel-like, contracted, subtended by bracts; bracts inconspicuous, less than 1.5 cm; spathaceous bracts absent; bracteoles persistent. **Flowers** bisexual (bisexual and pistillate in *C. repens*), radially symmetric; pedicels very short or well developed; sepals distinct, subequal; petals distinct, white or pink to rose [rarely blue], equal, not clawed; stamens 6 or 0–3, all fertile, equal; filaments glabrous or bearded; ovary 2–3-locular, ovules [1–]2 per locule, 1-seriate. **Capsules** 2–3-valved, 2–3-locular. **Seeds** [1-]2 per locule; hilum punctiform; embryotega abaxial. $x = 6$–8.

Species ca. 20 (7 in the flora): se United States, tropical America; major center of distribution in Mexico.

SELECTED REFERENCES Anderson, E. and R. E. Woodson Jr. 1935. The species of *Tradescantia* indigenous to the United States. Contr. Arnold Arbor. 9: 1–132. Lakela, O. 1972. Field observations of *Cuthbertia* (Commelinaceae) with description of a new form. Sida 5: 26–32.

1. Petals white; filaments glabrous; plants creeping, or ascending and stoloniferous.
  2. Robust, stoloniferous plants to 1 m; leaves oblong to lanceolate-oblong, 15–30 cm; flowers fragrant. . . . . .5. *Callisia fragrans*
  2. Weak, mat-forming plants; leaves lanceolate to ovate, 1–3.5 cm; flowers odorless.
    3. Inflorescences sessile in distal leaf axils; flowers sessile or subsessile; petals inconspicuous; stamens 0–6; ovary and capsule 2-locular. . . . . .6. *Callisia repens*
    3. Inflorescences pedunculate; flowers distinctly pedicellate; petals conspicuous; stamens 6; ovary and capsule 3-locular. . . . . .7. *Callisia cordifolia*
1. Petals pink to rose; filaments bearded; plants erect to ascending, or creeping.
  4. Plants creeping; leaves oblong-elliptic to lanceolate-oblong, 1–3.5 cm. . . . . 4. *Callisia micrantha*
  4. Plants erect to ascending; leaves linear, mainly 5–25 cm.
    5. Distal leaf blades as wide as opened, flattened sheaths or wider, 0.4–1.5 cm wide. . . . . 1. *Callisia rosea*
    5. Distal leaf blades much narrower than opened, flattened sheaths, 0.1–0.5 mm wide.
      6. Plants not cespitose or scarcely so; roots persistently woolly; bracts usually minute, scarious, 1–3(–7) mm. . . . . .2. *Callisia ornata*
      6. Plants cespitose; roots glabrous to sparsely puberulent; bracts often elongate, ± herbaceous, 2–14 mm. . . . . .3. *Callisia graminea*

1. Callisia rosea (Ventenat) D. R. Hunt, Kew Bull. 41: 409. 1986 [E]

Tradescantia rosea Ventenat, Descr. Pl. Nouv., plate 24. 1800; *Cuthbertia rosea* (Ventenat) Small

Herbs, perennial, erect. **Roots** sparsely pubescent to glabrescent. **Stems** 20–55 cm. **Leaves** often laxly spreading; basal leaf sheaths ± glabrous, margins ciliate; blade linear, 8–35 × 0.4–1.5 cm (distal leaf blades as wide as sheaths when sheaths opened, flattened). **Inflorescences:** bracts usually minute, 1–3(–5) mm, scarious. **Flowers** pedicellate; pedicels 0.5–1.2 cm; petals pink to rose, 7–12 mm; stamens 6; filaments bearded. **Capsules** 3-locular, 2–4 mm. **Seeds** 1.5–2 mm. $2n = 12$.

Flowering spring–early summer. Deciduous or pine-oak woods, sandy or shallow, rocky soils, occasionally in grassland, swamp forest, or along railroads and roadsides; Fla., Ga., N.C., S.C.

The evidence that this and the following two taxa are species, instead of varieties or subspecies, is hardly compelling. It is a matter of the author's preference. The work of N. H. Giles Jr. (1942) clearly demonstrated cytological diversity within *Callisia graminea*, but the remaining taxa have never been investigated in the same detail. The taxa appear to hybridize when they come in contact.

2. Callisia ornata (Small) G. C. Tucker, J. Arnold Arbor. 70: 118. 1989 [E]

Cuthbertia ornata Small, Man. S.E. Fl., 259. 1933; *Tradescantia rosea* Ventenat var. *ornata* (Small) E. S. Anderson & Woodson

Herbs, perennial, not or only scarcely cespitose, erect. **Roots** persistently densely woolly. **Stems** 20–50 cm. **Leaves** strongly ascending; basal leaf sheaths glabrous to puberulent; blade linear, (2–)10–25 × 0.1–0.4 cm (distal leaf blades much narrower than sheaths when sheaths opened, flattened). **Inflorescences:** bracts not distinct from bracteoles, usually minute, 1–3(–7) mm, scarious (rarely somewhat herbaceous). **Flowers** pedicellate; pedicels 1–1.5 cm; petals pink to rose, 9–13 mm; stamens 6; filaments bearded; ovary 3-locular. **Capsules** 3-locular, 3–5 mm. **Seeds** 1.5–2 mm.

Flowering spring, fall. Coastal strand, oak, pine, and

palmetto woods and scrub, occasionally fields or road-sides, usually in sandy soil; Fla.

A white-flowered form has been described from central Florida as *Callisia ornata* f. *leucantha* Lakela (O. Lakela 1972), but such color forms are hardly worth taxonomic recognition.

3. **Callisia graminea** (Small) G. C. Tucker, J. Arnold Arbor. 70: 118. 1989 [E] [F]

Cuthbertia graminea Small, Fl. S.E. U.S., 237, 1328. 1903; *Tradescantia rosea* Vent. var. *graminea* (Small) E. S. Anderson & Woodson

Herbs, perennial, cespitose, erect to ascending. **Roots** glabrous to sparsely puberulent. **Stems** (4–)15–25(–40) cm. **Leaves** ascending; basal leaf sheaths glabrous to pilose or puberulent; blade linear, 4–17 × 0.1–0.5 mm (distal leaf blades much narrower than sheaths when sheaths opened, flattened). **Inflorescences:** bracts often elongate, sometimes minute, 2–14 mm, if elongate ± herbaceous, if minute scarious. **Flowers** pedicellate; pedicels (0.7–)1.2–2 cm; petals pink to rose, 8–10 mm; filaments bearded. **Capsules** 2–3.5 mm. **Seeds** 1.5–2 mm. $n = 6, 12, 18$.

Flowering spring–fall. Sandy soil in pine-oak woods (especially longleaf pine and turkey oak) and pine barrens, often on sandhills, occasionally in thickets, old fields and roadsides; Fla., Ga., N.C., S.C., Va.

I have not been able to confirm the record of this species from Maryland in M. L. Brown and R. G. Brown (1984).

SELECTED REFERENCE Giles, N. H. Jr. 1942. Autopolyploidy and geographical distribution in *Cuthbertia graminea* Small. Amer. J. Bot. 29: 637–645.

4. **Callisia micrantha** (Torrey) D. R. Hunt, Kew Bull. 38: 131. 1983 [F]

Tradescantia micrantha Torrey in W. H. Emory, Rep. U.S. Mex. Bound. 2(1): 224. 1859; *Phyodina micrantha* (Torrey) D. R. Hunt

Herbs, perennial, creeping, succulent. **Stems** 3–30 cm. **Leaves** ± conduplicate; blade oblong-elliptic to lanceolate-oblong, 1–3.5 × 0.3–0.8 cm, margins ciliolate, glabrous. **Inflorescences** sessile or nearly sessile, subtended by 0–2 leaves that resemble spathaceous bracts. **Flowers** pedicellate; pedicels 0.8–1.2 cm, glabrous or nearly so; sepals strongly keeled, 4–5 mm, shortly, densely pubescent on keel; petals bright pink

to rose, ovate, 5–7 mm; stamens 6; filaments bearded. **Capsules** 3-locular, 2 mm. **Seeds** 1.5 mm. $2n = 24$.

Flowering spring–fall (May–Sep). Sandy or clayey soils in open oak or mesquite woods and prairies; Tex.; Mexico.

The generic placement of this species requires some explanation because the leaves subtending the inflorescence resemble the bracts of species of *Tradescantia*. The interpretation of D. R. Hunt (1986b), which I am following, is that the true bracts are small and are borne distal to those leaves.

5. **Callisia fragrans** (Lindley) Woodson, Ann. Missouri Bot. Gard. 29: 154. 1942 [I]

Spironema fragrans Lindley, Edwards's Bot. Reg. 26: plate 47. 1840

Herbs, perennial, robust, stoloniferous. **Stems** ascending, to 1 m. **Leaves** spirally arranged; blade oblong to lanceolate-oblong, 15–30 × 2.5–5 cm, apex acuminate, glabrous. **Inflorescences** terminal, panicles to 30 cm or longer. **Flowers** fragrant, subsessile; petals inconspicuous, white, lanceolate, 6 mm; stamens 6, long-exserted, connectives white, broad, flaglike; filaments glabrous; ovaries 3-locular, stigmas penicillate. **Capsules** 3-locular.

Flowering late winter (Feb). Pinelands and hammocks; introduced; Fla.; native, Mexico.

6. **Callisia repens** (Jacquin) Linnaeus, Sp. Pl. ed. 2, 1: 62. 1762 [F] [I]

Hapalanthus repens Jacquin, Enum. Syst. Pl., 1, 12. 1760

Herbs, perennial, mat-forming, repent (flowering stems ascending). **Leaves** 2-ranked, gradually reduced toward ends of flowering stems; blade ovate to lanceolate or lanceolate-oblong, 1–3.5 × 0.6–1 cm (distal leaf blades much narrower than sheaths when sheaths opened, flattened), margins scabrid, apex acute, glabrous. **Inflorescences** sessile in axils of distal leaves of flowering stems, composed of pairs of sessile cymes (sometimes reduced to single cymes). **Flowers** bisexual and pistillate, odorless, subsessile; petals inconspicuous, white, lanceolate, 3–6 mm; stamens 0–6, long-exserted; filaments glabrous; ovary 2-locular, stigma penicillate. **Capsules** 2-locular. **Seeds** 1 mm.

Flowering early spring (Tex.) or summer–fall (Fla.). Shady, rocky or gravelly places, and in citrus groves;

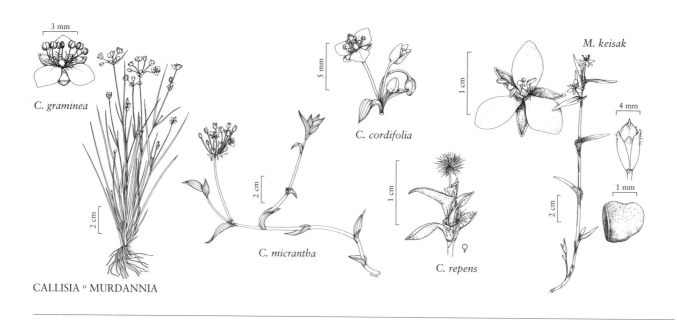

C. graminea

C. micrantha

C. cordifolia

C. repens ♀

M. keisak

CALLISIA ° MURDANNIA

introduced; Fla., La., Tex.; native, Mexico; West Indies; Central America; South America (to Argentina).

7. Callisia cordifolia (Swartz) E. S. Anderson & Woodson, Contr. Arnold Arbor. 9: 117. 1935 [F]

Tradescantia cordifolia Swartz, Prodr., 57. 1788; *Leiandra cordifolia* (Swartz) Rafinesque; *Phyodina cordifolia* (Swartz) Rohweder; *Tradescantella floridana* (S. Watson) Small; *Tradescantia floridana* S. Watson

Herbs, perennial, mat-forming. **Leaves** 2-ranked, gradually reduced toward end of flowering shoot; blade lanceolate or lanceolate-elliptic to ovate, 1–3 × 0.5–1.4 cm, mar-gins scabrous, glabrous. **Inflorescences** terminal and axillary from distal leaves, pedunculate, composed of pairs of sessile cymes; bracts linear to linear-lanceolate, 2–6 mm. **Flowers** odorless, 4–5 mm wide, pedicillate; sepals distinct, maroon, 2–3 mm; petals white, ovate, 2.5 mm; stamens 6, nearly equal or antipetalous stamens slightly longer than antisepalous; filaments glabrous; ovaries 3-locular. **Capsules** 3-locular. **Seeds** 0.6–0.7 mm. $2n = 14$ (as floridana).

Flowering spring–fall. Moist, usually shady places with calcareous soil, e.g., hammocks, fern grottoes, shell middens; Fla., Ga.; Mexico; West Indies; n South America.

The single Georgia record, a specimen labeled Rome, Floyd County (*Babcock s.n.*, MO), is considered credible by Dr. Nancy Coile.

5. MURDANNIA Royle, Ill. Bot. Himal. Mts., 403, plate 95, fig. 3. 1840, name conserved • [For Murdan Aly, plant collector and keeper of the herbarium at Saharunpore] [I] [W]

Herbs, annual or perennial. **Roots** thin [tuberous]. **Leaves:** blade sessile. **Inflorescences** terminal, terminal and axillary, or all axillary, thyrses to fascicles of 1-flowered cymes; spathaceous bracts absent; bracteoles persistent or caducous. **Flowers** bisexual or bisexual and staminate, radially or bilaterally symmetric; pedicels well developed; sepals distinct, subequal; petals distinct, white to purple or violet (rarely yellow), subequal, not clawed; stamens 2–3 fertile and antisepalous, 3–4 staminodial and antipetalous (if 4, then 1 antisepalous); filaments glabrous or bearded; antherodes usually 3-lobed; ovaries 3-locular, ovules 1–many per locule, 1[–2]-

seriate. **Capsules** 3-valved, 3-locular. **Seeds** 1–many per locule, 1[–2]-seriate; hilum punctiform to linear; embryotega abaxial to semilateral. $x = 6, 7, 9, 10, 11$.

Species ca. 50 (3 in the flora): introduced; pantropical and warm temperate.

SELECTED REFERENCES Faden, R. B. 1978. Review of the lectotypification of *Aneilema* R. Br. (Commelinaceae). Taxon 27: 289–298.   Shinners, L. H. 1962. *Aneilema* (Commelinaceae) in the United States. Sida 1: 100–101.

1. Flowers in 1-flowered cymes; cymes solitary or in fascicles; capsules (4–)5–9 mm. . . . . . .1. *Murdannia keisak*
1. Flowers in several-flowered cymes; cymes solitary or thyrses; capsules 2.5–5 mm.
  2. Bracteoles caducous; capsules with 2 seeds per locule; 2 stamens fertile, 4 staminodial
  . . . . .2. *Murdannia nudiflora*
  2. Bracteoles persistent; capsules with 3–7 seeds per locule; 3 stamens fertile, 3 staminodial. . . . . .3. *Murdannia spirata*

1. **Murdannia keisak** (Hasskarl) Handel-Mazzetti, Symb. Sin. 7: 1243. 1936 F I W

Aneilema keisak Hasskarl, Commelin. Ind., 32. 1870

Herbs, annual, with long-trailing, decumbent shoots. **Leaves:** blade linear-oblong to linear-lanceolate, 1.5–7 × 0.2–1 cm, glabrous. **Inflorescences** terminal and in distal leaf axils; cymes 1–several, 1-flowered, solitary or fascicled. **Flowers** bisexual, radially symmetric, 1 cm wide; sepals 5–6 mm; petals purplish lilac or purple to pink or white, 5–8 mm; stamens 3; filaments bearded; staminodes 3. **Capsules** (4–)5–9 mm. **Seeds** 2–6 per locule, 1.6–3 mm, faintly ribbed.

Flowering fall. Roadside ditches and swales, margins of lakes, creeks, rivers, swamps, bogs, swamp forest, and other aquatic habitats, often growing in water; introduced; Ark., Fla., Ga., Ky., La., Md., Miss., N.C., Oreg., S.C., Tenn., Va., Wash.; Europe; native, Asia.

I agree with C. P. Dunn and R. R. Sharitz (1990) that this species is still expanding its range since its introduction early this century. It tends to be very weedy.

SELECTED REFERENCE Dunn, C. P. and R. R. Sharitz. 1990. The history of *Murdannia keisak* (Commelinaceae) in the southeastern United States. Castanea 55: 122–129.

2. **Murdannia nudiflora** (Linnaeus) Brenan, Kew Bull. 7: 189. 1952 I W

Commelina nudiflora Linnaeus, Mant. Pl. 2: 177. 1771; *Aneilema nudiflorum* (Linnaeus) Sweet

Herbs, annual, unbranched to much branched, 8–30 cm. **Leaves** spirally arranged; blade linear or linear-lanceolate to lanceolate-oblong, 1.5–7(–18) × 0.3–0.8 cm. **Inflorescences** terminal or terminal and axillary; cymes few-flowered, solitary, long-pedunculate; bracteoles caducous, scars spaced less than 2 mm apart. **Flowers** bisexual, slightly bilaterally symmetric, 4–6 mm wide; sepals 2–3 mm; petals pinkish purple or violet, 3–3.5(–6) mm; fertile stamens 2; filaments bearded; staminodes 4. **Capsules** 2.5–5 mm. **Seeds** 2 per locule, deeply pitted, 1.3–1.8 mm. $2n = 20$ (Trinidad).

Flowering summer–fall. Weed in lawns, gardens, and other open or lightly shaded, disturbed sites; introduced; Ala., Fla., Ga., La., N.C., S.C., Tex.; Central America; South America; native, Asia.

3. **Murdannia spirata** (Linnaeus) Brückner in H. G. A. Engler and K. Prantl, Nat. Pflanzenfam. ed. 2, 15a: 173. 1930 I W

Commelina spirata Linnaeus, Mant. Pl. 2: 176. 1771

Herbs, annual, erect to decumbent, usually much branched with age, to 30 cm. **Leaves:** blade lanceolate-oblong to ovate, 1–4 × 0.3–1 cm. **Inflorescences:** cymes 1–2, several-flowered, elongate; bracteoles persistent, 2–6 mm apart. **Flowers** bisexual, radially symmetric, 8

mm wide; sepals 2.5–4 mm; petals rose or lavender with darker veins, 4–5 mm; fertile stamens 3; filaments bearded; staminodes 3. **Capsules** 3–5 mm. **Seeds** 3–7 per locule, less than 1 mm, warty.

Flowering fall–early winter. Palm hammocks, low prairies, glades, pastures, and roadsides; introduced; Fla.; native, Asia.

Murdannia spirata was first collected in Florida in 1965. It tends to be weedy.

6. COMMELINA Plumier ex Linnaeus, Sp. Pl. 1: 40. 1753; Gen. Pl. ed. 5: 25. 1754

• Dayflower, widow's-tears [for two Dutch botanists, Jan and Kaspar Commelijn, because of the two showy petals]

Herbs, perennial or annual. **Roots** thin or tuberous. **Leaves** 2-ranked or spirally arranged; blade sessile or petiolate. **Inflorescences** terminal, leaf-opposed; cymes 1–2, enclosed in spathes, proximal cyme several-flowered, distal cyme vestigial or with 1–several usually staminate flowers; spathes often filled with mucilaginous liquid, margins distinct or basally connate; bracteoles usually absent. **Flowers** bisexual and staminate, bilaterally symmetric; pedicels well developed; sepals distinct or proximal 2 connate, unequal; petals distinct, proximal petal often different color than distal 2, smaller or subequal, distal 2 blue (occasionally lilac, lavender, yellow, peach, apricot, or white), clawed; stamens (5–)6, proximal 3 fertile, medial different in form, size from others, distal (2–)3 staminodial; filaments glabrous; antherodes commonly 4–6-lobed; ovaries 2–3-locular, ovules 1–2 per locule, 1-seriate. **Capsules** 2–3-valved, 2–3-locular. **Seeds** 1–2 per locule; hilum linear; embryotega lateral. $x = 11$–15.

Species ca. 170 (9 in the flora): almost worldwide, mainly tropical.

SELECTED REFERENCES Brashier, C. K. 1966. A revision of *Commelina* (Plum.) L. in the U.S.A. Bull. Torrey Bot. Club 93: 1–19.   Faden, R. B. 1993b. The misconstrued and rare species of *Commelina* (Commelinaceae) in the eastern United States. Ann. Missouri Bot. Gard. 80: 208–218.   Pennell, F. W. 1916. Notes on plants of the southern United States—I. Bull. Torrey Bot. Club 43: 96–111.

1. Spathes with margins distinct to base.
  2. Perennials with erect to ascending stems; leaves linear to linear-lanceolate. . . . . 1. *Commelina dianthifolia*
  2. Annuals or perennials usually with decumbent to scandent stems; leaves narrowly lanceolate to ovate-elliptic.
    3. Spathes generally whitish or pale green toward peduncle with contrasting, dark green veins; proximal petal white or paler than others; capsules 2-locular; seeds rugose pitted-reticulate. . . . . 2. *Commelina communis*
    3. Spathes without contrasting veins; proximal petal ± concolorous with others; capsules 3-locular; seeds reticulate or smooth to faintly alveolate.
      4. Spathes not at all to slightly falcate; distal cyme usually vestigial (rarely well developed, 1-flowered); seeds smooth to faintly alveolate. . . . . 3. *Commelina caroliniana*
      4. Spathes usually distinctly falcate; distal cyme in larger spathes usually well developed, 1–several-flowered; seeds reticulate. 4. *Commelina diffusa*
1. Spathes with margins connate basally.
  5. Flowers apricot- or peach-colored. . . . . 5. *Commelina gambiae*
  5. Flowers blue (rarely lilac to lavender or white).
    6. Leaf sheaths with auricles at summit; roots stout; proximal petal white, minute; locules all 1-seeded. . . . . 6. *Commelina erecta*
    6. Leaf sheaths not auriculate; roots thin; proximal petal blue to lilac or lavender, conspicuous; some locules usually 2-seeded.
      7. Perennials with erect to ascending stems; leaf sheaths with red hairs at summit; leaf blades 6–20 cm; spathes 1.5–3.5 cm; subterranean, cleistogamous flowers absent. . . . . 7. *Commelina virginica*

7. Annuals or perennials with ascending to decumbent, repent or scrambling stems; leaf sheaths with or without red hairs at summit; leaf blades 1.5–7(–9) cm; spathes 0.5–2 cm; subterranean, cleistogamous flowers sometimes present.
  8. Leaf blades ovate to lanceolate-elliptic; sheaths often with red hairs at summit; lateral stamen filaments not winged; capsules usually 5-seeded. . . . . 8. *Commelina benghalensis*
  8. Leaf blades oblong to lanceolate-oblong or oblong-elliptic; sheaths without red hairs; lateral stamen filaments winged; capsules usually 1-seeded. . . . . 9. *Commelina forskaolii*

1. **Commelina dianthifolia** Delile in P. J. Redouté, Liliac., 7(65): plate 390. 1812 [F]

Herbs, perennial, unbranched to usually sparsely branched. **Roots** tuberous. **Stems** erect to ascending. **Leaves:** blade linear to linear-lanceolate, 4–15 × 0.4–1 cm, apex acuminate, glabrous to puberulent. **Inflorescences:** distal cyme usually 1-flowered, exserted; spathes solitary, green, often suffused and/or striped with purple, pedunculate, falcate or not, 2.5–8 × 0.7–1.7 cm, margins distinct, scabrous, not ciliate, apex acuminate, glabrous to puberulent; peduncles 1.5–9.5 cm. **Flowers** bisexual and staminate; pedicels puberulent; petals dark blue, proximal petal somewhat smaller; staminodes 3; antherodes yellow, cruciform. **Capsules** 3-locular, 2-valved, 5–6 mm, apiculate. **Seeds** 5, brown, 2.2–2.7 × 1.7–2.2 mm, rugose, pitted.

Flowering summer–fall. Rocky soils; Ariz., Colo., N.Mex., Tex.; Mexico.

Two varieties have been recognized: *Commelina dianthifolia* var. *dianthifolia* (Arizona, New Mexico, Texas), with the spathes gradually tapering into a long, acuminate apex, and *C. dianthifolia* var. *longispatha* (Torrey) Brashier (Arizona, Colorado, New Mexico), with the spathes abruptly narrowed below the middle into a long, attenuate tip (C. K. Brashier 1966). Although most U.S. specimens are readily separable into these taxa, their ranges and ecologies overlap very broadly in Arizona and New Mexico. Until their variation in Mexico is studied, I can see no useful purpose in maintaining these varieties.

2. **Commelina communis** Linnaeus, Sp. Pl. 1: 40. 1753
 • Asiatic dayflower, comméline commune [F][I][W]

Herbs, annual, erect to decumbent, rooting at proximal nodes. **Stems** diffusely branched. **Leaves:** blade narrowly lanceolate to ovate-elliptic, 5–12 × 1–4 cm, apex acute to acuminate. **Inflorescences:** distal cyme usually vestigial, included, sometimes 1-flowered, exserted; spathes soli-

tary, green, paler or whitish basally with contrasting, dark green veins, pedunculate, usually not falcate, 1.5–3(–3.5) × 0.8–1.3(–1.8) cm, margins distinct, scabrous, not ciliate, apex acute to acuminate, glabrous to puberulent; peduncles 0.8–3.5(–5) cm. **Flowers** bisexual (rarely staminate); proximal petal paler or white, very reduced, distal petals blue to bluish purple; staminodes 3; antherodes yellow sometimes with central maroon spot, cruciform. **Capsules** 2-locular, 2-valved, 4.5–8 mm. **Seeds** 4, brown, (2–)2.5–4.2 × 2.2–3 mm, rugose pitted-reticulate.

Flowering summer–fall. Weedy and waste places; edges of fields, woods, and marshes, often in thick herbaceous vegetation, occasionally in woods; introduced; Ont., Que.; Ala., Ark., Conn., Del., D.C., Ga., Ill., Ind., Iowa, Kans., Ky., La., Md., Mass., Mich., Minn., Miss., Mo., Nebr., N.J., N.Y., N.C., Ohio, Okla., Pa., R.I., S.C., S.Dak., Tenn., Tex., Va., W.Va., Wis.; native, Asia.

Commelina communis var. *ludens* (Miquel) C. B. Clarke is distinguished by its darker flowers, antherodes with maroon centers (instead of entirely yellow), distalmost cyme less well developed and usually not producing a flower, and spathes proportionally broader. I have not found it possible to separate this regularly from *C. communis* var. *communis*, which also occurs in the flora. A variegated form of *C. communis* var. *ludens*, f. *aureostriata* MacKeever, occurs spontaneously and has been noted from Arkansas, Kentucky, Louisiana, Maryland, North Carolina, Texas, and Virginia.

SELECTED REFERENCES Pennell, F. W. 1937. *Commelina communis* in the eastern United States. Bartonia 19: 19–22. Pennell, F. W. 1938. What is *Commelina communis?* Proc. Acad. Nat. Sci. Philadelphia 90: 31–39.

3. **Commelina caroliniana** Walter, Fl. Carol., 68. 1788 [I]

Herbs, annual, diffusely spreading, rooting at nodes. **Stems** decumbent to scandent. **Leaves:** blade lanceolate to lanceolate-elliptic or lanceolate-oblong, 2.5–10.5 × 0.7–2.4 cm, margins scabrous, apex acute to acuminate, glabrous. **Inflorescences:** distal cyme vestigial, included (rarely 1-flowered and exserted); spathes solitary, bright

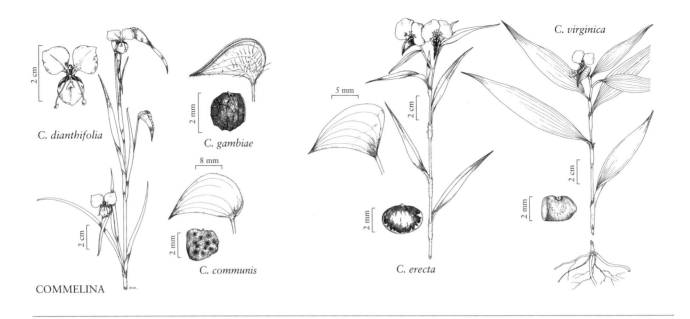

C. dianthifolia

C. gambiae

C. communis

COMMELINA

C. erecta

C. virginica

green, paler basally, without contrasting veins, pedunculate, not at all to slightly falcate, 1.2–3(–3.7) × 0.5–1 cm, margins distinct, usually ciliate, apex acuminate, glabrous or very sparsely pilose; peduncles 0.6–2.3 cm. **Flowers** bisexual; petals all blue, proximal petal white medially, smaller; medial stamen with white connective; staminodes 3; antherodes yellow, often with central maroon spot, cruciform. **Capsules** 3-locular, 2-valved, (5–)6–8 mm. **Seeds** 5, dark brown, 2.4–4.3(–4.6) × (1.6–)2–2.3 mm, smooth to faintly alveolate, mealy. $2n$ = ca. 86.

Flowering summer–fall (rarely winter). Fields, roadsides, railroad rights-of-way, yards, waste places, especially in moist situations, weed in crops, especially rice, sugar cane and corn, and rarely in forests; introduced; Ala., Ark., Fla., Ga., La., Md., Miss., Mo., N.C., S.C., Tex.; native, India.

SELECTED REFERENCE   Faden, R. B. 1989. *Commelina caroliniana* (Commelinaceae): A misunderstood species in the United States is an old introduction from Asia. Taxon 38: 43–53.

4. **Commelina diffusa** Burman f., Fl. Indica, 18, plate 7, fig. 2. 1768   I W

Herbs, perennial or annual, spreading. **Stems** decumbent to scandent. **Leaves:** blade narrowly lanceolate to lanceolate-oblong, lanceolate-elliptic or ovate, 1.5–14 × 0.5–3.3 cm, margins scabrous, apex acute to acuminate, glabrous. **Inflorescences:** distal cyme 1–several-flowered, usually

exserted; spathes solitary, bright green, without contrasting veins, pedunculate, usually distinctly falcate, (0.5–)0.8–4 × 0.4–1.2(–1.4) cm, margins distinct, glabrous or scabrous, sometimes also sparsely ciliate or ciliolate basally, apex usually acuminate, usually glabrous or nearly so; peduncles 0.5–2(–4) cm. **Flowers** bisexual and staminate; petals all blue (rarely all lavender), proximal petal smaller; medial stamen anther connective usually with transverse band of violet; staminodes 2–3; antherodes yellow, medial often absent or vestigial, cruciform. **Capsules** 3-locular, 2-valved, 4–6.3 mm. **Seeds** 5 (or less through abortion), brown, 2–2.8(–3.2) × 1.4–1.8 mm, deeply reticulate.

Varieties 4 (2 in the flora): tropical America, Asia, Africa, Pacific Islands.

The name *Commelina nudiflora* Linnaeus has been incorrectly used for this species.

1. Leaf blades 1.5–5(–8) × 0.5–1(–2.2) cm; medial anther connective with broad transverse violet band; capsules 5-seeded (occasionally less, by abortion). . . . . 4a. *Commelina diffusa* var. *diffusa*
1. Leaf blades 6–14 × 1–3.3 cm; medial anther connective without dark band; capsules typically 1–2-seeded. . . . . 4b. *Commelina diffusa* var. *gigas*

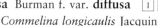

4a. Commelina diffusa Burman f. var. **diffusa** ⊡ ⊡

Commelina longicaulis Jacquin

Herbs, annual (sometimes perennial in south), diffusely spreading, rooting at nodes. **Leaves:** blade lanceolate to lanceolate-oblong, lanceolate-elliptic or ovate, 1.5–5(–8) × 0.5–1.8(–2.2) cm, apex acute to acuminate. **Inflorescences:** proximal cyme 2–4-flowered, distal cyme of larger spathes usually exserted, 1–several-flowered; spathes solitary, pedunculate, usually distinctly falcate, (0.5–)0.8–2.5(–3.7) × 0.4–1.2(–1.4) cm, apex usually acuminate; peduncles 0.5–2(–2.9) cm. **Flowers** blue (rarely lavender); medial stamen anther connective with broad, transverse band of violet. **Capsules** 3-locular, 2-valved, 4–6.3 × (2.1–)3–4 mm. **Seeds** 5, brown, 2–2.8(–3.2) mm × 1.4–1.8 mm, deeply reticulate. $2n = 30$.

Flowering spring–fall. Disturbed situations (lawns, gardens, and cultivated ground), moist places, and woods; introduced; Ala., Ark., D.C., Fla., Ga., Ill., Ind., Kans., Ky., La., Md., Miss., Mo., N.C., Ohio, Okla., S.C., Tenn., Tex., Va.; tropical America, Asia, Africa, Pacific Islands.

The report of this plant from Minnesota (H. A. Gleason and A. Cronquist 1991) is probably based on wrongly determined material.

4b. Commelina diffusa Burman f. var. **gigas** (Small) Faden, Ann. Missouri Bot. Gard. 80: 213. 1993 ⊡

Commelina gigas Small, Man. S.E. Fl., 264. 1933

Herbs, perennial, robust, spreading, sometimes scrambling in shrubs. **Leaves:** blade narrowly lanceolate to lanceolate-elliptic, 6–14 × 1–3.3 cm, apex acuminate. **Inflorescences:** distal cyme usually exserted, 1–3-flowered; proximal cyme 2–5-flowered; spathes pedunculate, falcate, 2.3–4 × 0.5–1.1 cm, apex acuminate; peduncles 1–2(–4) cm. **Flowers** blue; medial stamen anther connective without dark band. **Capsules** 3-locular, 2-valved. **Seeds** typically only 1–2 developing, dark brown, 2.1–2.8 mm, deeply reticulate. $2n = 90$.

Flowering spring–fall (perhaps all year round). Hammocks, streamsides, ditches, cypress swamps, wet woods, and lake shores; introduced; Fla.; Asia.

Commelina diffusa is a very variable species throughout its vast range. I have not been able to match *C. diffusa* var. *gigas* with specific herbarium specimens from elsewhere. It probably arrived as an introduction

instead of arising from diploid *C. diffusa* through *in situ* autopolyploidy.

SELECTED REFERENCE Austin, D. F. 1985. *Commelina gigas* rediscovered and lost. Palmetto 5(4): 11.

5. Commelina gambiae C. B. Clarke in A. L. P. de Candolle and C. de Candolle, Monogr. Phan. 3: 146. 1881 ⊡ ⊡

Commelina nigritana Bentham var. *gambiae* (C. B. Clarke) Brenan

Herbs, annual, 10–30 cm. **Stems** ascending to decumbent. **Leaves:** blade sessile, linear to linear-lanceolate, 2–15 × 0.3–1 cm, glabrous or sparsely hirsute. **Inflorescences:** proximal cyme ± 2-flowered, distal cyme absent; spathes solitary, whitish basally, pedunculate, usually slightly falcate, 0.9–2 × 0.4–1 cm, margins connate basally, apex acuminate, sparsely hirsute; peduncles 0.3–1.6 cm. **Flowers** bisexual, less than 1 cm wide; proximal sepals connate, forming cup; paired petals apricot- or peach-colored, proximal petal white, minute; staminodes 3; antherodes yellow, cruciform. **Capsules** 3-locular, 3-valved, 4–5.5 mm. **Seeds** 5, brown, 1.9–2.7 × 2–2.4 mm, reticulate. $2n = 56$.

Flowering fall. Roadsides, pastures, and levees; introduced; Fla.; native, w Africa.

Commelina gambiae was first found in Florida in 1976 and is now recorded from six counties. *Commelina nigratana* var. *nigritana*, which is unknown outside of Africa, differs by having three one-seeded locules, the seeds each with two pits and lacking the reticulation.

6. Commelina erecta Linnaeus, Sp. Pl. 1: 41. 1753 ⊡

Commelina angustifolia Michaux; *C. crispa* Wooton; *C. elegans* Kunth

Herbs, perennial. **Roots** fleshy, stout, tufted. **Stems** cespitose, usually erect to ascending (rarely decumbent, rooting at nodes). **Leaves:** leaf sheath auriculate at apex; blade sessile or petiolate, linear to lanceolate (rarely lanceolate-ovate), 5–15 × 0.3–4 cm, apex acuminate (rarely acute). **Inflorescences:** distal cyme absent; spathes solitary or clustered, green, pedunculate, not at all to strongly falcate, 1–2.5(–4) × 0.7–1.5(–2.5) cm, margins longly connate, glabrous except along connate edge, apex acute to acuminate, sometimes purple, usually variously pubescent; peduncles 0.5–1(–2) cm. **Flowers** bisexual and staminate, 1.5–4 cm wide; proximal petal minute, white, distal petals blue (rarely lavender or white); staminodes 3, staminodes and medial stamen entirely yellow; antherodes cruciform. **Capsules** 3-locular, 2-valved (very rarely 3-valved), 3.5–4.5 × 3–5 mm; abaxial locule warty, indehiscent (very rarely smooth and dehiscent); adaxial locules smooth, dehiscent. **Seeds** 3, brown, with soft, whitish tissue at both ends or in a band, 2.4–3.5 × 2.3–2.8 mm, nearly smooth. $2n = 60$.

Flowering spring–fall. Rocky woods and hillsides, scrub oak woods, pine woods and barrens, sand dunes, hammocks, shale barrens, roadsides, railroad rights-of-way, fields, and occasionally a weed in cultivated ground; Ala., Ark., Ariz., Colo., Del., D.C., Fla., Ga., Ill., Ind., Iowa, Kans., Ky., La., Md., Miss., Mo., Nebr., N.J., N.Mex., N.Y., N.C., Okla., Pa., S.C., Tenn., Tex., Va., W.Va., Wis.; Central America.

Commelina erecta grows in temperate regions of North and Central America, as well as in tropical regions.

This is by far the most variable species of *Commelina* in the flora. Three freely intergrading varieties may be recognized, although they are of questionable significance: *C. erecta* var. *erecta*, with larger leaves lanceolate to lanceolate-ovate, (1.5–)2–4 cm wide, and spathes (2.2–)2.5–3.6 cm, occurs throughout our region; *C. erecta* var. *angustifolia* (Michaux) Fernald, with leaves linear to narrowly lanceolate, 0.3–1.5 cm wide, and spathes 1–2 cm, is mainly southern but extends as far north as Virginia; and *C. erecta* var. *deamiana* Fernald, with leaves linear to narrowly lanceolate, 0.5–1.7 cm wide, and spathes 2–3.5 cm, occurs in midwestern United States south to Texas.

7. Commelina virginica Linnaeus, Sp. Pl. ed. 2, 1: 61. 1762 [E] [F]

Herbs, perennial. **Rhizomes** present. **Stems** erect to ascending, to 1 m. **Leaves** spirally arranged; leaf sheaths with red hairs at summit; blade lanceolate-elliptic to lanceolate-oblong, 6–20 × 1–5 cm, apex acuminate. **Inflorescences:** distal cyme vestigial, included (very rarely 1-flowered and exserted); spathes clustered, subsessile, funnelform, 1.5–3.5 × 1.2–2 cm, margins long-connate, glabrous (rarely puberulent). **Flowers** bisexual (rarely staminate); petals all pale blue, proximal one smaller; staminodes 3; antherodes entirely yellow, cruciform. **Capsules** 3-locular, 2-valved, (5.5–)6–9 × 3–6 mm. **Seeds** 5, brown, (2.4–)3–5(–6) × 2.4–3.1 mm, smooth with a few, shallow, irregular depressions, mealy. $n = 30$.

Flowering mid summer–fall. Wet places, especially swamps, river and stream banks, ditches, and bottom-lands, shade or full sun; Ala., Ark., Del., D.C., Fla., Ga., Ill., Ind., Ky., La., Md., Miss., Mo., N.J., N.C., Okla., Pa., S.C., Tenn., Tex., Va.

8. Commelina benghalensis Linnaeus, Sp. Pl. 1: 41. 1753 [I] [W]

Herbs, annual. **Roots** thin. **Rhizomes** short, subterranean, bearing cleistogamous flowers produced from base of plant. **Stems** ascending to decumbent or occasionally scrambling. **Leaves:** leaf sheaths, not auriculate, often with red hairs at summit; blade ovate to lanceolate-elliptic, (1–)2–9(–11) × 1–3(–4.5) cm, apex rounded, obtuse or acute, pubescent. **Inflorescences:** distal cyme often exserted and 1-flowered, sometimes vestigial; spathes often clustered, subsessile (peduncles 1–3.5 mm), funnelform, 0.5–1.5(–2) cm, margins long-connate, pubescent. **Flowers:** chasmogamous flowers bisexual and staminate, subterranean cleistogamous flowers bisexual; petals of chasmogamous flowers all blue (rarely lilac), proximal smaller; lateral stamen filaments not winged; staminodes 2–3; antherodes yellow, cruciform; lateral stamen pollen white; medial stamen pollen yellow. **Capsules** 3-locular, 2-valved, 4–6 mm. **Seeds** 5, brown or blackish, seeds of adaxial locule 1.7–2.5 mm, shallowly reticulate, mealy. $2n = 22$.

Flowering spring–fall. Citrus plantations, fields, yards, and other cultivated and disturbed sites; introduced; Calif., Fla., Ga., La.; tropical America; native, tropical Asia, Africa, Pacific Islands.

Commelina benghalensis was first recognized and collected in California in 1980. The California plants are approximately hexaploid and represent a separate and much later introduction than the southeastern plants. The species is federally listed as an Obnoxious Weed.

9. Commelina forskaolii Vahl, Enum. Pl. 2: 172. 1805 [I]

Herbs, annual or perennial, mat-forming. **Roots** thin. **Stems** erect to ascending, to 30 cm. **Leaves:** leaf sheaths not auriculate, with colorless hairs at summit; blade oblong or lanceolate-oblong to oblong-elliptic, 1.5–6 × 0.4–1.1 cm, margins strongly undulate, completely glabrous or adaxially sparsely pilose. **Inflorescences:** distal cyme usually exserted, 1-flowered; spathes solitary, pedunculate, strongly inflated, not falcate, 0.7–1.5 × 0.5–0.9 cm, margins connate, often violet, apex acute, sparsely hirsute; peduncles 0.4–1.1 cm. **Flowers** sometimes cleistogamous and subterranean (not yet seen in our area); chasmogamous flowers bisexual and staminate; petals blue, proximal one very reduced, conspicuous; antherodes entirely yellow, cruciform; lateral stamens with filaments laterally compressed and ± winged. **Capsules** 3-locular, 2-valved. **Seeds** sometimes 5, brown, usually only 1 abaxial locule seed developing, enclosed in decid-

uous, muricate, abaxial capsule valve; seeds of adaxial locule, when present, 2.5 mm, smooth. $2n = 30$.

Flowering summer–winter. Sanitary department landfill; introduced; Fla.; native, tropical Asia, Africa, Pacific Islands.

This species is known from a single population in Dade County, Florida, discovered in 1980. The population was sprayed with herbicide by the U. S. Department of Agriculture in 1984 because of the potential of the species to be a weed. Its current status is unknown.

208. ERIOCAULACEAE Desvaux

• Pipewort Family

Robert Kral

Herbs, annual or perennial, cespitose or solitary, rosulate, rarely caulescent, mostly scapose, glabrous or variously pubescent with simple or compound hairs. **Rootstocks** rhizomes or stems, thickened, short to variously elongate; roots fibrous, spongy, or spongy-septate (diaphragmatic). **Stems** erect to repent or prostrate, axis sympodial or monopodial, sometimes branching. **Leaves** mostly many ranked in rosettes, sometimes in loose spirals, mostly grasslike; blade linear to linear-triangular, lingulate, tapering, base mostly clasping; veins parallel. **Inflorescences** terminal and/or axillary, simple [compound], involucrate [proliferative], buttonlike or short-cylindric; scape sheaths spathelike, tubular, enclosing scape base, distally open; scapes 1–many, simple [compound], terete, usually twisted, mostly multiribbed; involucral bracts spirally imbricate series, usually chaffy or scarious, grading inward to receptacular bracts or these absent; receptacle glabrous or variously hairy. **Flowers** (florets) unisexual, staminate and pistillate on same [rarely different] plants, radially or bilaterally symmetric; sepals 2–3, distinct or variously connate, diverging from stipelike base or merely part of lobeless column; petals 0 or 2–3, diverging from short to elongate stipe (part of androphore or gynophore) or merely part of lobeless column; stamens 2–4(–6), often unequal; filaments arising from floral axis, rarely epipetalous; anthers mostly black, 1–2-locular, 2–4-sporangiate, versatile or basifixed, dehiscence longitudinal; pollen 1-grooved, 3-nucleate; appendages present in some flowers, glandlike or peglike, rarely bladelike, sometimes fringed or variously cleft; pistil compound, 2–3-carpellate; ovary superior, raised on gynophore, 1 locule per carpel; ovules 1 per locule, adaxial-apical, pendulous, orthotropous, bitegmic; style terminal, appendaged or unappendaged, 2–3-branched, branches simple or apex rebranched. **Fruits** capsules, thin-walled, loculicidal. **Seeds** translucent, ovoid, ellipsoid, or broadly fusiform, mostly 1 mm or shorter, variously ribbed or sculptured; endosperm copious, mealy-starchy, with compound starch grains; embryo apical.

Genera ca. 13, species 1200+ (3 genera, 16 species in the flora): nearly worldwide, primarily tropical and subtropical regions.

Most Eriocaulaceae grow in full sun in acidic wet soils or in aquatic situations.

SELECTED REFERENCES Kral, R. 1966. Eriocaulaceae of continental North America north of Mexico. Sida 2: 285–332. Kral, R. 1989. The genera of Eriocaulaceae in the southeastern United States. J. Arnold Arbor. 70: 131–142. Moldenke, H. N. 1937. Eriocaulaceae. In: N. L. Britton et al., eds. 1905+. North American Flora. . . . 47+ vols. New York. Vol. 19, pp. 17–50. Tomlinson, P. B. 1969. Eriocaulaceae. In: C. R. Metcalfe, ed. 1960+. Anatomy of the Monocotyledons. 8+ vols. Oxford. Vol. 3, pp. 146–192.

1. Lacunar tissue evident at leaf base; larger roots pale, thickened, septate, unbranched; perianth parts in 2s (except in *Eriocaulon cinereum* and *E. microcephalum*), petals with glands on adaxial surface; stamens 3–4 or 6; anthers 2-locular, apex of staminal column with 2–3 glands, glands unappendaged; pistil 2(–3)-carpellate; style unappendaged, 2(–3)-branched . . . . 1. *Eriocaulon*, p. 199

1. Lacunar tissue not evident at leaf base; larger roots either dark, fibrous, and evidently branched or pale, thickened, and spongy, neither septate nor branched; perianth parts in 3s (except in *Lachnocaulon digynum*), petals if present without glands; stamens (2–)3; anthers 1–2-locular; apex of staminal column usually with 3 glands, glands appendaged or unappendaged; pistil (2–)3-carpellate; style appendaged, 2–3-branched.

    2. Roots dark, slender, fibrous, evidently branched; scapes glabrous or hairy, hairs neither swollen nor glandular; expanded inflorescences ovoid to globose or short-cylindric, basal involucral bracts reflexed, obscured by inflorescence; hairs of perianth club-shaped; staminal filaments adnate to rim of androphore; anthers 1-locular; style branches 2-cleft . . . . 2. *Lachnocaulon*, p. 206

    2. Roots pale, thickened, spongy, appearing unbranched; scapes hairy, at least some hairs swollen basally, often glandular distally; expanded inflorescences hemispheric to globose, basal bracts not obscured by inflorescence; hairs of perianth tapering, acute, not club-shaped; staminal filaments low in corolla tube; anthers 2-locular; style branches undivided. . . . . 3. *Syngonanthus*, p. 209

1. ERIOCAULON Linnaeus, Sp. Pl. 1: 87. 1753; Gen. Pl. ed. 5, 38. 1754 • Pipewort, button-rods, hat-pins, ériocaulon [derived from Greek *erion*, wool, and *caulos*, stalk]

Herbs, annual or perennial, often cespitose, rosulate. **Roots:** larger roots unbranched, pale, septate, thickened, spongy. **Stems** rarely sparingly branched, short or elongate. **Leaves** many ranked in flat or high spiral; blade basally pale, distally greener, linear-attenuate or triangular-acuminate, lingulate, narrowing gradually or abruptly from base, base noticeably lacunate, less distinctly so distally. **Inflorescences:** scape sheaths tubular, orifice oblique (often 2–3-cleft); scapes 1–several per rosette, glabrous; heads pale to dark, white, gray, or gray-brown, hemispheric to globose or short-cylindric; receptacle hairy or glabrous; involucral bracts obscured or not obscured by inflorescence, pale to dark, chaffy or scarious; receptacular bracts narrower, thinner than involucral bracts, often scarious. **Flowers** mostly staminate and pistillate on same plants, 2–3-merous; sepals 2(–3), adnate to stipelike base, boat-shaped, scarious, apex often covered with multicellular hairs, hairs mealy white or translucent, frequently club-shaped; petals 2(–3), narrower, shorter than sepals, apex hairy, hairs club-shaped, glands adaxial, subapical, dark, rarely pale. **Staminate flowers:** androphore apically dilated stalk; petals separated from sepals by androphore, diverging as lobes from apex; stamens 3–4 or 6, 2–3 alternating with petals; apex of staminal column with 2–3 glands, glands unappendaged; filaments arising from androphore rim; anthers 2-locular, 4-sporangiate, dorsifixed, usually versatile, well exserted at anthesis, jet black (except in *E. cinereum*). **Pistillate flowers:** gynophore separating petals from sepals, stipelike; pistil 2(–3)-carpellate; style 1, unappendaged, style branches 2(–3).

Species ca. 400 (11 in the flora): mostly pantropic, mostly aquatic or on wet, mainly acidic substrates.

1. Receptacle and/or base of flowers copiously hairy; some or most receptacular bracts and
   perianth parts with chalk white hairs; heads white, 5 mm or more in full flower or in fruit.
   2. Heads hard, very slightly flattened when pressed; scape sheaths shorter than most leaves;
      involucral bracts straw-colored, apex acute; receptacular bracteoles pale, apex narrowly
      acuminate; pistillate flower petals adaxially glabrescent; terminal cells of club-shaped
      hairs of perianth whitened, basal cells often uncongested, transparent; plants of moist
      but seldom aquatic or permanently wet situations. . . . . 5. *Eriocaulon decangulare*
   2. Heads soft, much flattened when pressed; scape sheaths longer than most leaves; invo-
      lucral bracts gray or dark, apex rounded or obtuse; receptacular bracteoles gray to dark
      gray, apex acute; pistillate flower petals adaxially villous; all cells of club-shaped hairs
      on perianth mealy white; plants in aquatic or wet substrates.
      3. Mature heads 10–20 mm wide; leaves 5–30 cm; petals of staminate flower conspicu-
         ously unequal. . . . . 4. *Eriocaulon compressum*
      3. Mature heads 5–10 mm wide; leaves (1–)2–5(–7) cm; petals of staminate flower
         nearly equal. . . . . 6. *Eriocaulon texense*
1. Receptacle and/or base of flowers glabrous or sparingly hairy; receptacular bracteoles and/
   or perianth glabrous or hairy, hairs club-shaped, clear or white; heads dark gray or white,
   usually less than 7.5 mm in full flower or in fruit.
   4. Stamens 6, pistil 3-carpellate.
      5. Anthers yellow; apex of receptacular bracteoles acute. . . . . 10. *Eriocaulon cinereum*
      5. Anthers black, apex of receptacular bracteoles obtuse. . . . . 11. *Eriocaulon microcephalum*
   4. Stamens 4, pistil 2-carpellate
      6. Heads 4–10 mm wide at maturity; outer involucral bracts usually reflexed, obscured
         by bracteoles and flowers.
         7. All bracts of staminate and pistillate flowers straw-colored or pale with grayish
            midzone; sepals of pistillate flowers basally pale, darkening distally to grayish,
            gray-green, or gray-brown; heads (young or mature) very pale; seeds faintly
            rectangular-reticulate, often papillate in lines; s coastal plain. . . . . 7. *Eriocaulon lineare*
         7. Inner involucral bracts, receptacular bracts, and sepals darkened, usually gray to
            near black; young heads dark; seeds very faintly reticulate, not papillate; n and/or
            montane. . . . . 8. *Eriocaulon aquaticum*
      6. Heads seldom as wide as 5 mm; outer involucral bracts not reflexed, not obscured by
         bracteoles and flowers.
         8. Bracts straw-colored, greenish, or light gray to gray, dull, margins often erose or
            lacerate, apex blunt to obtuse; scapes linear; plants of brackish substrates. . . . 2. *Eriocaulon parkeri*
         8. Bracts dark, gray to blackish, very lustrous, margins all nearly entire (except
            Eriocaulon nigrobracteatum), apex acute; scapes filiform; plants of acidic sub-
            strates.
            9. Bracts narrowly ovate to oblong or spatulate, apex acute; bracts and perianth
               parts (except petals in some cases) glabrous; seed conspicuously pale-reticulate
               . . . . 1. *Eriocaulon ravenelii*
            9. Bracts wider in outline, apex rounded or apiculate; bracts (margins and apex)
               and perianth hairy; seed not pale-reticulate.
               10. Petals of pistillate flowers stipitate, suborbiculate-rhombic; outer involu-
                  cral bracts straw-colored, inner and receptacular bracts dark gray, gray-
                  green, or gray-brown. . . . . 3. *Eriocaulon koernickianum*
               10. Petals of pistillate flowers short-stipitate or nearly sessile, oblong; involu-
                  cral and receptacular bracts blackish or with pale base. . . . . . 9. *Eriocaulon nigrobracteatum*

1. **Eriocaulon ravenelii** Chapman, Fl. South. U.S., 503. 1860 [C][E][F]

Herbs, biennial or perennial, 4–20 cm. **Leaves** linear-attenuate, 3–7(–15) cm, apex subulate, flat. **Inflorescences:** scape sheaths shorter than leaves, loose; scapes filiform, 0.5 mm wide, 4–5(–6)-ribbed; heads gray-brown to dark gray, nearly globose or ovoid, 3–4 mm wide, relatively soft; receptacle glabrous or with sparse, clear hairs; outer involucral bracts usually not reflexed, not obscured by bracteoles and perianth, gray, ovate-oblong or broadly spatulate, 2 mm, margins nearly entire, apex narrowly rounded to acute, surfaces glabrous; inner bracts, receptacular bracts gray, very lustrous, narrowly ovate to oblong or spatulate, 2 mm, margins nearly entire, apex acute to acuminate or lacerate, glabrous. **Staminate flowers:** sepals 2, gray, oblong to oblanceolate, 1.5–2 mm, apex usually acute, glabrous; androphore narrowly club-shaped; petals 2, pale, scalelike, minute, glabrous, with inconspicuous glands; stamens 4; anthers black. **Pistillate flowers:** sepals 2, gray, narrowly oblong to lance-linear, 1.5–2 mm, apex acute, glabrous; petals 2, yellow-white, narrowly oblanceolate or oblong, 1.5–2 mm, apex acute, glabrous or with a few hairs at apex or adaxially, with inconspicuous glands; pistil 2-carpellate. **Seeds** dark brown, somewhat lustrous, broadly ellipsoid, 0.5 mm, conspicuously irregularly pale-reticulate, alveolae mainly rectangular.

Flowering late summer–winter. Mildly acid sandy pineland swamps, particularly on wet fluctuating shores of shallow ponds toward coasts; of conservation concern; 0–100 m; Fla., S.C.

2. **Eriocaulon parkeri** B. L. Robinson, Rhodora 5: 175. 1903 • Ériocaulon de Parker [E]

Herbs, perennial, 10–20(–30) cm. **Leaves** linear-attenuate, 2–6(–9) cm, apex filiform-terete. **Inflorescences:** scape sheaths slightly longer or slightly shorter than leaves, loose; scapes linear, 0.5–1 mm wide, 4–5-ribbed; mature heads dull gray or lead-colored, rarely straw-colored, hemispheric to subglobose, 3–4 mm wide, mostly nearly glabrous; receptacle glabrous; outer involucral bracts usually not reflexed, not obscured by braceteoles and perianth, straw-colored, greenish, or light gray to gray, dull, ovate to suborbiculate or obovate, 2 mm, margins often erose or lacerate, apex blunt, glabrous; inner bracts, receptacular bracteoles grayish, cuneate to narrowly obovate, 2 mm, margins often erose or lacerate, apex obtuse, glabrous or with a few white hairs abaxially at apex. **Staminate flowers:** sepals 2, gray, linear to oblong or oblanceolate, 2 mm, apex obtuse, glabrous or with a few white hairs abaxially at apex; androphore club-shaped; petals 2, triangular, minute, white-hairy; stamens 4; anthers black. **Pistillate flowers:** sepals 2, gray, oblong or oblanceolate, 2 mm, scarious, apex obtuse, abaxially hairy apically; petals 2, yellow-white, spatulate, 2 mm, apex obtuse, glabrous or with a few white, club-shaped hairs apically, adaxially; pistil 2-carpellate. **Seeds** red-brown, ovoid to broadly ellipsoid, 0.5(–7) mm, with delicate reticulum of horizontally oriented alveolae.

Flowering summer–fall. Muddy tidewater banks, brackish marsh, mud flats; 0–100 m; Que.; Conn., Del., Maine, Md., Mass., N.J., N.Y., N.C., Va.

A considerable amount of transitional material occurs between *Eriocaulon parkeri* and *E. aquaticum* at places along coastal streams where brackish habitat meets more acid habitat upstream.

3. **Eriocaulon koernickianum** Van Heurck & Müller-Argoviensis in H. F. Van Heurck, Observ. Bot. 1: 101. 1870 [C][E][F]

Herbs, perennial, 5–8 cm. **Leaves** linear-attenuate, 2–5 cm, apex subulate to blunt. **Inflorescences:** scape sheaths as long as leaves, inflated; scapes filiform, 0.5 mm wide, 3–4-ribbed; heads dark gray or gray-green with rims of bracts and perianth pale, nearly globose or short-oblong, 3–4 mm wide, soft; receptacle glabrous; outer involucral bracts usually not reflexed, not obscured by bracteoles and perianth, straw-colored, very lustrous, broadly oblong to suborbiculate, 1–1.25 mm, margins nearly entire, apex rounded, glabrous; inner bracts, receptacular bracteoles dark gray, gray-green, or gray-brown, very lustrous, oblong to cuneate, obliquely keeled, 1.5 mm, margins slightly erose, apex acute to obtuse, apiculate, with a few white, club-shaped hairs. **Staminate flowers:** sepals 2, grayish, linear-curvate, 1–1.5 mm, apex with a few white, club-shaped hairs abaxially, marginally; androphore broadly club-shaped; petals 2, low, tooth-like, nearly equal, apex with club-shaped hairs; stamens 4; anthers black. **Pistillate flowers:** sepals 2, gray, linear-curvate, 1 mm, apex with scattered hairs abaxially, hairs pale, club-shaped, otherwise glabrous; petals 2, yellow-white, stipitate, broadly suborbiculate-rhombic, 1 mm, apex with white, club-shaped hairs abaxially;

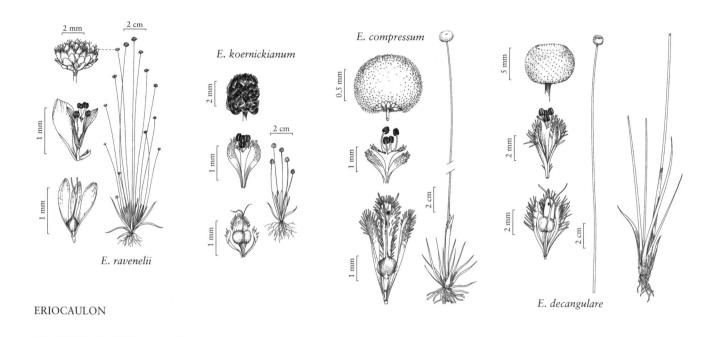

E. koernickianum

E. compressum

E. ravenelii

E. decangulare

ERIOCAULON

pistil 2-carpellate. **Seeds** deep reddish brown, broadly ovoid or ellipsoid, 0.5 mm, often indistinctly reticulate or rugulose, papillate.

Flowering spring–early fall. Moist to wet sands and sandy silts of seeps, particularly over and around arenaceous outcrops; of conservation concern; 0–1000 m; Ark., Ga., Okla., Tex.

Eriocaulon koernickianum is the most diminutive of our eriocaulons, widely disjunct in Georgia but seemingly most abundant on Piedmont granites.

4. **Eriocaulon compressum** Lamarck in J. Lamarck et al., Encycl. 3: 276. 1789 E F

Eriocaulon decangulare Walter 1788, not Linnaeus 1753; *E. cephalotes* Michaux; *Sphaerochloa compressa* Palisot de Beauvois

Herbs, perennial, 20–70 cm. **Leaves** linear-attenuate, 5–30 cm, apex subulate. **Inflorescences:** scape sheaths mostly longer than principal leaves, loose; scapes linear, 1–3 mm wide, multiribbed (ribs lacunar); heads chalk white except for dark gray or near black exserted tips of receptacular bracts, anthers, hemispheric to subglobose, 10–20 mm wide, soft, much flattened when pressed; receptacle pilose; involucral bracts frequently squarrose, later obscured by mature bracteoles and flowers, gray, broadly ovate to oblong or elliptic, 2–3 mm, margins entire, apex rounded or obtuse,

glabrous; inner bracts, receptacular bracteoles dark gray, spatulate-linear to oblong, 2–3 mm, margins entire, apex acute with white, club-shaped hairs. **Staminate flowers:** sepals 2, pale or with dark apex, linear or linear-spatulate, 2–4 mm, apex acute to blunt with mealy white, club-shaped hairs; androphore broadly club-shaped; petals 2, pale, oblong, conspicuously unequal, larger lobe apically fringed with pale, club-shaped hairs, smaller lobe glabrous or with a few hairs at apex; stamens 3–4(–6); anthers black. **Pistillate flowers:** sepals 2, dark at apex, oblong-spatulate, 2.5–3 mm, abaxially with mealy white, club-shaped hairs, adaxially with translucent hairs; petals 2, pale, oblong-spatulate, apex acute, abaxially with mealy white, club-shaped hairs, adaxially with translucent hairs; pistil 2-carpellate. **Seeds** dark lustrous brown, broadly ovoid to near round but asymmetric, 0.5 mm, mostly minutely spiny-papillate.

Flowering late winter–spring. Sands and peats of shallow pineland ponds, savanna, seeps, ditches, or low flatwoods; 0–300 m; Ala., Del., Fla., Ga., La., Md., Miss., N.J., N.C., S.C., Tex.

Eriocaulon compressum is polymorphic in habit. Male flowers vary considerably in length and degree of connation, and female flowers are often sterile. Of the southeastern coastal plain species this and the similar, but proportionately smaller, *E. lineare*, are the most aquatic, the former most common around clay-based ponds, the latter around karst ponds. Unlike the more northern *E. aquaticum*, these two species seldom frequent the shallows of flowing streams.

5. Eriocaulon decangulare Linnaeus, Sp. Pl. 1: 87. 1753

F

Eriocaulon serotinum Walter; *E. statices* Crantz; *Randalia decangularis* Palisot de Beauvois; *Symphachne xyrioides* (Linnaeus) Palisot de Beauvois

Herbs, perennial, 30–110 cm. **Leaves** linear or linear-attenuate, abruptly, then gradually, narrowing, 10–40(–50) cm, apex acute or obtuse. **Inflorescences:** scape sheaths shorter than principal leaves; scapes linear, distally 1–2(–3) mm wide, multiribbed; heads dull white, hemispheric to nearly globose, 7–15 mm wide, hard, very slightly flattened when pressed; receptacle copiously hairy; involucral bracts reflexed, obscured by reflexed proximal bracteoles and flowers, straw-colored, narrowly ovate to lanceolate, 2–4 mm, margins erose to entire, apex acute, glabrous or apex with white, club-shaped hairs; inner bracts and receptacular bracteoles pale, linear to oblong or lanceolate, 3–4 mm, margins entire, sometimes becoming erose, apex narrowly acuminate, glabrous or apex with white, club-shaped hairs. **Staminate flowers:** sepals 2, yellow-white, linear, curved, 3 mm, distal surface abaxially, adaxially with white, club-shaped hairs; androphore club-shaped; petals 2, yellow-white, translucent, triangular to linear, nearly equal, with small tuft of white, club-shaped hairs abaxially at apex; stamens 4; anthers black. **Pistillate flowers:** sepals 2, yellow-white, linear, 2–3 mm, apex acute, abaxially with white, club-shaped hairs at apex; petals 2, pale, spatulate or narrowly elliptic, 1–2 mm, abaxially with translucent hairs proximally, white, club-shaped hairs distally, or adaxially glabrescent; pistil 2-carpellate. **Seeds** pale brown, ellipsoid, 0.75–1 mm, very finely cancellate, sometimes with cancella concealed by rows of delicate nearly appressed hairs.

Flowering late spring–summer. Moist to wet sands or peats of shores, pine savanna, ditches, edges of cypress domes or savanna; 0–300 m; Ala., Ark., Del., D.C., Fla., Ga., La., Md., Miss., N.J., N.C., S.C., Tenn., Tex., Va.; Mexico; Central America (Nicaragua).

A possible variety, *Eriocaulon decangulare* var. *latifolium* Chapman ex Moldenke [in N. L. Britton et al., N. Amer. Fl. 19(1): 21. 1937], has yet to be thoroughly investigated. This plant typically is in the taller range and has thicker stems and scapes than in the type; it has very stiff blunt leaves to 50 cm × 13–20 mm with thicker scapes, heads 13–20 mm wide, and floral parts in the high range for the species. It occurs in wetter situations than usual for the species, and (fide R. R. Haynes, pers. comm.) flowers later. This morphology occurs in northwestern Florida and southern Alabama within boggy edges of cypress-titi-*Nyssa* on permanently wet substrates.

6. Eriocaulon texense Körnicke, Linnaea 27: 494. 1856

E

Herbs, perennial, 5–30 cm. **Leaves** linear-attenuate, (1–)2–5(–7) cm, apex subulate. **Inflorescences:** scape sheaths mostly longer than principal leaves, inflated; scapes filiform to linear, 0.5–1 mm wide, 4–7-ribbed; mature heads white or gray, hemispheric to nearly globose, 5–10 mm wide, soft, much flattened when pressed; receptacle pilose; involucral bracts becoming reflexed, obscured by bracteoles and proximal flowers, straw-colored to gray, suborbiculate to broadly obovate, 1.5 mm, margins entire, apex rounded to apiculate, glabrous; inner bracts, receptacular bracteoles gray to dark gray, obovate to cuneate, 1.5 mm, apex obtuse to acute, blade densely ciliate, hairs white, club-shaped. **Staminate flowers:** sepals 2, proximally pale, distally dark gray, opaque, linear-spatulate, 1.5 mm, apex acute, distal adaxial surfaces, margins with white, club-shaped hairs; androphore club-shaped; petals 2, unequal, narrowly triangular, short, apically ciliate with white, club-shaped hairs; stamens 4; anthers black. **Pistillate flowers:** sepals 2, dark gray, narrowly oblong-obovate, curved, keeled, 1.5 mm, keels with white, club-shaped hairs; petals 2, yellow-white, linear to oblong or obovate, 1–1.5 mm, apex acute, blade ciliate with white, club-shaped hairs, adaxial surfaces pilose with mixture of translucent, tapering and white, club-shaped hairs; pistil 2-carpellate. **Seeds** brownish, ovoid or broadly ellipsoid, 0.6 mm, indistinctly alveolate, proximal ribs often with fine pale papillae.

Flowering in spring. Sphagnous bogs and low pine savanna, seeps; 0–150 m; Ala., Ga., La., Miss., N.C., Tex.

In habit *Eriocaulon texense* is much like a diminutive version of *E. compressum*, likewise preferring deep bogs and also just as variable in the color of its bracts, bracteoles, and perianth.

7. Eriocaulon lineare Small, Fl. S.E. U.S., 236, 1328. 1903 E

Herbs, perennial, 6–20 cm (–80 cm when submersed). Leaves linear-attenuate, 1–10 cm (–20 cm when submersed), apex filiform. Inflorescences: scape sheaths mostly shorter than leaves in emergents, exceeding them in drier sites; scapes linear, 1 mm wide, 4–7-ribbed; heads (young or mature) very pale, hemispheric to nearly globose, rarely short-cylindric, 4–10 mm wide, soft, flattened when pressed; receptacle glabrous; involucral bracts sometimes squarrose, obscured by reflexed bracteoles and flowers of mature heads, straw-colored, orbiculate to ovate, 2–2.5 mm, margins entire, apex rounded, with white, club-shaped hairs; receptacular bracts and bracteoles pale except for grayish midzone, obovate to cuneate, 2 mm, margins entire, apex acute, ciliate, distal abaxial surfaces with white, club-shaped hairs. Staminate flowers: sepals 2, grayish, oblong-linear, curvate, 1.5–2 mm, apex acute with white, club-shaped hairs; androphore club-shaped; petals 2, pale, triangular, 0.5 mm, ciliate, hairs club-shaped; stamens 4; anthers black. Pistillate flowers: sepals 2, basally pale, darkening distally to grayish, gray-green, or gray-brown, narrowly oblong-obovate, curved, 2 mm, apex rounded, abaxially with white, club-shaped hairs; petals 2, yellow-white, broadly spatulate, flat, 1.5–2 mm, apex rounded, abaxially with white hairs, adaxially with white or clear hairs; pistil 2-carpellate. Seeds dark red-brown, ovoid or ellipsoid, 0.5–0.75 mm, faintly rectangular-reticulate, often papillate in lines.

Flowering mostly summer–fall. Sandy or peaty shores, hypericum ponds, wet savanna, southern coastal plain terraces; 0–100 m; Ala., Fla., Ga., N.C.

Eriocaulon lineare closely resembles *E. texense*, although it has paler bracts and flowers and a glabrous (rather than pilose) receptacle. *Eriocaulon lineare* blooms later and is most common in the margins or shallows of ponds, rather than in the sphagnous bogs favored by *E. texense*.

8. Eriocaulon aquaticum (Hill) Druce, Pharm. J. 83: 700. 1909 • Ériocaulon aquatique

Cespa aquatica Hill, Herb. Brit., 96*, plate 66bis. 1769; *Eriocaulon articulatum* (Hudson) Morong; *E. pellucidum* Michaux; *E. pumilum* Rafinesque; *Nasmythia articulata* Hudson; *N. septangularis* (Withering) Martius

Herbs, perennial, 4–21 cm (–100 cm when submersed). Leaves linear-attenuate, 1–10 cm (–40 cm when submersed), apex often subulate. Inflorescences: scape sheaths longer (shores) or shorter (submersed) than principal leaves, inflated; scapes linear to filiform, 1 mm wide, (4–)5–7-ribbed; mature heads white to pale gray, young heads dark, hemispheric to globose, 4–10 mm wide, soft, flattened when pressed; receptacle glabrous or rarely with a few clear hairs; involucral bracts becoming reflexed, obscured by proximal bracts and flowers, dark gray, broadly oblong to narrowly ovate or broadly obovate, 1–1.5 mm, margins entire, apex rounded, abaxially with white, club-shaped hairs, otherwise glabrous; inner and receptacular bracts gray to near black, oblanceolate or cuneate, 1.5 mm, margins entire, apex acute, distal margins and abaxial apical surface with white, club-shaped hairs. Staminate flowers: sepals 2, grayish, oblong-linear or linear-oblanceolate, curved, 1.5 mm, apex acute to rounded, margins and abaxial surface with white, club-shaped hairs; androphore club-shaped; petals 2, triangular, nearly equal, 0.5 mm, ciliate, hairs white, club-shaped; stamens 4; anthers black. Pistillate flowers: sepals 2, gray, oblong to narrowly obovate, curved, keeled, 1.5 mm, blade usually ciliate, distal abaxial surfaces with white hairs; petals 2, pale, oblong-linear to linear oblanceolate, 1.5 mm, apex acute to obtuse, apically ciliate, adaxially with white, club-shaped hairs; pistil 2-carpellate. Seeds light brown or red-brown, ovoid to broadly ellipsoid, 0.5 mm, very faintly reticulate, not papillate.

Flowering spring–fall. Sandy or peaty, often sphagnous, shores, bogs, muskegs, shallows; 0–600 m; N.B., Nfld. and Labr. (Nfld.), N.S., Ont., Que.; Ala., Conn., Del., Ind., Maine, Md., Mass., Mich., Minn., N.H., N.J., N.Y., N.C., Ohio, Pa., R.I., S.C., Vt., Va., Wis.; Europe (Great Britain, Ireland).

The name *Eriocaulon septangulare* Withering, widely used for this species, is invalid (H. N. Moldenke 1937). Some (T. G. Tutin et al. 1964–1980, vol. 5) retain *E. aquaticum* (Hill) Druce as the valid name if one accepts both North American and European plants as the same species (the alternative taken here). If North

American plants are considered to be distinct from Eurasian ones, then the appropriate binomial for ours becomes *E. pellucidum* Michaux.

9. Eriocaulon nigrobracteatum Orzell & E. L. Bridges, Phytologia 74: 105, figs. 1–4. 1993 [C][E][F]

Herbs, perennial, 5–15(–19) cm. Leaves linear-attenuate, 0.5–1.5 (–4) cm, apex subulate. Inflorescences: scape sheaths longer than principal leaves; scapes filiform, 0.3–0.4 mm, 4-ribbed; mature heads blackish at level of involucre, white above it, hemispheric to globose, 3–4(–5) mm wide, soft; receptacle glabrous; involucral bracts spreading, not obscured by bracteoles and flowers, blackish, very lustrous, broadly ovate to orbiculate, 1–3 mm, margins entire to erose, apex broadly rounded, glabrous; inner bracts, receptacular bracteoles blackish or with pale base, broadly to narrowly ovate, 2–3 mm, margins somewhat erose with age, apex rounded or apiculate, margins and abaxial surfaces with white, club-shaped hairs, apically ciliate. Staminate flowers: sepals 2, gray-brown, broadly spatulate to oblong-obovate, curved, keeled, 1–1.2 mm, keel and margin with white, club-shaped hairs; androphore club-shaped to campanulate, flaring; petals 2, oblong-obovate, nearly equal, 1 mm, apex ciliate and abaxially with white, club-shaped hairs; stamens 4; anthers black. Pistillate flowers: sepals 2, gray, oblong, curved, keeled, 1 mm, adaxial surfaces pilose, hairs translucent, sharp, apically with white, club-shaped hairs; petals 2, pale, short-stipitate or nearly sessile, oblong, 1 mm (narrower and longer than those of *E. koernickianum*), apex obtuse, adaxial surface and margins with white, club-shaped hairs; pistil 2-carpellate. Seeds brown, nearly round, 0.3 mm, obscurely reticulate, alveolae irregularly and horizontally rectangular.

Flowering spring. Muck of deep sphagnous seep bogs in longleaf pinelands; of conservation concern; 0–100 m; Fla.

Eriocaulon nigrobracteatum is a very limited, local endemic whose relationships with *E. aquaticum* should be clarified.

SELECTED REFERENCE   Orzell, S. L. and E. L. Bridges. 1993. *Eriocaulon nigrobracteatum* (Eriocaulaceae), a new species from the Florida panhandle with a characterization of its poor fen habitat. Phytologia 74: 104–124.

10. Eriocaulon cinereum R. Brown, Prodr., 254. 1810 [I]

Eriocaulon sieboldianum Siebold & Zuccarini ex Steudel

Herbs, annual or short-lived perennial, 15–30 cm. Leaves linear-attenuate, to 10 cm, apex subulate-filiform. Inflorescences: scape sheaths shorter than principal leaves, loose; scapes filiform, 0.5 mm wide, 6–8-ribbed; mature head silvery-gray, ovoid to nearly globose, 4–5 mm wide, soft; receptacle sparsely pilose, hairs translucent; involucral bracts spreading-ascending, not obscured by bracteoles and flowers of mature heads, pale, ovate to oblong, oblanceolate or broadly obovate, grading narrower inward, 2 mm, margins often lacerate or erose, apex acute, glabrous; receptacular bracts pale except for gray midzone, linear-oblong, 2 mm, margins erose or entire, apex acute, glabrous. Staminate flowers: sepals 3, dark, either deeply and narrowly lobed or broadly spatulate, 3-dentate, 1.5 mm, distally with scattered white, linear, short hairs; androphore flaring; petals 3, oblong, apex with white, tapering hairs; stamens 6; anthers yellow. Pistillate flowers: sepals 3, 2 linear-elliptic, 1 broadly elliptic-oblanceolate, concave, 1.5 mm, apex acute, sometimes with few translucent hairs; gynophore 1 mm with 3 spreading, peglike, minute appendages; pistil 3-carpellate. Seeds pale brown, ovoid or broadly ellipsoid, 4.5 mm, finely reticulate, alveolae horizontally rectangular.

Flowering summer–fall. Muck of rice fields; 0–200 m; introduced; Calif., La.; Europe (Italy); e Asia (Southeast Asia); Australia.

11. Eriocaulon microcephalum Kunth in A. von Humboldt et al., Nov. Gen. Sp. 1: 253. 1816

Herbs, perennial, to 5 cm. Leaves: narrowly triangular-acuminate, 1.5–3 cm, apex narrow, blunt, calloused. Inflorescences: scape sheaths mostly shorter than longer leaves, dilated apically; scapes linear, 1 mm wide, distally 4–5-ribbed; mature heads pale, hemispheric to globose, 3–4 mm wide, soft; receptacle glabrous or sparsely pilose; involucral bracts erect or ascending, nearly covering head, yellow-white or yellow-brown, broadly obovate to suborbiculate, grading narrower inward, 2 mm, margins entire, erose with age, apex rounded, surfaces glabrous; receptacular bracteoles dark brown, narrowly obovate, 2 mm, margins entire, erose with age, apex obtuse, margins and abaxial sur-

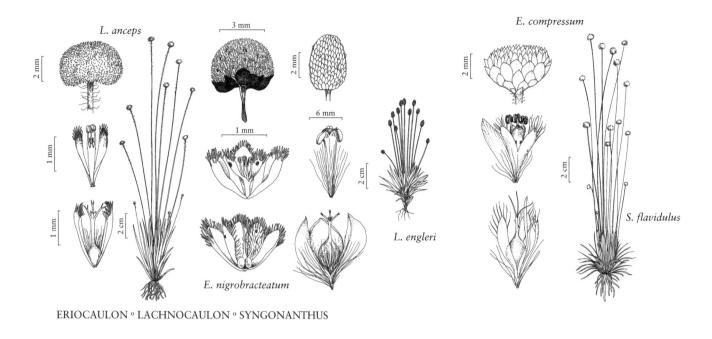

ERIOCAULON ∘ LACHNOCAULON ∘ SYNGONANTHUS

face with scattered, stubby, white, multicellular hairs. **Staminate flowers:** sepals 3, gray, cuneate-spatulate, 2 mm; apex shallowly 3-cleft, surfaces distally hairy, ciliolate, hairs white, stubby; androphore cylindric; petals 3, minute, unequal, white-ciliate; stamens 6; anthers nearly black. **Pistillate flowers:** sepals 3, gray, translucent, curved, keeled, 2 mm, apex rounded, apiculate, ciliate, hairy with white, blunt hairs; petals 3, yellow-white, oblong or narrowly spatulate, flat, 2 mm, apex rounded, blade adaxially pilose with translucent hairs, apically ciliate with white hairs; pistil 3-carpellate. **Seeds** rich red-brown, ovoid or ellipsoid, 0.6–0.8 mm, faintly reticulate with horizontally aligned rectangular alveolae.

Flowering summer–fall. Moist boggy upland meadows (paramos), western-montane; above 1000 m; Calif.; Mexico; Central America; South America.

Eriocaulon microcephalum is known in the flora only from one locality, Kern County, California; this was a collection by L. J. Xantus de Vesey in 1857–58 from "vicinity of Fort Tijon." It has not been found since. The specimen differs, however, in no way from those I have seen throughout the range, and in light of the number of Mexican western cordilleran plant species that have been continuously discovered in California, it seems best to include the species here.

2. LACHNOCAULON Kunth, Enum. Pl. 3: 497. 1841 • Hat-pins, bog bachelor's-buttons [Greek *lachnos*, wool, and *chaulos*, stem, in reference to the long, soft, upwardly pointed hairs on scapes of the type]

Herbs, perennial, cespitose, rosulate. **Roots** branched, dark, not septate, slender, fibrous. **Stems** sparingly branched, short or elongate. **Leaves** crowded, in spirals; blade mostly linear, lacunar tissue not evident, base pale, dilated. **Inflorescences:** scape sheaths tubular, orifice oblique, acute, or 2-cleft; scapes usually several per stem, filiform, glabrous or hairy, hairs neither swollen nor glandular apically; heads white, gray, gray-brown, or brown, ovoid to globose or short-cylindric; receptacle densely pale-pilose; involucral bracts pale to dark, gradate, broad, chaffy, basal ones often reflexed, obscured by inflorescence; receptacular bracts as long as involucral bracts. **Flowers:** staminate and pistillate on same plants, 3-merous (2-merous in *Lachnocaulon digynum*); sepals 3, connivent, forming club-shaped flower, nearly distinct, spatulate, scarious, apex with club-shaped hairs, surface glabrous or sparsely hairy; petals absent or reduced to small scales or hairs. **Staminate flowers:** androphore cylindric; stamens (2–)3; apex of staminal column with 2–3 lance-ovoid or peglike, often appendaged glands; filaments adnate to rim of androphore, alternating with glands; anthers 1-locular, 2-sporangiate, dorsifixed, versatile, exserted at anthesis, yellowish or pale. **Pistillate flowers:** gynophore short; pistil 3-carpellate (2-carpellate in *L. digynum*); style 1, appendaged at apex, style branches 2–3, 2-cleft, alternating with appendages similar to those of staminate flower.

Species ca. 10 (5 in the flora): se United States, West Indies (Cuba).

1. Apical hairs of receptacular bracts and perianth white, mealy, opaque; heads pale gray to white; scapes hairy (except for glabrous-scaped, s Florida extreme of *Lachnocaulon anceps*).
  2. Leaves narrowly linear, abruptly attenuate; mature heads seldom wider than 4 mm, dull gray-brown or pale gray; scapes glabrous or nearly so distally; seeds dark red-brown, very lustrous, longitudinal ribs faint, transverse ribs forming finely cross-striolate pattern. . . . . .1. *Lachnocaulon beyrichianum*
  2. Leaves linear, gradually attentuate; mature heads seldom as narrow as 4 mm, whitish to pale gray; scapes pilose from base to apex; seeds pale to dark brown, not lustrous, longitudinal ribs conspicuous, transverse ribs less conspicuous than longitudinal ribs, but coarser than in *Lachnocaulon beyrichianum*. .2. *Lachnocaulon anceps*
1. Apical hairs of bracts and perianth translucent, not white, the brown color of bract and perianth showing through; heads brown or gray-brown; scapes glabrous or with ascending hairs.
  3. Scapes with ascending hairs; heads dull gray-brown; hairs of receptacle copious, partly obscuring flowers (old heads may lose some hairs); gynoecium 3-carpellate. . . . . .3. *Lachnocaulon minus*
  3. Scapes glabrous; heads either chocolate brown or dull brown, if dull brown, with sepals of pistillate flowers yellow-white, hardly obscured by receptacular hairs; gynoecium 2–3-carpellate.
    4. Heads dark brown or reddish brown, usually short cylindric by seeding time; gynoecium 3-carpellate; leaves 2–4 cm; scape sheaths shorter than or as long as leaves. . . . . .4. *Lachnocaulon engleri*
    4. Heads gray or gray-brown, usually globose by seeding time; gynoecium 2-carpellate; leaves 0.5–2 cm; scape sheaths longer than or at least rising above leaves. . . .5. *Lachnocaulon digynum*

1. **Lachnocaulon beyrichianum** Sporleder ex Körnicke, Linnaea 27: 567. 1856 [C] [E]

Herbs, perennial, cespitose, forming domes of rosettes, 15–23 cm. **Leaves** narrowly linear, abruptly attenuate, 1.5–4 cm, apex subulate. **Inflorescences:** scape sheaths shorter than leaves; scapes filiform, 0.5 mm wide, obscurely ridged, proximally densely pilose with ascending hairs, distally gla-brous or nearly so; mature heads dull gray-brown or pale gray, globose to short cylindric, 3.5–4(–5) mm wide; receptacle pilose with translucent hairs; outer involucral bracts reflexed, brown to chestnut brown, ovate, 1 mm, apex obtuse, abaxially pilose; inner bracts distally brown, broadly spatulate or pandurate, 1–1.5 (–2) mm, apex obtuse, pilose-ciliate, adaxially pilose with whitish, mealy, opaque, club-shaped hairs. **Staminate flowers:** sepals 3, brown to chestnut brown, oblanceolate-spatulate, 1–1.5(–2) mm, apex obtuse, ciliate, abaxially with white, mealy, opaque hairs;

androphore pale, club-shaped, as long as sepals, glabrous; stamens 3, appendages 3. **Pistillate flowers:** sepals 3, tan or pale brown, oblong to linear, 2–3 mm, apex acute, glabrous or abaxially pilose toward apex, hairs white; gynoecium 3-carpellate; stylar column pale, dilated apically, appendages 3. **Seeds** dark red-brown, very lustrous, ellipsoid, 0.5 mm, apex apiculate, longitudinal ribs faint, transverse ribs forming finely cross-striolate pattern.

Flowering summer–fall. Sands and sandy peats, frequently dry pond shores, flatwoods, pine savanna, sandhills ecotones; of conservation concern; 0–100 m; Fla., Ga., N.C., S.C.

Through its range *Lachnocaulon beyrichianum* is sympatric with *L. anceps*; it has narrower, more glabrous foliage, smaller heads, darker bracts, and smoother, more lustrous seeds. Where the two are in the same site, *L. beyrichianum* takes the drier substrata.

2. **Lachnocaulon anceps** (Walter) Morong, Bull. Torrey Bot. Club 18: 360. 1891   F

Eriocaulon anceps Walter, Fl. Carol., 82. 1788; *E. villosum* Michaux; *Lachnocaulon floridanum* Small; *L. glabrum* Körnicke

Herbs, perennial, cespitose, forming domes of rosettes, 15–40 cm. **Leaves** narrowly to broadly linear, gradually attenuate, 2.5–6 (–12) cm. **Inflorescences:** scape sheaths as long as leaves; scapes filiform, 0.5 mm thick distally, obscurely 4–5-ribbed, pilose from base to apex (s Florida extreme of species glabrous-scaped); mature heads whitish to pale gray, globose to short-cylindric, 4–7(–9) mm wide; receptacle densely pilose; outer involucral bracts becoming reflexed, brownish, oblong to triangular, 1–1.5 mm, apex acute or obtuse, surfaces hairy abaxially and at margins; receptacular bracts dark brown, mostly spatulate, 1.5–2 mm, apex obtuse, distally with white, club-shaped hairs. **Staminate flowers:** sepals 3, dark brown, narrowly spatulate, 1.5–2 mm, apex acute, abaxially with white hairs; androphore pale, narrowly club-shaped, 1.5 mm, glabrous; stamens 3, appendages 3. **Pistillate flowers:** sepals 3, pale, oblong to broadly linear, 2 mm (enlarging to 3 mm), apex acute, distal abaxial surface pubescent, hairs white; gynoecium 3-carpellate; stylar column pale, dilated apically, appendages 3. **Seeds** pale to dark brown, not lustrous, ellipsoid, 0.5–0.55 mm, longitudinal ribs conspicuous, pale, transverse ribs less conspicuous, pattern fine but coarser than in *L. beyrichianum*.

Flowering spring–summer. Moist sands and peats of

shores, pine savanna, bog and seep edges, flatwoods clearings; 0–400 m; Ala., Fla., Ga., La., Miss., N.C., S.C., Tenn., Tex., Va.; West Indies (Cuba).

3. **Lachnocaulon minus** (Chapman) Small, Fl. S.E. U.S., 235, 1328. 1903   E

Lachnocaulon michauxii Kunth var. *minor* Chapman, Fl. South. U.S. ed. 3, 531. 1897; *L. eciliatum* Small

Herbs, short-lived perennial, solitary or cespitose, forming low domes of rosettes, 6–15(–20) cm. **Leaves** linear-triangular, 2–4 cm, apex narrowly acute to subulate. **Inflorescences:** scape sheaths shorter than principal leaves; scapes filiform, 0.3–0.4 mm thick, 4–5-ribbed, pilose from base to apex, with ascending hairs; mature heads dull gray-brown, globose to short-cylindric, 3–4 mm wide; receptacle copiously hairy, partly obscuring flowers (old heads may lose some hairs); involucral bracts soon reflexed, pale to dark brown, mostly ovate or broadly oblong, 1 mm, apex broadly acute, distal margins pilose, hairs translucent, capitate, abaxial surface distally pilose, hairs translucent; receptacular bracts pale brown to chestnut brown, broadly spatulate or pandurate, concave, 1 (–1.5) mm, apex rounded or obtuse, abaxial surface pilose toward apex, hairs translucent. **Staminate flowers:** sepals 3, chestnut brown, base greenish, spatulate, blade concave, 1–1.5 mm, apex rounded or obtuse, abaxial surface pilose toward apex, hairs translucent; androphore yellow-white, narrowly obconic, as long as sepals, glabrous, rarely with apex hairy, hairs club-shaped, short; anthers 3, appendages 3. **Pistillate flowers:** sepals 3, pale, spatulate, concave, 1–1.5 mm, apex obtuse or short-acuminate, abaxial surface apically with translucent hairs; gynoecium 3-carpellate; stylar column pale, apex 3-cleft, appendages 3. **Seeds** deep clear brown, ellipsoid, 0.5 mm, apex apiculate, body longitudinally low-ribbed, ribs pale, transverse striae fine.

Flowering summer–fall. Sands and peats of pond edges, ditchbanks, lake shores, drawdowns or moist exposed seeps; 0–200 m; Ala., Fla., Ga., N.C., S.C.

The favorite habitat of *Lachnocaulon minus* is the fluctuating shores of karst ponds where it may be aspect dominant along the edge of the maximum pool.

4. Lachnocaulon engleri Ruhland in H. G. A. Engler, Pflanzenr. 13[IV,30]: 241. 1903 [C] [E] [F]

Herbs, short-lived perennial, solitary or cespitose, forming low domes of rosettes, 6–15 cm. **Leaves** mostly linear-triangular, 2–4 cm, apex acute. **Inflorescences:** scape sheaths shorter than or as long as principal leaves; scapes filiform, distally 0.4 mm thick, 4–5-ribbed, glabrous; mature heads dark brown or reddish brown, nearly globose, usually short cylindric by seeding time, 3–4 mm; receptacle densely pilose; involucral bracts soon reflexed, brown, mostly ovate, 1(–1.5) mm, apex broadly acute to obtuse, surfaces glabrous or apex sparsely pilose, hairs translucent, club-shaped; receptacular bracts dark brown, broadly spatulate or obovate-cuneate, concave, with broad, low, pale keel, 1–1.5 mm, apex short-acuminate, glabrous or abaxial surface with translucent, club-shaped hairs distally. **Staminate flowers:** sepals 3, dark brown, narrowly spatulate to oblanceolate-cuneate, concave, 1 mm, glabrous or abaxially with scattered translucent, club-shaped hairs distally; androphore pale, narrowly obconic, as long as sepals, glabrous, rarely apex with few club-shaped hairs; stamens 2–3, appendages (2–)3. **Pistillate flowers:** sepals 3, dark brown, spatulate-oblong, concave, 1 mm, abaxially sparsely pubescent distally, hairs translucent; gynoecium 3-carpellate; styles: apex 3-cleft, appendages 3. **Seeds** rich translucent brown, broadly ovoid to ellipsoid, 0.4–0.5 mm, longitudinal ribs conspicuous and pale, transverse striae fine, forming linear cancella.

Flowering summer–fall. Moist sandy-peaty lake and pond shores, drawdowns, ditchbanks, low sandy clearings in savanna and flatwoods; of conservation concern; 0–100 m; Ala., Fla.

5. Lachnocaulon digynum Körnicke, Linnaea 27: 570. 1856 [E]

Lachnocaulon diandrum Van Heurck & Müller-Agroviensis

Herbs, perennial, densely cespitose, forming rosettes, 5–15 cm. **Leaves** linear or linear-triangular, mostly 0.5–2 cm, apex narrowly acute to acute, rarely blunt. **Inflorescences:** scape sheaths longer than or at least rising above leaves; scapes linear to filiform, distally 0.3–0.4 mm wide, glabrous; mature heads gray or gray-brown, hemispheric, usually globose by seeding time, 2–3.5 mm wide; receptacle densely pilose; involucral bracts soon reflexed, brown, ovate to triangular, 0.5–1 mm, apex acute, surfaces abaxially pilose; receptacular bracts brown, mostly narrowly obovate-cuneate or spatulate, concave, 1–1.5 mm, apex acute, abaxially pilose distally, hairs translucent, club-shaped. **Staminate flowers:** sepals 3, deep brown, spatulate, 1 mm, apex obtuse, distally pilose, hairs translucent; androphore pale, claviform, as long as sepals; stamens 2, appendages 2. **Pistillate flowers:** sepals 2, pale, broadly spatulate to narrowly oblong-obovate, concave, oblique, keeled, 1 mm, apex obtuse, apex ciliate, abaxially pilose distally, hairs translucent; gynoecium 2-carpellate; styles apically dilated, apex 2-cleft, appendages 2, deeply 2-cleft. **Seeds** brown, ovoid to ellipsoid, 0.5 mm, apex apiculate, longitudinal ribs fine, pale, transverse striae fine, indistinct.

Flowering summer–fall. Moist to wet sandy peats of "slick" seeps, bogs, ditchbanks, streambanks and low places in pine savanna; 0–100 m; Ala., Fla., La., Miss., Tex.

Lachnocaulon digynum is difficult to find throughout most of its known range. Various encroachments upon its habitat are causing concern.

3. SYNGONANTHUS Ruhland in I. Urban, Symb. Antill. 1: 487. 1900 • Shoe-buttons

[Greek *syngonos*, joined together, and *anthos*, flower, from connate petals of pistillate flowers]

Herbs, perennial [annual], often densely cespitose, rosulate [caulescent]. **Roots** appearing unbranched, pale, not septate, thickened, spongy. **Stems** simple [branched], short [to very elongate]. **Leaves** spreading in flat spiral in rosettes [spreading to erect in high spiral]; blade mostly slender, linear, not evidently lacunate [lacunate], base pale. **Inflorescences:** scape sheaths narrowly tubular, orifice diagonal, surfaces tomentose to scattered-pilosulous, glandular-hairy [glabrous]; scapes 1–several per stem, slender, terete [compressed], pubescent [glabrous or variously pubescent], at least some hairs swollen basally, often glandular distally; mature heads white [dark], hemispheric to globose [turbinate, urceolate]; receptacle pale, pilose with sharp hairs; involucral bracts spreading to ascending, mostly not obscured by expanding flowers, spirally imbricate, pale [brown or black], often gradate, chaffy-papery; receptacular bracts absent [few or reduced]. **Flowers** with staminate and pistillate on same plants [staminate and pistillate on different plants], 3-merous [2-merous]. **Staminate flowers:** sepals 3, nearly distinct, plane or concave, chaffy, often with tapering, acute hairs; androphore club-shaped or funnel-form; petals 3, connate at apex of androphore, narrowly campanulate, 3-lobed, lobes low-triangular [to variously elongate]; stamens 3, low in corolla tube alternating with 3 sessile glands at base of tube; filaments low in corolla tube (at disc, midway), alternating with glands; anthers 2-locular, locules divergent, [2–]4-sporangiate, dorsifixed, exserted, yellowish or white. **Pistillate flowers:** gynophore short; sepals distinct [connate], papery; petals connate, connivent or connate over ovary; pistil 3-carpellate; style1; style branches 3, undivided.

Species ca. 200 (1 in the flora): mostly in tropical and subtropical regions of Africa, West Indies, Central America, and South America, the primary center in the planalto of Brazil.

1. Syngonanthus flavidulus (Michaux) Ruhland in H. G. A. Engler, Pflanzenr. 13[IV,30]: 256. 1903 [E] [F]

Eriocaulon flavidulum Michaux, Fl. Bor.-Amer. 2: 166. 1803; *Paepalanthus flavidulus* (Michaux) Kunth

Herbs, perennial, cespitose, rosulate, 5–30 cm. **Stems** contracted or elongate to 5 cm, chaffy. **Leaves:** principal leaves mostly spreading-recurved, narrowly linear, 2–6 cm, base pale, apex narrowly acute to subulate, abruptly dilated, surfaces nearly glabrous or grading to copiously hairy. **Inflorescences:** scape sheaths erect or narrowly ascending, as long as leaves, proximal surfaces glabrous or scattered glandular-pilosulous; scapes filiform, to 1 mm wide, pilose, hairs variably ascending to spreading, translucent, pubescence increasingly glandular distally; mature heads white or yellow-white (aging dull white or pale gray), hemispheric to globose, 5–10 mm wide; receptacle densely pilose, hairs translucent, sharp; involucral bracts ascending, later spreading to recurved, lance-oblong or narrowly ovate, larger ones 3 mm, margins entire, apex acute, glabrous; receptacular bracts absent. **Staminate flowers:** sepals pale translucent, spatulate to oblanceolate, 2–3 mm, apex acute, adaxially pilose distally; androphore 2.5 mm, hairy at base; petals pale yellow, short-triangular. **Pistillate flowers:** sepals pale, translucent, linear or narrowly oblong, 3 mm, apex acuminate, surfaces glabrescent; petals connivent or connate above ovary forming beaklike apex, pale, narrowly linear, 3 mm; gynophore hairy; style with apex unappendaged; stigma branches exsert, linear. **Seeds** translucent brown, ovoid, 0.5 mm, finely longitudinally lined, very finely cross-lined, apex apiculate.

Flowering spring–early summer. Moist sands of bog and pond edges, ditchbanks, pine savanna, sandhills ecotones; 0–200 m; Ala., Fla., Ga., N.C., S.C.

209. JUNCACEAE Jussieu

• Rush Family

Ralph E. Brooks

Steven E. Clemants

Herbs, perennial, occasionally annual, usually rhizomatous, sometimes cespitose. **Culms** round or flat. **Leaves** mostly basal; sheath margins fused or overlapping, often with 2 earlike extensions (auricles) at blade junction; blade flat or round, glabrous or margins hairy. **Inflorescences** of headlike clusters or single flowers variously arranged; bracts subtending inflorescence 1 or more, mostly leaflike; bracts subtending inflorescence branches 1–2, reduced; bracteoles subtending solitary flower 0–2, translucent, reduced. **Flowers** usually bisexual, radially symmetric; sepals and petals similar, persistent, green to brown or purplish black; stamens usually 3 or 6; anthers persistent, linear; pistils 1; ovaries superior, locules 1 or 3, placentas 1 and basal or 3 and axile or parietal; stigmas generally longer than styles. **Fruits** capsules, loculicidal. **Seeds** 3–many, often with white appendages on 1 or both ends.

Genera 9, species ca. 350 (2 genera, 118 species in the flora): arctic and temperate regions, tropical mountains.

SELECTED REFERENCES Buchenau, F. 1890. Monographia Juncacearum. Bot. Jahrb. Syst. 12: 1–495, 622–623, plates 1–3. Buchenau, F. 1906. Juncaceae. In: H. G. A. Engler, ed. 1900–1953. Das Pflanzenreich. . . . 107 vols. Berlin. Vol. 25[IV,26], pp. 1–284.

1. Leaves glabrous; sheaths open; fruits 1- or 3-chambered; seeds many. . . . . 1. *Juncus*, p. 211
1. Leaves generally with hairy margins; sheaths closed; fruits 1-chambered; seeds 3. . . . . 2. *Luzula*, p. 255

1. JUNCUS Linnaeus, Sp. Pl. 1: 325. 1753; Gen. Pl. ed. 5, 152. 1754 • Rush, jonc [classical name for the genus]

Ralph E. Brooks*

Steven E. Clemants*

Herbs, perennial or rarely annual, rhizomatous or cespitose. **Culms** round or flattened in cross section. **Cataphylls** often present at culm base. **Leaves:** sheaths open; blade flat, channeled,

*SEC authored subg. Ensifolia and subg. Septati. All other subgenera were written by REB.

ensiform or terete, sometimes septate, margins involute. **Inflorescences** terminal or pseudoaxillary, monochasia or dichasia, usually with monochasial branches, cymes or 1–many heads in racemes or panicles; bracteoles 2 or absent. **Flowers:** tepals (4–)6 in 2 whorls; stamens (2–)3–6. **Capsules** 1-locular or 3-locular, septicidal. **Seeds** many, ellipsoid to ovoid, sometimes tailed.

Species ca. 300 (95 in the flora): worldwide except Antarctica.

SELECTED REFERENCES Engelmann, G. 1866–1868. Revision of the North American species of the genus *Juncus*, with a description of new or imperfectly known species. Trans. Acad. Sci. St. Louis. 2(2, 3): 424–498. Hermann, F. J. 1975. Manual of the Rushes (*Juncus* spp.) of the Rocky Mountains and Colorado Basin. Fort Collins, Colo. [U.S.D.A. Forest Serv., Gen. Techn. Rep. RM-18.]

1. Flowers borne singly; bracteoles present (except *J. pelocarpus*, *J. subtilis*).
  2. Inflorescences appearing lateral, inflorescence bract erect, terete, appearing to be continuation of culm; basal leaves bladeless, cauline leaves absent. . . . . 1a. *Juncus* subg. *Genuini*, p. 212
  2. Inflorescences appearing terminal, inflorescence bract erect or ascending, flat, involute or terete; basal leaves (at least some) usually with blade, cauline leaves present or absent.
    3. Leaves terete, septate; capsules beaked. . . . .
    . . . . 1g. *Juncus* subg. *Septati* (in part; *J. pelocarpus*, *J. subtilis*), p. 240
    3. Leaves flat, involute or rarely nearly terete, not septate; capsule rarely beaked. . . . .
    . . . . 1b. *Juncus* subg. *Poiophylli*, p. 217
1. Flowers solitary or in heads; bracteoles absent or partially absent.
  4. Leaves flat, terete and channeled, or ensiform.
    5. Leaves ensiform, imperfectly septate. . . . . 1d. *Juncus* subg. *Ensifolii*, p. 233
    5. Leaves flat, terete and channeled, not septate. . . . . 1c. *Juncus* subg. *Graminifolii*, p. 225
  4. Leaves terete or compressed.
    6. Capsules large; seeds large, long tailed; leaves not noticeably septate. . . . 1f. *Juncus* subg. *Alpini*, p. 238
    6. Capsules smaller; seeds not tailed or if tailed not long; leaves septate or not.
      7. Leaves not septate; plants of salty or alkaline habitats. . . . . 1e. *Juncus* subg. *Juncus*, p. 236
      7. Leaves septate; plants usually of fresh water habitats. . . . . 1g. *Juncus* subg. *Septati* (in part), p. 240

1a. JUNCUS Linnaeus subg. GENUINI Buchenau, Abh. Naturwiss. Vereine Bremen 4: 406. 1875

Juncus sect. *Genuini* (Buchenau) Dumortier; *Juncus* sect. *Juncotypus* Dumortier; *Juncus* subg. *Juncotypus* (Dumortier) Kreczetowicz & Gontscharow

Herbs, perennial, rhizomatous. **Culms** terete. **Leaves** basal; blade usually absent, rarely present (*J. hallii* and *J. parryi*), if present terete, channeled, not septate. **Inflorescences** lateral cymes, sympodial; bracts erect, terete, appearing to be continuation of culm; bracteoles 2, at base of perianth. **Flowers** borne singly, not in heads. **Capsules** 3-locular. **Seeds** usually not tailed.

Species ca. 25 (11 in the flora): worldwide.

SELECTED REFERENCE Stace, C. A. 1970. Anatomy and taxonomy in *Juncus* subgenus *Genuini*. Bot. J. Linn. Soc. 63(suppl. 1): 75–81.

1. Flowers 1–3(–5).
  2. Leaf blade reduced to bristle.. . . . . 4. *Juncus drummondii*
  2. Leaf blade well developed.

1. Juncus gymnocarpus Coville, Mem. Torrey Bot. Club 5: 106. 1894 • Pennsylvania rush E

Juncus smithii Engelmann, Trans. Acad. Sci. St. Louis 2: 444. 1866, not Kunth 1841

Herbs, perennial, 5–10 dm. **Rhizomes** widely creeping, 2–4 mm diam. **Culms** terete, 1.5–2.5 mm diam. **Cataphylls** several. **Leaves:** blade absent. **Inflorescences** 8–30(–50)-flowered, open, 1.5–4 cm; primary bract terete, 1–2.5 dm, much longer than inflorescence. **Flowers** pedicellate; bracteoles broadly ovate; tepals light brown, ovate-lanceolate, 1.8–2.5 mm, apex acuminate; inner series shorter, margins scarious, apex acute; stamens 6, filaments 1–1.3 mm, anthers 0.5–0.7 mm; style 0.5 mm. **Capsules** reddish tan to brown, lustrous, 3-locular, widely ellipsoid, 2–3 × 1.8–2.2 mm, exceeding perianth. **Seeds** dark amber, obovoid, 0.7–1 mm, not tailed.

Flowering and fruiting summer. Sphagnous swamps, low woods, edges of lakes; 600–1500 m; Fla., Miss., N.C., Pa., S.C., Tenn.

2. Juncus parryi Engelmann, Trans. Acad. Sci. St. Louis 2: 446. 1866 • Parry's rush E

Juncus drummondii E. Meyer var. *parryi* (Engelmann) M. E. Jones

Herbs, perennial, strongly tufted, 0.5–3 dm. **Rhizomes** densely short-branched. **Culms** terete. **Cataphylls** several. **Leaves:** auricles 0.2–0.3 mm, apex acute to rounded, scarious. **Inflorescences** 1–3-flowered, open; primary bract terete, 2–4 cm, usually longer than inflorescence.

Flowers pedicellate; bracteoles broadly ovate; tepals light brown with green midstripe, lanceolate, 5.5–9 mm, margins scarious; inner series loosely subtending capsule at maturity, shorter; stamens 6, filaments 0.7–1 mm, anthers 1.1–1.6 mm; style 0.2 mm. **Capsules** tan, 3-locular, narrowly oblong, 6–9 × 1.5–2 mm, usually exceeding perianth. **Seeds** amber, body 0.6 mm, tails 0.4 mm.

Flowering and fruiting summer. Exposed rocky slopes and stream banks in montane and alpine areas, conifer forests; 1500–4000 m; Alta., B.C.; Calif., Colo., Idaho, Mont., Nev., Oreg., Utah, Wash., Wyo.

3. Juncus hallii Engelmann, Trans. Acad. Sci. St. Louis 2: 446. 1866 • Hall's rush E

Herbs, perennial, strongly tufted, to 4 dm. **Rhizomes** densely short-branched. **Culms** terete. **Cataphylls** several. **Leaves:** auricles 0.2 mm, apex acutish to rounded, scarious; blade 4–15 cm. **Inflorescences** 2–7-flowered, loose to congested; primary bract usually longer than inflorescence.

Flowers pedicellate; bracteoles ovate; tepals light brown with green midstripe, lanceolate or widely so, 4–5 mm, margins scarious; inner series loosely subtending capsule at maturity, shorter; stamens 6, filaments 0.8 mm, anthers 0.8–1.1 mm; style 0.2 mm. **Capsules** brown, 3-locular, oblong-ovoid, 3.5–5 × 1.5–1.9 mm, slightly exceeding perianth. **Seeds** amber, body 0.5 mm, tails 0.3 mm.

Flowering and fruiting summer. Exposed slopes, stream banks, and meadows in montane and alpine areas; 1600–3000 m; Colo., Idaho, Mont., Utah, Wash., Wyo.

4. Juncus drummondii E. Meyer in C. F. von Ledebour, Fl. Ross. 4: 235. 1853 · Drummond's rush [E] [F]

Juncus compressus Jacquin var. *subtriflorus* E. Meyer; *J. drummondii* var. *longifructus* H. St. John; *J. drummondii* var. *subtriflorus* (E. Meyer) C. L. Hitchcock; *J. pauperculus* Schwarz; *J. subtriflorus* (E. Meyer) Coville

Herbs, perennial, strongly tufted, to 4 dm. **Rhizomes** densely short-branched. **Culms** terete. **Cataphylls** several. **Leaves:** blade absent or rarely present, to 1 cm. **Inflorescences** 2–5-flowered, loosely compact; primary bract usually longer than inflorescence. **Flowers** pedicellate; tepals brown to chestnut brown with green midstripe, lanceolate or widely so, (4–)5–8 mm, margins clear; inner series loosely subtending capsule at maturity, shorter; stamens 6, filaments 0.7–1 mm, anthers 1.1–1.6 mm; style 0.2 mm. **Capsules** brown to chestnut brown, 3-locular, oblate, 4.5–7(–8) × 1.8–2.2 mm, nearly equal to or exceeding perianth. **Seeds** amber, body oblate, 0.5–0.6 mm.

Flowering and fruiting summer. Exposed slopes, stream banks, and meadows in montane and alpine areas; 1600–4000 m; Alta., B.C., N.W.T., Yukon; Alaska, Calif., Colo., Idaho, Mont., Nev., N.Mex., Oreg., Utah, Wash., Wyo.

Plants with capsules distinctly longer than the perianth have been referred to as *Juncus drummondii* var. *subtriflorus.* Those plants frequently occur sympatrically with *J. drummondii* (strict sense) through most of its range, leaving considerable doubt as to the value of recognizing such variation.

5. Juncus filiformis Linnaeus, Sp. Pl. 1: 326. 1753 · Thread rush [F]

Herbs, perennial, 0.2–3.5 dm. **Rhizomes** widely creeping, sparingly branched, 1.5–2 mm diam., nodes closely set. **Culms** terete, 1 mm diam. **Cataphylls** several. **Leaves:** blade absent. **Inflorescences** 3–10(–12)-flowered, loosely congested, 1–2 cm; primary bract terete, nearly equaling to much longer than culm. **Flowers** pedicellate; bracteoles broadly ovate, tepals light brown or green, lanceolate, 2.5–4.2 mm; inner series loosely subtending capsule at maturity, slightly shorter, margins scarious; stamens 6, filaments 0.5–0.9 mm, anthers 0.5–0.7 mm; style 0.2 mm. **Capsules** tan, 3-locular, nearly globose, 2.5–3 × 1.8–2.1 mm, shorter than perianth. **Seeds** amber, 0.5–0.6 mm, not tailed. $2n = 40, 70, 80, 84.$

Flowering and fruiting summer. Usually sandy, moist or wet soil along stream banks, pools, lakes or in meadow depressions, rarely in bogs, frequently hidden by larger vegetation; 0–3000 m; Greenland; St. Pierre and Miquelon; Alta., B.C., Man., Nfld. and Labr., N.B., N.S., N.W.T., Nunavut, Ont., P.E.I., Que., Sask., Yukon; Alaska, Colo., Idaho, Maine, Mass., Mich., Minn., Mont., Nebr., N.H., N.J., N.Y., Oreg., Pa., Utah, Vt., Wash., W.Va., Wis., Wyo.; Eurasia; Atlantic Islands (Iceland).

6. Juncus inflexus Linnaeus, Sp. Pl. 1: 326. 1753 · Blue rush [I] [W]

Juncus glaucus Ehrhart ex Sibthorp

Herbs, perennial, cespitose, stooling, 4–10 (–12) dm. **Rhizomes** 3–5 mm diam. **Culms** terete, 1.5–3 mm diam. **Cataphylls** several. **Leaves:** blade absent. **Inflorescences** many flowered, open, 2–7 cm; primary bract terete, 10–25 cm. **Flowers** pedicellate; bracteoles ovate; tepals straw-colored to reddish brown, lanceolate, 2.7–3.5 mm, margins scarious; inner series loosely subtending capsule at maturity, shorter; stamens 6, filaments 0.8–1.5 mm, anthers 0.8–1 mm; style 0.3 mm. **Capsules** reddish brown to chestnut brown, 3-locular, 3-gonous-ovoid to widely ellipsoid, 3–4 mm, exceeding perianth. **Seeds** amber, obovoid, 0.3–0.5 mm, not tailed. $2n = 20, 38, 40.$

Flowering and fruiting summer. Wet soils along streams, ditches, and on wet, sandy and peaty hillsides; introduced; Ont.; Mich., N.Y., Pa., Va.; Europe; Asia; Africa.

7. Juncus effusus Linnaeus, Sp. Pl. 1: 326. 1753 · Soft rush [E] [F] [W]

Juncus conglomeratus Linnaeus; *J. effusus* var. *brunneus* Engelmann; *J. effusus* var. *caeruleomontanus* H. St. John; *J. effusus* var. *conglomeratus* (Linnaeus) Engelmann; *J. effusus* var. *costulatus* Fernald; *J. effusus* var. *dicipiens* Buchenau; *J. effusus* var. *exiguus* Fernald & Wiegand; *J. effusus* var. *gracilis* Hooker, *J. effusus* var. *pacificus* Fernald & Wiegand; *J. effusus* var. *pylaei* (Laharpe) Fernald & Wiegand; *J. effusus* var. *solutus* Fernald & Wiegand; *J. effusus* var. *subglomeratus* Lamarck & de Candolle; *J. griscomii* Fernald, *J. pylaei* Laharpe

Herbs, perennial, 4–13 dm. **Rhizomes** short-branched, forming distinct, often large clumps. **Culms** erect, terete, 1–2.5 mm diam. at top of sheaths. **Cataphylls** several. **Leaves:** blade absent. **Inflorescences** lateral, compound

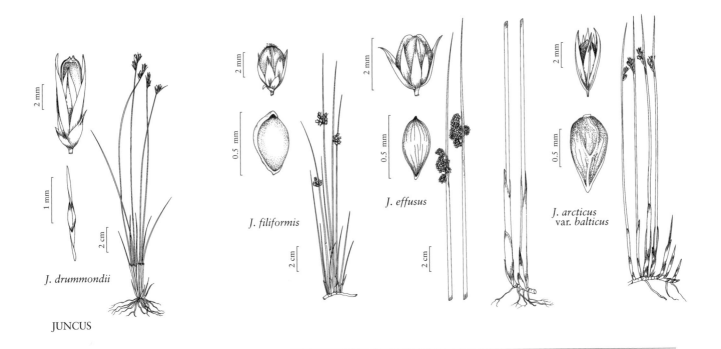

J. drummondii

J. filiformis

J. effusus

J. arcticus var. balticus

JUNCUS

dichasia, many flowered; primary bract erect, terete, extending well beyond dichasium. **Flowers:** tepals tan or darker, usually with greenish midstripe, lanceolate, 1.9–3.5 mm; inner slightly shorter; stamens 3, filaments 0.5–0.8 mm, anthers 0.5–0.8 mm; style 0.2 mm. **Capsules** greenish tan or darker, 3-locular, broadly ellipsoid to oblate, 1.5–3.2 mm. **Seeds** amber, (0.3–)0.4–0.5 mm. $2n = 40, 42$.

Flowering summer, fruiting summer–fall. Swamps and their edges, marshes, moist meadows, and moist or saturated soils, often conspicuous in pasture meadows where it is shunned by grazing animals; 0–2500 m; B.C., Man., N.B., Nfld. and Labr. (Nfld.), N.S., Ont., P.E.I., Que.; Ala., Alaska, Ark., Calif., Colo., Conn., Del., D.C., Fla., Ga., Idaho, Ill., Ind., Kans., Ky., La., Maine, Md., Mass., Mich., Minn., Miss., Mo., Mont., Nebr., Nev., N.H., N.J., N.Y., N.C., Ohio, Okla., Oreg., Pa., R.I., S.C., Tenn., Tex., Vt., Va., Wash., W.Va., Wis.

The *Juncus effusus* complex has been variously recognized as containing several species or a single species with numerous infraspecific taxa. Unfortunately, North American treatments have dealt primarily with taxa in either the eastern or western portions of the continent. In considering the continent as a whole, little sense can be made of these treatments. The North American *J. effusus* complex is one that is in obvious need of modern systematic scrutiny.

SELECTED REFERENCES Fernald, M. L. and K. M. Wiegand. 1910. The North American variation of *Juncus effusus*. Rhodora 12: 81–93. Hämet-Ahti, L. 1980. The *Juncus effusus* aggregate in eastern North America. Ann. Bot. Fenn. 17: 183–191.

8. Juncus patens E. Meyer, Syn. Luzul., 28. 1823

Herbs, perennial, occasionally tufted, occasionally rhizomatous, 3–9 dm. **Rhizomes,** if present, short. **Culms** green or glaucous, 1.5–2.5 mm diam. **Cataphylls** several. **Leaves:** blade absent. **Inflorescences** lateral, many flowered, loose to congested, 2–9 cm; primary bract exceeding inflorescence. **Flowers** variously pedicellate; bracteoles membranous; tepals greenish, light brown or reddish, lanceolate, 2.3–3 mm; inner series nearly equal, margins scarious; stamens 6, filaments 0.6–1 mm, anthers 0.4–0.6 mm; style 0.1 mm. **Capsules** 3-locular or pseudo-3-locular, nearly globose, 2–2.7 mm, equal to or exceeding perianth. **Seeds** dark amber, oblate, 0.4–0.5 mm.

Flowering and fruiting summer. Stream banks, lake or pond shores, ditches, and other wet places in sandy or clayey soils; 0–1600 m; Calif., Oreg.; nw Mexico.

9. Juncus arcticus Willdenow, Sp. Pl. 2: 206. 1799

Arctic rush

Herbs, perennial, 2–10 dm. **Rhizomes** long-creeping. **Culms** erect, 1–3 mm diam. **Cataphylls** several. **Leaves:** blade usually absent (present in var. *mexicanus*). **Inflorescences** lateral, 3–many-flowered, loose to congested; primary bract barely exceeding to many times longer than inflorescence. **Flowers** variously pedicellate; bracteoles membranous; tepals chestnut brown or paler, lanceolate, (2.5–)3.3–5.5(–6) mm, margins clear; inner series loosely subtending capsule at maturity; usually slightly shorter, margins scarious to clear, apex acutish to obtuse; stamens 6, filaments 0.2–1.1 mm, anthers 0.9–2.2 mm; style 0.9–1.5 mm. **Capsules** 3-locular or infrequently pseudo-3-locular, oblate to narrowly ovoid, 3.5–4(–4.5) mm, equal to or exceeding perianth. **Seeds** dark amber, oblate to ellipsoid, 0.6–0.8 mm.

Varieties 7 (3 in the flora): North America; Mexico; South America; Asia.

Numerous entities have been circumscribed and recognized at various nomenclatural ranks by a plethora of authors addressing state or regional floras. In considering the *Juncus arcticus-balticus* complex as a whole in North America, one is soon confronted with a wide-ranging and obviously polymorphic complex that has not read the literature. It is abundantly clear that the systematics of the group will not be solved on the basis of morphology alone and that resolution of the problem is ripe for molecular investigations.

1. Anthers ± as long as filaments; blades absent
   . . . . 9a. *Juncus arcticus* var. *alaskanus*
1. Anthers at least 2 times as long as filaments.
   2. Blade-bearing leaves absent. . . . .
      . . . . 9b. *Juncus arcticus* var. *balticus*
   2. Distal 1 or 2 leaves with obvious blade. . . .
      . . . . 9c. *Juncus arcticus* var. *mexicanus*

9a. Juncus arcticus Willdenow var. **alaskanus** (Hultén) S. L. Welsh, Anderson's Fl. Alaska, 611. 1974

• Alaskan rush

Juncus arcticus subsp. *alaskanus* Hultén, Fl. Alaska Yukon 3: 418. 1943; *J. arcticus* subsp. *sitchensis* Engelmann; *J. balticus* Willdenow var. *alaskanus* (Hultén) A. E. Porsild; *J. balticus* var. *haenkei* (E. Meyer) Buchenau; *J. balticus* subsp. *sitchensis* (Engelmann) Hultén; *J. haenkei* E. Meyer

Herbs, 1–3 dm. **Culms** 0.7–1.2 mm diam. **Leaves:** blade

absent. **Inflorescences** 3–8-flowered, slightly compact, 1–1.5 cm; primary bract 3–8 cm. **Flowers:** tepals shining chestnut brown with lighter midstripe, lanceolate, 4–5(–6) mm, apex acuminate; inner series shorter, margins clear, obtuse to acute; stamens 2 mm, filaments 1 mm, anthers 1 mm. **Capsules** dark tan, oblate, 3–4 mm diam., usually slightly exceeding perianth.

Flowering spring–summer. Moist to dry places, wet meadows, stream banks; 0–2200 m; Greenland; B.C., Man., Nfld. and Labr. (Labr), N.W.T., Nunavut, Ont., Que., Yukon; Alaska; Asia.

Perhaps *Juncus arcticus* var. *alaskanus* is not distinct from the Eurasian *J. arcticus* var. *arcticus*.

9b. Juncus arcticus Willdenow var. **balticus** (Willdenow) Trautvetter, Trudy Imp. S.-Petersburgsk. Bot. Sada 5: 119. 1878 • Baltic rush [F]

Juncus balticus Willdenow, Ges. Naturf. Freunde Berlin Mag. Neuesten Entdeck. Gesammten Naturk. 2: 298. 1809; *J. arcticus* subsp. *ater* (Rydberg) Hultén; *J. arcticus* subsp. *littoralis* (Engelmann) Hultén; *J. arcticus* var. *gracilis* Hooker; *J. arcticus* var. *littoralis* (Engelmann) B. Boivin; *J. arcticus* var. *montanus* (Engelmann) S. L. Welsh; *J. arcticus* var. *montanus* (Engelmann) Balslev; *J. ater* Rydberg; *J. balticus* subsp. *littoralis* (Engelmann) Hultén; *J. balticus* subsp. *vallicola* (Rydberg) Lint; *J. balticus* Willdenow var. *condensatus* Suksdorf.; *J. balticus* var. *eremicus* Jepson; *J. balticus* var. *littoralis* Engelmann; *J. balticus* var. *melanogenus* Fernald & Wiegand; *J. balticus* var. *montanus* Engelmann; *J. balticus* var. *vallicola* Rydberg; *J. breweri* Engelmann; *J. vallicola* (Rydberg) Rydberg

Herbs, 2–10 dm. **Leaves:** blade absent. **Inflorescences** 6–many-flowered, loose to congested; primary bract usually several times longer than inflorescence. **Flowers:** tepals blackish to light brown, occasionally paler, (2.5–)3.5–5.5(–6) mm; filaments 0.2–0.5 mm, anthers 1.2–2.2 mm.

Flowering spring–early summer. Shores, stream margins, wet meadows, fens, marshes, often slightly alkaline soils; Greenland; Alta., B.C., Man., N.B., Nfld. and Labr., N.W.T., N.S., Nunavut, Ont., P.E.I., Que., Sask., Yukon; Alaska, Ariz., Ark., Calif., Colo., Idaho, Ill., Ind., Iowa, Kans., Maine, Mass., Mich., Minn., Mo., Mont., Nebr., Nev., N.H., N.Mex., N.Dak., N.Y., Ohio, Okla., Oreg., Pa., S.Dak., Tex., Utah, Vt., Va., Wash., W.Va., Wis., Wyo.; Mexico; South America; Europe; Asia.

9c. Juncus arcticus Willdenow var. **mexicanus**
(Willdenow ex Roemer & Schultes) Balslev, Brittonia
35: 308. 1983 • Mexican rush

Juncus mexicanus Willdenow ex
Roemer & Schultes, Syst. Veg. 7(1):
178. 1829; *J. balticus* Willdenow
var. *mexicanus* (Willdenow ex
Roemer & Schultes) Kuntze

Herbs, 2–8 dm. **Culms** usually
compressed. **Leaves:** blade distinct, somewhat flattened, to 20
cm or more. **Flowers:** tepals usu-
ally brown or blackish, 3.5–4.5(–5.5) mm; filaments
0.2–0.4 mm, anthers 1.2–2.2 mm.

Intermittent or permanent water courses, around
springs, and in meadows in sandy or gravelly soils; 500–
3200 m; Ariz., Calif., Colo., N.Mex., Nev., Utah; Mexico; Central America; South America.

Juncus arcticus var. *mexicanus* is the common phase
of the species in the extreme southwest United States
and Mexico; northward the variety becomes sporadic in
its occurrence. Rare individuals having leaves with
blades are sometimes found in otherwise bladeless populations in more northern areas but it is doubtful that
these oddities represent an equivalent variant.

10. Juncus lesueurii Bolander, Proc. Calif. Acad. Sci. 2:
179. 1863 (as leseurii) • Lesueur's rush [E]

Juncus balticus Willdenow subsp.
pacificus Engelmann; *J. lesueurii* var.
tracyi Jepson

Herbs, perennial, 3–14 dm. **Rhizomes** long-creeping. **Culms**
erect, 1–3 mm diam. **Cataphylls**
several. **Leaves:** blade absent. **Inflorescences** lateral, 3–many-
flowered, mostly loose; primary

bract much longer than inflorescence. **Flowers** variously
pedicellate; bracteoles membranous; tepals green, lanceolate, 5–8 mm, margins brown; inner series loosely
subtending capsule at maturity, usually slightly shorter,
margins scarious to clear, acutish; stamens 6, filaments
0.2–1.1 mm, anthers 0.9–2.3 mm; style 1–1.5 mm. **Capsules** 3-locular, oblate to narrowly ovoid, 4–7 mm,
shorter than or nearly equal to perianth. **Seeds** dark
amber, oblate to ellipsoid, 0.4–0.7 mm.

Flowering and fruiting summer. Borders of salt or
freshwater marshes and usually near dunes along the
coast; 0–500 m; B.C.; Calif., Oreg., Wash.

11. Juncus textilis Buchenau, Abh. Naturwiss. Vereine
Bremen 17: 336. 1903 [E]

Juncus lesueurii Bolander var. *elatus*
S. Watson

Herbs, perennial, 10–20 dm. **Rhizomes** long-creeping. **Culms**
erect, 2–5 mm diam. **Cataphylls**
several. **Leaves:** blade absent. **Inflorescences** lateral, many flowered, loose; primary bract barely
exceeding to many times longer
than inflorescence. **Flowers** variously pedicellate; bracteoles membranous; tepals greenish to pale brown, lanceolate, 3.5–5 mm; inner series loosely subtending capsule at maturity, usually slightly shorter, margins
scarious to clear, acutish; stamens 6, filaments 0.3–0.9
mm, anthers 1–2.5 mm; style 1–1.5 mm. **Capsules** dark
brown, 3-locular, oblate to narrowly ovoid, 3–4 mm,
nearly equal to perianth. **Seeds** dark amber, oblate to
ellipsoid, 0.5–0.8 mm.

Flowering and fruiting summer. Moist or wet exposed areas; 100–1800 m; Calif.

1b. JUNCUS Linnaeus subg. **POIOPHYLLI** Buchenau, Abh. Naturwiss. Vereine Bremen 4:
406. 1875

Juncus sect. *Poiophylli* (Buchenau) Vierhapper; *Juncus* subg. *Pseudotenageia* Kreczetowicz &
Gontscharow; *Juncus* sect. *Tenageia* Dumortier; *Juncus* subg. *Tenageia* (Dumortier) Kuntze

Herbs, perennial, rarely annual, perennials rhizomatous. **Stems** terete. **Leaves** basal and cauline;
blade usually flat, involute, rarely nearly terete, channeled, not septate. **Inflorescences** terminal
cymes, usually sympodial (monopodial in *J. trifidus*); bracts erect or ascending, flat, involute,
or terete; bracteoles 2, at base of perianth. **Flowers** borne singly, not in heads. **Capsules** 1-
locular (placentae ½–¾ distance to central axis) or 3-locular, rarely beaked. **Seeds** not tailed
(except *J. vaseyi*).

Species 25 (18 in the flora): worldwide.

SELECTED REFERENCES Brooks, R. E. 1989. A Revision of *Juncus* subgenus *Poiophylli* (Juncaceae) in the Eastern United States. Ph.D. dissertation. University of Kansas. Catling, P. M. and K. W. Spicer. 1987. The perennial *Juncus* of section *Poiophylli* in the Canadian prairie provinces. Canad. J. Bot. 65: 750–760. Wiegand, K. M. 1900. *Juncus tenuis* Willd., and some of its North American allies. Bull. Torrey Bot. Club 27: 511–527.

1. Plants annual. . . . . 29. *Juncus bufonius*
1. Plants perennial.
  2. Leaves finely serrulate; auricles deeply lacerate. . . . . 12. *Juncus trifidus*
  2. Leaves entire; auricles not lacerate.
    3. Auricles at summit of leaf sheath 3–6 mm, membranous, transparent.
      4. Capsules less than ¾ length of tepals, borne widely spaced along usually diffuse branches of inflorescence; plants mostly 7–9 dm. . . . . 25. *Juncus anthelatus*
      4. Capsules ¾ or more length of tepals, borne congested or branch internodes ca. as long as tepals; plants mostly less than 7 dm. . . . . 24. *Juncus tenuis*
    3. Auricles absent or merely short, scarious or cartilaginous projections less than 2 mm.
      5. Leaf blade channeled to terete.
        6. Inflorescence falsely lateral, overtopped by terete bract that looks to be continuation of culm. . . . . 16. *Juncus coriaceus*
        6. Inflorescence terminal, involucre bract flat or much shorter than inflorescence.
          7. Perianth segments obtuse to nearly acute. . . . . 13. *Juncus squarrosus*
          7. Perianth segments acute to acuminate.
            8. Seeds distinctly tailed, tail at least 1/2 length of seed body; capsules golden tan. 17. *Juncus vaseyi*
            8. Seeds not tailed; capsules chestnut brown or paler.
              9. Capsules 1-locular or pseudo-3-locular, widely ellipsoid to globose, apex rounded; seeds 0.3–0.4 mm. . . . . 23. *Juncus dichotomus* (in part)
              9. Capsules 3-locular, ellipsoid, apex somewhat truncate; seeds 0.48–0.7 mm. . . . . 18. *Juncus greenei*
      5. Leaf blade flat.
        10. Capsules chestnut brown or darker; tepals apically obtuse.
          11. Anther 1–2 times length of filament, less than 1 mm; capsule usually exceeding perianth. . . . . 14. *Juncus compressus*
          11. Anther 2–4 times length of filament, longer than 1 mm; capsule usually nearly equal to perianth. . . . . 15. *Juncus gerardii*
        10. Capsules tan or light brown; tepals acute to acuminate.
          12. Capsules 3-locular.
            13. Tepals 2.5–3.5(–4.4) mm; capsule nearly globose to ellipsoid.
              14. Tepals green; flowers chiefly arranged along inner side of branches; plants of e North America. . . . . 22. *Juncus secundus*
              14. Tepals usually with brownish midstripe; flowers not secund; plants of w North America. . . . . 20. *Juncus confusus*
            13. Tepals (3.3–)4–5.5(–5.7) mm; capsule ellipsoid to narrowly so.
              15. Anthers 0.6–0.9 mm; auricles scarious, 0.4–1.1(–2) mm; plants of c United States. . . . . 19. *Juncus brachyphyllus*
              15. Anthers (0.8–)1.2–1.5(–1.7) mm; auricles scarious to membranous, 0.2–0.3 mm; plants endemic to s Appalachian Piedmont . . . . 21. *Juncus georgianus*
           12. Capsules 1- or 3-locular or pseudo-3-locular.
            16. Auricles 0.5–1(–1.5) mm, membranous. . . . . 26. *Juncus occidentalis*
            16. Auricles 0.1–0.3(–1) mm, scarious to leathery.
              17. Auricles leathery, yellowish; tepals spreading in fruit, nearly equal to or exceeding capsule; anthers 0.6–1 mm. . . . . 28. *Juncus dudleyi*
              17. Auricles scarious, whitish or sometimes purplish tinged, not leathery; tepals not spreading in fruit; anthers (0.2–)0.4–0.6 mm.

18. Mature capsules light brown or darker; plants of coastal plain of e, se North America. . . . . .23. *Juncus dichotomus* (in part)
18. Mature capsules light tan or darker; plants of interior continent. . . . . .27. *Juncus interior*

12. Juncus trifidus Linnaeus, Sp. Pl. 1: 326. 1753 [F]

Juncus monanthos Jacquin; *J. trifidus* subsp. *carolinianus* Hmit-Ahti; *J. trifidus* subsp. *monanthos* (Jacquin) Ascherson & Graebner; *J. trifidus* var. *monanthos* (Jacquin) Bluff & Fingerhuth

Herbs, perennial, tufted, 1–4 dm. **Rhizomes** densely branching. **Culms** few to many, erect to drooping. **Cataphylls** 2–4. **Leaves** basal and cauline, 2–4; auricles deeply lacerate, to 4 mm, membranous; blade dark green, flat to slightly channeled, 5–12 cm × 0.5–1 mm, margins finely serrulate. **Inflorescences** 1–3(–4)-flowered; primary bract usually exceeding inflorescence. **Flowers:** bracteoles 2, usually lacerate; tepals brownish, widely lanceolate, 2.1–3 mm; inner series slightly shorter, apex acute; stamens 6, filaments 0.3–0.7 mm, anthers 1.2–1.5 mm; stigma 1.5–3 mm. **Capsules** brown, ellipsoid, 2.2–3.5 × 1.3–1.7 mm, longer than tepals. **Seeds** brown, obliquely ovoid, 0.9–1.3 mm, not tailed. 2*n* = 40.

Flowering and fruiting late spring–early summer. Mostly in crevices of granitic (schistose) cliffs or rubble slopes in the higher regions of the Appalachians; 20–2000 m; Greenland; N.B., Nfld. and Labr., N.S., Que.; Maine, N.H., N.Y., N.C., Tenn., Vt., Va.; Europe; n Asia.

13. Juncus squarrosus Linnaeus, Sp. Pl. 1: 327. 1753 [I]

Herbs, perennial, cespitose to mat-forming, 2–5 dm. **Rhizomes** short, nearly erect, branching. **Culms** 1(–2). **Cataphylls** 2–4. **Leaves** basal, 3–6; auricles 0.2–0.4 mm, membranous to scarious; blade arcuate, spreading, grayish green, channeled, 7–30 cm × 1–2 mm, nearly leathery, margins entire. **Inflorescences** terminal, 7–40-flowered, somewhat loose, 3–10 × 1–2 cm; primary bract usually shorter than inflorescence. **Flowers:** pedicels 0.5–2 mm; bracteoles 2; tepals brown to blackish, 4–5.5 mm; outer and inner series nearly equal, apex obtuse to nearly acute; stamens 6, filaments 0.5 mm, anthers 1.5–2 mm; style 0.2 mm. **Capsules** brownish, 3-locular, ovoid to ellipsoid, 4–5 × 2.3–3 mm, nearly equal to perianth. **Seeds** brownish, ellipsoid, 0.6–0.9 mm, not tailed. 2*n* = 42.

Flowering and fruiting late spring–summer. Moist, open sites and bogs, occasionally disturbed areas; 20–200 m; introduced; Greenland; Wis.; native, Europe.

14. Juncus compressus Jacquin, Enum. Stirp. Vindob., 60, 235. 1762 [I]

Juncus bulbosus Linnaeus 1762, not 1753; *J. supinus* Moench

Herbs, perennial, to 8 dm. **Rhizomes** short-creeping or densely branching, if densely branching herb appearing cespitose. **Cataphylls** 1–3. **Leaves** basal and cauline, 1–2; auricles 0.3–0.5 mm, scarious to membranous; blade flat to slightly channeled, 5–35 cm × 0.8–2 mm, margins entire. **Inflorescences** 5–60-flowered, lax, loose to moderately congested, 1.5–8 cm; primary bract usually exceeding inflorescence. **Flowers:** bracteoles 2; tepals brownish, ovate to oblong, 1.7–2.7 mm; inner and outer series nearly equal, apex obtuse; stamens 6, filaments 0.5–0.7 mm, anthers 0.6–1 mm; style 0.3 mm. **Capsules** brown or darker, pseudo-3-locular, widely ellipsoid to obovoid, 2.5–3.5 × 1.4–1.8 mm, exceeding tepals. **Seeds** light brown, ellipsoid to lunate, 0.3–0.6 mm, not tailed. 2*n* = 44.

Flowering and fruiting late spring–early summer. Disturbed ground, especially ditches, along railroads and banks of canals and roadsides; frequently in saline or alkaline soils; 1500–2100 m; introduced; Man., Nfld. and Labr. (Nfld.), N.S., Ont., Que.; Colo., Maine, Md., Mich., Minn., Mont., N.Y., N.Dak., Utah, Wis., Wyo.; Europe; w Asia.

15. Juncus gerardii Loiseleur-Deslongchamps, J. Bot. (Desvaux) 2: 284. 1809 [F]

Juncus bulbosus Linnaeus var. *gerardii* (Loiseleur-Deslongchamps) A. Gray; *J. fucensis* H. St. John; *J. gerardii* var. *pedicellatus* Fernald

Herbs, perennial, 2–9 dm. **Rhizomes** long-creeping. **Cataphylls** 1–3. **Leaves** basal, (1–)2–4; auricles 0.4–0.6(–0.8) mm, scarious; blade flat or somewhat channeled, 10–40 cm × 0.4–0.7 mm, margins entire. **Inflorescences** 10–30(–80)-flowered, usually loose and somewhat lax, 2–16 cm; primary bract rarely sur-

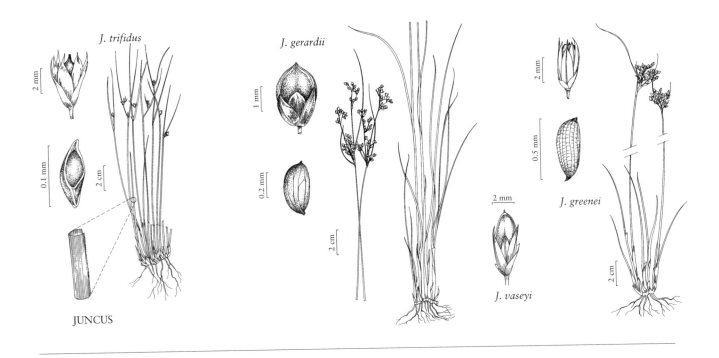

J. trifidus

J. gerardii

J. greenei

J. vaseyi

JUNCUS

passing inflorescence. **Flowers:** bracteoles 2; tepals dark brown or blackish, lanceolate-ovate to oblong, 2.6–3.2(–3.8) mm; inner and outer series nearly equal, apex obtuse; stamens 6, filaments 0.4–0.7 mm, anthers 1.1–1.6(–1.8) mm; style 0.4 mm. **Capsules** chestnut brown or brown, 3-locular, widely ellipsoid, (2.2–)2.5–3.2 (–3.5) × 1.3–1.9 mm. **Seeds** dark brown, ellipsoid to lunate, 0.5–0.6 mm, not tailed. $2n$ = ca. 80, 84.

Flowering and fruiting late spring–summer. Forming extensive colonies in exposed coastal estuary meadows and salt marshes just above high-tide line; also inland; Greenland; St. Pierre and Miquelon; B.C., Man., N.B., Nfld. and Labr. (Nfld.), N.S., Ont., P.E.I., Que.; Colo., Conn., Del., Ill., Ind., Kans., Ky., Maine, Md., Mass., Mich., Minn., Mo., Mont., N.H., N.J., N.Y., N.Dak., Ohio, Oreg., Pa., R.I., Utah, Vt., Va., Wash., Wis.; Europe; Asia

16. Juncus coriaceus Mackenzie, Bull. Torrey Bot. Club 56: 28. 1929 [E]

Herbs, perennial, tufted, 3–9 dm. **Rhizomes** densely branching. **Culms** few to many. **Cataphylls** 1–3. **Leaves** basal, 0–1(–2); auricles 0.2–0.4(–0.6) mm, scarious to leathery; blade nearly terete but slightly channeled adaxially, 10–50 cm × 0.7–1.8 mm, margins entire. **Inflorescences** falsely lateral, 5–35-flowered, loose to congested, 1–8 cm; pri-

mary bract exceeding inflorescence. **Flowers:** pedicels 0.5–2.5 mm; bracteoles 2; tepals dark green, lanceolate, 3.5–5 mm; inner series slightly shorter, apex acute; stamens 6, fewer in cleistogamous flowers, filaments 0.8–1.2 mm, anthers 0.5–1.1 mm; style 0.9–1.3 mm. **Capsules** light to dark brown, 1-locular, widely ovoid to nearly globose, 3.5–5 × 2.6–3.2 mm. **Seeds** light to dark brown, oblate, 0.5–0.7 mm, not tailed. $2n$ = 80.

Flowering and fruiting spring–early summer. Springy, wet woodlands, stream banks, marshy areas, flatwood depressions, and shaded or exposed disturbed, sites with poor drainage; Ala., Ark., Del., Fla., Ga., Ky., La., Md., Miss., N.J., N.Y., N.C., Okla., S.C., Tenn., Tex., Va.

17. Juncus vaseyi Engelmann, Trans. Acad. Sci. St. Louis 2: 448. 1866 · Vasey's rush [E] [F]

Juncus greenei Oakes & Tuckerman var. *vaseyi* (Engelmann) B. Boivin

Herbs, perennial, tufted, 2–7 dm. **Rhizomes** densely branching. **Culms** 1–15(–25). **Cataphylls** 1–2(–3). **Leaves** basal, (1–)2–3; auricles 0.2–0.4(–0.6) mm, scarious, rarely ± leathery; blade green, nearly terete, 10–30 cm × 0.5–1 mm, margins entire. **Inflorescences** terminal, 5–15(–30)-flowered, congested or not, 1–5 cm; primary bract usually much shorter than inflorescence. **Flowers:** bracteoles 2; tepals greenish to tan, lanceolate or widely so, 3.3–4.4 mm; outer and inner series nearly equal, apex acuminate; stamens 6, filaments 0.5–0.9 mm, anthers 0.4–0.8 mm; style 0.1–0.2 mm. **Capsules** golden tan or light brown, 3-locular, ellipsoid, (3.3–)3.8–4.7 × (1.1–)1.3–1.7 mm. **Seeds** tan, ellipsoid to lunate, body 0.5–0.7 mm, tails 0.2–0.5 mm. 2*n* = ca. 80.

Flowering and fruiting summer–early fall. Permanently moist, usually exposed areas including wet meadows, raised sites or margins of bogs, or depressions along sandy lakeshores; 200–1500 m; Alta., B.C., Man., Nfld. and Labr. (Labr.), N.W.T., Ont., Que., Sask.; Colo., Idaho, Ill., Iowa, Maine, Mich., Minn., Mont., N.Y., S.Dak., Vt., Wis., Wyo.

18. Juncus greenei Oakes & Tuckerman, Amer. J. Sci. Arts 45: 37. 1843 • Greene's rush [E] [F]

Herbs, perennial, tufted, to 7 dm. **Rhizomes** short, densely branched. **Culms** 1–30. **Cataphylls** 1–2. **Leaves** basal, (1–)2–3; auricles (0.2–)0.4–0.6 (–0.8) mm, scarious, rarely ± leathery; blade dark green, nearly terete, 5–30 cm × 0.4–0.8 mm, margins entire. **Inflorescences** terminal, 5–50-flowered, usually congested, 1–8 cm; primary bract usually surpassing inflorescence. **Flowers:** bracteoles 2; tepals dark green or darker, lanceolate, 2.5–4.2 mm; outer and inner series nearly equal, apex acute; stamens 6, filaments 0.5–0.8 mm, anthers 0.5–0.8 mm; style 0.2 mm. **Capsules** chestnut brown or dark brown, infrequently lighter, 3-locular, ellipsoid, (2.5–)2.9–3.5(–4) × (1.1–)1.5–1.8 mm, slightly exceeding tepals, apex somewhat truncate. **Seeds** dark tan, ellipsoid to lunate, 0.48–0.65–(0.7) mm, not tailed. 2*n* = ca. 80.

Flowering and fruiting summer. Usually dry, well-drained, sandy soil in pine lands, near lake shores, or among sand dunes and often associated with disturbance; N.B., N.S., Ont., P.E.I., Que.; Conn., Ill., Ind., Iowa, Maine, Mass., Mich., Minn., N.H., N.J., N.Y., Pa., R.I., Vt., Wis.

19. Juncus brachyphyllus Wiegand, Bull. Torrey Bot. Club 27: 519. 1900 [E]

Juncus kansanus F. J. Hermann
Herbs, perennial, cespitose, to 8 dm. **Rhizomes** densely branching. **Cataphylls** 1–2. **Leaves** basal, 2–5; auricles 0.4–1.1(–2) mm, scarious; blade flat, 9–25 cm × 0.8–1.5(–2.4) mm, margins entire, becoming involute under xeric conditions. **Inflorescences** 10–150-flowered, usually congested, 1.5–8 cm; primary bract usually longer than inflorescence. **Flowers:** brac-

teoles 2; tepals green to tannish, lanceolate, 3.5–5.7 mm; outer and inner series nearly equal, in fruit apex appressed to capsule; stamens 6, filaments 0.6–1.1 mm, anthers 0.6–0.9 mm; style 0.2 mm. **Capsules** dark tan to reddish brown, 3-locular, ellipsoid to narrowly so, 2.6–4.7 × 1.3–2 mm, barely shorter than tepals. **Seeds** tan, ellipsoid to lunate, (0.3–)0.4–0.5(–0.6) mm, not tailed. 2*n* = ca. 80.

Flowering and fruiting late spring–early summer. Exposed moist or wet soils associated with depressions in temporal wetlands, along stream banks or lake shores especially in sandy soils; 900–1650 m; Ark., Calif., Idaho, Kans., Mo., Mont., Nebr., Okla., Oreg., Tex., Wash.

20. Juncus confusus Coville, Proc. Biol. Soc. Wash. 10: 127. 1896 [E] [F]

Juncus exilis Osterhout
Herbs, perennial, cespitose, 3–5 dm. **Rhizomes** densely branched. **Culms** (1–)5–15(–25). **Cataphylls** 1–3. **Leaves** basal, 2–4; auricles 0.3–0.7 mm, apex usually rounded, scarious to membranaceous; blade flat, 3–15 cm × 0.4–1 mm, margins entire. **Inflorescences** 3–25-flowered, congested, 1–2.5 × 1–2 cm; primary bract usually exceeding inflorescence. **Flowers** not secund; bracteoles 2; tepals dark green to blackish, usually with brownish midstripe, lanceolate to lanceolate-ovate, 3.5–4.3 mm, margins clear; outer and inner series nearly equal; stamens 6, filaments 0.6–0.9 mm, anthers 0.3–0.5 mm; style 0.1 mm. **Capsules** tan or darker, 3-locular, nearly globose to widely obovoid, 2.5–3.5 × 1.3–1.8 mm, shorter than perianth. **Seeds** yellowish, obovoid to ellipsoid, 0.4–0.5 mm, not tailed. 2*n* = 80.

Flowering and fruiting late spring–summer. Moist, open grasslands and meadows; 700–3400 m; Alta., B.C., Sask.; Ariz., Calif., Colo., Idaho, Nev., Mont., Oreg., S.Dak., Utah, Wash., Wyo.

21. Juncus georgianus Coville, Bull. Torrey Bot. Club 22: 44. 1895 • Georgia rush [E]

Herbs, short-lived perennial, cespitose, to 4 dm. **Culms** 3–40. **Leaves** basal, 2–3; auricles 0.2–0.3 mm, scarious to membranous; blade flat, 5–15 cm × 0.4–0.7 mm, margins entire. **Inflorescences** (3–)8–30(–45)-flowered, diffuse, 3–11 cm; primary bract rarely surpassing inflorescence. **Flowers:** bracteoles 2; tepals greenish to tan, lanceolate,

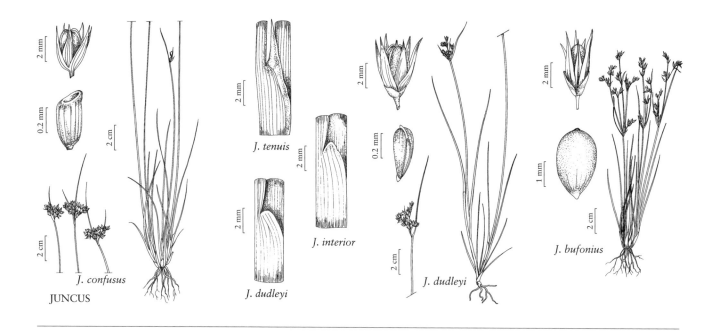

J. tenuis

J. interior

J. dudleyi

J. confusus

JUNCUS

J. dudleyi

J. bufonius

(3.3–)3.6–5.1(–5.7) mm; outer and inner series nearly equal, apex acuminate; stamens 6, filaments 0.2–0.4 mm, anthers (0.8–)1.2–1.5(–1.7) mm; style 0.6–0.8 mm. **Capsules** tan, 3-locular, ellipsoid to narrowly so, 2.7–4 × 1.2–1.7 mm. **Seeds** tan, ellipsoid or widely so, 0.4–0.5 mm, not tailed. $2n$ = ca. 80.

Flowering and fruiting spring. Exposed sites with thin, usually sandy soil over surfacing granite (flat-rocks); the soil in these areas may be moist in the spring from seepage or for a short period after rainfalls; 300–600 m; Ala., Ga., N.C., S.C.

22. Juncus secundus P. Beauvois ex Poiret in J. Lamarck et al., Encycl., suppl. 3: 160. 1813   E

Herbs, short-lived perennial, cespitose, 1.5–7 dm. **Culms** 1–30. **Leaves** basal, (1–)2–3; auricles 0.2–0.4(–0.6) mm, scarious; blade flat, 10–30 cm × 0.5–1 mm, margins entire. **Inflorescences** 5–15(–30)-flowered, usually somewhat loose, 1–5 cm; primary bract usually shorter than inflorescence. **Flowers** chiefly along inner side of branches; bracteoles 2; tepals greenish, lanceolate, 3.3–4.4 mm; outer and inner series equal, apex acuminate; stamens 6, filaments 0.5–0.9 mm, anthers 0.4–0.8 mm; style 0.1–0.2 mm. **Capsules** tan or light brown, 3-locular, ellipsoid, (3.3–)3.8–4.7 mm, nearly equal to tepals. **Seeds** tan, ellipsoid to lunate, 0.5–0.6(–0.7) mm, not tailed. $2n$ = ca. 80.

Flowering and fruiting spring. Exposed sites, usually with well-drained sandy soil, often associated with shallow bedrock; N.S., Ont.; Ark., Conn., Del., Ga., Ill., Ind., Ky., La., Maine, Md., Mass., Mo., N.H., N.J., N.Y., N.C., Ohio, Pa., R.I., Tenn., Vt., Va., W.Va.

23. Juncus dichotomus Elliott, Sketch Bot. S. Carolina 1: 406. 1817

Juncus dichotomus var. *platyphyllus* Wiegand; *J. tenuis* Willdenow var. *dichotomus* (Elliott) A. W. Wood; *J. tenuis* var. *platyphyllus* (Wiegand) Cory; *J. platyphyllus* (Wiegand) Fernald

Herbs, perennial, to 10 dm. **Rhizomes** densely branched to short-creeping. **Cataphylls** 1–3. **Leaves** basal, (1–)2–3; auricles 0.2–0.5(–0.6) mm, scarious to leathery; blade nearly terete, channeled or flat, 10–25 (–40) cm × (0.5–)0.7–1(–1.2) mm, margins entire. **Inflorescences** terminal, (5–)10–85(–100)-flowered, congested to somewhat loose, (1–)2.5–10–13 cm; primary bract usually exceeding inflorescence. **Flowers:** bracteoles 2; tepals green, lanceolate, (3–)3.3–4.5(–5.5) mm; outer and inner series nearly equal; stamens 6, filaments 0.6–1.2 mm, anthers 0.4–0.8(–1) mm; style 0.2 mm. **Capsules** tan to brown, 1-locular to pseudo-3-locular, ellipsoid to widely so, (2.5–)2.8–3.5(–4.5) × 1.6–2.2 mm. **Seeds** brownish to amber, ellipsoid to lunate, 0.3–0.4 mm, not tailed. $2n$ = ca. 80.

Flowering and fruiting late spring–summer. Ditches, shores, clearings, and other typically open areas, usually in sandy, well-drained (but frequently wet) soil; Ala., Ark., Colo., Conn., Del., D.C., Fla., Ga., La., Md., Mass., Miss., N.J., N.Y., N.C., Ohio, Pa., R.I., S.C., Tenn., Tex., Va., W.Va.; Mexico; Central America; South America.

24. Juncus tenuis Willdenow, Sp. Pl. 2: 214. 1799
• Path rush F W

Juncus bicornis Michaux; *J. bicornis* var. *williamsii* (Fernald) Victorin; *J. macer* Gray; *J. macer* var. *williamsii* (Fernald) Fernald; *J. tenuis* var. *bicornis* (Michaux) E. Meyer; *J. tenuis* var. *multicornis* E. Meyer; *J. tenuis* var. *williamsii* Fernald

Herbs, perennial, tufted, 1.5–5 dm. **Rhizomes** densely branching. **Culms** few–20. **Leaves** basal, (1–)2–3; auricles 2–5 mm, apex acute, membranous; blade flat, 3–12 cm × 0.5–1 mm, margins entire. **Inflorescences** 5–40-flowered, borne congested or branch internodes ca. as long as tepals, somewhat loose, 1–5 cm; primary bract usually longer than inflorescence. **Flowers:** bracteoles 2; tepals greenish, lanceolate, 3.3–4.4 mm; outer and inner series nearly equal; stamens 6, filaments 0.5–0.9 mm, anthers 0.1–0.2 mm; style 0.1–0.2 mm. **Capsules** tan or light brown, 1-locular to pseudo-3-locular, ellipsoid, (3.3–)3.8–4.7 × (1.1–)1.3–1.7 mm, nearly equal to tepals. **Seeds** tan, ellipsoid to lunate, 0.5–0.7 mm, not tailed. $2n = 80$.

Flowering spring–early summer. Exposed or shaded sites in soils ranging from sandy to clayey under moist or drier conditions, oftentimes these sites naturally or otherwise disturbed (e.g., game or human trails); Alta., B.C., Man., N.B., N.S., Ont., P.E.I., Que., Sask.; Ala., Alaska, Ariz., Ark., Calif., Colo., Conn., Del., D.C., Fla., Ga., Idaho, Ill., Ind., Iowa, Kans., Ky., La., Maine, Md., Mass., Mich., Minn., Miss., Mo., Mont., Nebr., Nev., N.H., N.J., N.Mex., N.Y., N.C., N.Dak., Ohio, Okla., Oreg., Pa., R.I., S.C., S.Dak., Tenn., Tex., Utah, Vt., Va., Wash., W.Va., Wis., Wyo.; introduced worldwide.

Juncus tenuis occurs throughout North America. It is particularly abundant in northeastern United States and eastern Canada, although infrequent in the south and west.

Through the use of isozyme electrophoresis, hybridization can be demonstrated between various members of the *Juncus tenuis* complex, including *Juncus tenuis*, *J. anthelatus*, *J. interior*, *J. secundus*, and *J. dichotomus* (R. E. Brooks, unpubl.). *Juncus ×oronensis* is thought to be a hybrid between *J. tenuis* and *J. vaseyi* in the northeast.

25. Juncus anthelatus (Wiegand) R. E. Brooks, Novon 9: 11. 1999 E

Juncus tenuis Willdenow var. *anthelatus* Wiegand, Bull. Torrey Bot. Club 27: 523. 1900; *J. macer* Gray var. *anthelatus* (Wiegand) F. J. Hermann

Herbs, perennial, cespitose, to (3–)7–9 dm. **Cataphylls** 1–2. **Leaves** basal, 2–3(–5); auricles transparent, 2–3.5 mm at summit of leaf sheath, apex acutish, membranous; blade flat, (10–)20–30 cm × 0.5–2.3 mm, margins entire. **Inflorescences** 10–100-flowered, diffuse; internodes of monochasia greater than 6 mm, primary bract usually exceeding inflorescence. **Flowers:** pedicels 0.1–0.3(–3) mm; bracteoles 2; tepals green, lanceolate, 3.2–4.5 × 0.7–1 mm; outer and inner series nearly equal, in fruit apically erect; stamens 6, filaments 0.8–1.1 mm, anthers 0.3–0.7 mm; style 0.2 mm. **Capsules** tan, 1-locular, widely ellipsoid to obovoid, 2–3.2 × 1.1–1.6 mm. **Seeds** tan, ellipsoid, 0.3–0.6 mm, not tailed. $2n = 80$.

Flowering and fruiting spring. Exposed or partially shaded sites in moist or seasonally wet, sandy or clay soils; N.B., Ont., Que.; Ala., Ark., Ga., Ill., Ind., Iowa, Kans., Ky., La., Maine, Md., Mich., Minn., Miss., Mo., N.H., N.Y., N.C., Ohio, Okla., Pa., S.C., Tenn., Tex., Vt., Va., W.Va., Wis.

26. Juncus occidentalis (Coville) Wiegand, Bull. Torrey Bot. Club 27: 521. 1900 • Western rush E

Juncus tenuis Willdenow var. *occidentalis* Coville, Proc. Biol. Soc. Wash. 10: 129. 1896; *J. tenuis* var. *congestus* Engelmann

Herbs, perennial, cespitose, 3–6 dm. **Rhizomes** densely branching. **Culms** few–20. **Leaves** basal, (1–)2–3; auricles 0.5–1(–1.5) mm, apex acutish, membranous; blade flat, 5–15 cm × 0.5–1.3 mm, margins entire. **Inflorescences** 5–50-flowered, somewhat loose, 1–7 cm; primary bract usually longer than inflorescence. **Flowers:** bracteoles 2; tepals greenish, lanceolate, (3.5–)5 mm; outer and inner series nearly equal; stamens 6, filaments 0.5–1 mm, anthers 0.1–0.3 mm; style 0.1–0.2 mm. **Capsules** light brown, 1-locular to pseudo-3-locular, ellipsoid, (3–)5–4.5 × 1.2–1.8 mm, nearly equal to shorter than tepals. **Seeds** tan, ellipsoid to lunate, 5.5–0.7 mm, not tailed. $2n = 80$.

Flowering spring–early summer. Moist, usually exposed sites in clayey or sandy soil around springs, along rivers and streams, and around lakes; below 2300 m; Ariz., Calif., Idaho, Oreg., Wash.

Juncus occidentalis should perhaps be considered a robust variant of *J. tenuis*.

27. Juncus interior Wiegand, Bull. Torrey Bot. Club 27: 516. 1900 • Interior rush E F

Juncus arizonicus Wiegand; *J. interior* var. *arizonicus* (Wiegand) F. J. Hermann; *J. interior* var. *neomexicanus* (Wiegand) F. J. Hermann; *J. monostichus* Bartlett; *J. neomexicanus* Wiegand

Herbs, perennial, tufted, 2–6 dm. **Rhizomes** densely branching. **Culms** 1–10. **Cataphylls** 1–3. **Leaves** basal, 1–2(–3); auricles whitish or purplish tinged, 0.2–0.4(–0.6) mm, scarious; blade flat, 5–15 cm × 0.5–1.1 mm, margins entire. **Inflorescences** usually somewhat compact, 1.5–7 cm; primary bract usually shorter than inflorescence. **Flowers:** bracteoles 2; tepals greenish, lanceolate, 3.3–4.4 mm, apex acuminate; stamens 6, filaments 0.5–0.9 mm, anthers 0.4–0.6 mm; style 0.1–0.2 mm. **Capsules** light tan or darker, 1-locular to pseudo-3-locular, ellipsoid to nearly globose, (3.3–)3.8–4.7 mm, nearly equal to or longer than tepals. **Seeds** tan, ellipsoid to lunate, 0.4–0.7 mm, not tailed. $2n = 80$.

Flowering and fruiting late spring–early summer. Dry, often upland sites in prairies, exposed disturbed sites, and ditches in sandy or clayey soils; Man., Ont., Sask.; Ark., Colo., Ill., Ind., Iowa, Kans., Ky., La., Minn., Mont., Mo., Nebr., N.Mex., N.Dak., Ohio, Okla., S.Dak., Tenn., Tex., Wis., Wyo.

28. Juncus dudleyi Wiegand, Bull. Torrey Bot. Club 27: 524. 1900 • Dudley's rush F

Juncus tenuis Willdenow var. *dudleyi* (Wiegand) F. J. Hermann; *J. tenuis* var. *uniflorus* Farwell

Herbs, perennial, 2–10 dm. **Rhizomes** densely branching. **Culms** 1–20. **Cataphylls** 1–3. **Leaves** basal, 2–3; auricles yellowish, 0.2–0.4 mm, leathery; blade flat, 5–30 cm × 0.5–1 mm, margins entire, turned up, occasionally involute. **Inflorescences** compact and few flowered to loose and lax with to 80 flowers, 1.5–5(–9) cm; primary bract usually exceeding inflorescence. **Flowers:** bracteoles 2; tepals greenish, lanceolate, 4–5 mm; inner series nearly equal, spreading in fruit, nearly equal to or exceeding capsule; stamens 6, filaments 0.8–1.2 mm, anthers 0.6–1 mm; style 0.2 mm. **Capsules** tan, 1-locular to pseudo-3-locular, ellipsoid, 2.9–3.6 × 1.5–1.9 mm. **Seeds** tan to amber, ellipsoid to lunate, 0.4–0.7 mm, not tailed. $2n = $ ca. 84.

Flowering and fruiting spring–early summer. Exposed or shaded sites in sandy to clayey soils, usually moist areas such as along stream banks, ditches, around springs; Alta., B.C., Man., N.B., Nfld. and Labr., N.W.T., N.S., Ont., P.E.I., Que., Sask., Yukon; Alaska, Ariz., Ark., Calif., Colo., Conn., Del., D.C., Idaho, Ill., Ind., Iowa, Kans., Ky., Maine, Md., Mass., Mich., Minn., Mo., Mont., Nebr., N.H., N.J., N.Mex., N.Y., N.Dak., Ohio, Okla., Oreg., Pa., R.I., S.Dak., Tenn., Tex., Utah, Vt., Va., Wash., W.Va., Wis., Wyo.; Mexico.

29. Juncus bufonius Linnaeus, Sp. Pl. 1: 328. 1753 F W

Juncus bufonius var. *congestus* (S. Watson) Fernald; *J. bufonius* var. *halophilus* Buchenau & Fernald; *J. bufonius* var. *hybridus* Farwell; *J. bufonius* var. *occidentalis* F. J. Hermann; *J. bufonius* var. *ranarius* Farwell; *J. congestus* S. Watson; *J. ranarius* Songeon & E. Perrier

Herbs, annual, cespitose, 0.5–4 dm. **Culms** 1–many, occasionally becoming decumbent. **Cataphylls** 0–2. **Leaves** basal and cauline; auricles rudimentary or absent; blade flat, 3–13 cm × 0.3–1.1 mm. **Inflorescences** loose and diffuse or less often compact, usually at least 1/2 total height of plant; primary bract shorter than inflorescence. **Flowers:** bracteoles 2; tepals greenish, lanceolate, 3.8–7(–8.5) mm; inner series slightly shorter, apex sometimes obtuse; stamens 3–6, filaments (0.7–)1–1.8 mm, anthers 0.3–0.8 mm; style 0.1–0.2 mm. **Capsules** tan to reddish brown, 3-locular, ellipsoid to narrowly so, slightly truncate, 2.7–4 × 1–1.5 mm, sometimes exceeding inner tepals but usually not outer series. **Seeds** yellowish, widely ellipsoid to ovoid, 0.26–0.49, not tailed. $2n = 27$–37, 58–81, 108–115.

Flowering and fruiting spring–early fall. Moist soils in meadows, along lakeshores or stream banks, ditches, or roadsides, especially frequent in drawdown areas; usually in open sites and often becoming weedy; Greenland; Alta., B.C., Man., N.B., Nfld. and Labr., N.W.T., N.S., Ont., P.E.I., Que., Sask., Yukon; Ala., Alaska, Ariz., Ark., Calif., Colo., Conn., Del., D.C., Fla., Ga., Idaho, Ill., Ind., Iowa, Kans., Ky., La., Maine, Md., Mass., Mich., Minn., Miss., Mo., Mont., Nebr., Nev., N.H., N.J., N.Mex., N.Y., N.C., N.Dak., Ohio, Okla., Oreg., Pa., R.I., S.C., S.Dak., Tenn., Tex., Utah, Vt., Va., Wash., W.Va., Wis., Wyo.; nearly worldwide.

Nearly worldwide, *Juncus bufonius* is found essentially throughout North America except north of the Alaskan and Canadian taiga. *Juncus bufonius* is a highly polymorphic complex that is poorly understood systematically. Insufficient evidence exists upon which to base the segregation of the plethora of taxa that have been recognized out of this group in the past.

1c. JUNCUS Linnaeus subg. GRAMINIFOLII Buchenau, Abh. Naturwiss. Vereine Bremen 4: 407, 441. 1875

Juncus sect. *Graminifolii* Engelmann

Herbs, annual or perennial, perennials rhizomatous. **Stems** terete. **Leaves** basal or cauline; blade flat or terete and channeled, not septate. **Inflorescences** terminal panicles or racemes of 2–many heads or single terminal head, sympodial; bracteoles absent below perianth. **Flowers** solitary or in heads. **Capsules** 1-locular or usually 3-locular. **Seeds** usually not tailed.

Species ca. 25 (20 in the flora): North America and south temperate areas of the world.

The relationships of both the perennial and annual species in this subgenus are poorly understood as evidenced by the variety of opinions expressed in numerous floras. The need for a multidisciplinary investigation of this subgenus, including morphometric and chromosomal evaluations, became obviously evident during the preparation of this manuscript.

SELECTED REFERENCE   Ertter, B. 1986. The *Juncus triformis* complex. Mem. New York Bot. Gard. 39: 1–90.

1. Plants annual.
    2. Proximal bracts foliose, clearly surpassing inflorescence; outer tepals acute, 2 times as long as delicate obtuse inner tepals. . . . . .49. *Juncus capitatus*
    2. All bracts commonly inconspicuous, membranous; inner and outer tepals similar in size and shape.
        3. Style and anthers more than 1 mm; culms to 13(–17) cm.
            4. Culms capillary, 0.1–0.4 mm diam.; capsules usually shorter than perianth. . . . 40. *Juncus triformis*
            4. Culms stouter, (0.2–)0.4–0.8 mm diam.; capsules usually nearly equal to tepals . . . . .41. *Juncus leiospermus*
        3. Style and anthers less than 0.5(–0.8) mm; culms rarely to 6.5 cm.
            5. Flowers mostly several per culm; bracts acute to acuminate.
                6. Capsules usually shorter than perianth, greenish or tan. . . . . .45. *Juncus capillaris*
                6. Capsules usually nearly equal to or longer than perianth, brownish.
                    7. Flowers 2-merous; capsules longer than tepals. . . . . .44. *Juncus tiehmii*
                    7. Flowers usually 3-merous; capsules nearly equal to perianth.
                        8. Flowers several per culm; capsules and tepals commonly turning dark reddish; seeds 0.4–0.5 mm. . . . . .42. *Juncus kelloggii*
                        8. Flowers solitary; capsules and tepals commonly remaining pale yellow-green until seeds ripen; seeds 0.3–0.4 mm. . . . . .43. *Juncus luciensis*
            5. Flowers uniformly solitary; bracts acute to truncate or absent.
                9. Tepals turning inward to enwrap shorter capsule at maturity, mostly less than 2.3 mm; flowers usually 3-merous; culms 0.1–0.2 mm diam. . . . . .46. *Juncus bryoides*
                9. Tepals erect to recurved at maturity, nearly equal to or shorter than capsule, 2–4 mm; flowers 2–3-merous; culms usually more than 0.2 mm diam.
                    10. Flowers 3-merous; bracts solitary, truncate, completely sheathing culm   47. *Juncus uncialis*
                    10. Flowers 2-merous; bracts 0–2, rounded to ovate, not sheathing culm 48. *Juncus hemiendytus*
1. Plants perennial.
    11. Culms arcuate-stoloniferous and creeping or floating, or growing submersed along bottom; nodes rooting. . . . . .39. *Juncus repens*
    11. Culms erect, never rooting at nodes.

12. Seeds tailed.
    13. Auricles absent or essentially so; anthers 1–1.5 mm. 32. *Juncus regelii*
    13. Auricles 1–3 mm, rounded to acutish; anthers 1.8–2.6 mm. . . . . 33. *Juncus howellii*
12. Seeds not tailed.
    14. Culms borne in tufts, rhizomes poorly developed.
        15. Stamens 6. . . . . 34. *Juncus macrophyllus*
        15. Stamens 3.
            16. Tepals straw-colored with green midstripe; capsules tan to reddish
                brown; glomerules mostly 1–5(–10) per culm. . . . . 37. *Juncus filipendulus*
            16. Tepals dark brown; capsules brown; glomerules more than 10 per
                culm. . . . . 38. *Juncus marginatus* (in part)
    14. Culms borne singly or in small groups along creeping rhizome.
        17. Stamens 3. . . . . 38. *Juncus marginatus* (in part)
        17. Stamens 6.
            18. Auricles absent or rudimentary, less than 1 mm.
                19. Anthers 1.6–2.4 mm. 31. *Juncus orthophyllus* (in part)
                19. Anthers 0.8–1.5 mm.
                    20. Tepals 3–4 mm. . . . . 36. *Juncus covillei* (in part)
                    20. Tepals 4.5–6 mm. . . . . 35. *Juncus falcatus*
            18. Auricles rounded or acute, 1–3 mm.
                21. Tepals 3–4 mm. . . . . 36. *Juncus covillei* (in part)
                21. Tepals 5–6 mm.
                    22. Tepals brown with green midstripe. . . . . 30. *Juncus longistylis*
                    22. Tepals brown. . . . . 31. *Juncus orthophyllus* (in part)

30. Juncus longistylis Torrey in W. H. Emory, Rep. U.S. Mex. Bound. 2(1): 223. 1859 • Longstyle rush E F

Herbs, perennial, 2–6 dm. **Rhizomes** long creeping. **Culms** slightly compressed. **Leaves:** basal 2–5, cauline 1–3; auricles 1–2.5 mm, apex truncate to obtuse; blade flat, 1–3 dm × 1.5–3 mm, the cauline shorter. **Inflorescences** glomerules 1–4(–8), each with 3–12 flowers, open or aggregate, 2–6(–10) cm; primary bract shorter than inflorescence. **Flowers:** tepals brown with green midstripe, lanceolate, 5–6 mm, margins scarious, sometimes papillose; outer series slightly shorter; stamens 6, filaments 0.5–1 mm, anthers 1.2–2 mm; style 0.6 mm. **Capsules** tan, 3-locular, obovoid, 3–5 mm, shorter than perianth. **Seeds** ovoid, 0.4–0.6 mm, not tailed. $2n = 40$.

Flowering and fruiting summer. Moist ground in mountain meadows; 1000–3300 m; Alta., B.C., Man., Nfld. and Labr. (Nfld.), Ont., Que., Sask.; Ariz., Calif., Colo., Idaho, Minn., Mont., Nebr., Nev., N.Dak., N.Mex., S.Dak., Oreg., Utah, Wash., Wyo.

31. Juncus orthophyllus Coville, Contr. U.S. Natl. Herb. 4: 207. 1893 • Straight-leaf rush E

Juncus longistylis Torrey var. *latifolius* Engelmann, Trans. Acad. Sci. St. Louis 2: 496. 1868; *J. latifolius* (Engelmann) Buchenau 1890, not Wulfen 1789

Herbs, perennial, 2–4 dm. **Rhizomes** creeping. **Culms** compressed. **Leaves:** basal several, cauline 0–3; auricles, if present, 0.5–1 mm, apex acutish; blade flat, basal 10–40 cm × 1–6 mm, cauline blade reduced. **Inflorescences** glomerules, usually 3–12, each with 5–10 flowers, open; primary bract much shorter than inflorescence. **Flowers:** tepals brown, lanceolate, 5–6 mm, apex obtuse; outer series shorter, margins scarious, apex acute, minutely papillose; inner series with margins clear; stamens 6, filaments 0.5–1 mm, anthers 1.6–2.4(–3) mm; style 0.5–2 mm. **Capsules** tan, 3-locular, obovoid, 3–5 mm, shorter than perianth. **Seeds** ovoid, 0.6 mm, not tailed.

Flowering and fruiting late spring to summer. Moist ground in mountain meadows; 1200–3500 m; B.C.; Calif., Idaho, Mont., Nev., Oreg., Wash.

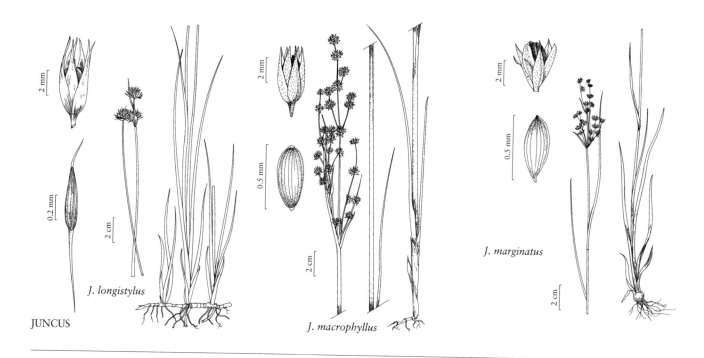

J. longistylus

JUNCUS

J. macrophyllus

J. marginatus

32. Juncus regelii Buchenau, Bot. Jahrb. Syst. 12: 414. 1890 • Regel's rush E

Juncus jonesii Rydberg

Herbs, perennial, 2–6 dm. **Rhizomes** long creeping. **Culms** erect slightly compressed, never rooting at nodes. **Leaves:** basal several, cauline 1–3; auricles absent or essentially so; blade flat, 0.5–3 dm × 1.5–4 mm. **Inflorescences** glomerules 1–5, each with (8–) 10–30 flowers, congested to open, 2–6 cm; primary bract shorter than inflorescence. **Flowers:** tepals dark brown with green midstripe, lanceolate, 4–6 mm, margins scarious, occasionally papillose; inner series slightly shorter; stamens 6, filaments 1–1.3 mm, anthers 1–1.5 mm; style 0.6 mm. **Capsules** tan or darker, 3-locular, obovoid, 3–5 mm, shorter than perianth. **Seeds** ovoid, body 0.6–0.8 mm, tails 0.2–0.4 mm.

Flowering and fruiting summer. Moist to wet meadows in montane or subalpine areas; 540–3000 m; B.C.; Calif., Idaho, Mont., Nev., Oreg., Utah, Wash., Wyo.

33. Juncus howellii F. J. Hermann, Leafl. W. Bot. 5: 182. 1949 • Howell's rush E

Herbs, perennial, 2–6 dm. **Rhizomes** long creeping. **Culms** erect, slightly compressed, never rooting at nodes. **Leaves:** basal 2–4, cauline 2–3; auricles 1–3 mm, apex rounded to acutish, membranous; blade flat, 10–30 cm × 2–4 mm, reduced distally, margins occasionally papillose. **Inflorescences** glomerules, usually 3–9, each with 3–8 (–10) flowers, open, 2–9 cm; primary bract much shorter than inflorescence. **Flowers:** tepals yellow-brown with green midstripe, lanceolate, 5–6.5 mm, margins clear; outer and inner series nearly equal, adaxially papillose; stamens 6, filaments 0.5–1 mm, anthers 1.8–2.6 mm; style 0.6 mm. **Capsules** tan, 3-locular, obovoid, 3–5 mm, shorter than perianth. **Seeds** ovoid, body 0.5–0.7 mm, tails 0.2–0.4 mm.

Flowering and fruiting summer. Moist ground in mountain meadows; 850–2500 m; Calif., Idaho, Oreg., Wash.

34. Juncus macrophyllus Coville, Univ. Calif. Publ. Bot. 1: 65. 1902 • Long-leaf rush [C] [E] [F]

Juncus canaliculatus Engelmann, Bot. Gaz. 7: 6. 1882, not Liebmann 1850; *J. longistylis* Torrey var. *scabratus* Hermann

Herbs, perennial, tufted, 2–10 dm. **Rhizomes** poorly developed. **Culms** erect, nearly terete to slightly compressed, never rooting at nodes. **Leaves:** basal several, cauline 1–3; auricles rounded, 1–3 mm, apex acutish, membranous; blade pale green, basal blade striate, channeled, basal ½–1 times length of culm; cauline 8–15 cm × 1–3 mm. **Inflorescences** glomerules, 8–25, each with 3–5 flowers, open; primary bract much shorter than inflorescence. **Flowers:** tepals greenish with reddish or brown tinge, lanceolate, 5–6 mm, outer series shorter; stamens 6, filaments 0.5–1 mm, anthers 1.8–2.6 mm; style 0.5–1 mm. **Capsules** tan, 3-locular, obovoid, 3–4.5 mm, shorter than perianth. **Seeds** ovoid, 0.6 mm, not tailed.

Flowering and fruiting summer. Wet banks and meadows in chaparral and low mountains; 700–2600 m; Ariz., Calif., Nev.

35. Juncus falcatus E. Meyer, Syn. Luzul., 34. 1823

Herbs, perennial, rhizomatous, 0.5–3 dm. **Culms** erect, slightly compressed. **Leaves:** basal 2–4, cauline 1–3; auricles absent or rudimentary; blade flat, 4–15 cm × 1.5–3 mm. **Inflorescences** glomerules, (1–)2–5, each with 2–15 flowers, open or aggregate; primary bract usually exceeding inflorescence. **Flowers:** tepals brown, ovate-lanceolate, 4.5–6 mm, margins scarious, minutely papillate; inner series slightly shorter. **Capsules** brown, 3-locular, obovoid, 3–5 mm, shorter than perianth. **Seeds** ovoid to pyriform, 0.7 mm, not tailed.

Varieties 2 (2 in the flora): North America, Pacific Islands.

1. Filaments much shorter than anthers. . . . .
. . . . .35a. *Juncus falcatus* var. *falcatus*
1. Filaments ca. as long as anthers. . . . .
. . . . .35b. *Juncus falcatus* var. *sitchensis*

35a. Juncus falcatus E. Meyer var. **falcatus**

Filaments 0.5 mm, anthers 1–1.3 mm; style 1 mm.

Flowering and fruiting summer–fall. Moist, sandy coastal areas; Calif.; Asia (Japan); Pacific Islands; Australia (including Tasmania).

35b. Juncus falcatus E. Meyer var. **sitchensis** Buchenau, Bot. Jahrb. Syst. 12: 428. 1890

Juncus falcatus subsp. *sitchensis* (Buchenau) Hultén; *J. falcatus* var. *alaskensis* Coville; *J. menziesii* R. Brown ex Hooker

Filaments 0.8–1.3 mm, anthers 1–1.5 mm; style 0.5 mm. $2n =$ ca. 40.

Flowering and fruiting summer–fall. B.C.; Alaska; Pacific Islands (Kuriles).

36. Juncus covillei Piper, Contr. U.S. Natl. Herb. 11: 182. 1906 • Coville's rush

Juncus falcatus E. Meyer var. *paniculatus* Engelmann, Trans. Acad. Sci. St. Louis 2: 495. 1868, not *J. paniculatus* Hoppe ex Mertens & W. D. J. Koch 1826

Herbs, perennial, strongly rhizomatous, 0.5–2.5 dm. **Culms** erect, slightly flattened. **Leaves:** basal 2–5, cauline 2–4; auricles absent or, if present, 0.5–1.5 mm, apex acute; blade flat, 5–15 cm × 2–3 mm. **Inflorescences** glomerules, 1–6, each with 3–7 flowers, open to aggregate, 2–6 cm; primary bract or distal leaves usually exceeding inflorescences. **Flowers:** tepals usually brown, ovate-oblong, 3–4 mm, margins scarious; inner series slightly shorter, minutely papillate; stamens 6, filaments 0.8–1.4 mm, anthers 0.8–1.4 mm; style 1 mm. **Capsules** 3-locular, narrowly ovoid, 4–5 mm. **Seeds** narrowly ovoid, 0.3 mm, not tailed.

Varieties 2 (2 in the flora): coastal to montane areas.

1. Perianth dark brown, apex of inner series rounded; capsule dark brown, 1 mm longer than perianth.. . . . . .36a. *Juncus covillei* var. *covillei*
1. Perianth pale brown, apex of inner series acute to obtuse; capsule pale brown, barely exceeding perianth.. . . . . . . .36b. *Juncus covillei* var. *obtusatus*

36a. Juncus covillei Piper var. **covillei**

J. falcatus E. Meyer var. *prominens* Buchenau; *J. latifolius* var. *paniculatus* Buchenau, not Hoppe; *J. prominens* Miyabe & Kudô

Perianth dark brown, apex of inner series rounded. **Capsules** dark brown, 1 mm longer than perianth. 2*n* = 36.

Flowering and fruiting summer–early fall. Wet areas along lakes, streams and rivers, especially on spray areas or occasionally flooded outcrops, coastal to low montane areas; below 300 m; B.C.; Calif., Idaho, Mont., Oreg., Wash.; Asia (Japan); Pacific Islands (Kuriles).

36b. Juncus covillei Piper var. **obtusatus** C. L. Hitchcock in C. L. Hitchcock et al., Vasc. Pl. Pacif. N.W. 1: 193. 1969 [E]

Juncus obtusatus Engelmann, Trans. Acad. Sci. St. Louis 2: 495. 1868, not Kitaibel 1863

Perianth pale brown, apex of inner series acute to obtuse. **Capsules** pale brown, barely exceeding perianth.

Flowering and fruiting summer–early fall. Wet areas along lakes, streams and rivers, especially on spray areas or occasionally flooded outcrops, mostly in montane areas; 500–3000 m; B.C.; Calif., Idaho, Mont., Oreg., Wash.

37. Juncus filipendulus Buckley, Proc. Acad. Nat. Sci. Philadelphia 14: 8. 1862 [E]

Juncus leptocaulis Torrey & A. Gray ex Engelmann

Herbs, perennial, tufted, 1.5–3.5 dm. **Rhizomes** poorly developed. **Culms** erect, compressed, bases often swollen. **Leaves:** basal 2–4, cauline 1–3; auricles 0.5–1 mm, apex rounded to nearly acute; blade flat, 3–15 cm × 1–2.5 mm. **Inflorescences** glomerules, (1–)2–5(–10), each with (3–)6–15 flowers, open; primary bract shorter than inflorescence. **Flowers:** tepals straw-colored with green midstripe, lanceolate or widely so, 3.5–5 mm, margins sometimes clear; outer and inner series nearly equal; stamens 3, filaments 1.5 mm, anthers 0.5 mm; style 0.5 mm. **Capsules** tan to reddish brown, 3-locular, obovoid, 2.6–3.2 mm, shorter than perianth. **Seeds** fusiform, 0.5–0.6 mm, not tailed.

Flowering and fruiting spring–early summer. Moist,

usually calcareous soils of swales or glades, occasionally in shallow water along streams; Ala., Ga., Ky., Okla., Tenn., Tex.

38. Juncus marginatus Rostkovius, De Junco, 38, plate 2, fig. 3. 1801 • Grass-leaf rush [F]

Juncus aristulatus Michaux; *J. aristulatus* var. *pinetorum* Coville; *J. biflorus* Elliott; *J. longii* Fernald; *J. marginatus* var. *aristulatus* (Michaux) Coville; *J. marginatus* var. *biflorus* (Elliott) Chapman; *J. marginatus* var. *odoratus* Torrey; *J. marginatus* var. *paucicapitatus* Engelmann; *J. marginatus* var. *setosus* Coville; *J. marginatus* var. *vulgaris* Engelmann; *J. odoratus* (Torrey) Steudel; *J. setosus* (Coville) Small

Herbs, perennial, occasionally tufted, sometimes rhizomatous, 3–13 dm. **Rhizomes** short, knotty. **Culms** compressed. **Leaves** basal and cauline; auricles 0.5–1.5 mm, apex rounded, membranous; basal blade flat, 20–45 cm × 1.5–5 mm, cauline reduced. **Inflorescences** glomerules, (2–)5–200, each with (1–)2–10(–20) flowers, mostly open, 3–10(–15) cm; primary bract shorter than inflorescence. **Flowers:** tepals dark brownish, usually with green midstripe, outer series ovate-lanceolate, 1.8–3.2 mm, margins broad, clear, awned or not, apex acutish; inner series ovate to lanceolate, 2–3.5 mm, slightly longer than outer series, apex obtuse to acute, awned or not; stamens 3, opposite outer tepals, shorter to longer than tepals, filaments 1.1–2.5 mm, anthers 0.3–1.2 mm; style 0.3 mm. **Capsules** brown and sometimes dark spotted, 3-locular, obovoid to nearly globose, 1.8–2.9 mm, shorter to longer than perianth. **Seeds** yellow to light brown, fusiform, 0.4–0.7 mm, not tailed. 2*n* = 38, 40.

Flowering and fruiting late spring–fall. Moist to wet sandy, peaty, or clayey soils, usually in open areas including bogs, shores, marshes, and ditches; N.S., Ont.; Ala., Ark., Calif., Colo., Conn., Del., D.C., Fla., Ga., Ill., Ind., Iowa, Kans., Ky., La., Maine, Md., Mass., Mich., Minn., Miss., Mo., Nebr., N.H., N.J., N.Y., N.C., Ohio, Okla., Oreg., Pa., R.I., S.C., S.Dak., Tenn., Tex., Vt., Va., W.Va., Wis.; Mexico; West Indies (Cuba); Central America.

The number of glomerules per inflorescence, stamen length versus perianth length, and tepal shape have separately and in combination been used to distinguish a number of taxa at various nomenclatural ranks. These characters, however, vary considerably across the distribution of the species (broad sense) and do so independently of one another to the point that if separate taxa are recognized, they pass insensibly among each other.

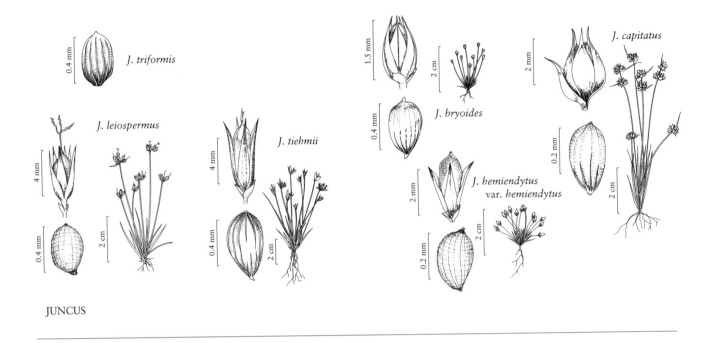

JUNCUS

39. Juncus repens Michaux, Fl. Bor.-Amer. 1: 191. 1803 · Creeping rush

Herbs, perennial, cespitose, floriferous culm 0.5–3 dm. Culms first ascending, soon arcuate-stoloniferous and creeping or floating, or growing submersed along bottom, each node with cluster of basal leaves and fibrous roots, eventually each emergent terrestrial node with floriferous culm. Leaves basal; auricles 0.5–1 mm, apex acutish, membranous or thicker; blade spreading, flat, 2–10 (–15) cm × 1–3 mm. Inflorescences glomerules, (1–)2–10, each with 3–12 flowers, open; primary bract usually shorter than inflorescence. Flowers: tepals green, margins scarious; inner series narrowly lanceolate, 5–9 mm, apex usually recurved; outer series obviously shorter, apex usually erect; stamens 3, filaments 1.5–3 mm, anthers 0.5–0.8 mm; style 0.5 mm. Capsules tan, 3-locular, narrowly ellipsoid, 3.5–5.5 × 0.8–1.2 mm. Seeds brown, ovoid, 0.3–0.4 mm, not tailed.

Flowering and fruiting summer–fall. Shores of ponds, lakes, and borrow pits, flatwood depressions, ditches, and drainage canals; Ala., Ark., Fla., Ga., La., Md., Miss., N.C., Okla., S.C., Tenn., Tex., Va.; Mexico (Tabasco); West Indies (Cuba).

40. Juncus triformis Engelmann, Trans. Acad. Sci. St. Louis 2: 492. 1868 · Long-styled dwarf rush [C][E][F]

Juncus megaspermus F. J. Hermann; *J. triformis* var. *stylosus* Engelmann

Herbs, annual, cespitose, 0.5–1.5 dm. Culms 2–80, capillary, 0.1–4 mm diam. Leaves 1/5–1/3 height of plant. Inflorescences headlike clusters, each with 1–7 flowers, bracts subtending inflorescence 2–10, ovate to broadly lanceolate, inconspicuous, 0.7–2.4 mm, membranous, apex acute. Flowers: tepals reddish with green central band, linear to lanceolate, 1.9–4.5 × 0.5–1.1 mm; inner series slightly longer than outer, apex acuminate to attenuate; stamens 3, filaments 0.4–1.1 mm, anthers 1–2.2 mm; style persistent, 1–3.2 mm, stigma 0.5–2.5 mm. Capsules reddish, 3-locular, ovoid to ellipsoid or oblate, 1.6–2.5 × 1.1–1.8 mm, usually 1/2 length of tepals. Seeds nearly globose to ellipsoid, 0.3–0.7 mm. $n = 18$.

Flowering and fruiting spring–mid summer. Shallow soil of seepage areas on granite outcrops, such as along stream banks, ditches, around springs; of conservation concern; to 2500 m; Calif.

41. Juncus leiospermus F. J. Hermann, Leafl. W. Bot. 5: 113. 1948 • Smooth-seeded rush [C] [E] [F]

Juncus leiospermus var. *ahartii* Ertter

Herbs, annual, cespitose, 0.2–1.2 dm. **Culms** to 100, (0.2–)0.4–0.8 mm diam. **Leaves** ⅓–¾ height of plant. **Inflorescences** terminal clusters, each with 1–7 flowers; bracts subtending inflorescence 2–8, round to acutely ovate, inconspicuous 0.7–2.4 mm, membranous. **Flowers:** tepals reddish to black, narrowly lanceolate to oblong, 2–4.6 × 0.5–1.4 mm; inner series usually slightly longer than outer, apex obtuse to acuminate; stamens 3, filaments 0.5–1.4 mm, anthers 1–3.2 mm; style persistent, 1.4–4 mm, stigma 1.3–4 mm. **Capsules** reddish, 3-locular, globose to ellipsoid or oblong, 2–4.5 × 1–3 mm, nearly equal to sepals. **Seeds** ovoid, 0.35–0.45 mm. $n = 16$.

Flowering and fruiting in spring. Margins of vernal pools; of conservation concern; to 500 m; Calif.

Plants from Butte and Calaveras Counties, California, tend to be smaller and consistently develop only one flower per culm; they may be recognized as *Juncus leiospermus* var. *ahartii* Ertter.

42. Juncus kelloggii Engelmann, Trans. Acad. Sci. St. Louis 2: 494. 1868 [E]

Juncus triformis Engelmann var. *brachystylus* Engelmann

Herbs, annual, cespitose, 0.1–0.55(–0.65) dm. **Culms** to 100. **Leaves** to 2.5 cm × 0.1–0.3 mm. **Inflorescences** terminal clusters, each with 1–4 flowers; bracts subtending inflorescence 2–5, ovate to lanceolate, inconspicuous, 1–2.5 mm, membranous, apex acute to obtuse. **Flowers** 3-merous; tepals dark reddish brown, 2–3.2 × 0.4–1 mm, margins unpigmented; inner series nearly equal to or slightly shorter than outer series, acute to acuminate; stamens 3, filaments 0.7–1.3 mm, anthers 0.3–0.5 mm; style 0.2–0.5 mm, stigma 0.4–1.3 mm. **Capsules** dark reddish, 3-locular, obovoid to ellipsoid, nearly equal to tepals, 1.8–2.9 × 1.2–1.6 mm. **Seeds** ovoid, 0.4–0.5 mm, apiculate. $n = 17$.

Flowering and fruiting spring–early summer. Sandy and clayey damp soils around vernal pools, seepage areas, and low spots in fields and meadows; to 800 m; B.C.; Calif., Nev., Oreg., Wash.

43. Juncus luciensis Ertter, Mem. New York Bot. Gard. 39: 58, figs. 13c–e, 14. 1986 • Santa Lucia dwarf rush [E]

Herbs, annual, cespitose, 0.07–0.6 dm. **Culms** to 160. **Leaves** to 1.5 cm × 0.1–0.3 mm. **Inflorescences** 1(–2) flowers; bracts subtending inflorescence 2, round to acutely ovate, inconspicuous, 0.4–1.6 mm, membranous. **Flowers:** tepals pale yellow-green until seeds ripen, tip darker, 1.6–3 (–4.2) mm; outer and inner series nearly equal or outer slightly longer; stamens 2–3, filaments 0.6–0.9 mm, anthers 0.3–0.5 mm; style 0.2–0.4 mm, stigma 0.6–1.1 mm. **Capsules** pale yellow-green to reddish tinged until seeds ripen, 3-locular, ovoid to ellipsoid, 1.3–2.9 × 0.9–1.6 mm. **Seeds** globose-ovoid, 0.3–0.4 mm. $n = 16$.

Flowering and fruiting spring–early summer. Wet sandy soil of seepage areas on sandstone, depressions in meadows, vernal pools, and streamsides; 300–1900 m; Calif.

Juncus luciensis occurs in California in the Diamond and Santa Lucia mountains and the Transverse and Peninsular ranges.

44. Juncus tiehmii Ertter, Mem. New York Bot. Gard. 39: 60, figs. 13f–g, 14. 1986 • Tiehm's dwarf rush [C] [F]

Herbs, annual, cespitose, 0.1–0.6 dm. **Culms** to 150, 0.1–0.2 mm diam. **Leaves** to 2.5 cm × 0.1–0.3 mm. **Inflorescences** headlike clusters, each with 1–4(–7) flowers; bracts subtending inflorescence 2–4(–8), ovate, inconspicuous, 0.6–1.5 mm, membranous, apex acute. **Flowers** 3-merous; tepals 4(–6), green or darker, acute to acuminate, 1.9–2.9 × 0.4–0.6 mm, nearly equal; stamens 2(–3), filaments 0.5–0.8 mm, anthers 0.3–0.4 mm; style 0.2–0.3 mm, stigma 0.2–0.7 mm. **Capsules** pink-tinged, 2- or 3-locular, ellipsoid to narrowly oblate, 1.9–2.9 × 1.1–1.5 mm, nearly equal or slightly longer than tepals. **Seeds** oblate to ovoid, 0.35–0.55 mm. $n = 17$.

Flowering and fruiting mid spring–early fall. Bare, moist granitic sand along streams, seepage areas around outcrops, and depressions in meadows; 300–3100 m; Calif., Idaho, Nev., Wash.; Mexico.

45. Juncus capillaris F. J. Hermann, Leafl. W. Bot. 5: 116. 1948 · Hair-stemmed dwarf rush E

Herbs, annual, cespitose, 0.09–0.6 dm. Culms to 20. Leaves to 2.2 cm. Inflorescences headlike clusters, each with 1–2 flowers; bracts subtending inflorescence 2–4, ovate, inconspicuous, 0.8–1.5 mm, membranous, apex acute. Flowers: tepals 4–6, chestnut brown to black, 1.8–2.8 × 0.8–1.5 mm; inner series usually slightly longer than outer, apex acuminate to attenuate; stamens 2–3, filaments 0.6–1.1 mm, anthers 0.3–0.4 mm; style 0.1–0.3 mm, stigma 0.4–0.6 mm. Capsules tan or apex reddish, 2–3-locular, globose to obovoid, 1.2–2 × 1.1–1.5 mm, usually shorter than tepals. Seeds ellipsoid-ovoid, 0.5–0.8 mm. *n* = 18.

Flowering spring–mid summer. Moist, bare flats, short turf, and mossy areas in meadows, stream banks, and seepage areas on outcrops (usually granite); 1200–3200 m; Calif., Oreg.

Juncus capillaris occurs in California in the Sierra Nevada and in Oregon in the Steens Mountains.

46. Juncus bryoides F. J. Hermann, Leafl. W. Bot. 5: 117. 1948 · Mosslike dwarf rush F

Herbs, annual, cespitose, 0.03–0.25 dm. Culms to 150, 0.1–0.2 mm wide. Leaves to 0.9 cm, ¼–1 times height of plant. Inflorescences terminal solitary flowers; bracts subtending inflorescence (1–)2, ovate, inconspicuous, 0.3–0.9 mm, membranous, apex acute. Flowers 3-merous; tepals (4–)6(–8), turning inward to enwrap shorter capsule at maturity, chestnut brown to black, lanceolate to oblong, 1.2–2.3(–2.8) × 0.4–0.6 mm; outer and inner series nearly equal, acute to acuminate; stamens 3, filaments 0.3–0.6 mm, anthers 0.15–0.25 mm; style 0.1 mm, stigma 0.2–0.3 mm. Capsules pale reddish, 3-locular, ovoid to ellipsoid, 1–1.9 mm × 0.5–1 mm. Seeds ovoid to globose, 0.3–0.5 mm. 2*n* = ca. 38.

Flowering and fruiting spring–mid summer. Usually fine, sandy soil of washes, swales in meadows, and seepage areas on rock outcrops; 600–3600 m; Calif., Colo., Idaho, Nev., Oreg., Utah; Mexico.

47. Juncus uncialis Greene, Pittonia 2: 105. 1890 · Inch-high rush E

Herbs, annual, cespitose, 0.8–3.5 cm. Culms to 70, more than 0.2 mm diam. Leaves to 2.2 cm, ½–¾ height of plant. Inflorescences terminal solitary flowers; bracts subtending inflorescence 1, inconspicuous, 0.25–0.9 mm membranous, apex widely truncate, completely sheathing culm. Flowers 3-merous; tepals erect to recurved at maturity, greenish or tinged red, 2–4 × 0.6–0.9 mm; outer and inner series nearly equal, apex acutish; stamens usually 3, filaments 0.9–1.6 mm, anthers 0.3–0.4 mm, 1/3 length of filaments; style 0.1–0.3 mm, stigma 0.4–1.3 mm. Capsules reddish to maroon, 3-locular, ovoid to ellipsoid, 1.8–3.2 × 1–2.5 mm, nearly equal or shorter than the tepals. Seeds ovoid, 0.3–0.4 mm. *n* = 16.

Flowering and fruiting spring–early summer. Margins of vernal pools and ponds; to 1700 m; Calif., Nev., Oreg.

48. Juncus hemiendytus F. J. Hermann, Leafl. W. Bot. 5: 118. 1948 E

Herbs, annual, cespitose, 0.03–0.36 dm. Culms to 5. Leaves to 1.8 cm, ½–⅓ height of plant. Inflorescences solitary flowers. Flowers 2-merous; tepals 4(–6), erect to recurved at maturity, reddish, 1.5–3 × 0.5–1.2 mm; outer and outer series nearly equal, apex rounded to acute; stamens 2(–3), filaments 0.5–1.4 mm, anthers 0.3–0.7 mm; style 0.1–0.4 mm, stigma 2(–3), 0.3–1 mm. Capsules reddish, 2–3-locular, obovoid to ellipsoid or oblate, 1.9–3.5 × 1–2 mm. Seeds ovoid to oblate, 0.3–0.55 mm.

Varieties 2 (2 in the flora).

1. Bracts proximal to inflorescence, 1–2. . . . . . .
. . . . 48a. *Juncus hemiendytus* var. *hemiendytus*
1. Bracts proximal to inflorescence absent. . . . . .
. . . . 48b. *Juncus hemiendytus* var. *abjectus*

48a. Juncus hemiendytus F. J. Hermann var.
hemiendytus · Hermann's dwarf rush E F

Juncus brachystylus (Engelmann)
Piper var. *uniflorus* (Engelmann) M.
Peck; *J. triformis* Engelmann var.
uniflorus Engelmann

Culms not dilated directly proximal to flower, 0.1–0.3 mm diam.
Inflorescences: bracts subtending inflorescence 1–2, rounded to obtusely ovate, 1–2 mm. **Capsules** usually longer than tepals. *n* = 16.

Flowering and fruiting spring–early fall. Various damp open habitats including vernal depressions, streambeds, swales in sagebrush flats, forest clearing and alpine meadows; 400–3200 m; Calif., Idaho, Nev., Oreg., Utah, Wash.

48b. Juncus hemiendytus F. J. Hermann var. **abjectus**
(F. J. Hermann) Ertter, Mem. New York Bot. Gard.
39: 76. 1986 · Least dwarf rush E

Juncus abjectus F. J. Hermann,
Leafl. W. Bot. 5: 120. 1948

Culms dilated directly proximal to flower, 0.2–0.5 mm diam. **Inflorescences:** bracts not subtending inflorescence. **Capsules** usually not exceeding tepals.

Flowering and fruiting spring–mid summer. Various damp open habitats including vernal depressions, streambeds, swales in sagebrush flats, forest clearing and alpine meadows; 1400–3400 m; Calif., Idaho, Nev., Oreg.

49. Juncus capitatus Weigel, Observ. Bot., 28. 1772 F
I

Herbs, annual, cespitose, 0.3–1 dm. **Culms** to 20. **Leaves** basal; auricles absent; blade 0.5–2.5 cm × 0.5 mm. **Inflorescences:** glomerules 1–2, each with 2–10(–14) flowers; bracts subtending inflorescence 1(–2), foliose, clearly surpassing inflorescence. **Flowers:** tepals 6, tan to brownish, subulate, lanceolate-ovate, 3.5–4.5(–5) mm, 2 times length of inner tepals, margins scarious; inner series delicate, shorter, apex acute; stamens 3, filaments 1 mm, anthers 0.5 mm; style 0.4–0.7 mm. **Capsules** tan to reddish brown, 3-locular, globose to obovoid, 1.2–1.7 mm, shorter than tepals. **Seeds** ellipsoid-ovoid, 0.3–0.4 mm. $2n = 18$.

Flowering and fruiting spring. Moist to wet areas, usually in sandy, often disturbed soil such as roadsides or along trails; below 1000 m; introduced; Calif., La., Okla., Tex.; native, Africa (north); Europe, Asia (Near East).

1d. JUNCUS Linnaeus subg. **ENSIFOLII** (Snogerup) Snogerup, Bot. Not. 131: 187. 1978

Juncus Linnaeus sect. *Ensifolii* Snogerup, Bot. Not. 116: 151. 1963

Herbs, perennial, rhizomatous, sometimes cespitose. **Culms** flattened, 2-edged, smooth. **Leaves** basal and cauline; blade ensiform (flattened with 1 edge facing culm), imperfectly septate. **Inflorescences** terminal panicles of 2–many heads, sympodial; bracteoles absent proximal to perianth. **Flowers** in multiflowered heads. **Capsules** 1-locular, apex separating at dehiscence. **Seeds** not tailed or occasionally tailed, tegmen reticulate and usually lined. $2n = 20$.

Species ca. 7 (6 in the flora): w North America, Mexico, e Asia (s to Japan).

Juncus subg. *Ensifolii* may be polyphyletic. *Juncus polycephalus* shows affinities with *J. scirpoides, J. megacephalus,* and *J. validus;* the remaining members show affinities with *J. nevadensis, J. megacephalus,* and *J. dubius.*

1. Capsules cylindric to ovoid, acute to obtuse proximal to beak, 2.7–3.8 mm (including beak).
    2. Plants robust, usually more than 6 dm; leaves 7–12 mm wide (rarely to 3 mm wide); capsules acuminate, without beak; stamens 6. . . . . 54. *Juncus xiphioides*
    2. Plants not robust, 2–6 dm; leaves 1.5–6 mm wide; capsules obtuse proximal to beak; stamens 3 or 6. . . . . 55. *Juncus ensifolius*

1. Capsules lanceoloid to subulate, tapering to beak, 3.3–5.6 mm (including beak).
    3. Body of capsule included, beak may be slightly exserted.
        4. Capsules rounded proximal to beak; tepals 3–4 mm; inflorescences with 10 or more
            3–22-flowered heads. . . . . 51. *Juncus macrandrus*
        4. Capsules acute proximal to beak; tepals 3.8–5.6 mm; inflorescences variable. . . 50. *Juncus phaeocephalus*
    3. Body of capsule exserted.
        5. Heads spheric, 20–30-flowered; capsules narrowly ovoid, apex remaining united after
            dehiscence. . . . . 52. *Juncus polycephalus*
        5. Heads turbinate to hemispheric, 3–11-flowered; capsules broadly lanceoloid to nar-
            rowly oblong, apex usually separating at dehiscence. . . . . 53. *Juncus oxymeris*

50. Juncus phaeocephalus Engelmann, Trans. Acad. Sci. St. Louis 2: 484. 1868 W

Juncus phaeocephalus var. *paniculatus* Engelmann

Herbs, perennial, rhizomatous, 2–9 dm. **Rhizomes** 2–3 mm diam. **Culms** erect, 2–5 mm diam. **Cataphylls** 2, chestnut brown, apex acute. **Leaves:** basal 2, cauline 1–2, light green; auricles absent; blade 1–15 cm × 2–5 mm. **Inflorescences** panicles of (2–)10–77 heads, 2–9.5 cm with ascending to erect branches, or heads solitary; primary bract erect; heads 3–35-flowered, spheric to obovoid, 0.5–1.2 mm diam. **Flowers:** tepals dark brown to purplish brown, lanceolate, 3.3–4.8 mm, nearly equal; outer tepals acuminate, mucro subulate; inner tepals acuminate or cuspidate, mucro subulate; stamens 6, anthers 2–3 times length of filaments. **Capsules** included, chestnut brown or straw-colored, 1-locular, oblong, 3.8–5.6 mm (including beak), apex acute proximal to beak, beak slightly exserted. **Seeds** obovoid, 0.5–0.6 mm, not tailed.

Fruiting summer. Meadows and borders of swamps; 100–300 m; Calif., Oreg., Wash.; Mexico (Baja California).

Plants with more than 10 heads of 3–12 flowers have been separated as *Juncus phaeocephalus* var. *paniculatus* Engelmann. This variety is very similar to the next species (*Juncus macrandrus*) and is probably better treated as part of that species. Until a more thorough study has been made of the entire subgenus, we are hesitant to make such a transfer.

51. Juncus macrandrus Coville in L. Abrams and R. S. Ferris, Ill. Fl. Pacific States 1: 367, fig. 900. 1923 E

Herbs, perennial, rhizomatous, 2.6–6.5 dm. **Rhizomes** 1.5–3.5 mm diam. **Culms** erect, 1.5–3 mm diam. **Cataphylls** 2, chestnut brown, apex acute. **Leaves:** basal 2, cauline 1–2, light green; auricles absent; blade 2.5–20.5 cm × 2–3 mm. **Inflorescences** panicles of 10–50 heads, 4–10 cm, ascending to erect branches; primary bract erect; heads 3–22-flowered, obovoid to hemispheric, 5–11 mm diam. **Flowers:** tepals dark brown to purplish brown, lanceolate, 3–4 mm, nearly equal; outer tepals acuminate, mucro subulate; inner tepals acuminate or cuspidate, mucro subulate; stamens 6, anthers 2.5–3 times filament length. **Capsules** included, chestnut brown, 1-locular, oblong, 2.3–3.2 mm (including beak), apex acuminate proximal to beak, beak slightly exserted. **Seeds** obovoid, 0.5–0.6 mm, not tailed.

Fruiting summer. Wet places, montane conifer forests; 1700–2900 m; Calif.

52. Juncus polycephalus Michaux, Fl. Bor.-Amer. 1: 192. 1803 (as polycephalos) E

J. engelmannii Buchenau

Herbs, perennial, cespitose, 6–10 dm. **Culms** erect, 3–14 mm diam. **Cataphylls** absent. **Leaves:** basal 2–3(–6), cauline 1–4, brownish green; auricles absent; blade 8–70 cm × 4–8 mm. **Inflorescences** panicles of 16–82 heads, 10–30 cm; primary bract erect; heads 20–30-flowered, spheric, 8–12 mm diam. **Flowers:** tepals green to reddish, lance-subulate, 3–4 mm, nearly equal, apex acuminate; stamens 3, anthers ⅓–½ length of filaments. **Capsules** exserted, straw-colored, 1-locular, narrowly ovoid, 4–5 mm, apex tapering to beak, remaining after dehiscence. **Seeds** lance-ellipsoid, 0.5–0.6 mm, not tailed.

Fruiting spring–fall. Wet or seasonally wet shores,

depressions, occasionally in fairly deep water of streams, usually with a peaty or mucky substrate, occasionally sandy to gravelly; 0–100 m; Ala., Fla., Ga., La., Md., Miss., N.C., S.C., Tex.

53. Juncus oxymeris Engelmann, Trans. Acad. Sci. St. Louis 2: 483. 1868 [E]

Juncus acutiflorus Bentham, not Ehrhart

Herbs, perennial, rhizomatous, 3–6 dm. **Rhizomes** 1–2 mm diam. **Culms** erect, 2–4 mm diam. **Cataphylls** 0–1, straw-colored, apex narrowly acute. **Leaves:** basal 1–3, cauline 3–4, straw-colored; auricles absent; blade 3–20 cm × 3–7 mm. **Inflorescences** panicles of 10–50 heads, 6–20 cm, erect to ascending branches; primary bract erect; heads 3–11-flowered, turbinate to hemispheric, 4–8 mm diam. **Flowers:** tepals straw-colored, lanceolate, 2.5–3.2 mm, nearly equal, apex acute to narrowly acuminate, mucronate; stamens 6, anthers 0.5–1.5 times length of filaments. **Capsules** exserted, chestnut brown, 1-locular, broadly lanceoloid to narrowly oblong, 3.3–4.7 mm, apex tapering to beak, separating at dehiscence. **Seeds** obovoid, 0.5 mm, not tailed.

Fruiting late spring–fall. Stream and lake shores, montane meadows and seasonally emergent wetlands; 100–2000 m; B.C.; Calif., Oreg., Wash.

Juncus oxymeris should be expected in Mexico (Baja California).

54. Juncus xiphioides E. Meyer, Syn. Junc., 50. 1822

Juncus xiphioides var. *auratus* Engelmann; *J. xiphioides* var. *littoralis* Engelmann

Herbs, perennial, rhizomatous, 5–9 dm. **Rhizomes** 2–3 mm diam. **Culms** erect, 2–6 mm diam. **Cataphylls** 0 or 1–2, straw-colored, apex narrowly acute. **Leaves:** basal 1–3, cauline 2–6, straw-colored; auricles absent; blade 10–40 cm × (3–)7–12 mm. **Inflorescences** panicles or racemes of 20–50 heads, 2–14 cm, erect or ascending branches; primary bract erect; heads 15–70-flowered, obovoid to globose, 7–11 mm diam. **Flowers:** tepals green to brown or reddish brown, lanceolate, 2.4–3.7 mm, nearly equal, apex acuminate; stamens 6; anthers ½ to equal filament length. **Capsules** slightly exserted, chestnut to dark brown, 1-locular, ellipsoid, 2.4–3.8 mm, apex acuminate, not beaked. **Seeds** elliptic to obovate, 0.4–0.6 mm, not tailed. $2n = 40$.

Fruiting early summer–fall. Salt marshes, moist ar-

eas, ditches, springs, lake and stream shores; 500–1600 m; Ariz., Calif., Nev., N.Mex., Utah; Mexico (Baja California).

This species and the next (*Juncus ensifolius*) are closely related and have been treated as members of a single species (*J. xiphioides*) by Engelmann. Until a study of the complete subgenus is done, we are hesitant to use a varietal name (*J. xiphioides* var. *triandrus*) for the widespread western taxon *J. ensifolius*.

55. Juncus ensifolius Wikström, Kongl. Vetensk. Acad. Handl. 2: 274. 1823

Herbs, perennial, rhizomatous, 2–6 dm. **Rhizomes** 2–3 mm diam. **Culms** erect, 2–6 mm diam. **Cataphylls** 0 or 1–2, straw-colored, apex narrowly acute. **Leaves:** basal 1–3, cauline 2–6, straw-colored; auricles absent; blade 2–25 cm × 1.5–6 mm. **Inflorescences** panicles or racemes of 2–50 heads or heads solitary, 2–14 cm, erect or ascending branches; primary bract erect; heads 3–70-flowered, obovoid to globose, 7–11 mm diam. **Flowers:** tepals green to brown or reddish brown, lanceolate; outer tepals 2.7–3.6(–4) mm, apex acuminate; inner tepals 2.2–3(–3.5) mm, nearly equal, apex acuminate; stamens 3 or 6; anthers ½ to equal filament length. **Capsules** included to slightly exserted, chestnut to dark brown, 1-locular, oblong, 2.4–4.3 mm, apex obtuse proximal to beak. **Seeds** elliptic to obovate, 0.4–1 mm, occasionally tailed.

Varieties 2 (2 in the flora): North America; introduced in Europe and East Asia.

1. Stamens 3. . . . . . 55a. *Juncus ensifolius* var. *ensifolius*
1. Stamens 6. . . . . . 55b. *Juncus ensifolius* var. *montanus*

55a. Juncus ensifolius Wikström var. ensifolius

Leaves 2–15 cm × 3–6 mm. **Inflorescences** 1–5(–11) heads; heads usually globose. **Flowers:** outer tepals 2.7–3.6(–4) mm; inner tepals 2.2–3(–3.5) mm; stamens 3. **Capsules** included to slightly exserted, oblong-ovoid to ellipsoid, 2.4–3.5 mm. **Seeds** 0.4–0.6, not tailed. $2n = 40$.

Fruiting summer. Wet meadows, marshy areas, wet granite areas, shores, banks and ditches, often montane; 400–3000 m; Alta., B.C., Ont., Que., Sask.; Alaska, Calif., Colo., Idaho, Mont., Nev., N.Y., N. Dak., Oreg., S.Dak., Utah, Wash., Wis., Wyo.; Mexico (s to Veracruz, Puebla, Guerrero); Europe; e Asia.

This species is probably introduced in Wisconsin and New York as well as in northern Europe.

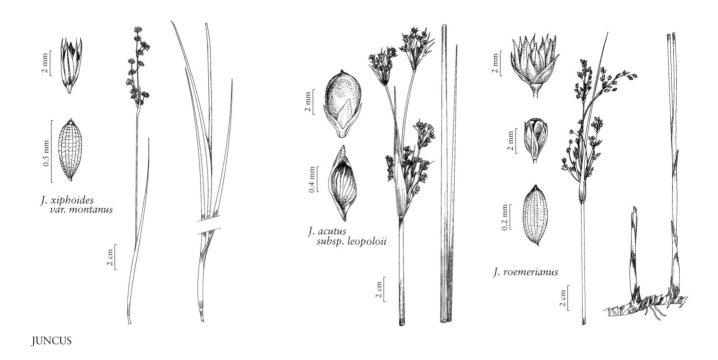

J. xiphoides
var. montanus

J. acutus
subsp. leopoloii

J. roemerianus

JUNCUS

55b. Juncus ensifolius Wikström var. **montanus** (Engelmann) C. L. Hitchcock in C. L. Hitchcock et al., Vasc. Pl. Pacif. N.W. 1: 195. 1969   F

Juncus xiphioides E. Meyer var. *montanus* Engelmann, Trans. Acad. Sci. St. Louis 2: 481. 1868; *J. brunnescens* Rydberg; *J. ensifolius* var. *brunnescens* (Rydberg) Cronquist; *J. parous* Rydberg; *J. saximontanus* A. Nelson; *J. tracyi* Rydberg; *J. utahensis* R. F. Martin; *J. xiphioides* var. *triandrus* Engelmann

Leaves 2–25 cm × 1.5–4 mm. **Inflorescences** 3–20 heads; heads obovoid to globose. **Flowers:** outer tepals 2.7–3.6 mm; inner tepals 2.7–3 mm; stamens 6. **Capsules** slightly exserted, ellipsoid, 2.4–4.3 mm. **Seeds** 0.4–1 mm, tailed or more often not tailed.

Fruiting early summer–fall. Wet meadows, bogs, springy woods, stream and lake shores; 400–3000 m; Alta., B.C.; Ariz., Calif., Colo., Idaho, Mont., Nev., N.Mex., Oreg., S.Dak., Tex., Utah, Wash., Wyo.; Mexico (s to Puebla, Veracruz).

1e. JUNCUS Linnaeus subg. JUNCUS

Juncus sect. *Acuti* Rouy; *Juncus* sect. *Juncastrum* Kuntze; *Juncus* sect. *Maritimi* (Engelmann) Rouy; *Juncus* sect. *Thalassii* (Buchenau) Vierhapper; *Juncus* subg. *Juncastrum* (Kuntze) Kreczetowicz & Gontscharow; *Juncus* subg. *Thalassii* Buchenau

Herbs, perennial, rhizomatous. **Stems** terete. **Leaves** basal; blade terete, not septate. **Inflorescences** terminal (sometimes appearing lateral) panicles of few to many heads, sympodial; bracteoles absent proximal to perianth. **Flowers** in few-flowered heads. **Capsules** 3-locular or pseudo-3-locular. **Seeds** usually tailed (not tailed in *J. roemerianus*).

Species 9 (4 in the flora): worldwide; coastal in North America.

SELECTED REFERENCE   Snogerup, S. 1993. A revision of *Juncus* subgen. *Juncus* (Juncaceae). Willdenowia 23: 23–73.

1. Capsules 3.5–5 mm; plants mostly of sw United States.
   2. Capsules nearly globose, obviously longer than perianth. . . . . *56a. Juncus acutus* var. *leopoldii*
   2. Capsules ellipsoid, shorter than or nearly equal to perianth. . . . . *57. Juncus cooperi*
1. Capsules less than 3.5 mm; plants mostly of ne United States Atlantic Coast and Gulf of
Mexico.
   3. Capsules nearly equal to or exceeding perianth; seeds tailed. . . . . *58. Juncus maritimus*
   3. Capsules mostly distinctly shorter than perianth; seed not tailed. . . . . *59. Juncus roemerianus*

56. Juncus acutus Linnaeus, Sp. Pl. 1: 325. 1753

Subspecies 2 (1 in the flora): North America; Mexico (Baja Calif.); South America; South Africa; Atlantic Islands.

56a. Juncus acutus Linnaeus subsp. **leopoldii** (Parlatore) Snogerup, Bot. Not. 131: 187. 1978 • Sharp rush [F]

Juncus leopoldii Parlatore, Giorn. Bot. Ital. 2: 324. 1845; *J. acutus* var. *sphaerocarpus* Engelmann; *J. robustus* S. Watson

Herbs, perennial, robust, tufted, often forming clumps to 1 m diam., 6–12 dm. **Culms** 3–4 mm diam. **Cataphylls** 1–2. **Leaves** basal, 2–4; sheaths brownish, inflated; auricles absent or prolonged 1–3 mm, cartilaginous; blade terete, 30–100 cm × 2–3 mm. **Inflorescences** sympodial or appearing lateral, glomerules numerous, each with 2–5 flowers, branches unequal, 5–20 cm; primary bract inflated, usually shorter than inflorescence. **Flowers:** tepals straw-colored, brown-tipped, 2–4 mm; outer tepals widely lanceolate, margins scarious, apex obtuse to acutish; inner tepals rounded, shorter; stamens 6, filaments 0.2–0.4 mm, anthers 1.2–1.8 mm; style 1–1.2 mm. **Capsules** yellow-brown, pseudo-3-locular, nearly globose, 3.5–5 × 2.4–3.3 mm, obviously exceeding perianth, mucronate. **Seeds** brownish, obliquely obovoid, body 0.6–0.7 mm, tails 0.2–0.3 mm.

Flowering and fruiting late spring–summer. Moist saline habitats and alkaline seeps; below 300 m; Ariz., Calif., Nev.; Mexico (Baja California); South America; South Africa; Atlantic Islands.

57. Juncus cooperi Engelmann, Trans. Acad. Sci. St. Louis 2: 590. 1868 • Cooper's rush

Herbs, perennial, robust, tufted, 4–10 dm. **Rhizomes** short. **Culms** terete, 1.5–2.5 mm diam. **Cataphylls** several, reddish brown, often apiculate. **Leaves** basal, 1–2; auricles essentially absent; blade terete, 10–40 cm, shorter than culm. **Inflorescences** glomerules 3–12, each with (1–)2–5 flowers, open, branches unequal, 3–12 cm, primary bract terete or slightly compressed, shorter to longer than inflorescence. **Flowers:** tepals straw-colored to pale green, ovate-lanceolate, 4–5 mm, apex acuminate or setaceous; inner series elliptic, nearly equal, margins wide, scarious, apiculate; stamens 6, filaments 1 mm, anthers 1.5–2 mm; style 1 mm. **Capsules** tan to light reddish brown, 3-locular, ellipsoid, 3.7–4.5 × 1.6–2 mm, mostly shorter than or nearly equal to perianth. **Seeds** dark amber, obovoid, body 0.7–1 mm, tails 0.1–0.5 mm.

Flowering and fruiting spring–summer. Saline flats and meadows or edges of salt marshes; below 600 m; Ariz., Calif., Nev.; Mexico.

58. Juncus maritimus Lamarck in J. Lamarck et al., Encycl. 3: 264. 1789 • Seaside rush [I]

Herbs, perennial, 5–10 dm. **Rhizomes** thick. **Culms** closely set along rhizomes, 2–3.5 mm diam. **Cataphylls** 3–5. **Leaves** basal, 2–4; auricles absent; blade terete, 40–60 cm × 1–2 mm. **Inflorescences** glomerules, to 100, each with 2–4 flowers, congested to open, 5–19 × 2–5 cm; primary bracts somewhat inflated, usually surpassing inflorescence. **Flowers:** tepals straw-colored; outer series 2.8–2.9 mm, apex acute; inner series elliptic, 2.3–2.5 mm, apex obtuse; stamens 6, 1.4–1.9 mm, anthers 2 times length of filaments; style 1–1.2 mm. **Capsules** straw-colored, 3-locular, ovoid to ellipsoid, 2.5–3.5 × 1.2–1.5 mm, nearly equal to or slightly exceeding perianth. **Seeds** brown, ellipsoid, body 0.6–0.7 mm, tails 0.5–1.2 mm. $2n = 40, 48$.

Flowering and fruiting summer. Coastal salt marshes, saline meadows, and sand dunes; introduced; N.Y.; Europe; Asia; Africa.

It is believed that this species has not been collected in North America since the late 1800s, when it was known to occur on Long Island, New York.

59. Juncus roemerianus Scheele, Linnaea 22: 348. 1849 · Needle rush E F

Herbs, perennial, 5–15 dm. **Rhizomes** long, thick, scaly. **Culms** gray-green, 2–5 mm diam. **Cataphylls** 2–5. **Leaves** basal, 1–3; auricles absent; blade terete, 40–80 cm × 1.5–2.5 mm. **Inflorescences** sympodial or appearing lateral, glomerules to 150, each with 2–6 perfect or pistillate flowers, congested to open, 4–20 × 5–10 cm; primary bracts terete, longer than inflorescence. **Flowers:** tepals pale brown, 2.5–3.3 mm, apex acuminate; outer series acuminate; inner series slightly shorter, acutish; stamens 6, sometimes rudimentary, filaments 0.2–0.7 mm, anthers 1.3–2.5 mm; style 0.3–1 mm. **Capsules** brown, 3-locular, ovoid to oblate, 1.7–3 × 1.3–1.6 mm, mostly distinctly shorter than perianth, apiculate. **Seeds** dark brown, ellipsoid, 0.4–0.6 mm, not tailed.

Flowering and fruiting spring–early fall. Coastal tidal marshes in dense zonal stands; Ala., Del., Fla., Ga., La., Md., Miss., N.C., S.C., Tex., Va.

1f. JUNCUS Linnaeus subg. ALPINI (Engelmann) Buchenau, Bot. Jahrb. Syst. 1: 118. 1880

Juncus [unranked] *Alpini* Engelmann; *Juncus* sect. *Alpini* (Engelmann) Vierhapper

Herbs, perennial, rhizomatous, usually freshwater habitats. **Stems** terete. **Leaves** basal or cauline; blade terete, channeled, imperfectly septate or not. **Inflorescences** terminal panicles or racemes of several heads or a single terminal head, sympodial or monopodial; bracteoles absent at base of perianth. **Flowers** in multiflowered heads. **Capsules** (1–)3-locular. **Seeds** long tailed, large.

Species 72 (4 in the flora); n North America, n Europe, n Asia.

1. Plants not cespitose, strongly rhizomatous.. . . . . .61. *Juncus castaneus*
1. Plants loosely cespitose, rhizomes (if present) short, sparingly branched.
  2. Leaves basal and cauline.. . . . . .60a. *Juncus stygius* var. *americana*
  2. Leaves all basal.
    3. Primary bract of infloresence much longer than inflorescence.. . . . . .62. *Juncus biglumis*
    3. Primary bract of inflorescence nearly equal to or shorter than inflorescence.. . . . . . .63. *Juncus triglumis*

60. Juncus stygius Linnaeus, Syst. Nat. ed. 10, 2: 987. 1759

Varieties 2 (1 in the flora): North America, Eurasia.

60a. Juncus stygius Linnaeus var. **americanus** Buchenau, Bot. Jahrb. Syst. 12: 393. 1890 · American moor rush

Juncus stygius subsp. *americanus* (Buchenau) Hultén

Herbs, perennial, loosely cespitose, 2–4 dm. **Rhizomes** short, sparingly branched. **Cataphylls** 0–1. **Leaves:** basal 1–3, cauline 1–2; auricles rounded, scarcely prolonged, scarious; blade imperfectly septate (often inconspicuous), ascending, terete to slightly flattened, 8–18 cm × 0.5–1 mm, much reduced distally. **Inflorescences** glomerules, 1–3, each with 1–3 flowers; peduncles 0.2–1 cm, primary bracts nearly equal to or slightly surpassing glomerules. **Flowers:** bracteoles absent; tepals pale to reddish brown, lanceolate to lanceolate-ovate, 4–5 mm × 1 mm, apex acutish; outer and inner series nearly equal; stamens 6, filaments 2.5–3.5 mm, anthers 0.4–0.5 mm; style 1–1.2 mm. **Capsules** greenish to tan, pseudo-3-locular, ellipsoid, 5.5–9 × 1.8–2.6 mm. **Seeds** pale yellow, fusiform, body 0.8–1.1 mm, tails 1–1.4 mm.

Flowering and fruiting mid–late summer. Wet moss, bogs, and bog-pools; Alta., B.C., N.B., Nfld. and Labr., Man., N.S., N.W.T., Nunavut, Ont., P.E.I., Que., Sask., Yukon; Alaska, Maine, Mich., Minn., N.Y.; e Asia.

61. Juncus castaneus Smith, Fl. Brit. 1: 383. 1800 ☐F

Juncus castaneus var. *pallidus* (Hooker ex Buchenau) B. Boivin; *J. castaneus* subsp. *leucochlamys* (N. W. Zinger ex V. I. Kreczetowicz) Hultén; *J. leucochlamys* N. W. Zinger ex V. I. Kreczetowicz

Herbs, perennial, strongly rhizomatous, 1–4 dm. **Culms** solitary, 1–2 mm diam. **Cataphylls** 1–2. **Leaves** partially cauline, 3–5, auricles absent distally, rounded proximally; blade channeled, to 20 cm, reduced distally. **Inflorescences** glomerules, 1–3(–5), each with 2–10 flowers; peduncles 0.4–1.5 cm; primary bracts somewhat inflated, usually surpassing inflorescence. **Flowers:** bracteoles absent; tepals brown or occasionally paler, lanceolate, 4.5–6.6 mm, apex acute to obtuse; inner series slightly shorter; stamens 6, filaments 2.5–3.5 mm, anthers 0.6–1.3 mm; style 1–1.3 mm. **Capsules** chestnut brown, 3-locular, narrowly oblong, 6.5–8.5 × 1.8–2.3 mm. **Seeds** pale yellow, fusiform, body 0.6–0.7 mm, tails 0.8–1.1 mm. $2n = 60$, 90, 120.

Flowering and fruiting late spring–summer. Tundra, subalpine and alpine bogs and meadows, and along streams in gravelly or clayey soils; 10–3700 m; Greenland; Alta., B.C., Man., Nfld. and Labr. (Labr.), N.W.T., Nunavut, Ont., Que., Sask., Yukon; Alaska, Colo., Mont., Nev., N.Mex., Utah, Wyo.; Europe; Asia.

In southern Alaska some plants with several many-flowered heads and capsules about double the length of the perianth have been referred to the Asian *Juncus castaneus* subsp. *leucochlamys*. The distinction, however, seems dubious without further investigation.

62. Juncus biglumis Linnaeus, Sp. Pl. 1: 328. 1753

Herbs, perennial, loosely cespitose, 0.25–1.6 dm. **Culms** nearly terete. **Cataphylls** 2–4. **Leaves** basal, 1–4; sheaths loose; auricles absent or rounded, 0.5 mm; blade imperfectly septate, ascending, nearly terete, 2–7 cm × 0.5–1.5 mm. **Inflorescences** heads, 1–2(–4)-flowered; primary bracts much longer than inflorescence. **Flowers:** tepals brown to blackish, oblong, 2.5–4 mm, apex obtuse; outer and inner series nearly equal; stamens 6, filaments 1–1.5 mm, anthers 0.5–0.7 mm; style deciduous, 0.3–0.4 mm. **Capsules** pale with dark purplish valve margins, pseudo-3-locular, oblate to narrowly ovoid, 4–5.5 × 1.7–2.3 mm, exceeding perianth, apex retuse. **Seeds** yel-

lowish tan, fusiform-ovoid, 0.7–0.9 mm, short tailed. $2n = 120$.

Flowering and fruiting summer. Wet tundra and mossy margins of ponds and streams, wet gravel and open, rocky slopes in alpine zones; 10–3400 m; Greenland; Alta., B.C., Man., Nfld. and Labr. (Labr.), N.W.T., Nunavut, Ont., Que., Yukon; Alaska, Colo., Mont., Wyo.; n Europe; Asia.

63. Juncus triglumis Linnaeus, Sp. Pl. 1: 328. 1753

Herbs, perennial, cespitose, 0.3–3.5 dm. **Culms** 1–8, 0.3–0.5 mm diam. **Cataphylls** 1–2. **Leaves** basal, 2–4; auricles slightly prolonged, rounded, scarious to ± leathery; blade deeply channeled, 2–10 cm, mostly shorter than culms. **Inflorescences** solitary heads, each with 2–3(–50) flowers; primary bracts brownish, nearly equal to or slightly shorter than inflorescence. **Flowers:** tepals pale brown or darker, oblong-lanceolate, 3–5 mm, outer and inner series nearly equal; stamens 6, filaments 2.5–4 mm, anthers 0.6–1 mm; styles 0.5–0.8 mm. **Capsules** tan, pseudo-3-locular, 3-gonous–cylindric, apex obtuse, mucronate. **Seeds** tan or darker, fusiform, body 0.5–1 mm, tails 0.6–1 mm.

Varieties 3 (2 in the flora): North America; Eurasia.

1. Most proximal bracts of inflorescence apex usually much shorter than inflorescence, obtuse to mucronate; capsules well exserted from perianth, 4.5–7 mm, apex conic or rounded proximal to persistent style.. . . . 63a. *Juncus triglumis* var. *triglumis*
1. Most proximal bracts of inflorescence equal to or longer than inflorescence, apex long acuminate or awned; capsules included or barely exserted from perianth, 3–5 mm, apex nearly truncate proximal to persistent style.. . . . . 63b. *Juncus triglumis* var. *albescens*

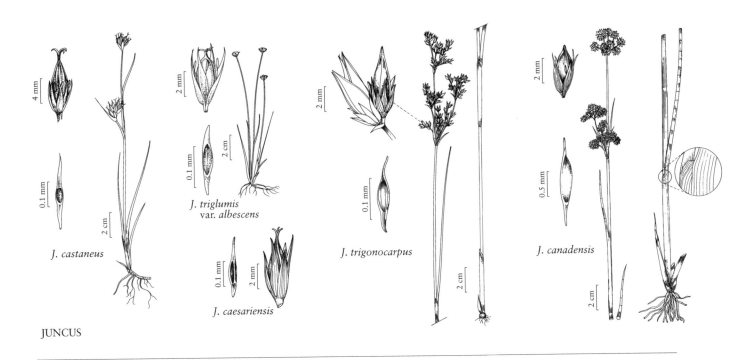

JUNCUS

63a. Juncus triglumis Linnaeus var. **triglumis**

Inflorescences: most proximal bracts of inflorescence usually much shorter than inflorescence, apex obtuse to mucronate. **Capsules** well exserted from perianth, 4.5–7 mm, apex conic or rounded proximal to persistent style.

Flowering and fruiting late spring–early summer. Wet gravelly soils of snowmelt basins; 400–3800 m; Greenland; Nunavut, Yukon; Alaska, Colo., Mont., Wyo.; Eurasia.

63b. Juncus triglumis Linnaeus var. **albescens** Lange, Consp. Fl. Groenland. 1: 123. 1880 [F]

Juncus albescens (Lange) Fernald; *J. triglumis* subsp. *albescens* (Lange) Hultén

Inflorescences: most proximal bracts of infloresecence equal to or longer than inflorescence, apex long acuminate or awned. **Capsules** included or barely exserted from perianth, 3–5 mm, apex nearly truncate proximal to persistent style. $2n = 132$.

Flowering and fruiting summer. Peat bogs; 0–820 m; Greenland; B.C., Man., Nfld. and Labr., N.W.T., Nunavut, Ont., Que., Sask., Yukon; Alaska, Colo., Mont., N.Mex., Oreg., Utah, Wyo.; e Asia.

1g. JUNCUS Linnaeus subg. **SEPTATI** Buchenau, Abh. Naturwiss. Vereine. Bremen 4: 406. 1875

Juncus sect. *Septati* (Buchenau) Vierhapper

Herbs, perennial (rarely annual), rhizomatous. **Culms** terete. **Leaves** basal and cauline: blade terete, nearly terete, or flattened, perfectly septate. **Inflorescences** terminal panicles or racemes of 2–many heads, a single terminal head, or rarely a cyme of heads, sympodial; bracts erect or ascending, flat, involute or terete; bracteoles absent. **Flowers** in multiflowered heads (rarely only 2–3 flowers per head). **Capsules** 1- or 3-locular, beaked. **Seeds** tailed or not tailed. $x = 20$.

Species ca. 80 (32 in the flora): worldwide.

1. Seeds tailed, 0.7–2.6 mm, including tails; seed body covered with whitish translucent veil.
  2. Seeds 1.8–2.6 mm, including tails; stamens usually 6.
    3. Culms and leaves scabrous; auricles 1–7.5 mm. . . . . .64. *Juncus caesariensis*
    3. Culms and leaves glabrous; auricles 0.2–0.4 mm. . . . . .65. *Juncus trigonocarpus*
  2. Seeds 0.7–1.9 mm, including tails; stamens usually 3.
    4. Seeds 1.1–1.9 mm; heads 5–50-flowered; inflorescence branches erect to ascending.
    . . . . .66. *Juncus canadensis*
    4. Seeds 0.7–1.2 mm; heads 2–8-flowered or if to 20-flowered then proximal inflorescence branches widely spreading (*J. subcaudatus*).
      5. Outer tepals obtuse to nearly acute. . . . . .67. *Juncus brachycephalus*
      5. Outer tepals acuminate to acute.
        6. Proximal inflorescence branches widely spreading. . . . . .68. *Juncus subcaudatus*
        6. Proximal inflorescence branches erect. . . . . .69. *Juncus brevicaudatus*
1. Seeds not tailed, 0.3–0.7 mm; seed body clear yellow-brown (seeds not seen in *J. pervetus*).
  7. Flowers 1–2(–4) at each node, not in heads; capsules fertile only proximal to middle.
    8. Culms erect, 0.3–7 dm, not forming mats. . . . . .70. *Juncus pelocarpus*
    8. Culms repent, 0.1–1 dm, forming mats. . . . . .71. *Juncus subtilis*
  7. Flowers in heads of (1–)3–60; capsules fertile throughout or only proximal to middle.
    9. Heads spheric or nearly spheric, (2–)15–60(–100)-flowered
      10. Stamens 6.
        11. Tepals lanceolate to lance-ovate, light green to light pink or white or dark reddish or purplish brown; capsule ovoid to obovoid, 1.9–3.5 mm; heads 1–2 per inflorescence (24–60 for *J. pervetus*).
          12. Heads 24–60. . . . . .74. *Juncus pervetus*
          12. Heads 1–2.
            13. Tepals dark purplish brown to black; anthers ¼ to equal length of filament. . . . . .90. *Juncus mertensianus*
            13. Tepals light green, or light pink to white; anthers 2–3 times length of filaments. . . . . .91. *Juncus chlorocephalus*
        11. Tepals lance-subulate, green to straw-colored; capsule lance-subulate, 3.2–6.5 mm; heads 1–23 per inflorescence.
          14. Outer tepals 2.4–4.1 mm, equaling inner tepals; auricles 0.5–1.7 mm
          . . . . .80. *Juncus nodosus* (in part)
          14. Outer tepals 4–6 mm, outer and inner tepals unequal in length; auricles 1–4 mm.
            15. Outer tepals longer than inner; capsule 4.3–5.7 mm. . . . . .81. *Juncus torreyi*
            15. Outer tepals shorter than inner; capsules 5.2–6.5(–8) mm. . . . . . .82. *Juncus texanus*
      10. Stamens 3 or 6.
        16. Plants cespitose, not rhizomatous; tepals lanceolate. . . . . .85. *Juncus acuminatus* (in part)
        16. Plants rhizomatous, not cespitose; tepals lance-subulate.
          17. Capsules included. . . . . .83. *Juncus brachycarpus*
          17. Capsule exserted to slightly included.
            18. Capsules remaining united at apex after dehiscence.
              19. Leaves laterally compressed; culms 3–5 mm diam. . . . .
              . . . . .79b. *Juncus validus* var. *fascinatus*
              19. Leaves terete; culms 1–4 mm diam.
                20. Heads sometimes spheric, usually lobed; tepals green to straw-colored, nearly equal; uppermost cauline leaf blade (below inflorescence bract) equaling or longer than its sheath. . . . . .78. *Juncus scirpoides*
                20. Heads spheric; tepals straw-colored to reddish brown, inner tepals shorter than outer; uppermost cauline leaf blade much shorter than its sheath. . . . . .77. *Juncus megacephalus*
            18. Capsule valves separating at apex during dehiscence.
              21. Tepals 4–5 mm; leaves laterally compressed. . . .79a. *Juncus validus* var. *validus*
              21. Tepals 2.9–4 mm; leaves terete.

22. Capsules 3.5–5 mm, exserted; culms 0.4–3 dm   80. *Juncus nodosus* (in part)
22. Capsules 3–3.5 mm; equaling perianth or slightly included; culms 2.5–8.5 dm. . . . . .84. *Juncus bolanderi*
9. Heads obconic to hemispheric, (2–)3–15-flowered.
  23. Stamens 3 (or 6).
    24. Culms creeping or floating, capillary; filiform, submersed leaves often formed before flowering; inflorescences with 1–9 heads.
      25. Capsule rounded at apex, 2.5–4 mm; tepals 2–3.6 mm, obtuse to acute . . . . .76. *Juncus bulbosus* (in part)
      25. Capsule acute to acuminate at apex, 3.5–6.1 mm; tepals 2.8–5.5 mm, acute to acuminate-subulate. .75. *Juncus supiniformis* (in part)
    24. Stems erect, not floating; filiform leaves absent; inflorescences with 3–200 heads.
      26. Capsule 1.3–2 times perianth length.
        27. Capsules beaked; plants often decumbent or matted. . . . . .75. *Juncus supiniformis* (in part)
        27. Capsules not beaked; plants erect.
          28. Capsule ⅓ longer than perianth, 2.8–4.2 mm. . . . . .86. *Juncus debilis*
          28. Capsule twice as long as perianth, 4–5.2 mm. . . . . .87. *Juncus diffusissimus*
      26. Capsule equaling or just exceeding perianth.
        29. Heads 5–50; tepals 2.6–3.5 mm, nearly equal; capsules 2.8–3.5 mm. . . . . .85. *Juncus acuminatus* (in part)
        29. Heads 30–250; tepals 1.7–2.9 mm, inner tepals shorter than outer tepals; capsules 1.9–2.9 mm.
          30. Leaf blade with prominent and conspicuous ringlike bands at position of cross partitions; roots without terminal tubers; inner tepals 1.7–2.1 mm. . . . . .88. *Juncus nodatus*
          30. Leaf blade with faint ringlike bands at position of septa; roots often with terminal tubers; inner tepals (1.8–)2.4–2.8 mm. . . . .89. *Juncus elliottii*
  23. Stamens 6.
    31. Culms creeping or floating, capillary; filiform, submersed leaves often formed before flowering; inflorescences with 1–9 heads.
      32. Auricles 0.8–2.1 mm; capsules 3.5–6.1 mm. . . . . .75. *Juncus supiniformis* (in part)
      32. Auricles 0.4–1 mm; capsules 2.5–4 mm.
        33. Capsules exserted about 1 mm beyond perianth; apex acute proximal to beak; outer tepals acute to acuminate. . . . . . . .95. *Juncus articulatus* (in part)
        33. Capsules as long as tepals or exserted to 0.5 mm beyond tepals, apex obtuse proximal to apex; outer tepals acute to obtuse. . . . . .76. *Juncus bulbosus* (in part)
    31. Culms erect; filiform submersed leaves not formed except in *J. militaris*; heads 1–66.
      34. Proximal culm leaf overtopping inflorescences, distal culm leaf usually inflated, bladeless sheath; often developing filiform (hairlike) leaves off rhizome in running water. . . . . .72. *Juncus militaris*
      34. Proximal culm leaf shorter than inflorescence, distal culm leaf usually with blade, sheath not inflated; not developing filiform leaves off rhizome.
        35. Tepals 1.6–3 mm; apex obtuse to acuminate, usually not subulate mucronate; anthers shorter than to equaling filaments.
          36. Inner tepals obtuse; inflorescence stiffly erect. . . . . . .94. *Juncus alpinoarticulatus*
          36. Inner tepals acuminate; inflorescence spreading. . .95. *Juncus articulatus* (in part)
        35. Tepals 2.3–6.2 mm; apex usually acuminate with subulate mucro (acute in *J. acutiflorus*); anthers equaling or longer than filaments.
          37. Capsule abruptly contracted at apex, usually not exserted   92. *Juncus nevadensis*
          37. Capsule narrowed to beak, exserted at least slightly.

38. Heads 2–9; tepals 2.8–5.5 mm. . . . . *75. Juncus supiniformis* (in part)
38. Heads 25–120; tepals (2–)2.5–3.5 mm.
    39. Tepals nearly equal; heads with 6–10 flowers. . . . . . *93. Juncus dubius*
    39. Tepals unequal, outer shorter than inner; heads with
        5–8 flowers. . . . . *73. Juncus acutiflorus*

64. Juncus caesariensis Coville, Mem. Torrey Bot. Club
5: 106. 1894 [E] [F]

Juncus asper Engelmann, Trans.
Acad. Sci. St. Louis 2: 478. 1868,
not Sauze 1864

Herbs, perennial, cespitose, 3–9
dm. **Culms** erect, 2–4 mm diam.,
scabrous. **Cataphylls** 0 or 1–2,
straw-colored, apex acuminate.
Leaves: basal 1–2, cauline 1–2;
auricles 1–7.5 mm, apex
rounded, scarious; blade terete, 4–25 cm × 1–1.8 mm,
scabrous. **Inflorescences** panicles of 5–30 heads, 12–15
cm, erect to ascending branches; primary bract erect;
heads 2–6-flowered, broadly obovoid, 5–10 mm diam.
Flowers: tepals green to reddish brown, lanceolate;
outer tepals 3.3–3.9 mm, apex acuminate; inner tepals
3.9–4.7 mm, acuminate; stamens 6, anthers ½ filament
length. **Capsules** exserted, chestnut brown, 3-locular,
ovoid, 4.5–5.3 mm, apex acuminate, valves separating
at dehiscence. **Seeds** fusiform, 2.2–2.6 mm, tailed; body
covered with whitish translucent veil.

Fruiting mid summer–early fall. Wet springy bogs,
swamps, and borders of wet woods; 0–100 m; N.S.;
Del., D.C., Md., N.J., N.C., Pa., Va.

65. Juncus trigonocarpus Steudel, Syn. Pl. Glumac. 2:
308. 1855 [E] [F]

Juncus caudatus Chapman

Herbs, perennial, cespitose, 5–10
dm. **Culms** erect, 2–5 mm diam.,
smooth.
Cataphylls 1–2, straw-colored,
apex rounded. **Leaves:** basal 0–1,
cauline 2–3; auricles 0.2–0.4 mm,
apex obtuse, leathery; blade terete, 2.5–22 cm × 1.3–4 mm,
smooth. **Inflorescences** panicles of 30–100 heads, 3–9
cm, stiffly erect or ascending branches; primary bract
erect; heads 2–8-flowered, obconic, 4–8 mm diam.
Flowers: tepals green to reddish brown, lanceolate;
outer tepals 2.5–2.8 mm, acuminate; inner tepals 2.8–
3.5 mm, acute; stamens 3–6, anthers ½ filament length.
Capsules exserted, chestnut brown, 3-locular, ovoid-
pyramidal, 4–5 mm, apex acuminate, valves separating
at dehiscence. **Seeds** fusiform, 1.8–2.6 mm, tailed; body
covered with whitish translucent veil.

Fruiting early fall–early winter (fruit often persisting

until spring). Wet pinewoods, pine barrens, bogs, hammocks, roadsides, and seepy areas; 0–200 m; Ala., Fla.,
Ga., La., Miss., N.C., S.C., Tex.

66. Juncus canadensis J. Gay in J. J. C. de Laharpe,
Essai Monogr. Jonc., 46. 1825 • Canada rush [E] [F]

Juncus canadensis var.
longicaudatus Engelmann; *J.
canadensis* var. *sparsiflorus* Fernald;
J. longicaudatus (Engelmann)
Mackenzie; *J. polycephalus* Michaux
var. *paradoxus* Torrey

Herbs, perennial, cespitose, 3–10
dm. **Culms** erect, 1–5 mm diam.,
smooth. **Cataphylls** 0 or 1–2,
straw-colored, apex rounded. **Leaves:** basal 1, cauline 2–
3; auricles 1–1.2 mm, apex rounded, scarious; blade
terete, 7–22 cm × 1.2–3 mm. **Inflorescences** panicles or
racemes of 3–50 heads, 2–20 cm, branches erect to
ascending; primary bract erect; heads 5–50-flowered,
obconic to spheric, 3–10 mm diam. **Flowers:** tepals
green or straw-colored to reddish brown, lanceolate;
outer tepals 2.7–3.8 mm, apex acuminate; inner tepals
2.9–4 mm, apex acuminate; stamens 3(–6), anthers
½ filament length. **Capsules** equaling perianth or exserted, chestnut brown, imperfectly 3-locular, lanceoloid, 3.3–4.5 mm, acute proximal to beak, valves separating at dehiscence. **Seeds** fusiform, 1.1–1.9 mm,
tailed; body covered with whitish translucent veil. $2n =$
80.

Fruiting mid summer–fall. Salt, brackish, and calcar-
eous marshes, acid bogs, roadsides, tidal flats, swamps,
patterned fen, lake shores, beaches; 0–200 m; St. Pierre
and Miquelon; Man., N.B., Nfld. and Labr. (Nfld.),
N.S., Ont., Oreg., P.E.I., Que.; Ala., Ark., Conn., Del.,
D.C., Fla., Ga., Ill., Ind., Iowa, Ky., La., Maine, Md.,
Mass., Mich., Minn., Miss., Mo., Nebr., N.H., N.J.,
N.Y., N.C., Ohio, Pa., R.I., S.C., S.Dak., Tenn., Vt.,
Va., W.Va., Wis.

Two varieties and two forms occurring within the
flora have been recognized (M. L. Fernald 1945b). *Juncus canadensis* var. *sparsiflorus* has stiffly erect inflorescence branches, and the flowers are generally longer
than those of var. *canadensis*. These varieties simply
serve to give name to parts of the broad morphologic
range of variation encountered in *J. canadensis* and do
not appear to represent any distinct biological entities.

Juncus canadensis and the following three species

form a distinctive group: they have been variously treated as species (as here), varieties of *J. canadensis*, or as two species, *J. canadensis* and a polymorphic species, *J. brachysephalus*, encompassing the other three species (B. Boivin 1967–1979, part IV). Most of the species are easily recognized at their extremes but show a fair amount of overlap.

67. Juncus brachycephalus (Engelmann) Buchenau, Bot. Jahrb. Syst. 12: 268. 1890 [E]

Juncus canadensis J. Gay var. *brachycephalus* Engelmann, Trans. Acad. Sci. St. Louis 2: 474. 1868

Herbs, perennial, cespitose, 2–7 dm. **Culms** erect, 1–4 mm diam., smooth. **Cataphylls** 1–2, straw-colored to pink, apex acute. **Leaves:** basal 1–3, cauline 1–2; auricles 0.6–1.5 mm, apex rounded, scarious; blade terete to compressed, 0.2–12 cm × 0.5–2 mm. **Inflorescences** panicles of 5–80 heads, 5–25 cm, branches ascending; primary bract erect; heads 2–6-flowered, ellipsoid to obovoid, 2–5 mm diam. **Flowers:** tepals green to light brown, lanceolate; outer tepals 1.8–2.5 mm, apex obtuse to nearly acute; inner tepals 2–2.8 mm, apex obtuse to nearly acute; stamens 3 or 6, anthers ½ filament length. **Capsules** exserted, chestnut to dark brown, imperfectly 3-locular, obconic, 2.4–3.8 mm, apex acute proximal to beak, valves separating at dehiscence. **Seeds** ellipsoid to fusiform, 0.8–1.2 mm, tailed; body covered with whitish translucent veil. 2*n* = 80.

Fruiting summer–early fall. Calcareous marshes, wet meadows, and wetland shores; 100–200 m; N.B., N.S., Ont., Que.; Ala., Colo., Ga., Ill., Ind., Maine, Mass., Mich., Minn., N.H., N.J., N.Y., N.C., Ohio, Okla., Pa., Tenn., Vt., Va., W.Va., Wis.

68. Juncus subcaudatus (Engelmann) Coville & S. F. Blake, Proc. Biol. Soc. Wash. 31: 45. 1918 [E]

Juncus canadensis J. Gay var. *subcaudatus* Engelmann, Trans. Acad. Sci. St. Louis 2: 474. 1868; *J. subcaudatus* var. *planisepalus* Fernald

Herbs, perennial, cespitose, 1.5–6 dm. **Culms** erect, 1–3 mm diam., smooth. **Cataphylls** 0–1, straw-colored, apex acute. **Leaves:** basal 1, cauline 1–3; auricles 0.5–1 mm, apex rounded, scarious; blade terete, 4.5–15 cm × 1–2 mm. **Inflorescences** panicles or racemes of 3–35 heads, 2–16 cm, branches widely spreading (at least the proximalmost); primary bract erect; heads 2–10(–20)-flowered,

obconic to nearly spheric, 3–9 mm diam. **Flowers:** tepals greenish, becoming straw-colored, lanceolate; outer tepals 1.8–3 mm, apex acuminate; inner tepals 2–3.2 mm, apex acuminate; stamens 3, anthers ⅓ filament length. **Capsules** exserted, straw-colored, imperfectly 3-locular, ovoid to prismatic, 3–3.8 mm, apex acute proximal to beak, valves separating at dehiscence. **Seeds** ellipsoid, 0.7–1.2 mm, tailed; body covered with whitish translucent veil.

Fruiting late summer–fall. Stream banks, lake and pond shores, bogs, and other wet places; 0–1000 m; Nfld. and Labr. (Nfld.), N.S.; Ala., Ark., Conn., Del., D.C., Ga., Ill., Ky., Maine, Md., Mass., Mo., N.J., N.Y., N.C., Ohio, Pa., Tenn., Va., W.Va.

69. Juncus brevicaudatus (Engelmann) Fernald, Rhodora 6: 35. 1904 • Narrow-panicled rush [E] [F]

Juncus canadensis J. Gay var. *brevicaudatus* Engelmann, Trans. Acad. Sci. St. Louis 2: 436. 1866; *J. canadensis* var. *coarctatus* Engelmann; *J. canadensis* var. *kuntzei* Buchenau; *J. coarctatus* (Engelmann) Buchenau; *J. kuntzei* (Buchenau) Vierhapper; *J. tweedyi* Rydberg

Herbs, perennial, cespitose, 1.4–5.5(–7) dm. **Culms** erect, terete, 1–3 mm diam., smooth. **Cataphylls** 0–1, straw-colored to pink, apex acute. **Leaves:** basal 1–3, cauline 1–2; auricles 0.5–3 mm, apex rounded to truncate, scarious; blade terete, 1.5–25 cm × 0.5–2.5 mm. **Inflorescences** terminal panicles or racemes of 2–35 heads, 1–12 cm, branches erect; primary bract erect; heads 2–8-flowered, ellipsoid to narrowly obconic, 2–9 mm diam. **Flowers:** tepals green to light brown, lanceolate; outer tepals 2.3–3.1 mm, apex acuminate to rarely obtuse; inner tepals 2.5–3.2 mm, apex acuminate; stamens 3 (or 6), anthers ¼–½ filament length. **Capsules** exserted, chestnut brown, imperfectly 3-locular, narrowly ellipsoid to prismatic, 3.2–4.8 mm, apex acute proximal to beak, valves separating at dehiscence. **Seeds** fusiform, 0.7–1.2 mm, tailed; body covered with whitish translucent veil. 2*n* = 80.

Fruiting mid summer–fall. Generally in acidic or peaty moist sites, including emergent shorelines and around hot springs; 100–2500 m; Alta., B.C., Man., N.B., Nfld. and Labr., N.S., Ont., P.E.I., Que., Sask.; Ariz., Colo., Conn., Maine, Mass., Mich., Minn., N.H., N.Y., N.C., Oreg., Pa., R.I., Tenn., Utah, Vt., Va., W.Va., Wis., Wyo.

Populations from around hot springs in the west have been separated as *Juncus tweedyi* Rydberg, but no morphologic distinction appears to exist between *J. tweedyi* and *J. brevicaudatus*.

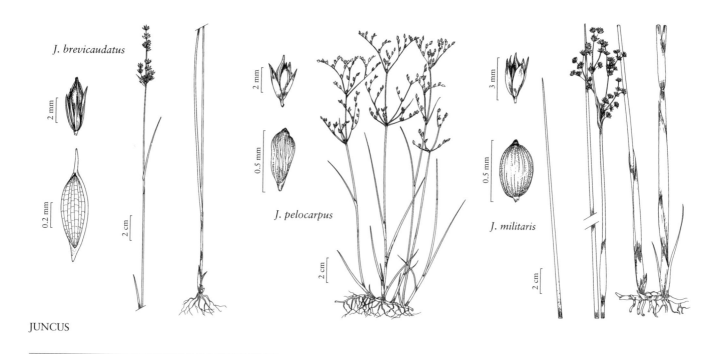

J. brevicaudatus

J. pelocarpus

J. militaris

JUNCUS

70. Juncus pelocarpus E. Meyer, Syn. Luzul., 30. 1823
  • Brown-fruited rush, jonc à fruits bruns ⒠ Ⓕ

Juncus abortivus Chapman; *J. pelocarpus* var. *crassicaudex* Engelmann; *J. pelocarpus* var. *sabulonensis* H. St. John

Herbs, perennial, rhizomatous, 0.3–7 dm. **Rhizomes** 1–3 mm diam., nodes not swollen. **Culms** erect, terete, 1–2 mm diam., smooth. **Cataphylls** 0 (rarely 1, straw-colored, apex obtuse). **Leaves:** basal 0–2, cauline 1–4, small fascicles of short capillary leaves often on rhizomes and stems; auricles 0.3–1 mm, apex rounded, membranaceous; blade terete, 1.5–11 cm × 0.8–1.1 mm. **Inflorescences** terminal cymes, flowers single or paired at nodes, (rarely in 3s), 2–25 cm, branches spreading to erect; primary bract erect. **Flowers:** tepals straw-colored, oblong; outer tepals 1.6–2.3 mm, apex obtuse; inner tepals 1.8–2.8 mm, apex obtuse; stamens 6, anthers 2–3 times filament length. **Capsules** included to exserted, chestnut brown, 1-locular, narrowly ovoid, 1.5–3.5 mm, apex acute proximal to beak, valves separating at dehiscence, fertile only proximal to middle. **Seeds** ovoid, 0.3–0.5 mm, not tailed; body clear yellow-brown. 2*n* = 40.

Fruiting late summer–fall. Shores, peat bogs, sandy soils, pools, occasionally submersed in lakes, rarely in salt water; 0–600 m; B.C., N.B., Nfld. and Labr. (Nfld.), N.S., Ont., P.E.I., Que.; Conn., Del., Fla., Ga., Ind.,

Maine, Md., Mass., Mich., Minn., N.H., N.J., N.Y., N.C., Pa., R.I., S.C., Vt., Va., Wis.

Populations from Virginia and south have been separated as *Juncus pelocarpus* var. *crassicaudex* (*J. abortivus*) based on their taller culms and thicker rhizomes. The evidence used to support the varietal status for the southeastern plants (N. A. Murray and D. M. Johnson 1987) clearly demonstrates that these plants are the southern end of a cline.

The flowers are often replaced by bulbils.

71. Juncus subtilis E. Meyer, Syn. Luzul., 31. 1823
  • Creeping rush, jonc délié ⒠

Juncus pelocarpus var. *fluitans* (Michaux) Buchenau; *J. pelocarpus* E. Meyer var. *subtilis* (E. Meyer) Engelmann; *J. uliginosus* Kunth var. *subtilis* (E. Meyer) Hooker

Herbs, perennial, emergent, rhizomatous (forming mats to 5 dm), 0.5–1 dm. **Rhizomes** 1 mm diam., nodes not swollen. **Culms** repent, floating or submersed, usually profusely branched, terete, 0.5–1 mm diam. **Cataphylls** absent. **Leaves:** basal 1–5, cauline 1–4, small fascicles of short capillary leaves often on rhizomes and stems; auricles 0.1–0.8 mm, apex acute, membranaceous; blade terete, 1.6–3 cm × 0.1–0.3 mm. **Inflorescences** cymes, flowers 1–3 at 1–2 nodes, 1–4 cm, branches spreading to erect. **Flowers:** tepals reddish, oblong; outer tepals 1.8–2.8

mm, obtuse; inner tepals 2.2–4.4 mm, obtuse; stamens 6, anthers ¾ to equal filament length. **Capsules** exserted, chestnut brown, 1-locular, ovoid to narrowly ovoid, 2.4–5 mm, apex acute proximal to beak, valves separating at dehiscence. **Seeds** ovoid, 0.3–0.5 mm, not tailed; body clear yellow-brown. $2n = 40$.

Fruiting late summer–early fall. Muddy, sandy or gravelly shores, fens in decomposed peat, fine muds rich in humus, and submersed in brackish pools; 0–200 m; Greenland; N.B., Nfld. and Labr., Ont., Que.; Maine.

Flowers are rarely replaced by bulbils. Most Greenland collections are sterile.

72. Juncus militaris Bigelow, Fl. Boston. ed. 2, 139. 1824 · Bayonet rush, jointed bog rush E F

Herbs, perennial, rhizomatous, 3–15 dm. **Rhizomes** 3–4 diam., nodes not swollen, smooth (often developing filiform leaves in running water). **Culms** erect, terete to compressed, 5–12 mm diam. **Cataphylls** 1–3, straw-colored to pink, apex acute. **Leaves:** basal 0, cauline 2, long capillary leaves often found in fascicles on rhizomes; auricles 0.3–0.5 mm, apex rounded, scarious; blade terete, 50–70(–100) cm × 2–5 mm, those of proximal leaves usually overtopping inflorescences, distal leaves usually inflated bladeless sheaths, occasionally absent or with well-developed blades. **Inflorescences** terminal panicles of 20–100 heads, 4–15 cm, branches erect to ascending; primary bract erect; heads (3–)5–13(–25)-flowered, hemispheric to turbinate, 6–8 mm diam. **Flowers:** tepals straw-colored or reddish, lanceolate, 2.3–3.2(–4) mm, nearly equal, apex acuminate to awned; stamens 6, anthers 1.5–2 times filament length. **Capsules** straw-colored, 1-locular, ovoid, 2.3–3.3 mm, equaling perianth, tapering to subulate tip, valves separating at dehiscence. **Seeds** obovoid, 0.5–0.6 mm, not tailed; body clear yellow-brown.

Fruiting late summer–fall. Mucky bottoms of shallow lakes and rivers, wet shores; 0–500 m; St. Pierre and Miquelon; N.B., Nfld. and Labr. (Nfld.), N.S., Ont., P.E.I.; Conn., Del., Ind., Maine, Md., Mass., Mich., N.H., N.J., N.Y., Pa., R.I., Vt.

The proximal culm leaf of *Juncus militaris* usually has a single well-developed blade that overtops the inflorescence; the distal leaf usually is an inflated bladeless sheath above it. An Alabama report, based on a single specimen collected by Drummond (not seen), at GH and MO, was discounted by Coville (and here). Coville believed the location and possibly the collector were wrongly attributed to this collection (see letter from Coville to Small at NY).

73. Juncus acutiflorus Ehrhart in G. F. Hoffmann, Deutschl. Fl. 1: 125. 1791

Herbs, perennial, rhizomatous, to 8 dm. **Rhizomes** 5–6 mm diam., nodes not swollen. **Culms** erect, terete, 3 mm diam., smooth. **Cataphylls** 1, straw-colored, apex acute. **Leaves:** basal 1, cauline 2; auricles 0.5 mm, apex rounded, cartilaginous; blade 7–45 cm × 1–2 mm, terete, not scabrous. **Inflorescences** terminal panicles of 70–120 heads, 6–10 cm, branches ascending; primary bract erect; heads 3–5(–8)-flowered, obconic, 2–4 mm diam. **Flowers:** tepals dark reddish brown, lanceolate, apex acuminate to subulate tip; outer tepals 1.9–2.2 mm; inner tepals 2–2.3 mm; stamens 6, anthers 2 times filament length. **Capsules** inserted with beak exserted, straw-colored, 1-locular, narrowly ovoid, 2.3–2.4 mm, tapering to subulate tip, valves separating at dehiscence. **Seeds** narrowly ovoid, 0.5–0.6 mm, not tailed; body clear yellow-brown.

Fruiting fall. Wet woods; 0–10 m; St. Pierre and Miquelon; Nfld. and Labr. (Nfld.); Europe; Asia; Africa.

74. Juncus pervetus Fernald, Rhodora 19: 17. 1917 E

Herbs, perennial, rhizomatous, 7–10 dm. **Rhizomes** 3–7 mm diam., nodes not swollen. **Culms** erect, terete, 2–4 mm diam. **Cataphylls** 2–3, straw-colored, apex rounded. **Leaves:** basal 0, cauline 1; auricles 0.5–1.5 mm, apex blunt, cartilaginous; blade 25–35 cm × 1–2 mm. **Inflorescences** terminal panicles of 24–60 heads, 5–10 cm, branches spreading; primary bract erect; heads 15–50-flowered, spheric, 4.5–6 mm diam. **Flowers:** tepals light brown, lanceolate to oblong, 2–2.5 mm, nearly equal; stamens 6, anthers longer than filament. **Capsules** included with beak slightly exserted, chestnut brown, 3-locular, ovoid, 2.5 mm, apex acute proximal to beak, valves separating at dehiscence, fertile throughout or only proximal to middle. **Seeds** not seen.

Fruiting fall. Upper border of salt marsh; 0 m; Mass.

All sheets that I have seen have aborted seeds. *Juncus pervetus* was thought to be *Juncus subnodulosus*, a European native (H. Weimarck 1946). It has been shown that this is a distinct species, however (S. Snogerup and B. Snogerup 1996). It would appear that this species is now extinct.

SELECTED REFERENCES   Snogerup, S. and B. Snogerup. 1996. *Juncus pervetus* Fernald, a misunderstood N. American species. Ann. Naturhist. Mus. Wien, B 98(suppl.): 423–426.   Weimarck, H. 1946. Studies in the Juncaceae. With special reference to the species in Ethiopia and the Cape. Svensk Bot. Tidskr. 40: 141–178.

75. Juncus supiniformis Engelmann, Trans. Acad. Sci. St. Louis 2: 461. 1868 • Hairy-leaved rush   E

Juncus oreganus S. Watson; *J. paucicapitatus* Buchenau

Herbs, perennial, cespitose or matted, often decumbent, rooting at proximal nodes or floating, 0.3–5 dm. **Culms** erect, decumbent and rooting at nodes, or floating, terete, 1–2 mm diam., smooth. **Cataphylls** 0 or 1–2, straw-colored, apex acute. **Leaves:** basal 1–3, cauline 1–4; auricles 0.8–2.1 mm, apex rounded to acute, membranaceous; blade terete, 3.7–15 cm × 0.4–1.3 mm, occasionally with filiform, flaccid, and floating leaves to 60 cm. **Inflorescences** terminal racemes of 2–9 heads, 2–10 cm, branches erect; primary bract erect; heads 2–12-flowered, obconic or rarely hemispheric, 4–13 mm diam. **Flowers:** tepals light brown or greenish to reddish brown, linear to narrowly lanceolate, apex acute to acuminate-subulate; outer tepals (2.1–)2.8–4.9 mm; inner tepals (2.1–)2.8–5.5 mm; stamens 3 or 6, anthers ½–¾ filament length. **Capsules** usually exserted, dark brown, 1-locular, ovoid to oblong, (3.2–)3.5–6.1 mm, apex acute to acuminate proximal to beak, valves separating at dehiscence. **Seeds** narrowly obovoid to obovoid, 0.6–0.7 mm, not tailed; body clear yellow-brown. n = ca. 30, ca. 50–60, $2n$ = ca. 112.

Fruiting mid summer–fall. Pond, lake and river shores, marshes, bogs, and ditches; 0–1000 m; B.C.; Alaska, Calif., Oreg., Wash.

The northern California and southern Oregon populations (*Juncus supiniformis* in the strict sense) form long filiform leaves before flowering, are shorter, and have smaller flowers than the northern populations. Except for the filiform leaves, the variation in sizes appears to follow a rough latitudinal cline with the largest plants and largest flowers in Alaska.

Flowers of *Juncus supiniformis* often form bulbils.

76. Juncus bulbosus Linnaeus, Sp. Pl. 1: 327. 1753 • Bulbous rush   I

Juncus kockii F. W. Schultz; *J. supinus* Moench

Herbs, perennial, cespitose, often with basal bulblike swellings, occasionally appearing rhizomatous, 0.3–3 dm, to 10 dm when floating or submersed. **Culms** erect or decumbent and rooting at nodes, or floating, terete, 1–2 mm diam., smooth. **Cataphylls** 1, straw-colored, apex acute. **Leaves:** basal 1, cauline 0–1; auricles 0.4–1 mm, apex acute, scarious; blade terete, occasionally filiform, flaccid, forming carpets, 2–10 cm × 0.8–1.4 mm. **Inflorescences** terminal racemes of 1–8(–30) heads, or single head, 2–10 cm, branches erect; primary bract erect; heads 2–6(–15)-flowered, obconic, 4.5–6.5 mm diam. **Flowers** often forming bulbils; tepals pale brown, ovate to lanceolate or inner oblong, 2–3.6 mm, nearly equal, apex acute to obtuse; stamens 3 or 6, anthers ¾ to equal filament length. **Capsules** equaling perianth or exserted (to 0.5 mm beyond tepals), chestnut brown, 1-locular, cylindric, 2.5–4 mm, apex obtuse proximal to beak, valves separating at dehiscence. **Seeds** ellipsoid, 0.5–0.6 mm, not tailed; body clear yellow-brown.

Fruiting mid summer–early fall. Margins and siliceous or peaty shores of pools and streams, often floating; 0–200 m; probably introduced in North America; St. Pierre and Miquelon; B.C., Nfld. and Labr. (Nfld.), N.S., Oreg.; Wash.; Europe; n Africa.

77. Juncus megacephalus M. A. Curtis, Boston J. Nat. Hist. 1: 132. 1835   E

Juncus scirpoides Lamarck var. *carolinianus* Coville; *J. scirpoides* var. *echinatus* Engelmann

Herbs, perennial, rhizomatous, 3–11 dm. **Rhizomes** 3–4 diam. **Culms** erect, terete, 3–4 mm diam., smooth. **Cataphylls** 1–2, purple, apex acute. **Leaves:** basal 0–1, cauline 2–3; auricles 0.5–2 mm, apex acute, membranaceous; blade terete, 0–24 cm × 0.5–1.7 mm, most distal cauline leaf blade 0–2 cm, shorter than sheath. **Inflorescences** panicles of (1–)3–21 heads, 1–8 cm, branches erect to spreading; primary bract erect; heads 40–60-flowered, spheric, 8–12 mm diam. **Flowers:** tepals straw-colored to reddish brown, lanceolate-subulate; outer tepals 2.9–4.1 mm, apex acuminate; inner tepals 2.2–3.7 mm, apex acuminate; stamens 3, anthers ¼–½ filament length. **Capsules** exserted, straw-colored, 1-locular, subulate, 2.5–4.2 mm, apex tapering to subulate beak, valves not separating at dehiscence, fertile throughout or only proximal to middle. **Seeds** ellipsoid to ovoid, 0.4 mm, not tailed; body clear yellow-brown.

Fruiting summer. Fresh marshes, moist hollows of sand dunes, swales, roadside ditches, and dry fertile soil; 0–100 m; Ala., Fla., Ga., La., Md., Miss., N.C., S.C., Tex., Va.

78. **Juncus scirpoides** Lamarck in J. Lamarck et al., Encycl. 3: 267. 1789 • Scirpuslike rush [E] [F]

Juncus echinatus Muhlenberg; *J. scirpoides* var. *compositus* Harper; *J. scirpoides* var. *genuinus* Buchenau; *J. scirpoides* var. *macrostemon* Engelmann; *J. scirpoides* var. *meridionalis* Buchenau

Herbs, perennial, rhizomatous, sometimes nearly cespitose, 0.8–7 dm. **Rhizomes** usually tuberous, 2–4 mm diam. **Culms** erect, terete, 1–3 mm diam., smooth. **Cataphylls** 0–1. **Leaves:** basal 1–2, cauline 2–3, green; auricles 1–2 mm, apex rounded, membranaceous; blade terete, 2–23 cm × 1–2 mm, distal cauline leaf blade 1.6–26 cm, equaling or longer than sheath. **Inflorescences** terminal panicles of 1–23(–32) heads, 2.5–9 cm, branches ascending to erect; primary bracts erect; heads 20–60-flowered, spheric or usually lobed, 6–11 mm diam. **Flowers:** tepals green to straw-colored, lance-subulate, 2–3.5 mm, nearly equal, apex acuminate; stamens 3, anthers ⅓ filament length. **Capsules** exserted, straw-colored, 1-locular, lance-subulate, 3–4 mm, apex tapering, remaining attached at tip, valves not separating at dehiscence, fertile throughout or only proximal to middle. **Seeds** oblong, 0.4 mm, not tailed; body clear yellow-brown.

Fruiting early summer–fall. Wet sandy soil, salt marshes, lake shores, ditches, meadows, wet woods; 0–1400 m; Ala., Ark., Del., D.C., Fla., Ga., Ill., Ind., Kans., Ky., La., Md., Mich., Miss., Mo., Nebr., N.J., N.Y., N.C., Okla., Pa., S.C., Tenn., Tex., Va., W.Va.

79. **Juncus validus** Coville, Bull. Torrey Bot. Club 22: 305. 1895 [E]

Juncus crassifolius Buchenau, Bot. Jahrb. Syst. 12: 326. 1890, not Bosc ex Laharpe 1827

Herbs, perennial, rhizomatous, sometimes nearly cespitose, 4–10 dm. **Rhizomes** not tuberous, 2 mm diam. **Culms** erect, terete, 3–5 mm diam., smooth. **Cataphylls** 0. **Leaves:** basal 2–3(–6), cauline 1–4; auricles 1–3 mm, apex acute, membrana-

ceous, absent on proximal leaves; blade green, laterally compressed, 9–70 cm × 2–6 mm. **Inflorescences** terminal panicles of 10–30 heads, 10–30 cm, branches spreading; primary bract erect; heads 20–30-flowered, spheric, (10–)12–15 mm diam. **Flowers:** tepals green to reddish, lance-subulate, 4–5 mm, apex acuminate; stamens 3, anthers 1/3–1/2 filament length. **Capsules** exserted, straw-colored, 1-locular, subulate, 4–5,5 mm, tapering to subulate tip, valves separating or not at dehiscence, fertile throughout or only proximal to middle. **Seeds** broadly ellipsoid, 0.5–0.6 mm, not tailed; body clear yellow-brown.

Varieties 2 (2 in the flora): United States.

1. Inflorescence 10–30 cm; heads 20–30-flowered; capsule valves fully separating at apex after dehiscence. . . . . 79a. *Juncus validus* var. *validus*
1. Inflorescence 2–5 cm; heads 6–15-flowered; capsule valves remaining united at apex after dehiscence. . . . . . 79b. *Juncus validus* var. *fascinatus*

79a. **Juncus validus** Coville var. **validus** [E]

Inflorescences 10–30 cm; heads 20–30-flowered. **Capsules** exserted, valves separating at dehiscence. $2n = 40$.

Fruiting late spring–fall. Moist to wet open sites, usually sandy, sometimes peaty, exposed or marshy shores of ponds, lakes, and streams, ditches, and fields; 0–100 m; Ala., Ark., Fla., Ga., Kans., La., Miss., Mo., N.C., Okla., S.C., Tenn., Tex., Va.

Juncus validus var. *validus* closely resembles *J. polycephalus* but differs in having complete septa, long-exserted capsules, and terete stems.

79b. **Juncus validus** Coville var. **fascinatus** M. C. Johnston, SouthW. Naturalist 9: 313. 1968 [E]

Inflorescences 2–5 cm; heads 6–15-flowered. **Capsules** exserted, valves not separating at dehiscence.

Fruiting summer. Moist to wet sandy areas, shores of ponds, lakes, and streams, sometimes in woodlands; 0–100 m; Tex.

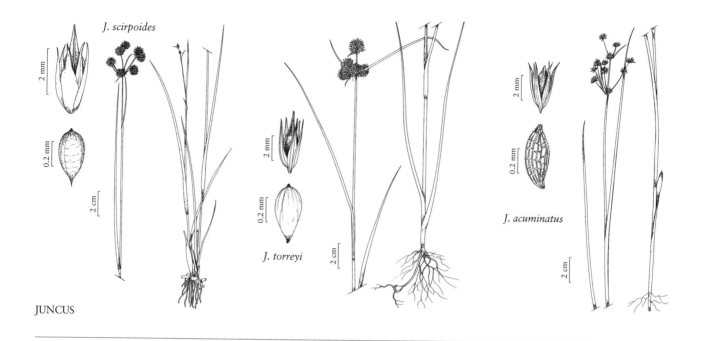

J. scirpoides

Juncus nodosus var. *meridionalis*

J. torreyi

J. acuminatus

JUNCUS

---

80. Juncus nodosus Linnaeus, Sp. Pl. ed. 2, 1: 466. 1762

Juncus nodosus var. *meridionalis* F. J. Hermann; *J. rostkovii* E. Meyer

Herbs, perennial, rhizomatous, 0.4–5.5(–7) dm. **Rhizomes** with swollen nodes, 1 mm diam. **Culms** erect, terete, 1–3 mm diam., smooth. **Cataphylls** 0 or 1–2, pink to gray, apex acute. **Leaves:** basal 1, cauline 2–4, green to pink; auricles 0.5–1.7 mm, apex rounded, membranaceous to cartilaginous; blade terete, 6–30 cm × 0.5–1.5 mm. **Inflorescences** terminal racemes of 3–15 heads, 0.6–6 cm, branches ascending to erect; primary bract erect; heads 6–30-flowered, spheric, 6–10(–12) mm diam. **Flowers:** tepals green to light brown, lance-subulate, 2.4–4.1 mm, nearly equal, apex acuminate; stamens 3 or 6, anthers ½ to equal filament length. **Capsules** exserted, chestnut brown, 1-locular, lance-subulate, 3.2–5 mm, apex tapering, valves separating at dehiscence, fertile throughout or only proximal to middle. **Seeds** oblong, ellipsoid, or obovoid, 0.4–0.5 mm, not tailed. $2n = 40$.

Fruiting early summer–fall. Sandy and muddy shores of lakes, streams, rivers, and estuaries (both freshwater and brackish), swamps, fens, salt marshes, and wet fields, often calcareous; 0–2200 m; Alta., B.C., Man., N.B., Nfld. and Labr., N.W.T., N.S., Ont., P.E.I., Que., Sask., Yukon; Alaska, Ariz., Colo., Conn., Del., Idaho, Ill., Ind., Iowa, Kans., Maine, Md., Mass., Mich., Minn., Mo., Mont., Nebr., Nev., N.H., N.J., N.Mex., N.Y., N.Dak., Ohio, Okla., Pa., R.I., S.Dak., Tex., Utah, Vt., Va., Wash., W.Va., Wis., Wyo.; Mexico (s to Puebla).

The Texas populations and some of the northern Mexican populations consistantly have 3 stamens. These populations have been separated as *Juncus nodosus* var. *meridionalis*. Plants with 3 stamens are found elsewhere, however, and other significant characters do not appear to separate these populations.

81. Juncus torreyi Coville, Bull. Torrey Bot. Club 22: 303. 1895 • Torrey's rush [F]

Juncus nodosus Linnaeus var. *megacephalus* Torrey, Fl. New York 2: 326. 1843; *J. megacephalus* (Torrey) A. W. Wood 1862, not M. A. Curtis 1835

Herbs, perennial, rhizomatous, (3–)4–10 dm. **Rhizomes** 1–3 mm diam., with swollen nodes. **Culms** erect, terete, 3–5 mm diam., smooth. **Cataphylls** 0. **Leaves:** basal 1–3, cauline 2–5, auricles 1–4 mm, apex rounded, scarious; blade strawberry-colored, green, or pink, terete, 13–30 cm × 1–5 mm. **Inflorescences** terminal clusters of 1–23 heads or single heads, 2–5.5 cm, branches spreading; primary bract erect to ascending; heads 25–100-flowered, globose, 10–15 mm diam. **Flowers:** tepals green to straw-colored, occasionally reddish, lanceolate-subulate; outer tepals (3.7–)4–6 mm, apex acuminate; inner tepals (3–)

3.4–4.6 mm, apex acuminate; stamens 6, anthers ½ filament length. **Capsules** equaling perianth or slightly exserted, straw-colored or chestnut brown, 1-locular, lance-subulate, 4.3–5.7 mm, apex tapering to subulate tip, valves separating at dehiscence, fertile throughout or only proximal to middle. **Seeds** oblong to ellipsoid, 0.4–0.5 mm, not tailed; body clear yellow-brown. $2n = 40$.

Fruiting early summer–fall. Wet sandy shores, edges of sloughs, along slightly alkaline watercourses, swamps, sometimes on clay soils, calcareous wet meadows, and alkaline soils; 0–600; Alta., B.C., Man., N.B., Ont., Que., Sask.; Ala., Ariz., Ark., Calif., Colo., Ga., Idaho, Ill., Ind., Iowa, Kans., Ky., La., Maine, Md., Mass., Mich., Minn., Miss., Mo., Mont., Nebr., Nev., N.H., N.J., N.Mex., N.Y., N.Dak., Ohio, Okla., Oreg., Pa., S.Dak., Tenn., Tex., Utah, Vt., Va., Wash., W.Va., Wis., Wyo.; Mexico (Baja California, probably elsewhere in n Mexico).

82. **Juncus texanus** (Engelmann) Coville in J. K. Small, Fl. S.E. U.S., 259. 1903 · Texas rush [E]

Juncus nodosus Linnaeus var. *texanus* Engelmann, Trans. Acad. Sci. St. Louis 2: 471. 1868

Herbs, perennial, rhizomatous, 2.5–6 dm. **Rhizomes** 1 mm diam., with swollen nodes. **Culms** erect, terete, 1–3 mm diam., smooth. **Cataphylls** 0 or 1–2, straw-colored, apex rounded. **Leaves:** basal 0–1, cauline 3–5, auricles 1.5–2.5 mm, apex rounded to acutae, scarious; blade straw-colored to green, terete, 5.5–20 cm × 1–2.5 mm. **Inflorescences** terminal panicles of 2–16 heads, 2.5–6 cm, branches ascending, spreading, or reflexed; primary bract erect to spreading; heads 10–40-flowered, spheric, 11–15 mm diam. **Flowers:** tepals green or straw-colored to reddish brown, lance-subulate, apex acuminate; outer tepals (3.5–)4–4.9 mm; inner tepals 3.9–5.4 mm; stamens 6, anthers 2–3 times filament length. **Capsules** exserted, chestnut brown, 1-locular, lance-subulate, 5.2–6.5(–8) mm, apex tapering, valves separating at dehiscence, fertile throughout or only proximal to middle. **Seeds** obovoid, 0.45–0.5 mm, not tailed; body clear yellow-brown.

Fruiting summer. Swamps, depressions, seeps, sand or gravel bars, and wet mud; 400–600 m; Tex.

83. **Juncus brachycarpus** Engelmann in A. Gray, Manual ed. 5, 542. 1867 [E]

Herbs, perennial, not cespitose, rhizomatous, (3–)4.5–8(–9) dm. **Rhizomes** tuberous, 3–4 mm diam. **Culms** erect, terete, 2–3 mm diam., smooth. **Cataphylls** 0(–1), straw-colored, apex acute. **Leaves:** basal 1–2, cauline 2–4, auricles 0.5–3.5 mm, apex rounded, scarious; blade green to straw-colored, terete, 3–50 cm × 1–2 mm diam. **Inflorescences** terminal panicles of 2–10(–20) heads or a single head, 1–4(–10) cm, branches ascending; primary bract erect; heads 30–100-flowered, spheric, 8–10 mm diam. **Flowers:** tepals green to straw-colored, often red-tinted, lanceolate-subulate, apex acuminate; outer tepals 2.5–3.8 mm; inner tepals 2–3.2 mm; stamens 3, anthers ¼–½ filament length. **Capsules** included, chestnut brown, 1-locular, obconic or ovoid, 1.8–2.7 mm, apex obtuse, valves separating at dehiscence, fertile throughout or only proximal to middle. **Seeds** ellipsoid to oblique-oblong, 0.3–0.4 mm, not tailed; body clear to yellow-brown. $2n = 44$.

Fruiting mid summer–fall. Damp clayey, peaty, or sandy soils, swamps, ditches, ponds, wet woods, wet prairies; 0–500 m; Ont.; Ala., Ark., Conn., Del., Ga., Ill., Ind., Kans., Ky., La., Md., Mass., Mich., Minn., Miss., Mo., N.J., N.Y., N.C., Ohio, Okla., Pa., S.C., Tenn., Tex., Va., W.Va.

84. **Juncus bolanderi** Engelmann, Trans. Acad. Sci. St. Louis 2: 470. 1868

Juncus bolanderi var. *riparius* Jepson

Herbs, perennial, not cespitose, rhizomatous, 2.5–8.5 dm. **Rhizomes** 2–3 mm, tuberous. **Culms** erect, terete, 1–3 mm diam., smooth. **Cataphylls** 1–2, straw-colored, apex acute. **Leaves:** basal 0–1, cauline 2–3, auricles 2–3 mm, apex acute, membranaceous; blade green to straw-colored, terete, 7–28 cm × 1–2 mm. **Inflorescences** terminal racemes of 1–8 heads or single head, 1–13 cm, branches ascending to spreading; primary bract erect; heads 40–70-flowered, spheric, 7–11 mm diam. **Flowers:** tepals dark reddish brown to light brown, lance-subulate, (2.6–)3–3.5(–4) mm, apex acuminate; stamens 3, anthers ¼ filament length. **Capsules** equaling perianth or slightly included, chestnut brown, 1-locular, ellipsoid to oblong, 3–3.5 mm, apex acuminate, valves separating at dehiscence, fertile throughout or only proximal to middle. **Seeds** oblong, 0.4–0.5 mm, not tailed; body clear yellow-brown.

Fruiting late spring–summer. Swamps, marshes, stream banks, beaches, roadside meadows, and other moist or wet areas; 0–1000 m.; B.C.; Calif., Idaho, Oreg., Wash.; nw Mexico.

85. Juncus acuminatus Michaux, Fl. Bor.-Amer. 1: 192. 1803 • Sharp-fruited rush [F]

Juncus acuminatus var. *legitimus* Engelmann; *J. pallescens* E. Meyer ex Buchenau; *J. pondii* A. W. Wood

Herbs, perennial, cespitose, not rhizomatous, 1.4–10 dm. **Culms** erect, terete, 1–3 mm diam., smooth. **Cataphylls** 1–2, gray, apex acute. **Leaves:** basal 1–2, cauline 1–2; auricles 1–1.5 mm, apex rounded, scarious; blade green, straw-colored, or pink, nearly terete, 1–40 cm. **Inflorescences** terminal panicles of 5–50 heads, 5–15 cm, branches ascending; primary bract erect; heads (2–)5–20(–50)-flowered, hemispheric (to spheric), 3–10 mm diam. **Flowers:** tepals light brown to greenish, lanceolate, 2.6–3.5(–3.9) mm, apex acuminate; stamens 3 (or 6), anthers ⅓ filament length. **Capsules** equaling perianth or slightly exserted, straw-colored, 1-locular, ellipsoid to narrowly ovoid, 2.8–3.5(–4) mm, apex acute, valves separating at dehiscence, fertile throughout or only proximal to middle. **Seeds** ellipsoid to oblong, 0.3–0.4 mm, not tailed; body clear yellow-brown. 2n = 40.

Fruiting early summer–fall. Shores, swamps, ditches, springs, wet meadows, and rock outcrops; 0–2200 m; B.C., N.S., Ont., Que.; Ala., Ariz., Ark., Calif., Colo., Conn., Del., D.C., Fla., Ga., Idaho, Ill., Ind., Iowa, Kans., Ky., La., Maine, Md., Mass., Mich., Minn., Miss., Mo., Mont., Nebr., N.H., N.J., N.Mex., N.Y., N.C., Ohio, Okla., Oreg., Pa., R.I., S.C., Tenn., Tex., Vt., Va., Wash., W.Va., Wis.; Mexico; Central America (Honduras).

86. Juncus debilis A. Gray, Manual, 506. 1848 • Weak rush

Juncus acuminatus Michaux var. *debilis* (A. Gray) Engelmann; *J. radicans* Schlechtendal

Herbs, perennial, cespitose, 1–2.5 dm. **Culms** erect, terete, 1–2 mm diam., smooth. **Cataphylls** 0–1, maroon or dark green, apex acute. **Leaves:** basal 0–1, cauline 1–3; auricles 1–1.5 mm, apex rounded, scarious; blade maroon or dark green, terete,

1–12.5 cm × 0.5–1.5 mm. **Inflorescences** terminal panicles of 3–50 heads, 2–8 cm, branches ascending to spreading; primary bract erect; heads 2–10-flowered, obpyramidal, 2–5 mm diam. **Flowers:** tepals green to straw-colored, lanceolate, 1.8–2.3(–2.5) mm, apex sharply acuminate; stamens 3, ½ filament length. **Capsules** exserted, straw-colored, 1-locular, narrowly ellipsoid to lanceoloid, 2.8–3.7(–4.2) mm, apex acute, valves separating at dehiscence. **Seeds** ellipsoid, 0.3–0.4 mm, not tailed; body clear yellow-brown.

Fruiting summer. Marshy shores, in small streams, swamps, wet clearings, spring runs, commonly in very soft mucky substrates; 0–700 m; Ala., Ark., Del., Fla., Ga., Ill., Ky., La., Md., Miss., Mo., N.J., N.Y., N.C., Pa., S.C., Tenn., Tex., Va.; Mexico (Chiapas); Central America (Honduras).

Juncus debilis A. Gray is a name being proposed for conservation.

87. Juncus diffusissimus Buckley, Proc. Acad. Nat. Sci. Philadelphia 14: 9. 1862 [E]

Herbs, perennial, cespitose, 2.5–6.5 dm. **Culms** erect, terete, 1–3 mm diam., smooth. **Cataphylls** 0–1, maroon or dark green, apex obtuse. **Leaves:** basal 1, cauline 2–3; auricles 1–2.1 mm, apex rounded, membranaceous; blade maroon or dark green, terete to compressed, 3–20 cm × 1–2.4 mm. **Inflorescences** terminal panicles of 30–70(–130), 5–20 cm, branches spreading; primary bract erect; heads (1–)2–10-flowered, hemispheric or narrower, 5–10 mm diam. **Flowers:** tepals green to straw-colored, lanceolate, apex acute; outer tepals (2–)2.6–3.2 mm; inner tepals (1.8–)2.3–3 mm; stamens 3, anthers ½–⅔ filament length. **Capsules** exserted, straw-colored, 1-locular, linear-lanceoloid, 4–5.2 mm, apex acute, valves separating at dehiscence. **Seeds** oblong-ellipsoid, 0.3–0.4 mm, not tailed; body clear yellow-brown.

Fruiting summer. In soft mucky substrates, marshy shores, sloughs, occasionally in wet wooded places, often in shallow water; commonly abundantly colonizing wet, sandy-alluvial outwash in ditches and clearings; 10–1000 m; Ala., Ark., Calif., Conn., Fla., Ga., Ill., Ind., Kans., Ky., La., Md., Miss., Mo., N.C., Ohio, Okla., S.C., Tenn., Tex., Va., Wash., W.Va.; probably introduced, South America.

88. Juncus nodatus Coville in N. L. Britton and A. Brown, Ill. Fl. N. U.S. ed. 2, 1: 482. 1913 [E]

Juncus acuminatus Michaux var. *robustus* Engelmann; *J. robustus* (Engelmann) Coville 1896, not S. Watson 1879

Herbs, perennial, cespitose, 3–10 dm. **Roots** without terminal tubers. **Culms** erect, terete, 4–6 mm diam., smooth. **Cataphylls** 1–2, straw-colored, apex acute. **Leaves:** basal 1–2, cauline 1–2; auricles 1.2–1.5 mm, apex rounded, scarious; blade straw-colored or green, terete, 20–65 cm × 1.1–3.5 mm, with prominent and conspicuous ringlike bands at position of cross partitions; distal cauline leaves reduced to 2.5 cm. **Inflorescences** terminal panicles of 30–250 heads, 8–12 cm, branches spreading; primary bract erect to ascending; heads 2–10-flowered, broadly obovoid to hemispheric, 0.3–0.5 mm diam. **Flowers:** tepals straw-colored, lance-subulate, apex acuminate; outer tepals 1.9–2.2 mm; inner tepals 1.7–2.1 mm; stamens 3, anthers equal filament length. **Capsules** exserted, straw-colored, 1-locular, ovoid, 1.9–2.5 mm, apex acute, valves separating at dehiscence. **Seeds** oblong or ellipsoid, 0.5–0.6 mm, not tailed; body clear yellow-brown.

Fruiting late spring–late summer. Commonly in shallow water, marshy shores, sloughs, wet flatwoods, and savannas, bogs, ditches, wet woods, shores, in standing water to 1 m deep; 100–200 m; Ala., Ark., Fla., Ill., Ind., Kans., Ky., La., Mo., Okla., Tenn., Tex.

89. Juncus elliottii Chapman, Fl. South. U.S., 494. 1860 [E]

Herbs, perennial, cespitose, 3–9 dm. **Roots** often with terminal tubers. **Culms** erect, terete, 1–3 mm diam., smooth. **Cataphylls** 1, maroon to brown, apex acute. **Leaves:** basal 1–3, cauline 1–2; auricles 0.5–2 mm, apex rounded, scarious; blade green or maroon, compressed, 2–16 cm × 1–2 mm, with faint ringlike bands at position of septa. **Inflorescences** terminal panicles of 40–100(–200) heads, 4–16 cm, branches ascending to spreading; primary bract erect to ascending; heads 2–10-flowered, hemispheric to obpyramidal, 0.3–0.5 mm diam. **Flowers:** tepals straw-colored, lanceolate, apex acuminate; outer tepals (2.2–)2.6–2.9 mm; inner tepals (1.8–)2.4–2.8; stamens 3, anthers ⅔ to equal filament length. **Capsules** exserted, chestnut brown, 1-locular, narrowly obpyriform to narrowly ovoid, 2.4–2.9 mm, apex acute,

valves separating at dehiscence. **Seeds** ellipsoid, 0.3–0.5 mm, not tailed; body clear yellow-brown. 2*n* = 40.

Fruiting summer. Wet sands, peaty sands, or peat, exposed shores of ponds and lakes, depressions in savannas and flatwoods, moist to wet, much disturbed clearings, roadsides and ditches; 0–700 m; Ala., Ark., Del., Fla., Ga., La., Miss., N.J., N.C., S.C., Tenn., Va.

Juncus elliottii has tubers at the ends of the roots.

90. Juncus mertensianus Bongard, Mém. Acad. Imp. Sci. St.-Pétersbourg, Sér. 6, Sci. Math. 2: 167. 1833 [E][F]

Juncus duranii Ewan; *J. mertensianus* var. *duranii* (Ewan) F. J. Hermann; *J. mertensianus* var. *filifolius* Suksdorf ; *J. slwookoorum* S. Young

Herbs, perennial, rhizomatous to cespitose, 0.5–4 dm. **Rhizomes** 1–2 mm diam., not swollen. **Culms** erect, terete, 1–3 mm diam., smooth. **Cataphylls** 0–1, straw-colored to chestnut brown, apex acute. **Leaves:** basal 1–2, cauline 0–1; auricles 1–1.2 mm, apex rounded to acute, membranaceous or scarious; blade green to straw-colored, terete, 3–15 cm × 0.3–0.6 mm. **Inflorescences** terminal single head (rarely cluster of 2 heads), 0.5–1.6 cm; primary bract erect; heads 12–60-flowered, spheric (to hemispheric), 4.5–15 mm diam. **Flowers:** tepals dark purplish brown to black, lanceolate to lance-ovate, apex acute, mucro subulate; outer tepals 2.4–4.9 mm; inner tepals 2.3–4.3 mm; stamens 6, anthers ¼ to equal filament length. **Capsules** included or slightly exserted, chestnut brown, 1-locular, obovoid, 1.9–3.5 mm, apex obtuse or rounded, valves separating at dehiscence, fertile throughout or only proximal to middle. **Seeds** ellipsoid, 0.4–0.5 mm, not tailed; body clear yellow-brown. 2*n* = 40.

Fruiting mid summer–fall. Montane to alpine meadows, stream banks, lake margins, and conifer woods; (400–)1900–3300 m; Alta., B.C., N.W.T., Sask., Yukon; Alaska, Ariz., Calif., Colo., Idaho, Mont., Nev., N.Mex., Oreg., Utah, Wash., Wyo.

Populations from southern California with brown tepals, anthers equaling filaments, and rounded to acute, translucent auricles have been separated as *Juncus duranii*; the typical form is so highly variable, however, that it can easily accommodate this local form. This species passes into *Juncus nevadensis* and has often been combined with that species. The two species can generally be separated, and we are following those treatments (F. J. Hermann 1975; A. Cronquist et al. 1972+, vol. 6).

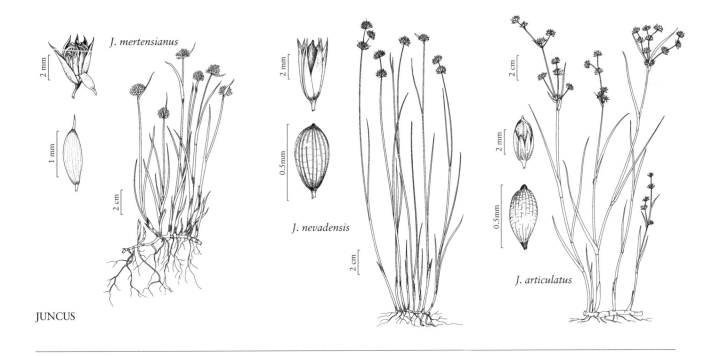

J. mertensianus

J. nevadensis

J. articulatus

JUNCUS

91. Juncus chlorocephalus Engelmann, Trans. Acad. Sci. St. Louis 2: 485. 1868 E

Herbs, perennial, cespitose, 2–4.5 dm. Culms erect, terete, 1–3 mm diam., smooth. Cataphylls 1, straw-colored or maroon, apex acute. Leaves: basal 1–2, cauline 2–3; auricles 2–3.5 mm, apex rounded, scarious; blade terete, 2–16 cm × 1–2 mm. Inflorescences single heads (rarely cluster of 2–3 heads), 0.5–2 cm; primary bract erect; heads 15–25-flowered, spheric, 11–14 mm diam. Flowers: tepals light green to light pink or white, lanceolate, 3.1–4.2 mm, apex obtuse; stamens 6, anthers 2–3 times filament length. Capsules included, straw-colored, 1-locular, broadly obovoid to ovoid, 2.2–2.5 mm, apex obtuse, valves separating at dehiscence, fertile throughout or only proximal to middle. Seeds oblong, 0.5 mm, not tailed; body clear yellow-brown.

Fruiting mid summer–fall. Sandbars, meadows, damp soil in rock outcrops, and talus; 1400–3000 m; Calif., Nev.

92. Juncus nevadensis S. Watson, Proc. Amer. Acad. Arts 14: 303. 1879 E F

Juncus badius Suksdorf; *J. columbianus* Coville; *J. mertensianus* Bongard subsp. *gracilis* (Engelmann) F. J. Hermann; *J. mertensianus* var. *badius* (Suksdorf) F. J. Hermann; *J. mertensianus* var. *columbianus* (Coville) F. J. Hermann; *J. mertensianus* var. *gracilis* (Engelmann) F. J. Hermann; *J. mertensianus* var. *suksdorfii* (Rydberg) F. J. Hermann; *J. nevadensis* var. *badius* (Suksdorf) C. L. Hitchcock; *J. nevadensis* var. *columbianus* (Coville) H. St. John; *J. nevadensis* var. *inventus* (L. F. Henderson) C. L. Hitchcock; *J. phaeocephalus* Engelmann var. *gracilis* Engelmann; *J. suksdorfii* Rydberg; *J. truncatus* Rydberg

Herbs, perennial, rhizomatous, 0.5–7 dm. Rhizomes 1 mm diam., not swollen. Culms erect, terete, 1.5–3 mm diam., smooth. Cataphylls 0–1, maroon or straw-colored, apex acute. Leaves: basal 1–3, cauline 1–2; auricles 1–3.2 mm, apex rounded to acute, membranaceous; blade green, laterally flattened, 1.5–31 cm × 0.5–2.2 mm. Inflorescences terminal panicles of 2–11 heads, 2–18 cm, branches erect to spreading; primary bract erect; heads 3–11-flowered, hemispheric to obpyramidal (rarely nearly spheric), 5–14 mm diam. Flowers: tepals dark brown to white, lanceolate, apex acute to acuminate, mucro subulate; outer tepals 2.8–6.2 mm; inner tepals 2.4–6 mm; stamens 6, anthers 1–2 times filament length. **Capsules** included to slightly exserted, chestnut brown, ellipsoid, 2.3–3.7 mm, abruptly contracted to apex, apex acute proximal to beak, valves separating at dehiscence. **Seeds** ellipsoid, 0.4–0.5 mm, not tailed.

Fruiting early summer–fall. Wet banks along margins of streams and lakes, marshes, bogs, sloughs, and wet or boggy prairies; 0–2600 m; Alta., Sask.; Ariz., Calif., Colo., Idaho, Mont., N.Mex., Nev., Oreg., Utah, Wash., Wyo.

This variable species has been divided into five separate taxa in the past (C. L. Hitchcock et al. 1955–1969, vol. 1), but more recently, most of the variants have not been separated (A. Cronquist et al. 1972+, vol. 6). The Oregon coastal endemic, *Juncus nevadensis* var. *inventus* L. F. Henderson) C. L. Hitchcock, is at one extreme of the variation, having only a single head, fairly dark brown tepals 3.5–4.5 mm, anthers only slightly longer than the filaments, and a distinct habitat. The *J. mertensianus* var. *columbianus* segregate, however, approaches var. *inventus* in number of heads, and the other characters are so highly variable in the complex, they cannot be used alone to separate this variety. Therefore, we are not recognizing any infraspecific taxa at this time.

93. **Juncus dubius** Engelmann, Trans. Acad. Sci. St. Louis 2: 459. 1868   [E]

Juncus rugulosus Engelmann

Herbs, perennial, rhizomatous, 3–7.5 dm. **Rhizomes** 2–3 mm diam., not swollen. **Culms** erect, terete, 2–4 mm diam., smooth or rugulose. **Cataphylls** 1–2, pink to straw-colored, apex acute. **Leaves:** basal 1–2, cauline 1–2; auricles 1–4.9 mm, apex rounded, membranaceous; blade green to straw-colored, terete, 20–30 cm × 1.5–4 mm, rugulose or smooth. **Inflorescences** terminal panicles of 25–66 heads, 7–13 cm, branches spreading; primary bract erect; heads 6–10-flowered, hemispheric to obovoid, 5–10 mm diam. **Flowers:** tepals straw-colored to brown, lanceolate, apex acuminate; outer tepals (2–)2.5–3.4 mm; inner tepals (2–)2.6–3.6 mm; stamens 6, anthers 1.5–2 times filament length. **Capsules** exserted, chestnut brown, 1-locular, subulate (2.4–)3–3.9 mm, apex tapering to subulate tip, valves separating at dehiscence. **Seeds** obovoid, 0.3–0.4 mm, not tailed. $2n = 40$.

Fruiting early summer–late fall. Montane meadows, riverbeds, canyons, aroyos; 100–1600 m; Calif.

Juncus dubius has rugulose stems and leaves throughout most of its range, but on the periphery (in Mariposa, San Diego, and Sonoma counties, California) the plants are smooth.

94. **Juncus alpinoarticulatus** Chaix in D. Villars, Hist. Pl. Dauphiné 1: 378. 1786 • Alpine rush

Juncus alpinoarticulatus subsp. *americanus* (Farwell) Hämet-Ahti; *J. alpinoarticulatus* subsp. *fuscescens* (Fernald) Hämet-Ahti; *J. alpinus* Villars; *J. alpinus* subsp. *nodulosus* (Wahlenberg) Lindman; *J. alpinus* var. *americanus* Farwell; *J. alpinus* var. *fuscescens* Fernald; *J. alpinus* var. *insignis* Fries ex Buchenau; *J. alpinus* var. *rariflorus* (Hartman) Hartman; *J. nodulosus* Wahlenberg; *J. rariflorus* Hartman; *J. richardsonianus* Schultes

Herbs, perennial, rhizomatous, 0.5–5 dm. **Rhizomes** 2–4 mm diam., not swollen. **Culms** erect, terete, 1–3 mm diam., smooth. **Cataphylls** 0–1, straw-colored or maroon, apex acute. **Leaves:** basal 0–2, cauline 1–2(–5); auricles 0.5–1.2 mm, apex rounded, scarious; blade green to straw-colored, terete, 1.5–12 cm × 0.5–1.1 mm. **Inflorescences** terminal panicles of 5–25 heads, 3–11 cm, branches erect to ascending; primary bract erect; heads 2–10-flowered, obpyramidal, usually with some flowers short-pedicellate, 2–6 mm in diam. **Flowers:** tepals greenish to straw-colored, lanceolate to oblong; outer tepals 1.8–3 mm, apex obtuse, mucronate; inner tepals 1.6–2.7 mm, apex obtuse; stamens 6, anthers ½ filament length. **Capsules** equaling perianth to usually exserted, chestnut brown to straw-colored, imperfectly 3-locular, oblong to oblong-ovoid, 2.3–3.5 mm, apex obtuse, valves separating at dehiscence. **Seeds** oblong to ovoid, 0.5–0.7 mm, not tailed. $2n = 40$.

Fruiting mid summer–fall. Wet meadows, sandy and gravelly, often calcareous shores, fens, and clayey pools over rock; 0–2600 m; Greenland; Alta., B.C., Man., N.B., Nfld. and Labr., N.S., N.W.T., Ont., P.E.I., Que., Sask., Yukon; Alaska, Colo., Idaho, Ill., Ind., Iowa, Maine, Mich., Minn., Mo., Mont., Nebr., N.Y., N.Dak., Ohio, Pa., S.Dak., Utah, Vt., Wash., Wis.; Eurasia.

Several attempts have been made to separate subspecies or varieties of this widespread and variable species. In one study, five varieties were recognized, with four in North America (B. Lindquist 1932). In another, at least six subspecies were recognized with two in North America (L. Hämet-Ahti 1986). The variation we have encountered does not fit nicely into the subspecies Hämet-Ahti has recognized, and until a full account of the variation throughout the range of the species is presented, we are not recognizing subspecific or varietal divisions of this species. Recent evidence suggests that this species may be one of the parents of the tetraploid *Juncus articulatus*. *Juncus alpinus* hybridizes with *J. brevicaudatus* (= *J.* ×*gracilescens* J. Hermann), *J. arti-*

culatus (= *J.* ×*alpiniformis* Fernald), *J. nodosus* (= *J.* ×*nodosiformis* Fernald), and *J. torreyi* (= *J.* ×*stuckeyi* Reinking).

95. Juncus articulatus Linnaeus, Sp. Pl. 1: 327. 1753
• Jointed rush   F

Juncus articulatus var. *obtusatus* Engelmann; *J. articulatus* var. *stolonifer* (Wohlleben) House; *J. lampocarpus* Ehrhart ex Hoffmann

Herbs, perennial, rhizomatous to nearly cespitose, 0.5–6(–10) dm. **Rhizomes** 2–3 mm diam., not swollen. **Culms** erect to decumbent (and floating), terete, 1–3 mm diam., smooth. **Cataphylls** 1, maroon to straw-colored, apex acute to obtuse. **Leaves:** basal 0–2, cauline (1–)3–6; auricles 0.5–1 mm, apex rounded, scarious; blade green to straw-colored, terete, 3.5–12 cm × 0.5–1.1 mm. **Inflorescences** terminal panicles of 3–30(–50) heads, 3.5–8 cm, branches spreading; primary bract erect; heads 3–10-flowered, obpyramidal to hemispheric, 6–8 mm diam. **Flowers:** tepals green to straw-colored or dark brown, ovate to lanceolate, 1.8–3 mm; outer tepals with apex acute or acuminate; inner tepals with apex acuminate to obtuse; stamens 6, anthers equal to filament length. **Capsules** exserted ca. 1 mm beyond perianth, chestnut brown to dark brown, imperfectly 3-locular, ellipsoid or ovoid, 2.8–4 mm, apex acute proximal to beak, valves separating at dehiscence. **Seeds** obovoid, 0.5 mm, not tailed. $2n = 80$.

Fruiting mid summer–fall. Wet ground in ditches, lake and stream margins, and a variety of other habitats, often a calciphile; 0–3000 m; St. Pierre and Miquelon; B.C., N.B., Nfld. and Labr. (Nfld.), N.S., Ont., Que., P.E.I.; Alaska, Ariz., Colo., Conn., Idaho, Ind., Ky., Maine, Mass., Mich., Minn., Nebr., Nev., N.H., N.J., N.Mex., N.Y., Ohio, Oreg., Pa., R.I., S.Dak., Utah, Vt., Va., Wash., W.Va.; Eurasia.

Juncus articulatus hybridizes with *J. brevicaudatus* (= *J.* ×*fulvescens* Fernald), *J. alpinus* (= *J.* ×*alpiniformis* Fernald), *J. nodosus*, and *J. canadensis*.

Juncus articulatus var. *obtusatus* Engelmann appears to be intermediate with *J. alpinus*. It has spreading inflorescence branches but obtuse inner tepals. This may represent a backcross with *J. alpinus*. Recent evidence suggests that *J. alpinus* is a polyploid species with *J. articulatus* as one of its parents.

2. LUZULA de Candolle in J. Lamarck and A. P. de Candolle, Fl. Franç. ed. 3, 1: 198; 3: 158. 1805, name conserved • Wood rush [possibly from Italian *lucciola*, to shine, sparkle, or Latin *gramen luzulae* or *luxulae*, diminutive of *lux*, light, because hairs of several species have shiny appearance when covered with dew]

Juncoides Adanson

Janice Coffey Swab

Herbs, perennial, usually cespitose, often with short, mostly vertical to running rhizomes and/or (less commonly) stolons. **Culms** round. **Cataphylls** absent. **Leaves:** sheaths closed, without auricles at throat (junction with blade), usually pilose; blade flat or channeled, never septate, margins with long, soft, multicellular hairs, apex often thickened (callous), veins commonly indistinct. **Inflorescences** terminal; flowers inserted individually or in dense clusters (glomerules) variously arranged; bracts subtending inflorescence (proximal inflorescence bracts) 2, mostly leaflike; bracts subtending inflorescence branches 1–2, reduced; bracteoles subtending flowers 2–3. **Flowers:** tepals 6, in 2 whorls; stamens 6. **Capsules** 1-locular, generally globose; beak often formed by persistent style base. **Seeds** 3, globose to ovoid, base often with tuft of fibrous hairs (vestige of funiculus); nutritive appendage from outer seed coat (caruncle) often present, white, barely visible to ± equaling seed body. $x = 6$.

Species ca. 108 (23 in the flora): temperate and arctic regions worldwide; tropical mountains.

The leaves of *Luzula* are primarily basal; cauline leaves are usually reduced.

Luzula species have diffuse centromeres and small chromosomes. That has resulted in much confusion in interpretation and reporting of chromosome counts. No attempt has been made to include reported counts that could not reasonably be verified by the author.

Excluded species: *Luzula sudetica* (Willdenow) de Candolle. Although reports of this European species appear frequently in the North American literature, I have seen no specimens that confirm its presence. No chromosome counts are published for North American material. Since this species has a distinct cytotype, $2n = 48$ (H. Nordenskiöld 1956), it should not be difficult to verify on this basis.

1. Flowers in dense clusters (glomerules); inflorescences spikelike or umbellate; seeds with caruncle conspicuous to barely visible.. . . . . .2c. *Luzula* subg. *Luzula*, p. 260
1. Flowers solitary or in small clusters of 2–4; inflorescences mostly unbranched or dichasial; seeds with caruncle conspicuous to absent.
   2. Flowers solitary; inflorescences corymbose, rarely to often branching; seeds with caruncle conspicuous.. . . . . .2a. *Luzula* subg. *Pterodes*, p. 256
   2. Flowers mostly in pairs, rarely in clusters of 3–4, or solitary; inflorescences paniculate or dichasial; seeds with caruncle inconspicuous or absent.. . . . . .2b. *Luzula* subg. *Anthelaea*, p. 257

2a. LUZULA de Candolle subg. PTERODES (Grisebach) Buchenau, Bot. Jahrb. Syst. 12: 76. 1890

Luzula sect. *Pterodes* Grisebach, Spic. Fl. Rumel. 2: 404. 1846

Leaves: apex callous. **Inflorescences** corymbose, simple or sparsely branched. **Flowers** solitary. **Seeds:** base without tuft of hairs; caruncle chalazal, conspicuous.

Species 8 (2 in the flora): North America, Eurasia, e Africa.

SELECTED REFERENCES Ebinger, J. E. 1964. Taxonomy of the subgenus *Pterodes*, genus *Luzula*. Mem. New York Bot. Gard. 10(5): 279–304.   Nordenskiöld, H. 1957. Hybridization experiments in the genus *Luzula*. III. The subg. *Pterodes*. Bot. Not. 110: 1–16.

1. Tepals 1.6–2.5 mm; inflorescences rarely branching; pedicels spreading.. . . . . .1. *Luzula rufescens*
1. Tepals 3–4.5 mm; inflorescences rarely branching or often branching; pedicels mostly erect to spreading.. . . . . .2. *Luzula acuminata*

1. Luzula rufescens Fischer ex E. Meyer, Linnaea 22: 385. 1849 · Rusty wood rush

Rhizomes absent. **Stolons** to 4 cm; scale leaves present; adventitious roots present. **Culms** loosely cespitose, 10–30 cm × 0.5–3.5 mm. **Leaves:** sheaths reddish, throats pilose; blade sparsely pubescent; basal leaves 6–10 cm × 2–5 mm; cauline leaves 2–3, to 4(–6) cm × 1–2.5 mm. **Inflorescences** simple or rarely branching; proximal inflorescence bract to ½ length of inflorescence; bracts brown, apex truncate, glabrous; bracteoles ovate, ½ length of tepals; pedicels spreading, 1–4 cm. **Flowers:** tepals brown, 1.6–2.5 mm, margins entire, clear; anthers 2 times filament length; stigmas 2 times style length. **Capsules** straw-colored, sometimes reddish, equal to or exceeding tepals; beak a mucro. **Seeds** dark brown to black, 1.4 mm; caruncle erect to slightly curved, ⅔ length of seed body. $2n = 52$ (Japan).

Flowering and fruiting summer. Dry to mesic open montane forests and forest margins to borders of bogs, marshes, and river bars; 0–1500 m; B.C., N.W.T., Yukon; Alaska; Asia.

2. Luzula acuminata Rafinesque, Autik. Bot. 3: 193. 1840 · Hairy wood rush [E]

Rhizomes similar to stolons. **Stolons** to 6 cm, scale leaves present. **Culms** loosely cespitose, 10–40 cm × 3–5 mm. **Leaves:** sheaths reddish, throat pilose; blade sparsely pubescent; basal leaves 32 cm × 12 mm; cauline leaves 2–4, 2–4 cm × 2–5 mm. **Inflorescences** cymose, rarely branching or often branching to 6 cm; proximal inflorescence bract ½–⅔ inflorescence length; bracts straw-colored, apex truncate, lacerate, glabrous or pubescent; bracteoles lance-ovate, ⅓–½ tepal length; pedicels single or paired, erect to spreading, 1–4 cm. **Flowers:** tepals pale to dark brown, 3–4.5 mm, margins entire, clear; anthers 2 times filament length; stigmas ± equaling style length. **Capsules** green to straw-colored, exceeding tepals; beak a mucro; valves widely spreading upon splitting. **Seeds** reddish to brown, 1–1.5 mm; caruncle curved, ± equaling length of seed body. $2n = 48$.

Varieties 2 (2 in the flora).

SELECTED REFERENCE   Ebinger, J. E. 1962. The varieties of *Luzula acuminata*. Rhodora 64: 74–83.

1. Inflorescences simple, with occasional branch from near base of flower. . . . .
. . . . .2a. *Luzula acuminata* var. *acuminata*
1. Inflorescences usually branching, with pedicels commonly paired. . . . .
. . . . .2b. *Luzula acuminata* var. *carolinae*

2a. Luzula acuminata Rafinesque var. **acuminata**
E F

Inflorescences simple, with occasional branch from near base of flower; secondary branches usually absent.

Flowering and fruiting early spring. Open woods, meadows, and hillsides; 0–500 m; St. Pierre and Miquelon; Alta., Man., N.B., Nfld. and Labr. (Nfld.), N.S., Ont., P.E.I., Que.; Ark., Conn., Ill., Ind., Iowa, Ky., Maine, Md., Mass., Mich., Minn., Mo., N.H., N.J., N.Y., N.C., Ohio, Pa., Vt., Va., W.Va., Wis.

2b. Luzula acuminata Rafinesque var. **carolinae** (S. Watson) Fernald, Rhodora 46: 5. 1944   E

Luzula carolinae S. Watson, Proc. Amer. Acad. Arts 14: 302. 1879

Inflorescences usually branching; pedicels commonly paired on primary branches.

Flowering and fruiting early spring. Mountain woods, clearings, and shaded stream banks to coastal plains; 0–500 m; Ala., Ark., Conn., Fla., Ga., Ky., La., Mass., N.C., Pa., S.C., Tenn., Va., W.Va.

2b. L U Z U L A de Candolle subg. **ANTHELAEA** (Grisebach) Buchenau, Bot. Jahrb. Syst. 12: 76, 86. 1890

Luzula sect. *Anthelaea* Grisebach, Spic. Fl. Rumel. 2: 404. 1846

Leaves: apex generally not callous. **Inflorescences** paniculate or dichasial. **Flowers** in pairs, rarely in clusters of 3–4 or solitary. **Seeds:** micropylar end with tuft of hairs; caruncle barely visible or absent.

Species ca. 50 (6 in the flora): Northern Hemisphere.

SELECTED REFERENCE   Häet-Ahti, L. 1971. A synopsis of the species of *Luzula*, subgenus *Anthelaea* Grisebach (Juncaceae) indigenous in North America. Ann. Bot. Fenn. 8: 368–381.

1. Tepals whitish to pinkish. . . . . .3a. *Luzula luzuloides* subsp. *luzuloides*
1. Tepals straw-colored to dark brown.
   2. Apex of tepals long-acuminate, reflexed; inflorescence branches widely spreading to 90°, stiff. . . . . .4. *Luzula divaricata*
   2. Apex of tepals acute, not reflexed; inflorescence branches spreading less than 90°, lax.
      3. Tepals 2.5–3.5 mm; capsule ± equaling tepal length, ovoid, distinct beak to 1 mm
      . . . . .5. *Luzula hitchcockii*
      3. Tepals 2.5 mm or shorter; capsule shorter than 2.5 mm, never ovoid, distinct beak absent.
         4. Culms commonly (20–)30–100 cm; margins of inflorescence bracts and bracteoles entire to lacerate. . . . . .8. *Luzula parviflora*

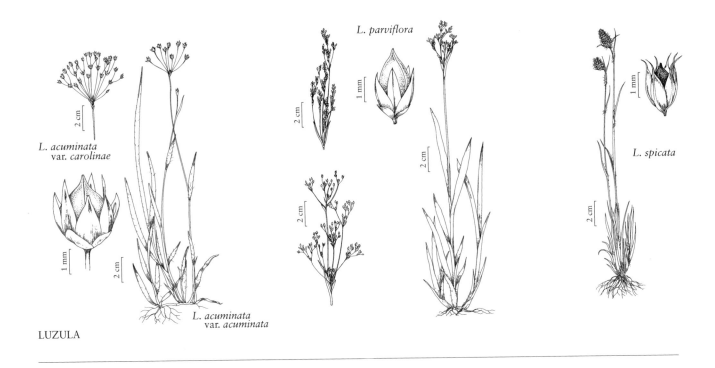

L. parviflora

L. acuminata
var. *carolinae*

L. acuminata
var. *acuminata*

L. spicata

LUZULA

4. Culms commonly less than 30 cm; margins of inflorescence bracts and bracteoles ciliate.
   5. Proximal inflorescence bract to 1 cm; stigmas 2 times style length; seeds dark reddish brown. . . . . 6. *Luzula wahlenbergii*
   5. Proximal inflorescence bract 0.8–1.5 cm; stigmas 5 times style length; seeds light yellow-brown. . . . . 7. *Luzula piperi*

3. **Luzula luzuloides** (Lamarck) Dandy & Wilmott, J. Bot. 76: 352. 1938 • Forest wood rush ⊡

Juncus luzuloides Lamarck in J. Lamarck et al., Encycl. 3: 272. 1789

Subspecies 2 (1 in the flora): introduced, North America; Europe.

3a. **Luzula luzuloides** (Lamarck) Dandy & Wilmott subsp. **luzuloides** ⊡

Rhizomes slender (1–1.5 mm wide). **Stolons** 1–1.5 mm wide. **Culms** cespitose, 45–70 cm. **Leaves:** sheath throat densely ciliate; basal leaf blade to 30 cm × 7 mm, densely ciliate; cauline leaf blade long-attenuate to thin apex, sparsely ciliate. **Inflorescences** anthelate; proximal inflorescence bract leaflike, ± equaling inflorescence length, apex attenuate; bracts brownish to clear, apex lacerate; bracteole margins entire to dentate to scarcely ciliate. **Flowers** in clusters of 2–8; tepals whitish to pinkish; outer whorl keeled, 1.7–2.1 mm; inner whorl 2.2–3 mm; anthers ca. 4 times filament length; stigmas shorter than styles. **Capsules** reddish, 1.5–1.8 mm; beak 0.4–0.6 mm. **Seeds** dark brown, shining, ellipsoid, 0.8–1.1 mm; caruncle 0.1–0.2 mm. 2*n* = 12 (Europe).

Flowering and fruiting late spring–summer. Open woods, fields, and lawns; 10–300 m; introduced; N.S., Ont., Que.; Conn., Maine, Mass., Minn., N.Y., Pa., Vt., Wis.; Europe.

4. **Luzula divaricata** S. Watson, Proc. Amer. Acad. 14: 302. 1879 ⊡

Rhizomes thick. **Culms** densely cespitose, reddish, 6–30 cm × 2 mm. **Leaves:** basal leaves numerous, blade to 20 cm × 4–6 mm, sometimes longer than stem, apex long-pointed to 12 mm, mostly glabrous; cauline leaves 2–3. **Inflorescences** 5–15 cm, width 1/2 to equaling length; branches widely spreading to 90°, stiff, not drooping; proximal bract inconspicuous, less than 2 cm; bracts and bracteoles clear, margins slightly lacerate, often with a few cilia. **Flowers** solitary; tepals pale brown with reddish tint, 1.8–2.4 mm, apex reflexed, long-acuminate; outer whorl slightly longer than inner whorl; anthers ± equaling filaments; stigmas 3 times style length. **Capsules** deep reddish brown, shorter to slightly longer than tepals. **Seeds** light brown, 1.2 mm.

Flowering and fruiting summer. Subalpine forest to alpine granitic slopes; 2100–3700 m; Calif., Idaho, Nev., Ore., Wash.

5. Luzula hitchcockii Hämet-Ahti, Ann. Bot. Fenn. 8: 368. 1971 • Hitchcock's wood rush [E]

Rhizomes short to long-running, thick. **Culms** cespitose, 15–50 cm. **Leaves:** basal blade flat, apex involute, mostly glabrous; cauline leaves 3–5, 2–6 cm × 3–10 mm. **Inflorescences** anthelate; branches spreading less than 90°, lax; proximal bracts much shorter than inflorescence; bracts brownish, clear toward apex, margins lacerate; bracteoles light brown, 1/2 length of tepals, margins dentate. **Flowers** solitary or in pairs; tepals reddish to dark brown, nearly equal, 2.5–3.5 mm, apex acute, not reflexed; anthers 3 times filament length; stigmas 2 times style length. **Capsules** blackish, ovoid, 2.5–3.5 mm; beak to 1 mm. **Seeds** dark brown, 1.2–1.8 mm; caruncle barely visible.

Flowering and fruiting summer. Montane forest to subalpine and alpine slopes, ridges, and meadows; 1700–2400 m; Alta., B.C.; Idaho, Mont., Oreg., Wash.

Although *Luzula hitchcockii* has been reported from California, no convincing specimens have been seen by this author. The species resembles the European *L. glabrata* (Hoppe) Desvaux very closely; however, it is readily distinguished from all other North American species. Plants may be heavily infested with *Ustilago vuijckii* Oudemans and Beijerinck, which considerably alters their usual appearance.

6. Luzula wahlenbergii Ruprecht, Fl. Samojed. Cisural., 58. 1845 • Wahlenberg's wood rush

Rhizomes short. **Culms** cespitose, 15–30(–35) cm. **Leaves:** sheath throat acutely angled, pilose; basal leaf blade 5–10 cm × 3–8 mm, apex short acuminate, mostly glabrous; cauline leaves (1–)2–4, 3–5 cm × 2–4 mm. **Inflorescences** anthelate, few flowered; main branches generally 2–4, spreading less than 90°, lax, usually arching; proximal inflorescence bract 1 cm or less; bracts and bracteoles brown; margins with long curly cilia. **Flowers** soli-

tary on short pedicels; tepals dark purplish brown, 2–2.5 mm, margins finely lacerate toward apex, apex acute, not reflexed; anthers ± equaling filament length; stigmas 2 times style length. **Capsules** dark brown, less than 2.5 mm, slightly longer than tepals; beak absent. **Seeds** dark reddish brown, cylindric, 1.2–1.6 mm; caruncle essentially absent. $2n = 24$.

Flowering and fruiting late summer. Wet mossy arctic and alpine tundra, lake shores, alluvial rivers, shores of alpine creeks, gneissic seashores; 40–600 m; Greenland; B.C., Man., N.W.T., Nunavut, Que., Yukon; Alaska; Eurasia.

7. Luzula piperi (Coville) M. E. Jones, Biol. Ser. Bull. State Univ. Montana 15: 22. 1910 • Piper's wood rush

Juncoides piperi Coville, Contr. U.S. Natl. Herb. 11: 185. 1906; *Luzula wahlenbergii* Ruprecht subsp. *piperi* (Coville) Hultén

Rhizomes horizontal, short. **Culms** densely cespitose, 10–30(–35) cm. **Leaves:** basal blade green, 5–10 cm × 2–4 mm, firm, essentially glabrous; cauline leaves 2–3, 3–7 cm × 3–5 mm. **Inflorescences:** branches spreading less than 90°, lax; proximal inflorescence bract leaflike, 0.8–1.5 cm; bracts and bracteoles brown, clear at apex, margins strongly ciliate. **Flowers** single or in clusters of 2–3; tepals dark brown, 1–2.5 mm, ± equal, apex acute, not reflexed; anthers ± equaling filament length; stigmas 5 times style length. **Capsules** dark brown, ellipsoid, shorter than 2.5 mm, longer than tepals; beak absent. **Seeds** light yellow-brown, lanceolate, narrowed at ends, 1.2 mm. $2n = 24$.

Flowering and fruiting summer. Snowbeds and mesic heaths in subalpine and oceanic zones; 400–2400 m; Alta., B.C., Yukon; Alaska, Idaho, Mont., Oregon, Wash.; e Asia.

SELECTED REFERENCE Hämet-Ahti, L. 1965. *Luzula piperi* (Cov.) M. E. Jones. An overlooked wood rush in western North America and eastern Asia. Aquilo, Ser. Bot. 3: 11–21.

8. Luzula parviflora (Ehrhart) Desvaux, J. Bot. (Desvaux) 1: 144. 1808 • Small-flowered wood rush [F]

Juncus parviflorus Ehrhart, Beitr. Naturk. 6: 139. 1791

Stolons to 5 cm or absent. **Culms** loosely cespitose, (20–)30–100 cm, base often reddish, often distinctly so at proximal internodes. **Leaves:** sheath throat with long, soft hairs; basal leaf blade 12–17 cm × 5–10 mm, mostly glabrous; cauline leaves 3–6, dull yellowish or bluish to gray-green to shiny, bright green, 7–9 cm × 3–5 mm, apex acute to acuminate. **Inflorescences** anthelate, few-to-many flowered, 4–20 × 4–12 cm; major branches spreading less than 90°, lax, often arching; proximal inflorescence bract inconspicuous to leaflike, to 5(–8) cm; bract margins entire to lacerate; bracteoles clear or brown, margins entire to lacerate. **Flowers** (1–)2–4, crowded or open; tepals pale brown to brown, broadly lanceolate, 1.8–2.5 mm, apex acute, not reflexed; anthers equaling to shorter than filaments; stigmas well exceeding style. **Capsules** straw-colored to dark brown to blackish, spheric, less than 2.5 mm, equal to gener-ally longer than tepals; beak absent. **Seeds** brown to brownish red or purple, ellipsoid, 1.1–1.5 mm. $2n = 24$.

Flowering and fruiting spring–late summer. Meadows in temperate to subalpine boreal forests, wet grasslands and tundra, willow copses, herb slopes; 0–3300 m.; Greenland; St. Pierre and Miquelon; Alta., B.C., Man., N.B., Nfld. and Labr., N.W.T., N.S., Nunavut, Ont., Que., Sask., Yukon; Alaska, Ariz., Calif., Colo., Idaho, Maine, Mass., Mich., Minn., Mont., Nev., N.H., N.Mex., N.Y., Oreg., S. Dak., Utah, Vt., Wash., Wis., Wyo.; Eurasia.

2c. LUZULA de Candolle subg. LUZULA

Luzula sect. *Gymnodes* Grisebach; *Luzula* subg. *Gymnodes* (Grisebach) Buchenau.

Leaves: apex generally callous. **Inflorescences** spikelike or umbellate. **Flowers** in glomerules. **Seeds** usually without tuft of hairs; caruncle conspicuous to barely visible.

Species ca. 50 (15 species in the flora): temperate and arctic regions and tropical mountains.

SELECTED REFERENCES Böcher, T. W. 1950. Contributions to the flora and plant geography of west Greenland II. Meddel. Grønland 147(7): 11–23. Kirschner, J. 1993. Taxonomic survey of *Luzula* sect. *Luzula* (Juncaceae) in Europe. Folia Geobot. Phytotax. 28: 141–182. Nordenskiöld, H. 1956. Cyto-taxonomical studies in the genus *Luzula*. II. Hybridization experiments in the *campestris–multiflora* complex. Hereditas (Lund) 42: 7–73. Novikov, V. S. 1990. Konspekt sistemy roda *Luzula* DC. (Juncaceae). (Survey of the system of the genus *Luzula*.) Byull. Moskovsk. Obshch. Isp. Prir., Otd. Biol. 95(6): 63–70.

1. Inflorescences dense, spikelike, though often interrupted, nodding. 9. *Luzula spicata*
1. Inflorescences of distinct glomerules, not spikelike, sometimes lax but not nodding.
  2. Inflorescences of sessile glomerules only or with smaller glomerules on short branches.
    3. Proximal inflorescence bract inconspicuous.
      4. Tepals shorter than capsule; bracteoles deep brown, dentate. . . . . 14. *Luzula arctica*
      4. Tepals and capsule about equal; bracteoles clear to deep brown, conspicuously fimbriate. . . . . 15. *Luzula confusa* (in part)
    3. Proximal inflorescence bract conspicuous, usually well exceeding inflorescence.
      5. Seeds 1.3 mm, caruncle absent. . . . . 10. *Luzula subcapitata*
      5. Seeds 0.8–1.5 mm, with caruncle.
        6. Seeds 0.8 mm, caruncle 0.2 mm. . . . . 11. *Luzula orestera*
        6. Seeds 0.9–1.5 mm, caruncle either barely visible or at least 0.3 mm.
          7. Capsules ovoid; caruncle barely visible. . . . . 12. *Luzula groenlandica*
          7. Capsules spheric; caruncle 0.3-0.8 mm. . . . . 13. *Luzula comosa* (in part)
  2. Inflorescences with central glomerule(s) sessile or nearly sessile, always with other pedunculate glomerules.
    8. Inflorescence branches straight.
      9. Inflorescence branches mostly erect; glomerules mostly cylindric.
        10. Tepals pale or with brown central area at maturity; seeds 0.7–1 mm; caruncle 0.2–0.3 mm. . . . . 16. *Luzula pallidula*
        10. Tepals pale brown to chestnut brown to blackish at maturity; seeds 0.9–1.7 mm; caruncle 0.2–0.7 mm.
          11. Seeds 1.1–1.7 mm; caruncle 0.2–0.6 mm. . . . . 17. *Luzula multiflora*
          11. Seeds 0.9–1.3 mm; caruncle 0.5–0.7 mm. . . . . 18. *Luzula bulbosa*
      9. Inflorescence branches divergent as much as 90°; glomerules not cylindric.
        12. Leaves without callous apex, glomerules on elongate peduncles. . . . . 20. *Luzula echinata*
        12. Leaves with callous apex; glomerules sessile to congested to spreading.

13. Stolons slender and short, conspicuous rhizomes present. . . . . 19. *Luzula campestris*
13. Stolons or conspicuous rhizomes absent. . . . . 13. *Luzula comosa* (in part)
8. Inflorescence branches arching.
  14. Inflorescence branches arching in all directions; basal leaf sheaths bluish green
     . . . . 23. *Luzula subcongesta*
  14. Inflorescence branches arching in 1 direction; basal leaf sheaths variously colored.
    15. Basal leaf sheaths dull brown-gray to green; cauline leaves as large or larger than basal leaves. . . . . 22. *Luzula kjellmaniana*
    15. Basal leaf sheaths shining reddish to brown to purple; cauline leaves usually smaller than basal leaves.
      16. Basal and cauline leaves not reaching inflorescence. . . . .
         . . . . 21a. *Luzula arcuata* subsp. *unalaschkensis*
      16. Basal and especially cauline leaves often reaching or exceeding inflorescence. . . . . 15. *Luzula confusa* (in part)

9. Luzula spicata (Linnaeus) de Candolle in J. Lamarck and A. P. de Candolle, Fl. Franç. ed. 3, 1: 161. 1805 • Spiked wood rush [F]

Juncus spicatus Linnaeus, Sp. Pl. 1: 330. 1753

Culms densely cespitose, reddish, 3–33 cm, base thick, extending 1–8 cm into soil. **Leaves:** sheath throats densely hairy; basal leaves erect, channeled, linear, 2–15 cm × 1–4 mm, apex not callous; cauline leaves 2–3. **Inflorescences** panicles of dense, nodding, spikelike clusters (each 1–25 mm), often interrupted by 10–70 mm; proximal inflorescence bract conspicuous, generally exceeding inflorescence; bracts clear; bracteoles clear, margins ciliate, apex narrow, extended. **Flowers:** tepals brown with clear margins or very pale throughout (outer whorl bristle-pointed), 2–2.5 mm; outer whorl longer than inner whorl; anthers ± equaling filaments. **Capsules** pale to dark brown or blackish, round, generally shorter than tepals, apex ± acute. **Seeds** brown, cylindric-ovoid, body 1–1.2 mm; caruncle 0.2 mm. $2n = 24$.

Flowering and fruiting summer. Alpine slopes and heaths, dry or damp situations among grasses, herbs, or lichens, and in subalpine forests; 0–3700 m; circumpolar; Greenland; St. Pierre and Miquelon; Alta., B.C., Man., Nfld. and Labr., N.W.T., Nunavut, Que., Yukon; Alaska, Calif., Colo., Idaho, Maine, Mont., Nev., N.H., N.Mex., N.Y., Oreg., Utah, Vt., Wash., Wyo.; Eurasia.

10. Luzula subcapitata (Rydberg) H. D. Harrington, Man. Pl. Colorado, 641. 1954 [E] [F]

Juncoides subcapitatum Rydberg, Bull. Torrey Bot. Club 31: 401. 1904

Rhizomes short, stocky. **Culms** cespitose, 8–40 cm × 1 mm, base thickened, glabrous. **Leaves:** basal leaves to 13 cm; cauline leaves 1–3, mostly less than 5 cm × 5–10 mm, apex involute, glabrous. **Inflorescences** compact, irregular in shape; glomerules 6–10, sessile; branches few or none; proximal inflorescence bract conspicuous, leaflike, lanceolate, equal to or exceeding inflorescences; bracteoles clear, ½ tepal length. **Flowers:** tepals shining brown, with thin clear margins, 1.5–2 mm; outer and inner whorls nearly equal, outer whorl slightly keeled; anthers longer than filaments; stigmas much longer than styles. **Capsules** deep purplish brown at maturity, globose, ± equaling tepals, apex rounded. **Seeds** brown, cylindric, 1.3 mm; caruncle absent.

Flowering and fruiting summer. Subalpine and alpine bogs; 3200–3700 m; Colo.

11. Luzula orestera Sharsmith, Aliso 4: 125. 1958 [E]

Luzula campestris (Linnaeus) de Candolle var. *sudetica* Linnaeus, not Celakovsky

Culms densely cespitose, stiffly erect, reddish brown, 3–26 cm. **Leaves:** basal leaves reddish, 2.5–7 cm × 2–5 mm, firm, apex callous, glabrous. **Inflorescences** pyramidal, 5–10 mm wide; glomerules 1–5, sessile, sometimes with 1–2 smaller glomerules on short peduncles; proximal inflorescence bract conspicuous, reddish, usually exceeding inflorescence, gen-

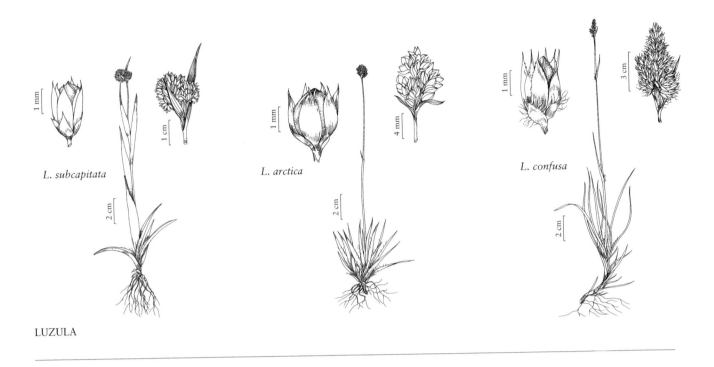

L. subcapitata

L. arctica

L. confusa

LUZULA

erally stiff; bracteoles clear. **Flowers:** tepals very dark with clear margins, 2–3 mm; outer whorl slightly exceeding inner whorl; anthers ± equaling filaments. **Capsules** dark brown to black in distal portion, much shorter than tepals; beak 0.3 mm. **Seeds** dark brown, oval, 0.8 mm; caruncle about 0.2 mm. $2n = 20, 22$.

Flowering and fruiting summer. Alpine and subalpine meadows, fell-fields; 2700–3600 m; Calif.

12. Luzula groenlandica Böcher, Meddel. Grønland 147: 18. 1950 · Greenland wood rush E

Culms cespitose, straight, 10–30 cm, stiff. **Leaves:** basal leaves with several outer dead ones usually present, to 9 cm × 3 mm, apex callous, sparingly pilose; cauline leaves to 5 cm × 2 mm. **Inflorescences:** glomerules 1(–3); peduncles not exceeding 5 mm; proximal inflorescence bract conspicuous, leaflike, usually conspicuously exceeding inflorescence; bracts mostly inconspicuous; bracteoles clear or light to dark brown, occasionally to ½ tepal length. **Flowers:** tepals with broad clear margins and apex, 1.9–2.5 mm, apex acute-acuminate; outer whorl ± equaling to exceeding inner whorl; anthers ± equaling filaments. **Capsules** dark reddish, shining, ovoid, generally shorter than tepals. **Seeds** translucent, brown, ellipsoid, 0.9–1.1 mm; caruncle barely visible. $2n = 24$.

Flowering and fruiting summer. Sandy sea shores with herbaceous vegetation to turfy tundra, often by water, to alpine flats; 0–400 m; Greenland; B.C., Man., Nfld. and Labr. (Labr.), N.W.T., Nunavut, Que., Yukon; Alaska.

13. Luzula comosa E. Meyer, Syn. Luzul., 21. 1823 E

Luzula campestris var. *columbiana* H. St. John; *L. campestris* (Linnaeus) de Candolle var. *comosa* (E. Meyer) Fernald & Wiegand; *L. campestris* var. *macrantha* (S. Watson) Fernald & Wiegand; *L. comosa* var. *congesta* (Thuillier) S. Watson; *L. comosa* var. *laxa* Buchenau; *L. comosa* var. *macrantha* S. Watson; *L. comosa* var. *subsessilis* S. Watson; *L. intermedia* (Thuillier) A. Nelson; *L. multiflora* (Erhard) Lejeune subsp. *comosa* (E. Meyer) Hultén; *L. multiflora* subsp. *congesta* (Thuillier) Hylander; *L. subsessilis* Buchenau

Stolons absent. **Culms** cespitose, 10–40 cm. **Leaves:** sheath throats with long, soft, wavy hairs; basal leaves 5–15 cm × 3–7 mm, margins with long, soft, wavy hairs, apex callous. **Inflorescences** umbellate or sessile glomerules, 5–15 × 5–7 mm; glomerules 1–6, spheric to nearly cylindric; branches straight, divergent by as much as 90°, 0.5–5 cm; proximal inflorescence bract conspicuous, shorter than to much longer than inflorescence; bracts clear, margins nearly entire to lacerate;

bracteoles clear, margins ciliate, especially in distal half. **Flowers:** tepals dark brown to pale with clear margins, 2–5 mm; outer whorl often longer than inner whorl; anthers much exceeding to shorter than filaments. **Capsules** greenish to dark brown, spheric, equal to or much shorter than tepals. **Seeds** red-brown to brown, cylindric, 1–1.5 mm; caruncle 0.3–0.8 mm. $2n = 12, 24$.

Flowering and fruiting spring–summer. Meadows, open woods, coniferous forest; 50–3200 m; Alta., B.C.; Alaska, Calif., Colo., Idaho, Mont., Nebr., N.Mex., Oreg., S.Dak., Utah, Wash., Wyo.

The highly variable nature of this species is indicated by the number of names applied to it.

14. **Luzula arctica** Blytt in M. N. Blytt and A. G. Blytt, Norges Fl. 1: 299. 1861 • Arctic wood rush [F]

Culms densely cespitose, 5–20 cm. **Leaves:** sheaths brown to straw-colored; basal leaves to 10 cm × 4 mm; cauline leaves usually 2, reduced. **Inflorescences:** glomerules 1–3, sessile; proximal inflorescence bract inconspicuous, brown, much shorter to ± equaling inflorescence, apex often clear, dentate; bracts deep brown, margins dentate; bracteoles deep brown, margins dentate. **Flowers:** tepals deep brown with narrow clear margins and apex, 1.7–2.1 mm; anthers ± equaling filament length. **Capsules** dark reddish to blackish, shining, spheric, 1.8–2.1 mm, usually exceeding tepals. **Seeds** translucent, clear brown, broadly elliptic, with few entangled hairs, 1–1.2 mm. $2n = 24$.

Flowering and fruiting summer. Wet, stony places on slopes and in dwarf shrub heaths in alpine and arctic tundra; circumpolar, 0–1200 m; Greenland; B.C., Man., Nfld. and Labr. (Labr.), N.W.T., Nunavut, Que., Yukon; Alaska; Eurasia.

15. **Luzula confusa** Lindeberg, Bot. Not. 1855: 9. 1856 • Northern wood rush [F]

Rhizomes 0.5–1 cm. **Culms** cespitose, stiffly erect, 3–28 cm. **Leaves:** sheaths reddish; basal leaves persisting for several seasons, reddish, exceeding inflorescence to as short as cauline leaves, glabrous; cauline leaves 2–3, reddish, rarely to 6 cm, often reaching or exceeding inflorescence, glabrous. **Inflorescences:** main branches 1–3(–4), arching in 1 direction, 1–4 cm, terminated by 2–several

peduncles each bearing 1–5 flowers; peduncles 2 mm; proximal inflorescence bract usually inconspicuous, though variable in size and shape; bracts and bracteoles conspicuous, clear to deep brown, margins lacerate to fimbriate to abundantly ciliate. **Flowers:** tepals dark brown with clear apex, 1.6–2.6 mm; outer whorl longer than inner whorl; anthers ca. 2 times filament length; stigmas much exceeding styles. **Capsules** dark brown, ovoid, ± equaling to shorter than tepals. **Seeds** dark brown, with tuft of fine tangled hairs, ellipsoid, 0.8–1.2 mm. $2n = 36$.

Flowering and fruiting late summer. Wet, exposed rocky and sandy hillsides, lichen tundra, and mountain summits; growing where it may be practically only vascular plant; circumpolar; 500–1300 m; Greenland; B.C., Man., Nfld. and Labr. (Labr.), N.W.T., Nunavut, Ont., Que., Sask., Yukon; Alaska, Maine, N.H., Vt.; Eurasia.

16. **Luzula pallidula** Kirschner, Taxon 39: 110. 1990 [I]

Juncus pallescens Wahlenberg; *Luzula pallescens* (Wahlenberg) Besser

Rhizomes thickened. **Culms** cespitose, 9–35 cm. **Leaves:** basal leaves 6–11.5 cm × 1.5–4 mm, apex not callous, sparingly ciliate. **Inflorescences** umbellate-paniculate; glomerules 4–30 (each with 9–24 flowers), central glomerules sessile or nearly sessile, cylindric, 6–10 × 4 mm; branches straight, erect, to 3 cm; proximal inflorescence bract conspicuous, leaflike, equal to much longer than inflorescence; bracts clear, sometimes variegated with purple; bracteole margins dentate to lacerate. **Flowers:** tepals clear to straw-colored throughout or centers brown with clear margins and apex, 1.5–2.6 mm; outer whorl exceeding inner whorl, outer whorl apex awned; anthers equaling to 1.5 times filament length. **Capsules** light or dark reddish, shining, spheric, usually equaling inner tepal whorl. **Seeds** translucent brown, ellipsoid, 0.7–1 mm; caruncle 0.2–0.3 mm. $2n = 12$.

Flowering and fruiting early–late summer. Moist to wet woods, grassy places, and clearings on rocky places and barrens; 0–1000 m; introduced; N.B., Nfld. and Labr. (Nfld.), Que.; N.Y., Vt.; Eurasia.

For discussion of the change of the widely known name for this species, see J. Kirschner (1990).

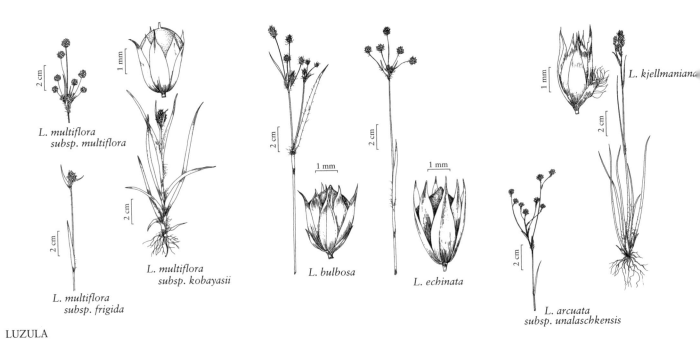

L. multiflora
subsp. multiflora

L. multiflora
subsp. frigida

L. multiflora
subsp. kobayasii

L. bulbosa

L. echinata

L. kjellmaniana

L. arcuata
subsp. unalaschkensis

LUZULA

17. Luzula multiflora (Ehrhart) Lejeune, Fl. Spa 1: 169. 1811

Juncus multiflorus Ehrhart, Calam. Gram. Tripet. Linn., no. 127. 1793, not Retzius 1795; *J. campestris* (Linnaeus) de Candolle var. *multiflorus* Ehrhart

Culms densely to loosely cespitose, 10–40 cm. **Leaves:** basal leaves 3.5–12 cm × 2–6 mm; cauline leaves equaling or exceeding inflorescences. **Inflorescences:** glomerules 3–16 (each with 8–16 flowers), 1–2 nearly sessile with others on evident peduncles, mostly cylindric; secondary branches sometimes present, usually straight, erect; proximal inflorescence bract barely as long as to exceeding inflorescence. **Flowers:** tepals pale brown to chestnut brown to blackish with clear margins, 2–4 mm; outer and inner whorl ± equal, or outer whorl slightly longer (outer whorl pointed, inner whorl pointed or truncate-mucronate); anthers not more than 2 times filament length; stigmas 0.8–1.5 mm; styles not persistent in fruit. **Capsules** pale to brown to black, globose, shorter than to ± equaling tepals. **Seeds** 1.1–1.7 mm; caruncles 0.2–0.6 mm.

Subspecies 6 (3 in the flora): North America; Eurasia.

Many names have been applied to members of this variable complex throughout the range of this flora. No monograph is yet available to enable a satisfactory treatment.

SELECTED REFERENCES Jarolimova, V. and J. Kirschner. 1995. Tetraploids in *Luzula multiflora* (Juncaceae) in Ireland. Folia Geobot. Phytotax. 30: 389–396. Kirschner, J. 1990. *Luzula multiflora* and allied species (Juncaceae): A nomenclatural study. Taxon 39: 106–114.

1. Tepals of outer and inner whorls not similar, outer whorl pointed, inner whorl truncate-mucronate; dark brown to chestnut to blackish . . . . 17c. *Luzula multiflora* subsp. *frigida*
1. Tepals of outer and inner whorls similar, pointed, straw-colored to chestnut.
  2. Tepals straw-colored to brown; cauline leaves not or only barely overlapping; proximal inflorescence bract equal to or barely exceeding inflorescence. . . . . 17a. *Luzula multiflora* subsp. *multiflora*
  2. Tepals chestnut; cauline leaves greatly overlapping; proximal inflorescence bract well exceeding inflorescence. . . . . 17b. *Luzula multiflora* subsp. *kobayasii*

17a. Luzula multiflora (Ehrhart) Lejeune subsp. **multiflora** [F] [I]

Luzula multiflora var. *acadiensis* Fernald

Cauline leaves 2–3, not or only barely overlapping; proximal inflorescence bract equal to barely exceeding inflorescence. **Tepals:** outer and inner whorl straw-colored to brown, 2.8–3.6 mm, apex pointed. **Capsules** pale to brown, shorter than tepals. **Seeds** 1.1–1.5 mm; caruncle 0.3–0.6 mm. $2n = 24$.

Flowering and fruiting spring–summer. Sparsely scattered in fields, meadows, open woods, ditches and clearings; 50–800 m; introduced; St. Pierre and Miquelon; Alta., B.C., Man., N.B., Nfld. and Labr. (Nfld.), N.S., Ont., P.E.I., Que., Sask.; Conn., Del., D.C., Ga., Idaho, Ill., Ind., Iowa, Ky., Maine, Md., Mass., Mich., Minn., Mo., Mont., N.H., N.J., N.Y., N.C., Ohio, Oreg., Pa., R.I., S.Dak., Tenn., Vt., Va., Wash., W.Va., Wis., Wyo.; Eurasia.

17b. Luzula multiflora (Ehrhart) Lejeune subsp. **kobayasii** (Satake) Hultén, Ark. Bot., n. s. 7: 32. 1968 [F]

Luzula kobayasii Satake, Bot. Mag. (Tokyo) 46: 544. 1932

Cauline leaves greatly overlapping; proximal inflorescence bract nearly as wide as leaves, well exceeding inflorescence. **Tepals:** outer and inner whorls chestnut, ca. 4 mm, apex pointed. **Capsules** chestnut brown, shorter than tepals. **Seeds** 1.5 mm; caruncle 0.2–0.3 mm.

Flowering and fruiting in summer. Grassy places; 0–500 m; Alaska; e Asia.

17c. Luzula multiflora (Ehrhart) Lejeune subsp. **frigida** (Buchenau) V. I. Krecztowicz, Bot. Zhurn. (Moscow & Leningrad) 12: 490. 1928 [F]

Luzula campestris (Linnaeus) de Candolle var. *frigida* Buchenau, Oesterr. Bot. Z. 48: 284. 1898; *L. frigida* (Buchenau) Samuelsson; *L. multiflora* (Erhard) Lejeune var. *fusconigra* Celakovsky; *L. sudetica* (Willdenow) de Candolle var. *frigida* (Buchenau) Fernald

Cauline leaves 2–3, not overlapping; proximal inflorescence bract not conspicuously exceeding inflorescence. **Tepals:** outer and inner whorl dark brown to chestnut to blackish, outer whorl pointed, inner whorl truncate-mucronate. **Capsules** dark chestnut to blackish, equal to slightly exceeding tepals. **Seeds** 1.1–1.4 mm; caruncle 0.2–0.3 mm. $2n = 36$.

Flowering and fruiting summer. Peaty barrens, slopes, fields, heaths, grasslands, and snowbeds; 50–1800 m; Greenland; B.C., Man., Nfld. and Labr. (Labr.), N.W.T., N.S., Nunavut, Ont., Que., Yukon; Alaska, Maine, Mass., N.H., Vt.; Eurasia.

18. Luzula bulbosa (A. Wood) B. B. Smyth & L. C. R. Smyth, Trans. Kansas Acad. Sci. 25: 107. 1913 [E] [F]

Luzula campestris (Linnaeus) de Candolle var. *bulbosa* A. Wood, Class-book Bot. ed. s.n.(k), 723. 1868

Rhizomes short, slender, bearing few to several white, swollen (storage) leaf bases. **Culms** weakly cespitose or solitary, 8–40 cm. **Leaves:** basal leaves few; cauline leaves 2–3, to 17 cm × 7 mm, margins scarcely to densely ciliate. **Inflorescences:** glomerules 3–20 (each with 6–20 flowers), central glomerules sessile or nearly sessile, cylindric, 5–12 × 5–7 mm; peduncles straight, erect, 0.5–7 cm; proximal inflorescence bract leaflike, shorter than inflorescence, margins pilose; bracts purplish with long clear apex, sheathing, margins sparsely to densely ciliate; bracteoles white-clear, shining, margins fimbriate. **Flowers:** tepals with shining chestnut centers and usually wide clear margins and apex, 2–3 mm; outer whorl usually exceeding inner whorl, at least by awned tip; anthers 1–2 times filament length; stigmas 3–4 times length of styles. **Capsules** brown, shining, obovoid, equal or longer than tepals, apex truncate. **Seeds** dark brown, ellipsoid, 0.9–1.3 mm; caruncle 0.5–0.7 mm. $2n = 12$.

Flowering and fruiting spring–early summer. Dry situations in woods and fields 50–600 m; Ala., Ark., Conn., Del., D.C., Fla., Ga., Ill., Ind., Kan., Ky., La., Md., Mass., Miss., Mo., N.J., N.Y., N.C., Ohio, Okla., Pa., S.C., Tex., Tenn., Va., W.Va.

19. **Luzula campestris** (Linnaeus) de Candolle in J. Lamarck and A. P. de Candolle, Fl. Franç. ed. 3, 3: 161. 1805   I   W

Juncus campestris Linnaeus, Sp. Pl. 1: 329. 1753

Rhizomes conspicuous. **Stolons** short, slender. **Culms** not cespitose, decumbent, 10–20 cm. **Leaves:** basal leaves few, 2.5–15 cm × 4 mm, apex callous, pilose. **Inflorescences** racemose; glomerules 2–6, central glomerules sessile or all congested, not cylindric; peduncles straight, divergent as much as 90°, to 3 cm; proximal inflorescence bract dark, often purplish, leaflike. **Flowers:** tepals dark reddish, shining, with wide clear margins and apex, apex acuminate, midrib extending as awned tip, 3–3.5 mm; outer and inner whorls equal; anthers ca. 2–6 times filament length; stigmas ± equal to style. **Capsules** brown, shining, usually lighter than tepals, conspicuously shorter than to nearly equal to tepals; beak obvious. **Seeds** reddish, globose, 1–1.3 mm; caruncle to ½ seed length. $2n = 12$.

Flowering and fruiting summer. Sunny clearings; 500–900 m; introduced; Nfld. and Labr. (Nfld.).

Luzula campestris may occur rarely elsewhere in Canada and the United States in lawns and cleared places (collected in Massachusetts in the 1920s). A common European species, the name is used in our floras for almost every species of the "*multiflora–campestris*" complex.

20. **Luzula echinata** (Small) Hermann, Rhodora 40: 84. 1938   E   F

Juncoides echinatum Small, Torreya 1: 74. 1901; *Luzula echinata* var. *mesochorea* Hermann

Rhizomes to 1.5 cm or shorter, knotty. **Culms** loosely cespitose, 15–45 cm, base sometimes swollen. **Leaves:** basal leaves 4–15 cm × 2–7 mm, margins pilose to sparsely hairy, apex not callous. **Inflorescences:** glomerules 4–15, central glomerules sessile or nearly sessile, broadly conic or globose (often loosely flowered); peduncles straight, divergent by as much as 90°, to 9 cm; proximal inflorescence bract leaflike, not longer than inflorescence; bracts and bracteoles clear, variously cut. **Flowers:** tepals greenish to pale or dark brown, usually with clear margins and apex, 2.8–4 mm; anthers ca. 2–5 times filament length; stigmas 2–3 times length of styles. **Capsules** pale to dark brown, obovoid to subglobose, usually shorter than tepals.

Seeds dark brown, globose, 1.2–1.6 mm; caruncle 0.5–0.6 mm. $2n = 12$.

Flowering and fruiting spring–early summer. Bluffs, wooded slopes, alluvial woods, streamsides, under hardwoods and occasionally in clearings; 50–800 m; Ala., Ark., Conn., Del., D.C., Fla., Ga., Ill., Ind., Iowa., Ky., La., Md., Mass., Miss., Mo., N.J., N.Y., N.C., Ohio, Okla., Pa., S.C., Tenn., Tex., Va., W.Va.

21. **Luzula arcuata** (Wahlenberg) Swartz, Summa Veg. Scand., 13. 1814 • Curved wood rush

Juncus arcuatus Wahlenberg, Fl. Lapp., 87. 1812

Subspecies 2 (1 in the flora): North America; Eurasia.

21a. **Luzula arcuata** (Wahlenberg) Swartz subsp. **unalaschkensis** (Buchenau) Hultén, Ark. Bot., n. s. 7: 32. 1968   F

Luzula arcuata var. *unalaschkensis* Buchenau, Bot. Jahrb. Syst. 12: 124. 1890; *L. arcuata* var. *kamtschadalorum* Samuelsson; *L. beringensis* Tolmachev; *L. kamtschadolarum* (Samuelsson) Gorodkov; *L. unalaschkensis* (Buchenau) Satake; *L. unalaschkensis* subsp. *kamtschdalorum* (Samuelsson) Tolmachev

Stolons short to (rarely) 15 cm. **Culms** loosely cespitose, usually slightly arched, 15–30 cm, thin. **Leaves:** sheaths brownish purple, throats rounded, densely pilose; basal leaves flat, 8–15 cm × 3–5 mm, not reaching inflorescences, margins pubescent; cauline leaves 1–3, smaller than basal, not reaching inflorescences. **Inflorescences** corymbose, proximal branch longest; glomerules 8–15 (each with 3–5 flowers, sometimes reduced to 1–2 flowers), pedunculate, capitate or spicate; branches mostly arching in same direction (often branching at apex); proximal inflorescence bract bladeless; bracts and bracteoles brown, ciliate. **Flowers:** tepals light brown, margins ± ciliate, apex acuminate, usually distinctly awned, 1.8–3 mm; anthers ± equaling filament length. **Capsules** brown, ellipsoid, shorter than tepals; beak short. **Seeds** brown, 1 mm, few hairs present; caruncle generally indistinct. $2n = 36$.

Flowering and fruiting summer. Rocky to gravelly snow patches on mountains, dry to mesic meadows, and patches of bare soil in heaths; 50–2500 m; Alta., B.C., N.W.T., Yukon; Alaska, Mont., Wash.; e Asia.

22. **Luzula kjellmaniana** Miyabe & Kudô, Trans. Sapporo Nat. Hist. Soc. 5: 38. 1913 [F]

Luzula beringiana Gjaerevoll; *L. multiflora* (Erhard) de Candolle subsp. *kjellmanniana* (Miyabe & Kudô) Tolmachev; *L. nivalis* (Laestadius) Beurling var. *latifolia* (Kjellmann) Samuelsson; *L. tundricola* Gorodkov

Culms loosely cespitose, light green, 15–25 cm. **Leaves:** sheaths dull brown-gray at base, becoming green distally, throats hairless; basal leaves flat toward base, channeled toward apex, 4–16 cm × 3–8 mm, apex abruptly pointed, callous, glabrous. **Inflorescences** asymmetric in appearance; glomerules (1–)2–5, central glomerules sessile or nearly sessile, 3–9 × 2.5–6 mm; capillary branches 2–8, arched in 1 direction, to 3.5 cm; proximal inflorescence bract bladeless. **Flowers:** tepals brownish translucent, 2–3 mm, apex lacerate to entire; anthers equal to or exceeding filaments. **Capsules** light chestnut brown, broadly 3-angled, shorter than tepals. **Seeds:** most failing to mature, brown, 3-angled, 0.9–1.1 mm, few hairs present; caruncle barely visible. $2n = 36$.

Flowering and fruiting summer. Tundra; 50–2000 m; B.C., N.W.T., Yukon; Alaska; ne Asia.

23. **Luzula subcongesta** (S. Watson) Jepson, Fl. Calif. 1: 258, fig. 45. 1921 [E]

Luzula spadicea (Allioni) de Candolle var. *subcongesta* S. Watson, Bot. California 2: 202. 1880

Rhizomes horizontal, stout. **Culms** 15–48 cm. **Leaves:** basal leaf sheaths bluish green; basal leaves bluish green, abruptly tapered, less than 19 cm × 3–5 mm, apex long. **Inflorescences** with glomerules crowded at ends of uneven peduncles; glomerules each with 3–12 flowers, central glomerules sessile or nearly sessile; peduncles arching in all directions; proximal inflorescence bract inconspicuous, 2–3 cm; bracteoles clear, margins ciliate. **Flowers:** tepals pale to dark brown, 1.5–2 mm, apex acute; anthers ± equaling filament length. **Capsules** pale to dark brown-purple, shorter than tepals; beak short. **Seeds** brown, oblong-ovoid, 1.1–1.2 mm; caruncle absent. $2n = 24$.

Flowering and fruiting summer. Alpine to subalpine moist or wet places; 2000–3500 m; Calif., Oreg.

210. CYPERACEAE Jussieu

• Sedge Family

[To be treated in volume 23]

211. POACEAE Barnhart

• Grass Family

[To be treated in volumes 24 and 25]

212. SPARGANIACEAE Rudolphi

• Bur-reed Family

Robert B. Kaul

Herbs, perennial, aquatic or paludal, emergent or leaves and inflorescence floating, rhizomatous, caulescent in flower, glabrous. **Leaves** basal and cauline, reduced to persistent foliaceous bracts on flowering stem, 2-ranked, erect or floating; sheaths open, sometimes inflated, especially on bracts, margins clear to scarious, not auriculate; blades flat to keeled, linear, apex obtuse to rounded or retuse, aerenchyma present. **Inflorescences** 1, terminal, erect and emergent or peduncle floating, usually not exceeding basal leaves; heads globose, sessile or peduncled, congested to remote along an often zigzag simple or branched rachis; staminate heads deciduous but rachis often persistent, distal to pistillate heads, sessile, proximal ones often subtended by foliaceous bract; pistillate heads sessile or peduncled, axillary or supra-axillary (peduncle partly adnate to rachis) to bract, or not bracteate. **Flowers** unisexual, staminate and pistillate on same plants, numerous, crowded in unisexual (occasionally bisexual) heads, sessile to stipitate, often subtended by tepaloid bract, wind-pollinated; tepals mostly 3–6, free, translucent to opaque, white to green when fresh, club-shaped to spatulate, entire to notched or erose-tipped, with or without subapical thickened dark area. **Staminate flowers** sessile, white or whitish; stamens 2–8, much exceeding tepals, distinct or filaments basally connate; anthers basifixed, connective widened. **Pistillate flowers** hypogynous, sessile to stipitate; pistil 1, exceeding tepals, syncarpous, 1- or 2(–3)-carpellate (functional), (0–)1–2-carpellate (abortive); ovary 1–2(–3)-locular; ovule 1 per locule, pendent; style 1, usually elongating in fruit, obsolete in 1 species; stigmas 1–2, white to greenish, linear to ovate or subcapitate; agamous flowers absent. **Fruits** achenelike drupes (often called achenes), obpyramidal with sides faceted and top truncate, to fusiform or ellipsoid with sides weakly or not faceted but tapering to beak, sometimes constricted at or near equator; dried exocarp persistent and spongy; endocarp hard; tepals persistent, attached at base except partly adnate to stipe in 1 species. **Seeds:** endosperm abundant, mealy; embryo linear, straight.

Genera 1, species 14 (9 species in the flora): mostly in north temperate zone, some circumboreal, a few south to Mexico, New Zealand, and Australia.

The flower heads are of two or more highly condensed racemes (D. Müller-Doblies 1970;

U. Müller-Doblies 1969). The staminate flowers are extremely crowded and difficult to differentiate. The drupes are crowded in globose head made burlike by the persistent styles (beaks) and stigmas.

All chromosome accounts published to 1985 are given by C. D. K. Cook and M. S. Nicholls (1986).

Sparganium is sometimes included in Typhaceae.

SELECTED REFERENCES Beal, E. O. 1960. *Sparganium* (Sparganiaceae) in the southeastern United States. Brittonia 12: 176–181. Cook, C. D. K. and M. S. Nicholls. 1986. A monographic study of the genus *Sparganium* (Sparganiaceae). Part 1. Subgenus *Xanthosparganium* Holmberg. Bot Helv. 96: 213–267. Cook, C. D. K. and M. S. Nicholls. 1987. A monographic study of the genus *Sparganium* (Sparganiaceae). Part 2. Subgenus *Sparganium*. Bot. Helv. 97: 1–44. Crow, G. E. and C. B. Hellquist. 1981. Aquatic vascular plants of New England. Pt. 2. Typhaceae and Sparganiaceae. New Hampshire Agric. Exp. Sta. Bull. 517: 1–21. Fernald, M. L. 1922b. Notes on *Sparganium*. Rhodora 24: 26–34. Harms, V. L. 1973. Taxonomic studies of North American *Sparganium*. I. *S. hyperboreum* and *S. minimum*. Canad. J. Bot. 51: 1629–1641. Hébert, L. P. 1973. Contribution à l'Étude des Sparganiaceae Rudolphi en Amerique du Nord. M.S. thesis. Université de Montréal. Müller-Doblies, D. 1970. Über die Verwandtschaft von *Typha* und *Sparganium* im Infloreszenz- und Blütenbau. Bot. Jahrb. Syst. 89: 451–562. Müller-Doblies, U. 1969. Über die Blütenstände und Blüten sowie zur Embryologie von *Sparganium*. Bot. Jahrb. Syst. 89: 359–450. Thieret, J. W. 1982. The Sparganiaceae in the southeastern United States. J. Arnold Arbor. 63: 341–356.

1. SPARGANIUM Linnaeus, Sp. Pl. 2: 971. 1753; Gen. Pl. ed. 5, 418. 1754 • Bur-reed, rubanier, ruban d'eau [probably Greek *sparganion*, a name used by Dioscorides for some plant, perhaps *Butomus umbellatus* Linnaeus; derived from *sparganon*, swaddling band, for strap-shaped leaves]

Herbs, erect or floating, or sometimes with some leaves floating and some emergent. **Leaves** flat, plano-convex, or abaxially keeled and V-shaped in section, spongy, margins entire. **Flowers** wind-pollinated, odorless, sessile. **Fruits** sessile or stipitate; tepals persistent, attached at base, in one species partially adnate to stipe. **Seeds** 1–2(–3), slender-ovoid; coat thin, appressed to endocarp. $x = 15$.

Species 14 (9 in the flora): mostly north temperate, some circumboreal, North America, a few s to Mexico, e Asia, Pacific Islands (New Zealand), Australia.

The plants flower in late spring to late summer, and the flowering season is shorter northward and at higher elevations. Fruiting is in late summer and fall, and some plants flower and fruit simultaneously in late season, especially northward and at higher elevations. Occasional plants have only staminate flowers.

Mature fruits are needed for identification of some species. The tepals, style, and often the stigmas remain attached to the mature fruits; the persistent style and stigma form the beak. The staminate rachises often persist, and the location and numbers of staminate heads can be determined from the scars of the fallen heads, even in fruiting specimens.

Some species show great variation, and specimens can be difficult to assign. Most characteristics depart at least occasionally from those given in the key, and foliar cross-sectional details are often distorted in dried specimens. In some erect species growing partially submersed, the first leaves of the season are submersed or floating and limp, and later leaves are emergent and stiffer. In species with normally erect leaves, deep or moving waters often suppress flowering and stimulate formation of only long, ribbonlike leaves that resemble those of species that produce only floating leaves. Species normally having floating leaves sometimes produce partially erect leaves when stranded. Depauperate or deep-water individuals of species that usually bear branched rachises sometimes have simple rachises with shortened internodes, and then the flowers and fruits are often reduced in number, but their morphology is little affected.

Many hybrids are reported in *Sparganium*, and they are discussed at length in a monograph by C. D. K. Cook and M. S. Nicholls (1986, 1987), but few are verified. For detailed discussion of the complex nature of the inflorescences, see U. Müller-Doblies (1969) and D. Müller-Doblies (1970); for extensive bibliographies, see Cook and Nicholls, and J. W. Thieret (1982); and for information about water quality and substrate preferences of the species, see E. O. Beal (1977), T. C. Brayshaw (1985), Cook and Nicholls, and G. E. Crow and C. B. Hellquist (1981).

The worldwide monograph by C. D. K. Cook and M. S. Nicholls (1986, 1987) is the most comprehensive treatment; most of their nomenclature is used herein. Nevertheless, the taxonomy must be regarded as tentative in the absence of detailed studies of any species over its full range. Phenotypic, ecotypic, and clinal variation are great in most species, the influence of hybridization is unclear, and some species are but dubiously differentiated from their Eurasian vicariants.

Collecting of *Sparganium* has been uneven in Alaska, Greenland, and northern Canada. The ranges there probably exceed those shown on the distribution maps.

1. Stigmas 2 on all or many pistillate flowers, 1 on others; fruits sessile, obpyramidal, depressed-truncate to somewhat rounded or tapering distally, not constricted at equator, body faceted; leaves erect, emergent. . . . . 1. *Sparganium eurycarpum*
1. Stigmas 1; fruits fusiform, ellipsoid, or obovoid, tapering at tip to beak (or beakless) and often to stipe below, usually constricted near equator; sides not or only weakly faceted; leaves erect and emergent or limp and floating.
   2. Leaves and inflorescences emergent, erect, stiff; leaves keeled throughout or at least toward base.
      3. Inflorescence rachis branched.
         4. Rachis branches, or at least some, with 1–3 pistillate heads; fruiting heads 1.5–2.5 cm diam.; fruits dull; bracts somewhat ascending. . . . . 2. *Sparganium americanum*
         4. Rachis branches without pistillate heads, rarely with 1; fruiting heads 2.5–3.5 cm diam; fruits distally shiny, proximally dull; bracts strongly ascending. . . . 3. *Sparganium androcladum*
      3. Inflorescence rachis unbranched.
         5. Staminate heads 1(–2); fruiting heads 1.2–1.6(–2) cm diam.; fruit beak 1.5–2 mm . . . . 4. *Sparganium glomeratum*
         5. Staminate heads 3–7(–10); fruiting heads 1.6–3.5 cm diam.; fruit beak 2–4.5 mm . . . . 5. *Sparganium emersum* (in part)
   2. Leaves and inflorescences floating, limp; leaves flat or plano-convex, unkeeled or keeled only near base.
      6. Staminate heads 1(–2); inflorescence rachis unbranched.
         7. Pistillate heads axillary and sessile or proximal head axillary and peduncled; fruit beak 0.5–1.5 mm. . . . . 8. *Sparganium natans*
         7. Pistillate heads, at least one of them, supra-axillary and peduncled; fruit beak less than 0.5 mm, or absent. . . . . 9. *Sparganium hyperboreum*
      6. Staminate heads 2 or more; inflorescence rachis branched or not.
         8. Rachis usually branched; tepals borne at middle of fruit stipe. . . . . 7. *Sparganium fluctuans*
         8. Rachis unbranched; tepals borne at or near base of fruit stipe.
            9. Staminate heads, or some of them, not contiguous; fruit beak 2–4.5 mm; leaves keeled toward base. . . . . 5. *Sparganium emersum* (in part)
            9. Staminate heads contiguous, appearing as single elongate head; fruit beak 1.5–2 mm; leaves unkeeled. . . . . 6. *Sparganium angustifolium*

1. **Sparganium eurycarpum** Engelmann in A. Gray, Manual ed. 2, 430. 1856 • Rubanier à gros fruits F

Sparganium californicum Greene; *S. erectum* Linnaeus subsp. *stoloniferum* (Hamilton ex Graebner) C. D. K. Cook & M. S. Nicholls; *S. eurycarpum* var. *greenei* (Morong) Graebner; *S. greenei* Morong

Plants robust, to 2.5 m; leaves and inflorescences emergent. **Leaves** erect, keeled but sometimes distally flattened, to 2.5 m × 6–20 mm. **Inflorescences:** rachis usually branched, erect; bracts ascending, not basally inflated; pistillate heads (1–)2–6(–8), axillary, not contiguous, peduncled on main rachis, sessile on branches, 1.5–5 cm diam. in fruit; staminate heads 10–40+, on main rachis and branches, proximal not contiguous, distal often crowded. **Flowers:** tepals often with dark subapical spot, entire to subentire; stigmas usually 2, often some flowers in head with 1, linear. **Fruits** straw-colored, darkening with age, somewhat lustrous, sessile, obpyramidal, body 3–7-faceted proximal to prominent shoulder, depressed-truncate to somewhat rounded or tapering distally, not constricted at equator, body 5–10 mm and often nearly as wide, gradually or abruptly beaked; beak straight, 2–4 mm, tepals attached at base, reaching to shoulder, about equaling body. **Seeds** 1–2 (–3), bistigmatic flowers often with 2, unistigmatic flowers with 1. $2n = 30$.

Flowering spring–fall (Apr–Oct southward, Jun–Jul northward). Lowland marshes, shores, and ditches, mostly in neutral-to-alkaline, hard, and even brackish waters on mud, sand, or gravel, sometimes among boulders on wave-washed shores, tolerant of some desiccation; 0–1600 m; St. Pierre and Miquelon; Alta., B.C., Man., N.B., Nfld. and Labr. (Nfld.), N.W.T., N.S., Ont., P.E.I., Que., Sask.; Ariz., Calif., Colo., Conn., Del., Idaho, Ill., Ind., Iowa, Kans., Ky., Maine, Md., Mass., Mich., Minn., Mo., Mont., Nebr., Nev., N.H., N.J., N.Mex., N.Y., N.Dak., Ohio, Okla., Oreg., Pa., R.I., S.Dak., Utah, Vt., Va., Wash., W.Va., Wis., Wyo.; Mexico (Baja California); e Asia.

Sparganium eurycarpum grows mostly near the coast (but not in salt marshes) in New England (G. E. Crow and C. B. Hellquist 1981), mostly in the interior in British Columbia (T. C. Brayshaw 1985), and in coastal and interior sites from Washington to Baja California. It is locally common to abundant in fresh to somewhat brackish waters across the continent but is less frequent toward its northern and southern limits.

Both stigmas of bistigmatic flowers often persist on the fruits, but sometimes one or both break away. The fruits of bistigmatic and 2-ovuled flowers are often larger than those of unistigmatic flowers.

From southern British Columbia to Baja California, especially near the coast but also at scattered sites inland, some plants having fewer bistigmatic flowers and smaller, rounded or tapering fruits have been called *Sparganium eurycarpum* var. *greenei* (Morong) Graebner or *S. greenei* Morong by some authors (e.g., H. L. Mason 1957; T. C. Brayshaw 1985) but not by others (e.g., C. L. Hitchcock et al. 1955–1969, vol. 1; A. Cronquist et al. 1972+, vol. 6). Such plants were placed in *S. erectum* Linnaeus, the strongly polymorphic and ecotypically variable Eurasian vicariant species, as *S. erectum* subsp. *stoloniferum* (Hamilton ex Graebner) C. D. K. Cook & M. S. Nicholls (C. D. K. Cook and M. S. Nicholls 1987). *Sparganium eurycarpum* is only weakly distinguished from *S. erectum*, largely on the basis of relative preponderance of bi- and unistigmatic flowers (± 50% in the two species, respectively), and by concomitant differences in fruit size and shape. In North America, however, some unistigmatic flowers occur among bistigmatic flowers on pistillate heads of many plants throughout the range of *S. eurycarpum*. Such unistigmatic flowers are common on some west-coast plants and on plants of alkaline marshes of the Great Basin and Great Plains, although there is considerable variation in all three areas. Because of these widespread variations, and in the absence of definitive, rangewide, comparative studies, our plants are here treated as *S. eurycarpum*. Comparative studies of Eurasian and North American plants, however, perhaps would show that all are best treated as *S. erectum*, the name with priority, but no such studies have been made.

2. **Sparganium americanum** Nuttall, Gen. N. Amer. Pl. 2: 203. 1818 • Rubanier d'Amérique

Plants slender and grasslike to usually robust, to 1 m; leaves and inflorescences usually emergent. **Leaves** erect but not especially stiff, usually keeled only near base, flattened distally, to 1 m × 6–12 mm. **Inflorescences:** rachis 0–3-branched, erect; bracts somewhat ascending, not basally inflated; pistillate heads 1–3 on branches, 2–6 on main rachis, axillary, not contiguous at anthesis, sessile, 1.5–2.5 cm diam. and often contiguous in fruit; staminate heads usually 3–7 on each branch, 4–10 on main rachis, not or barely contiguous. **Flowers:** tepals often with prominent subapical dark spot, entire to crenulate or emarginate; stigma 1, linear-lanceolate. **Fruits** tan to dark greenish brown, dull, subsessile to stipitate, fusiform, sometimes barely constricted near equator, body

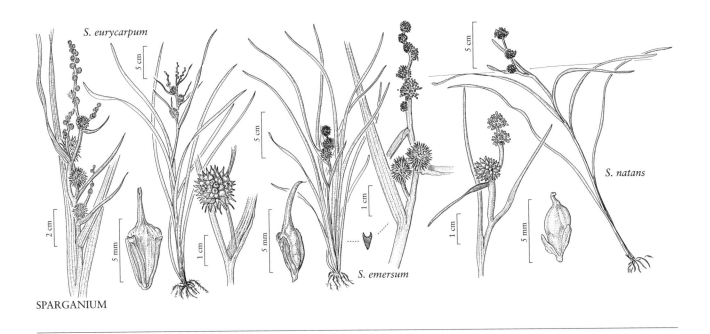

S. eurycarpum

S. natans

S. emersum

SPARGANIUM

not strongly faceted, 3.5–5(–7) mm, tapering to beak; beak usually curved, not hooked, 3–5 mm; tepals attached at base, reaching equator or slightly beyond. **Seeds** 1.

Flowering late spring–summer (Apr–Jun southwestward, May–Sep southeastward, Jun–Aug northward). Shores and shallow neutral-to-alkaline waters, sometimes forming large stands; 0–800 m; St. Pierre and Miquelon; Man., N.B., Nfld. and Labr. (Nfld.), N.S., Ont., P.E.I., Que.; Ala., Ark., Conn., Del., Fla., Ga., Ill., Ind., Iowa, Kans., Ky., La., Maine, Md., Mass., Mich., Minn., Miss., Mo., N.H., N.J., N.Y., N.C., Ohio, Okla., Pa., R.I., S.C., Tenn., Tex., Vt., Va., W.Va., Wis.; Mexico (Durango).

For differences between this highly polymorphic species and the similar but less variable *Sparganium androcladum*, see the discussion under that species. Also see the discussion under *S. emersum*.

E. O. Beal (1977) recognized three morphologically overlapping races of *Sparganium americanum*: the coastal race, growing in the lower coastal plain from Virginia to Florida and Louisiana, and north in the Mississippi Embayment to Oklahoma, Arkansas, and Missouri, has leaves wide for the species, rachises 2–5-branched, and stigmas 1.5+ mm; the Appalachian race, growing in the Appalachian region from Maine to western North Carolina and in the Ozark Mountains, has leaves narrow for the species, rachises simple to sparingly branched, and stigmas 0.9 mm or less; and the ubiquitous race, growing throughout the range of the species with increasing robustness southward, morphologically overlaps the others.

3. **Sparganium androcladum** (Engelmann) Morong, Bull. Torrey Bot. Club 15: 78. 1888 • Rubanier rameux [E]

Sparganium simplex Hudson var. *androcladum* Engelmann in A. Gray, Manual ed. 5, 481. 1867; *S. americanum* Nuttall var. *androcladum* (Engelmann) Fernald & Eames

Plants robust, to 1.2 m; leaves and inflorescences erect, emergent. **Leaves** stiff, keeled from base at least to middle, flattened distally, to 1.2 m × 5–15 mm. **Inflorescences:** rachis (0–)1–3 branched, erect, bracts strongly ascending, not basally inflated; pistillate heads (1–)2–4 on main rachis, 0(–2) on lateral branches, axillary, not contiguous, sessile or short-peduncled, 2.5–3.5 cm diam. in fruit; staminate heads 3–8 on main rachis, 1–6 on lateral rachises, most not contiguous. **Flowers:** tepals often with prominent subapical dark spot, subentire to entire; stigma 1, linear. **Fruits** brown, proximally dull, distally shiny, subsessile to short-stipitate, fusiform, usually constricted near equator, body not strongly faceted, 5–7 × 2.5–3 mm, tapering to beak; beak often curved and hooked, 4–7 mm; tepals attached at base, reaching to equator or somewhat beyond. **Seeds** 1.

Flowering late spring–summer (Apr–Jul). Shores and shallow, quiet, circumneutral waters; 0–800 m; Ont., Que.; Ark., Conn., Ill., Ind., Iowa, Ky., Maine, Mass., Mich., Minn., Mo., N.H., N.J., N.Y., Ohio, Okla., Pa., R.I., Tenn., Vt., W.Va., Wis.

Sparganium androcladum is less variable than the similar and more common *S. americanum*, from which it is distinguished by its generally larger size and more robust habit; leaves stiffer, wider, more strongly keeled; inflorescence branches usually without pistillate heads, the bracts ascending; fruiting heads larger; fruits distally shiny, the beak longer and hooked. Immature fruits of *S. androcladum* resemble mature fruits of *S. americanum*. Some specimens will not key readily to either species. The complex nomenclatural history is discussed by C. D. K. Cook and M. S. Nicholls (1987).

Sparganium androcladum has an unusual, discontinuous distribution, which is wholly within the range of *S. americanum*, except in the central Mississippi Valley (M. L. Fernald 1922b). In the absence of convincing specimens, the presence of *S. androcladum* in Virginia cannot be confirmed, although E. O. Beal (1960) reported it there, and it occurs nearby in West Virginia.

4. **Sparganium glomeratum** (Beurling ex Laestadius) L. M. Newman in C. J. Hartman et al., Handb. Skand. Fl. ed. 12: 111. 1889 • Rubanier aggloméré

Sparganium erectum Linnaeus var. *glomeratum* Beurling ex Laestadius, Årsberätt. Bot. Arbeten Upptäckter 1850(bihang 2): 2. 1853

Plants slender to robust, to 0.4 (–0.6) m; at least some leaves and inflorescences emergent, erect. **Leaves** stiff, weakly keeled, to 50 cm × 6 mm. **Inflorescences:** rachis unbranched, condensed, erect; bracts ascending, somewhat inflated near base; pistillate heads 2–6, mostly supra-axillary, sometimes opposite bract above, upper crowded, sessile, proximal head not contiguous with distal, peduncled, 1.2–1.6(–2) cm diam. and contiguous in fruit; staminate heads 1(–2), contiguous or not with distalmost pistillate head. **Flowers:** tepals without subapical dark spot, entire to erose; stigma 1, lanceolate. **Fruits** greenish brown, lustrous, stipitate, fusiform, body not faceted, slightly constricted near equator, 3–6 × 2–3 mm, tapering to beak; beak straight, 1.5–2 mm; tepals attached at base, reaching ⅓–½ length of fruit. **Seeds** 1. 2*n* = 30.

Flowering summer (Jul–Aug). Shallow, quiet, neutral, mesotrophic waters; 0–1000 m; Alta., B.C., Nfld. and Labr. (Labr.), Ont., Que., Sask.; Minn., Wis.; circumboreal.

Sparganium glomeratum is apparently rare, or perhaps is only rarely collected, in North America, except it is locally common in sedge-marshes and black-ash swamps near the western end of Lake Superior. The species is rather invariable throughout its circumboreal range (C. D. K. Cook and M. S. Nicholls 1986).

5. **Sparganium emersum** Rehmann, Verh. Naturf. Vereins Brünn 10: 80. 1872 • Rubanier émergent F

Sparganium angustifolium Michaux subsp. *emersum* (Rehmann) Brayshaw; *S. chlorocarpum* Rydberg; *S. chlorocarpum* var. *acaule* (Beeby ex Macoun) Fernald; *S. emersum* subsp. *acaule* (Beeby ex Macoun) C. D. K. Cook & M. S. Nicholls

Plants robust to slender, to 2 m; leaves and inflorescences emergent, stiff, or some or all leaves floating, limp. **Leaves** when erect rather stiff, partially to entirely keeled, flat, to 0.8 m × 4–10 mm; floating leaves limp, keeled at least near base, to 2 m × 4–18 mm. **Inflorescences:** rachis unbranched, erect; bracts ascending to erect, often somewhat inflated toward base; pistillate heads 1–6, contiguous or not, proximal axillary and peduncled, distal supra-axillary and sessile, 1.6–3.5 cm diam. in fruit; staminate heads 3–7(–10), contiguous or not with each other, sometimes some contiguous and some not, but not contiguous with distalmost pistillate head. **Flowers:** tepals without subapical dark spot, tip erose; stigmas 1, linear-lanceolate. **Fruits** green to reddish brown, lustrous, stipitate, fusiform, slightly constricted below equator, body not faceted, 3–4 × 1.5–2 mm excluding stipe, tapering to beak; beak straight to curved, 2–4.5 mm; tepals attached at base, reaching about to equator. **Seeds** 1. 2*n* = 30.

Flowering late spring–early fall (May–Oct southward, Jun–Aug northward). Still to flowing eutrophic and mesotrophic, circumneutral to somewhat alkaline waters, sometimes abundant; 0–3000 m; St. Pierre and Miquelon; Alta., B.C., Man., N.B., Nfld. and Labr., N.W.T., N.S., Ont., P.E.I., Que., Sask., Yukon; Alaska, Ariz., Calif., Colo., Conn., Idaho, Ill., Ind., Iowa, Maine, Md., Mass., Mich., Minn., Mont., Nebr., Nev., N.H., N.J., N.Y., N.C., N.Dak., Ohio, Oreg., Pa., S.Dak., Utah, Vt., Va., Wash., W.Va., Wis., Wyo.; Eurasia.

Less robust plants of *Sparganium emersum* are found from Manitoba, Minnesota, and Iowa, east to the Atlantic coast. They have erect leaves, 3–7 mm wide, much exceeding the short inflorescence; supra-axillary, sessile, and crowded pistillate heads; and 3–4 mm fruits with a 3–4 mm beak. Such plants have been called *S. emersum* subsp. *acaule* (C. D. K. Cook and M. S. Nicholls 1986), and they are widely known as *S. chlorocarpum* Rydberg var. *acaule* or *S. chlorocarpum* f. *acaule*. Inasmuch as they often grow with and are not always clearly distinct from typical plants, and in the absence of definitive, range-wide studies, they are recognized here only as a form (fide E. G. Voss 1966; G. E. Crow and C. B. Hellquist 1981).

Sparganium emersum is easily confused with *S. angustifolium,* especially when floating leaves are present, but it is distinguished by its leaves which are triangulate at least at the base, by its often more numerous staminate heads, at least some of which are not contiguous, and by its greenish fruits with longer beaks (C. D. K. Cook and M. S. Nicholls 1986). In northern and northwestern states and British Columbia, it resembles *S. angustifolium* in the distribution of the staminate heads, the distal of which tend to be confluent. Cook and Nichols noted that some characters of *S. angustifolium* might have been incorporated into *S. emersum* and our plants could be a stable hybrid that differs somewhat from *S. emersum* as it is known in Europe. G. E. Larson (1993) called for further study of the entire *S. emersum-S. angustifolium* complex, which is perhaps better treated as one variable species, *S. angustifolium* (T. C. Brayshaw 1985).

Hybrids with *Sparganium hyperboreum* and *S. natans* have been reported from Europe but are doubted by C. D. K. Cook and M. S. Nicholls (1986). In eastern provinces and states, *S. emersum* could be confused with *S. americanum* and *S. androcladum,* but it can be distinguished from them by its supra-axillary pistillate heads and its tepals lacking a subapical dark spot.

6. Sparganium angustifolium Michaux, Fl. Bor.-Amer. 2: 189. 1803 • Rubanier à feuilles étroites

Sparganium angustifolium var. *multipedunculatum* (Morong) Brayshaw; *S. emersum* Rehmann var. *multipedunculatum* (Morong) Reveal; *S. multipedunculatum* (Morong) Rydberg, *S. simplex* Hudson var. *multipedunculatum* Morong

Plants slender, to more than 2 m long; leaves and inflorescences usually floating. **Leaves** limp, unkeeled, flat to plano-convex, 0.2–0.8(–2.5) m × 2–5(–10) mm. **Inflorescences:** rachis unbranched, flexuous, its fertile part usually erect at water surface; bracts ascending, lower bracts inflated near base; pistillate heads 2–5, at least some supra-axillary, not contiguous, sessile or most proximal peduncled (often prominently so in deeper water), 1–3 cm diam. in fruit; staminate heads (1–)2–4, contiguous and appearing as one elongate head, not contiguous with distalmost pistillate head. **Flowers:** tepals without subapical dark spot, erose-tipped, stigmas 1, lance-ovate. **Fruits** reddish to brownish, dull, short-stipitate, ellipsoid to fusiform, not faceted, body constricted at equator, 3–7 mm, tapering to beak; beak straight, 1.5–2 mm; tepals attached at base, reaching about to equator. **Seeds** 1. 2*n* = 30.

Flowering summer–fall (Jun–Oct southwestward, Jul–Aug northward). Acid, oligotrophic waters of lakes, ponds, ditches, and streams, usually in shallow waters but to 2.5 m deep; 0–4000 m; Greenland; St. Pierre and Miquelon; Alta., B.C., Man., N.B., Nfld. and Labr., N.W.T., N.S., Ont., P.E.I., Que., Sask., Yukon; Alaska, Ariz., Calif., Colo., Conn., Idaho, Maine, Mass., Mich., Minn., Mont., Nev., N.H., N.Mex., N.Y., Oreg., Utah, Vt., Wash., Wis., Wyo.; circumboreal.

Sparganium angustifolium is sometimes abundant, its leaves then covering the surface. It is a relatively invariable species that forms fertile hybrids with *S. emersum* (C. D. K. Cook and M. S. Nicholls 1986), from which it is distinguished by its contiguous staminate heads and flat to plano-convex leaves. See the discussion under *S. emersum.*

7. Sparganium fluctuans (Engelmann ex Morong) B. L. Robinson, Rhodora 7: 60. 1905 • Rubanier flottant E

Sparganium androcladum (Engelmann) Morong var. *fluctuans* Engelmann ex Morong, Bull. Torrey Bot. Club 15: 78. 1888

Plants limp, to more than 1 m long; leaves and inflorescences floating. **Leaves** limp, unkeeled, flat, mostly 0.6–1 m × 4–10 mm. **Inflorescences:** rachis 0–2-branched, distal part erect at water surface; bracts ascending, not inflated at base; pistillate heads 0–2 on main rachis, 1–2 on secondary rachises, axillary or supra-axillary, often some contiguous or nearly so, especially in fruit, sessile, 1.5–2.3 cm diam. in fruit; staminate heads 3–6 on main rachis, 1–4 on secondary rachises, contiguous or not, but not contiguous with distalmost pistillate head. **Flowers:** tepals often with prominent apical dark spot, apically erose to fimbriate; stigmas 1, linear-lanceolate. **Fruits** eventually dark reddish brown, dull, stipitate, elliptic to obovoid or fusiform, body not faceted, sometimes constricted near equator, 2–5 × 1.5–2 mm, tapering to beak; beak curved, 2–3.5 mm; tepals borne at middle of fruit stipe, basally adnate to stipe, reaching nearly to equator. **Seeds** 1.

Flowering summer (Jul–Aug). Quiet, cold, acidic to neutral, oligotrophic waters to 2 m deep but usually less, abundant in some areas but not in others, sometimes covering the water with its strap-shaped leaves; 0–600 m; Alta, B.C., Man., N.B., Nfld. and Labr. (Nfld.), N.S., Ont., P.E.I., Que., Sask.; Conn., Maine, Mass., Mich., Minn., N.H., N.J., N.Y., Pa., Vt., Wis.

The tepals are usually adnate to the fruit stipe for about half their length, but in our other species they are free from it.

Sparganium fluctuans is not known to hybridize. When vegetative, it is sometimes confused with *S. angustifolium*, but the floating leaves of that species are usually less than 5 mm wide and plano-convex. *Sparganium fluctuans* is more robust than the grasslike *S. natans*, with which it sometimes grows.

8. **Sparganium natans** Linnaeus, Sp. Pl. 2: 971. 1753 • Rubanier nageant

Sparganium minimum (Hartman) Fries

Plants slender, grasslike, limp, to 0.6 m; leaves and inflorescences floating or, when stranded, more or less erect. **Leaves** limp in water, unkeeled, flat, 0.04–0.6 m × 2–6(–10) mm; leaves of stranded plants shorter, firmer. **Inflorescences:** rachis unbranched, flexuous; bracts ascending, basally inflated; pistillate heads 1–3, axillary, not contiguous, sessile or most proximal short-peduncled (often long-peduncled in Alaska and nw Canada), 0.5–1.2 cm diam. in fruit; staminate head 1 or apparently so, terminal, not contiguous with distalmost pistillate head. **Flowers:** tepals without subapical dark spot, erose; stigmas 1, lance-ovate. **Fruits** dark greenish or brownish, subsessile, body ellipsoid to obovoid, not faceted, barely or not constricted at equator, 2–4 × 1–1.5 mm, tapering to beak; beak curved, 0.5–1.5 mm; tepals attached at base, reaching about to equator. **Seeds** 1. $2n = 30$.

Flowering summer–fall (Jun–Sep southwestward, Jul–Aug northward). Cool, quiet, slightly acid to somewhat basic waters of bays, pools, ditches, and peat bogs, usually in shallow water but sometimes to 60 cm depth, where less floriferous, abundant in its northern range, less so southward; 0–3500 m; St. Pierre and Miquelon; Alta., B.C., Man., N.B., Nfld. and Labr., N.W.T., N.S., Nunavut, Ont., P.E.I., Que., Sask., Yukon; Alaska, Calif., Colo., Conn., Idaho, Ill., Ind., Maine, Mass., Mich., Minn., Mont., N.H., N.Y., Oreg., Pa., R.I., Utah, Vt., Wash., Wis., Wyo.; circumboreal (not in Greenland).

This species has long been known as *Sparganium minimum*, but the correct name is *S. natans* (C. D. K. Cook 1985).

The leaves of *Sparganium natans* are thinner and more translucent than those of the similar *S. hyperboreum*, and they lack the yellowish cast of that species. Its distalmost pistillate head is not contiguous with the staminate head, as is sometimes the case in *S. hyperboreum*, and its beaked fruit also distinguishes it from that species. See the discussion under *S. hyperboreum* for a description of *S. hyperboreum* × *S. natans*.

9. **Sparganium hyperboreum** Beurling ex Laestadius, Årberätt. Bot. Arbeten Upptäckter 1850(bihang 2): 4. 1853 • Rubanier hyperboréal

Plants slender, grasslike, to 0.8 m; leaves and inflorescences usually floating. **Leaves** limp, unkeeled, flat, 0.1–0.4(–0.8) m × 1–5 mm. **Inflorescences:** rachis unbranched, flexuous; bracts ascending, lower bracts slightly inflated near base; pistillate heads 1–4, axillary, contiguous, sessile, or most proximal peduncled and supra-axillary, 0.5–1.4 cm diam. in fruit; staminate heads 1(–2), terminal, contiguous or not with distalmost pistillate head. **Flowers:** tepals without subapical dark spot, erose; stigmas 1, ovate. **Fruits** brown or yellowish, dull, subsessile, body ellipsoid to obovoid, not faceted, ± constricted at equator, 2–5 × 1.5–2.5 mm; beak less than 0.5 mm, or absent; tepals attached at base, not reaching equator. **Seeds** 1. $2n = 30$.

Flowering summer (Jul–Aug). Cold, quiet, shallow, oligo- to mesotrophic arctic-alpine waters; 0–3000 m; Greenland; St. Pierre and Miquelon; Alta., B.C., Nfld. and Labr., N.W.T., N.S., Nunavut, Ont., P.E.I., Que., Yukon; Alaska; circumboreal.

Sparganium hyperboreum is distinguished from other floating-leaved species by its beakless fruits with sessile stigmas.

Putative hybrids between *Sparganium hyperboreum* and *S. natans* have been found in Manitoba, Newfoundland, Northwest Territories, and Alaska by V. L. Harms (1973), who discussed variation in both species. The hybrids have wider (2–5 mm) leaves, golden-brown fruits with short (1 mm) beaks, and supra-axillary pistillate heads (V. L. Harms 1973; C. D. K. Cook and M. S. Nicholls 1986).

213. TYPHACEAE Jussieu [as Typhae]

• Cat-tail Family

S. Galen Smith

Herbs, perennial, of fresh to slightly brackish wetlands, often emergent, rhizomatous, caulescent in flower, smooth, glabrous. **Leaves** basal and cauline, 2-ranked, mostly ascending; sheaths open, margins overlapping, clear, summit tapered into blade or auriculate; blades twisted into loose helix, narrowly linear-attenuate, apex acute, aerenchyma prominent. **Inflorescences** 1, terminal, erect, equaled or exceeded by cauline leaves, cylindric, spikelike (hereafter "spikes"); staminate spike distal to pistillate spike; young spikes subtended by early-deciduous bracts resembling reduced foliage leaves, 1 bract subtending and several within staminate spike, 1 bract subtending pistillate spike; staminate spike flowers deciduous but axis generally persistent; staminate axis with numerous simple or branched scales arising among flowers; pistillate axis with numerous projections ("compound pedicels"), evident on denuded fruiting spike, each bearing several flowers; in some species flowers subtended by slender bracteoles. **Flowers** unisexual, staminate and pistillate on same plants, numerous, densely packed in unisexual spikes, minute, wind-pollinated (stigmas receptive several days before pollen is shed); perianth probably represented by staminate scales and by hairs on stipes of pistillate flowers. **Staminate flowers** stipitate; stamens 1–several, filaments connate; anthers basifixed, connective distally extended. **Pistillate flowers** hypogynous, stipitate (stipe bearing numerous straight hairs, developing after flowering, acting in wind dispersal of fruits); pistils 1, 1-carpellate; ovaries 1-locular; placentation apical; ovules 1; styles 1, unbranched; stigmas 1, whitish or green, drying brown, 1-sided, smooth; agamous flowers numerous (ovaries modified after flowering as carpodia). **Fruits** follicles, fusiform; pericarp clear, transparent, splitting longitudinally in water to release seed. **Seeds:** endosperm starchy, oily; embryo cylindric.

Genera 1, species ca. 8–13 (3 in the flora): boreal to tropical regions worldwide.

The extensive literature on morphology and taxonomy of Typhaceae has been recently reviewed by U. Müller-Doblies and D. Müller-Doblies (1977); R. M. T. Dahlgren and H. T. Clifford (1982); R. M. T. Dahlgren et al. (1985); and J. W. Thieret and J. O. Luken (1996). The inflorescence is probably reduced from a compound structure.

Sparganium and *Typha* are very similar and perhaps should be placed in one family, as summarized by J. W. Thieret and J. O. Luken (1996):

> Pre-Englerian [authors] . . . placed *Typha* and *Sparganium* together in a single family, the Typhaceae. [H. G. A.] Engler (1886) put these genera in separate families, thus starting a tradition that has been followed by almost all subsequent authors until recently, when [D.] Müller-Doblies (1970) re-examined the relationships of the genera and concluded that "the five different characters by which Engler justified the family Sparganiaceae are wrong or, in two cases, without any significance . . . The few remaining but very obvious differences may be explained to a large extend [sic] by an adaptation of *Typha* to anemochory [wind-dispersal of propagules]. . . ."

The phylogenetic relationships of the Typhales with other families remain controversial, and it seems best to treat the taxon as an isolated order of uncertain relationships pending further research. Various authors have placed the Typhales close to or within the Pandanales, Arales, Poales, Liliales, Pontederiales, or Philydrales or in the Commeliniflorae generally close to the Cyperales and Juncales (J. W. Thieret and J. O. Luken 1996).

SELECTED REFERENCES Briggs, B. G. and L. A. S. Johnson. 1968. The status and relationships of the Australian species of *Typha*. Contr. New South Wales Natl. Herb. 4: 57–78. Eckardt, T. 1964. Pandanales. In: H. Melchior, ed. 1964. A. Engler's Syllabus der Pflanzenfamilien . . . , ed. 12. 2 vols. Berlin. Vol. 2, pp. 598–602. Engler, H. G. A. 1886. Ueber die Familie der Typhaceen. Bot. Centralbl. 25: 127. Fassett, N. C. and B. M. Calhoun. 1952. Introgression between *Typha latifolia* and *T. angustifolia*. Evolution 6: 367–379. Hotchkiss, N. and H. L. Dozier. 1949. Taxonomy and distribution of North American cattails. Amer. Midl. Naturalist 41: 237–254. Kronfeld, E. M. 1889. Monographie der Gattung *Typha* Tourn. (Typhinae Agdh., Typhaceae Schur-Engl.). Verh. Zool.-Bot. Ges. Wien 39: 89–192. Morton, J. F. 1975. Cattails (*Typha* spp.)—weed problem or potential crop? Econ. Bot. 29: 7–29. Müller-Doblies, D. 1970. Über die Verwandschaft von *Typha* und *Sparganium* im Infloreszenz- und Blütenbau. Bot. Jahrb. Syst. 89: 451–562. Müller-Doblies, U. and D. Müller-Doblies. 1977. Typhaceae. In: G. Hegi et al. 1964+. Illustrierte Flora von Mitteleuropa, ed. 3. 5+ vols. in 8+. Berlin and Hamburg. Vol. 2, part 1(4), pp. 275–317. Ramey, V. 1981. *Typha*—not just another weed. Aquaphyte 1: 1–2. Smith, S. G. 1967. Experimental and natural hybrids in North American *Typha* (Typhaceae). Amer. Midl. Naturalist 78: 257–287. Smith, S. G. 1987. *Typha*: Its taxonomy and the ecological significance of hybrids. Arch. Hydrobiol. 27: 129–138. Thieret, J. W. and J. O. Luken. 1996. The Typhaceae in the southeastern United States. Harvard Pap. Bot. 8: 27–56.

1. TYPHA Linnaeus, Sp. Pl. 2: 971. 1753; Gen. Pl. ed. 5, 418. 1754 • Cat-tail, cat-o'-nine-tails, cat-tail flag, bulrush, reed-mace, quenouille, massette, canne, tule, queue de rat Ⓦ [Greek, perhaps from *typhein*, to smoke or to emit smoke, in allusion either to the use of the spikes for maintaining smoky fires or to the smoky brown color of the fruiting spikes]

Plants of fresh to slightly brackish wetlands, often emergent. **Rhizomes** at base of erect shoots, mostly horizontal, unbranched, to 70 cm × 5–40 mm, starchy, firm, scaly. **Erect shoots** vegetative or flowering, single at rhizome apices or arising from shoot bases, thus clustered, unbranched, to 4 m, elliptic in cross section; stems often somewhat compressed distally, aerenchyma absent. **Foliage leaves** persistent, intergrading proximally with scale leaves, to 15 on each flowering shoot; blade twisted into loose helix, mostly slightly oblanceolate, thickly concave-convex or plano-convex proximally to thinly plane distally (abaxially keeled in the Old World *Typha elephantina* Roxburgh); mucilage-secreting glands numerous in adaxial surface of sheath and sometimes proximally on blade, colorless to brown, roughly rectangular. **Inflorescences:** staminate scales shorter than or exceeding flowers; pistillate spikes usually persisting into winter, when dry fruiting flowers often falling in masses; pistillate bracteoles absent or numerous, colorless except for brown apical blade at spike surface, filiform, blade club-shaped to lanceoloid. **Staminate flowers:** anthers dehiscing longitudinally, 4-sporangiate. **Pistillate flowers:** pistil hairs colorless and wholly filiform, or apically enlarged and brown, exceeded by stigmas; carpodia obovoid, spongy, bearing rudimentary styles. $x = 15$.

The extensive literature on *Typha* has been reviewed (C. M. Finlayson et al. 1983; J. B. Grace and J. S. Harrison 1986; J. W. Thieret and J. O. Luken 1996). A modern taxonomic revision is much needed, especially for eastern Asia, adjacent islands, and South America (S. G. Smith 1987); the latest worldwide monograph is by E. M. Kronfeld (1889). The center of diversity (ca. 6 species) is central Eurasia.

Typha is ecologically important in many fresh to slightly brackish wetlands, often emergent in up to 1.5 m of water. Each spike may produce hundreds of thousands of seeds, which are efficiently wind-dispersed and germinate on bare wet soils or under very shallow water. The seedlings rapidly form clones by means of rhizomes in the first season, flower the second season (R. R. Yeo 1964), and often form very large, persistent, often monospecific stands. Some species produce large amounts of biomass, comparable to the most productive agricultural crops. The three species are ecotypically well differentiated in North America (S. J. McNaughton 1966). Some mechanisms of competition between *Typha* species were studied by J. B. Grace and R. G. Wetzel (1982) and J. B. Grace (1988) (cf. Thieret and Luken 1996; J. B. Grace and J. S. Harrison 1986).

Common teratological forms are longitudinally split pistillate spikes (caused by parasitic insects), pistillate spikes interrupted by zones of naked axis, and partially merged pistillate and staminate spikes.

Typha species are or have been utilized in numerous ways worldwide (C. M. Finlayson et al. 1983; J. B. Grace and J. S. Harrison 1986; J. F. Morton 1975; V. Ramey 1981; J. W. Thieret and J. O. Luken 1996). Leaves are used for dwellings (walls, roof thatch, floor coverings); for mats, baskets, and other handicraft objects; for caning chairs; and for caulking barrels, boats, and houses. "Fluff" from fruiting spikes is used for tinder and insulation; for dressing burns; and for stuffing pillows, quilts, mattresses, life preservers, toys, and diapers. Young shoot bases, young rhizomes, starch from mature rhizomes, staminate flowers before anthesis, and pollen are all minor sources of food. *Typha* is valuable as habitat and food for many kinds of wildlife. It is useful for removal of various kinds of pollutants; a potential source of fiber for paper and other products; and a potential source of energy, e.g., for alcohol manufacture. The seeds comprise about 18–20% of an edible oil (69% linolenic acid). Several species are cultivated as ornamentals. The North American species are often sold commercially and planted for wildlife habitat and in wetland restoration.

The larger *Typha* species and *T. ×glauca* can be serious weeds in managed aquatic systems worldwide, where they can invade canals, ditches, reservoirs, cultivated fields, and farm ponds; they can be a nuisance in recreational lakes; and they can reduce biodiversity and displace species more desirable for certain kinds of wildlife (J. B. Grace and J. S. Harrison 1986; J. F. Morton 1975; J. W. Thieret and J. O. Luken 1996).

Users of this treatment should be aware of the following: 1) Leaves shrink considerably in width as they dry. 2) Leaf mucilage glands are usually colorless and difficult to see in fresh leaves of all three species early in the season and in *Typha latifolia* at all stages. They are brown and clearly evident to the unaided eye in mid- to late-season fresh or dried *T. angustifolia* and *T. domingensis* and are easily stained (with, e.g., safranin). Brown necrotic spots, apparently caused by feeding arthropods, may superficially resemble mucilage glands. 3) Spikes are commonly poorly developed as a result of drought or other causes; fruiting spike thicknesses given herein are for normal spikes. 4) Except for the presence of mucilage glands on the leaf blades, unique to *T. domingensis* and its hybrids, the microscopic flower and bracteole structures are generally essential for accurate identification of *Typha* species and hybrids. This is in part because of changes in the inflorescences during development and in part because of phenotypic plasticity, especially of leaf blade widths. It is often necessary to use forceps to pull a few pistillate flowers out of the spike and observe them with a dissecting microscope at 20× to 30×. 5) Pollen is often infested with fungi, which attach the grains together and simulate genetically aborted grains, and the grains of *T. angustifolia* and *T. domingensis* often adhere in small groups for no obvious reason.

Hybrids among the three North American species have been experimentally produced and occur in most regions of sympatry (S. G. Smith 1967, 1987). Local studies were provided (J. R. Dugle and T. P. Copps 1972; T. M. Tompkins and J. Taylor 1983). Protogyny and slight differences in flowering dates favor interspecific pollination. Hybrid seedlings are likely wherever two species form mixed stands and bare wet soil is available for seed germination and seedling establishment. 1) *T. latifolia* × *T. angustifolia* (=*T.* ×*glauca* Godr., pro sp.), often called "hybrid cattail," is abundant throughout most of the region of sympatry of the parents except along the southeast coast, where it is uncommon. Almost all plants are putative F$_1$s which are intermediate between the parental species in all morphologic characters studied and are highly sterile, producing very few or no seeds or viable pollen grains. Fertile or sterile intermediates between *T.* ×*glauca* and *T. angustifolia* occasionally occur, however. In spite of its sterility, *T.* ×*glauca* is remarkably successful ecologically. It often spreads by means of rhizomes to form very large clones and out-competes the parental species, especially in eutrophic, disturbed habitats with unstable water levels (S. W. Harris and W. H. Marshall 1963; S. G. Smith 1987). Unfortunately it has been treated as a species by many authors (e.g., N. Hotchkiss and H. L. Dozier 1949). 2) *Typha domingensis* × *T. latifolia* (= *T.* ×*provincialis* A. Camus, *T. bethulona* Costa) is known only from very few collections in Arkansas, California, Florida, Missouri, Nebraska, and North Carolina. All of these are highly sterile putative F$_1$s except for one putative F$_2$, in which the characteristics of the parental species are recombined, from southern California. 3) *Typha angustifolia* × *T. domingensis* is known from scattered specimens in Arkansas, California, Kansas, Kentucky, Missouri, and Nebraska. It is not known from the southeast coast, perhaps because of differences between the species in flowering dates. Most plants are highly fertile, and some may be F$_2$ or later generation hybrids. 4) Putative *T. angustifolia* × *T. domingensis* × *T. latifolia* trihybrids are locally common in California and rare in south-central United States. Introgression between the interfertile *T. angustifolia* and *T. domingensis* is probably locally common in the south-central U.S. and north-central California, while introgression between *T. latifolia* and the other two species is probably very uncommon because of hybrid sterility. Published research presumably demonstrating introgression (e.g., N.C. Fassett and B. M. Calhoun 1952) is faulty (S. G. Smith 1967, 1987). The tetraploid *T. orientalis* of the Pacific Basin may be of hybrid origin (B. G. Briggs and L. A. S. Johnson 1968; S. G. Smith 1967, 1987).

The following key includes the hybrids, which are not further described in the text:

1. Pistillate bracteoles absent, or if present then narrower than stigmas and generally not evident at spike surface; stigmas ovate to lanceolate, persistent on mature spikes; pistillate spikes green in flower when fresh, in fruit mostly 19–36 mm thick; carpodia concealed among pistil hairs; compound pedicels on denuded axis 0.6–3.5 mm; staminate scales colorless to brown.

   2. Pistillate bracteoles absent; stigmas ovate to ovate-lanceolate, often blackish when dry;

pistillate spikes contiguous with staminate spikes or sometimes separated by gap of to 4(–8) cm, in fruit mostly 28–36 mm thick; compound pedicels on denuded axis 1.5–3.5 mm; seeds usually numerous; staminate scales colorless to straw-colored; pollen in tetrads. . . . . .1. *Typha latifolia*

2. Pistillate bracteoles present (but generally evident only at 20–30× after removal from spike, resembling perigonial hairs, with brown, enlarged tips narrower than stigmas); stigmas lanceolate, brown when dry; pistillate spikes usually separated from staminate spikes by gap, in fruit mostly 19–25 mm thick; compound pedicels on denuded axis 0.6–2 mm; seeds absent or few; staminate scales brownish; pollen a mixture of tetrads, triads, dyads, and single grains, sometimes mostly single grains.

   3. Mucilage glands absent from blade; pistillate spikes after flowering medium to dark brown, rarely bright orange-brown. . . . . *Typha angustifolia* × *T. latifolia* (*T.* ×*glauca*)

   3. Mucilage glands usually present on adaxial surface of blade near sheath summit; pistillate spikes after flowering bright orange-brown. . . . . *Typha domingensis* × *T. latifolia*

1. Pistillate bracteoles present, many as wide as or wider than stigmas, evident at spike surface; stigmas linear (to narrowly lanceolate), sometimes deciduous and thus absent from mature spikes; pistillate spikes brown at all stages (or whitish when flowering and fresh; *T. angustifolia* sometimes greenish in fruit when fresh), in fruit mostly 13–25 mm thick; carpodia often evident at spike surface among pistil-hair tips; compound pedicels on denuded axis 0.5–0.9 mm; staminate scales brown or straw-colored.

   4. Mucilage glands absent from adaxial surface of blade and generally from central part of sheath near sheath summit; pistillate bracteole tips darker than (or as dark as) stigmas, very dark to medium brown, rounded (to acute), in mature spikes about equaling pistil hairs; pistil-hair tips medium brown, distinctly enlarged at 10–20× magnification; pistillate spikes medium to dark brown; leaf sheath summits with membranous auricles (often disintegrating late in season). . . . . .2. *Typha angustifolia*

   4. Mucilage glands present on adaxial surface of all of sheath and usually about 1–10 cm of adjacent blade; pistillate bracteole tips much paler than to about same color as stigmas, straw-colored to light brown, mostly acute to acuminate, in mature spikes exceeding pistil hairs; pistil-hair tips colorless to usually orangish (or slightly brownish in hybrids), not evidently enlarged, or often with 1 subapical, orange, swollen cell evident at 20–30×; pistillate spikes bright cinnamon- to orange- or medium brown; leaf sheath summits tapered to blade or sometimes with membranous auricles.

      5. Pistillate bracteole blades much paler than to nearly same color as stigmas, straw-colored to mostly bright orange-brown, usually many acuminate; pistillate spikes usually bright cinnamon- to orange-brown; mucilage glands numerous on proximal 1–10 cm of leaf blade. . . . . .3. *Typha domingensis*

      5. Pistillate bracteole blades usually about same color as stigmas, light- to medium brown, usually acute; pistillate spikes usually medium brown; mucilage glands often few or absent from leaf blade. . . . . *Typha angustifolia* × *T. domingensis*

1. Typha latifolia Linnaeus, Sp. Pl. 2: 971. 1753

• Broad-leaved cat-tail, tule espedilla, quenouille à. feuilles larges F W

Erect shoots 150–300 cm; flowering shoots 1–2 cm thick in middle, stems 3–7 mm thick near inflorescence. **Leaves:** usually glaucous when fresh; sheath sides papery or membranous, margins narrowly clear, summit tapered into blade to distinctly shouldered, or rarely with firm, papery auricles; mucilage glands at sheath-blade transition usually colorless, obscure, absent from sheath center and blade; widest blades on shoot 10–23(–29) mm wide when fresh, 5–20 mm when dry, distal blades about equaling inflorescence. **Inflorescences:** staminate spikes contiguous with pistillate or in some clones separated by to 4(–8) cm of naked axis, about as long as pistillate, ca. 1–2 cm thick at anthesis; staminate scales colorless to straw-colored, filiform, simple, ca. 4 × 0.05 mm; pistillate spikes in flower pale green drying brownish, later blackish brown or reddish brown, in fruit often mottled with whitish patches of pistil-hair tips, 5–25 cm × 5–8 mm in flower, 24–36 mm thick in fruit; compound pedicels in fruit bristlelike, variable in same

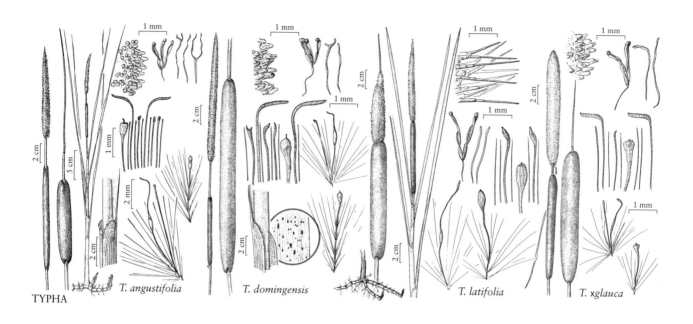

TYPHA   *T. angustifolia*   *T. domingensis*   *T. latifolia*   *T. xglauca*

spike, 1.5–3.5 mm; pistillate bracteoles absent. **Staminate flowers** 5–12 mm; anthers 1–3 mm, thecae yellow, apex dark brown; pollen in tetrads. **Pistillate flowers** 2–3 mm in flower, 10–15 mm in fruit; pistil-hair tips colorless, whitish in mass, not enlarged; stigmas persistent, forming solid layer on spike surface, pale green in flower, drying brownish, then reddish brown or usually distally blackish, spatulate, ovate to ovate-lanceolate, 0.6–1 × 0.2–0.25 mm; carpodia exceeded by and hidden among pistil hairs, straw-colored, apex rounded. **Seeds** numerous. $2n = 30$.

Flowering late spring–summer in north, spring–early summer in south. Fresh to slightly brackish water or wet soil; 0–2300 m; Alta., B.C., Man., N.B., Nfld. and Labr. (Nfld.), N.W.T., N.S., Ont., P.E.I., Que., Sask., Yukon; Ala., Alaska, Ariz., Ark., Calif., Colo., Conn., Del., Fla., Ga., Idaho, Ill., Ind., Iowa., Kans., Ky., La., Maine, Md., Mass., Mich., Minn., Miss., Mo., Mont., Nebr., Nev., N.H., N.J., N.Mex., N.Y., N.C., N.Dak., Ohio, Okla., Oreg., Pa., R.I., S.C., S.Dak., Tenn., Tex., Utah, Vt., Va., Wash., W.Va., Wis., Wyo.; Mexico; Central America; South America; Eurasia; Africa; introduced, Australia (Tasmania).

The erect shoots of *Typha latifolia* are more fanlike when young than in other North American species because the proximal leaves (dying by mid season) spread more widely. Undoubtedly native throughout its North American range, where it is often a codominant or minor component of marshes, wet meadows, fens, and other communities. In many places it is apparently being replaced by *T. angustifolia* and *T. angustifolia* × *T. latifolia* (*T.* ×*glauca*) at least partly due to human disturbance of habitats. Locally in California and perhaps

elsewhere where hybrids are common, the pollen grains of some *T. latifolia* plants separate slightly and may be shed partly as mixtures of triads, dyads, and monads, perhaps due to introgression (S. G. Smith, unpublished). See also hybrids in key and genus.

SELECTED REFERENCES Kaul, R. B. 1974. Ontogeny of foliar diaphragms in *Typha latifolia*. Amer. J. Bot. 61: 318–323. Rowlatt, U. and H. Morshead. 1992. Architecture of the leaf of the greater reed mace, *Typha latifolia* L. Bot. J. Linn. Soc. 110: 161–170. Yeo, R. R. 1964. Life history of common cattail. Weeds 12: 284–288.

2. **Typha angustifolia** Linnaeus, Sp. Pl. 2: 971. 1753
 · Narrow-leaved cat-tail, quenouille à feuilles étroites
F W

Erect shoots 150–300 cm, not glaucous; flowering shoots 5–12 mm thick in middle; stems 2–3 mm thick near inflorescence. **Leaves:** sheath sides membranous, margin broadly clear, summit with membranous auricles which often disintegrate late in season; mucilage glands at sheath-blade transition brown, absent from blade and usually from sheath center near summit; widest blades on shoot 4–12 mm wide when fresh, 3–8 mm when dry; distal blade usually markedly exceeding inflorescence. **Inflorescences:** staminate spikes separated from pistillate by 1–8(–12) cm of naked axis, ca. as long as pistillate, 1 cm thick in anthesis; staminate scales variable in same spike, straw-colored to medium brown, filiform, simple to bifid or sometimes cuneate and irregularly branched, to 6 × 0.1 mm; pistillate spikes in flower when fresh dark brown with whitish stigmas (drying brown), later medium brown, in fruit when fresh as stigmas wear off often greenish due to green carpodia, (4–)6–20 cm × 5–6 mm in flower, 13–22 mm in fruit; compound pedicels in fruit peglike, 0.5–0.7 mm; pistillate bracteole blades forming spike surface before flowering, later exceeded by stigmas and about equaling or slightly exceeded by pistil hairs, very dark to medium brown, much darker than (or sometimes as dark as) stigmas, irregularly spatulate, 0.6 × 0.1–0.2 mm, wider than or about as wide as stigmas, apex rounded (to acute). **Staminate flowers** 4–6 mm; anthers 1.5–2 mm, thecae yellow, apex dark brown; pollen in monads or some in irregular clusters. **Pistillate flowers** 2 mm in flower, 5–7 mm in fruit; pistil-hair tips medium brown, distinctly swollen at 10–20×; stigmas sometimes deciduous in fruit, in flower erect, elongating, bending to form surface mat, white in flower, drying brownish, later medium brown, narrowly linear-lanceolate, 0.6–1.4 × 0.1 mm; carpodia slightly exceeded by and visible among pistil hairs at mature spike surface, green when young and fresh, straw-colored with orange-brown spots when dry, apex nearly truncate. $2n = 30$.

Flowering late spring–summer. Often somewhat brackish or subsaline water or wet soil; 0–1900 m; B.C., Man., N.B., N.S., Ont., P.E.I., Que., Sask.; Ark., Calif., Colo., Conn., Del., Ill., Ind., Iowa, Kans., Ky., Maine, Md., Mass., Mich., Minn., Mo., Nebr., Nev., N.H., N.J., N.Mex., N.Y., N.C., N.Dak., Ohio, Okla., Oreg., Pa., R.I., S.C., S.Dak., Tenn., Vt., Va., Wash., W. Va., Wis., Wyo.; Eurasia.

Prior to N. Hotchkiss and H. L. Dozier (1949), *Typha domingensis* was generally included within *T. angustifolia* in North America. Because of many misidentified specimens, range expansion in recent years, and undercollecting, the distribution on the margins of the main range is somewhat uncertain. Many literature reports are based on misidentified specimens. Some workers suggested *T. angustifolia* was early introduced from Europe into Atlantic Coastal North America and migrated westward (R. L. Stuckey and D. P. Salamon 1987). In recent decades it has expanded its range in many regions and become much more abundant, especially in roadside ditches and other highly disturbed habitats. For example, although it was known only from one Wisconsin station in 1929 (N. C. Fassett 1930) and was very local in Iowa in 1939 (A. Hayden 1939), it is now common and widespread in both states. As it often out-competes many native marsh species to produce very dense, pure stands, and hybridizes with *T. latifolia* to form the probably even more competitive *T.* ×*glauca*, *T. angustifolia* and *T.* ×*glauca* should perhaps be classified as noxious weeds in parts

of North America. Beyond the main range of *T. angustifolia*, there are specimens of *T.* ×*glauca* from north-central Montana (Phillips County), west-central Manitoba (La Pas), and Anticosti Island, Quebec. Many erroneous reports have come from outside of Europe and North America. For hybrids see also genus and key.

SELECTED REFERENCE Stuckey, R. L. and D. P. Salamon. 1987. *Typha angustifolia* in North America: A foreigner masquerading as a native. [Abstract.] Ohio J. Sci. 87: 4.

3. **Typha domingensis** Persoon, Syn. Pl. 2: 532. 1807 • Southern cat-tail

Erect shoots 150–400 cm, not glaucous; flowering shoots 1–2 cm thick in middle; stems 3–4 mm thick near spike. **Leaves:** sheath sides membranous, margin broadly clear, summit tapered to blade or with persistent, membranous auricles; mucilage glands at sheath-blade transition orange-brown, numerous on entire sheath and proximal 1–10 cm of blade; widest blades on shoot 6–18 mm wide when fresh, 5–15 mm when dry; distal blade about equaling inflorescence. **Inflorescences:** staminate spike separated from pistillate by (0–)1–8 cm of naked axis, ca. 1.4 × longer than pistillate, 1 cm thick at anthesis; staminate scales straw-colored to mostly bright orange-brown, variable in same spike, linear to cuneate, often laciniate distally, to 3–4 × 0.3 mm; pistillate spikes in flower when fresh bright cinnamon-brown with whitish stigmas (drying brownish), later orange- (to medium) brown, in fruit generally paler as stigmas and often bracteole blades wear off, ca. 6–35 cm × 5–6 mm in flower, 15–25 mm in fruit; compound pedicels in fruit peglike, ca. 0.6–0.9 mm; pistillate bracteole blades forming spike surface before flowering, later slightly exceeded by stigmas and slightly exceeding pistil hairs, straw-colored to bright orange-brown, much paler than to nearly same color as stigmas, irregularly narrowly to broadly spatulate or lanceolate, 0.8 × 0.1–0.3 mm, mostly wider than stigmas, apex variable in same inflorescence or different plants, acute or acuminate. **Staminate flowers** 5 mm; anthers 2–2.5 mm, thecae yellow, apex bright orange-brown; pollen in single grains. **Pistillate flowers** 2 mm in flower, 8–9 mm in fruit; pistil-hair tips straw-colored to orange-brown in mass, usually with 1 subapical bright orange-brown, generally enlarged cell; stigmas often deciduous in fruit, in flower erect, elongating, bending to form surface mat, white in flower when fresh, later bright orange-brown, narrowly linear-lanceolate, ca. 1 × 0.1 mm; carpodia slightly exceeded by pistil hairs, usually evident at fruiting spike surface, straw-colored, orange-spotted, apex broadly rounded. $2n = 30$.

Flowering spring–summer. Often in brackish water or wet soil; 0–2000 m; Ala., Ariz., Ark., Calif., Colo., Del., Fla., Ga., Ill., Kans., Ky., La., Md., Miss., Mo., Nebr., Nev., N.Mex., N.C., Okla., S.C., Tex., Utah, Va., Wyo.; Mexico; Central America; South America; West Indies; Eurasia; Africa; Pacific Islands (New Zealand); Australia.

Typha domingensis aggressively invades and forms nearly pure stands in brackish or nutrient-enriched wetlands in the Florida Everglades and elsewhere. It is established but does not mature fruits on the cold coast of northern California. There are specimens of putative hybrids with *T. angustifolia* beyond the main range of *T. domingensis*, in southeastern and northwestern Nebraska and southeastern Kentucky, and with *T. latifolia* in southeastern Nebraska. Vegetative *T. domingensis* or hybrids occur on the Atlantic Coast north as far as Delaware. The northern Illinois locality is a power plant cooling pond. The Wyoming record is from a hot spring and may be a hybrid with *T. latifolia*. *Typha domingensis* probably should be treated as a highly variable pantropic and warm temperate species, occurring to 40° north and south latitude worldwide, and needing study to determine infraspecific taxa and delimitation from related species (B. G. Briggs and L. A. S. Johnson 1968; S. G. Smith 1987). For hybrids see also genus and key.

214. BROMELIACEAE Jussieu

• Bromeliad or Pineapple Family

Harry E. Luther

Gregory K. Brown

Herbs, perennial, terrestrial, among or on rocks, or epiphytic. **Roots** usually present, often poorly developed in epiphytic taxa. **Stems** very short to very elongate. **Leaves** usually spirally arranged, forming water-impounding rosette, occasionally lax and/or 2-ranked, simple, margins serrate or entire, trichomes nearly always covering surface, peltate, water-absorbing. **Inflorescences** terminal or lateral, sessile to scapose, simple or compound; bracts usually present, conspicuous. **Flowers** bisexual or functionally unisexual, radially symmetric to slightly bilaterally symmetric; perianth in 2 distinct sets of 3; stamens in 2 series of 3; ovary inferior or superior; placentation axile. **Fruits** capsules or berries. **Seeds** plumose, winged, or unappendaged.

Genera 56, species 2600+ (4 genera, 19 species, and 2 natural hybrids in the flora): widely distributed in the Neotropics (1 species in West Africa).

Bromeliaceae contain three subfamilies: Bromelioideae, Pitcairnioideae, and Tillandsioideae. Generic circumscriptions are problematic, especially in parts of the Bromelioideae and Tillandsioideae.

Pineapple, *Ananas comosus* (Linnaeus) Merrill, the only agriculturally important member of the family, is in worldwide cultivation in tropical climates. Horticultural interest in bromeliads is widespread among the public; the Bromeliad Society, Inc. caters to that interest.

SELECTED REFERENCES Brown, G. K. and A. J. Gilmartin. 1989. Chromosome numbers in Bromeliaceae. Amer. J. Bot. 76: 657–665. Brown, G. K. and A. J. Gilmartin. 1989b. Stigma types in Bromeliaceae—A systematic survey. Syst. Bot. 14: 110–132. Brown, G. K. and R. G. Terry. 1992. Petal appendages in Bromeliaceae. Amer. J. Bot. 79: 1051–1071. Smith, L. B. and R. J. Downs. 1974. Pitcairnioideae. In: Organization for Flora Neotropica. 1968+. Flora Neotropica. 75+ nos. New York. No. 14(1). Smith, L. B. and R. J. Downs. 1977. Tillandsioideae. In: Organization for Flora Neotropica. 1968+. Flora Neotropica. 75+ nos. New York. No. 14(2). Smith, L. B. and R. J. Downs. 1979. Bromelioideae. In: Organization for Flora Neotropica. 1968+. Flora Neotropica. 75+ nos. New York. No. 14(3).

1. Leaf margins spinose; flowers functionally unisexual, staminate and pistillate on different plants; seeds narrowly winged to almost wingless, plumose appendages absent. . . . . 1. *Hechtia*, p. 287
1. Leaf margins entire; flowers bisexual (in flora) or functionally unisexual; seeds not winged, plumose appendages basal or apical.

2. Inflorescences 2-ranked, 1–50(–200)-flowered. . . . . 2. *Tillandsia*, p. 288
2. Inflorescences many-ranked, 5–many-flowered.
   3. Floral bracts broad, conspicuous, mostly obscuring rachis, flowers laxly to densely
     arranged. . . . . 3. *Guzmania*, p. 296
   3. Floral bracts small, inconspicuous, not obscuring rachis, flowers laxly arranged. . . . 4. *Catopsis*, p. 297

1. HECHTIA Klotzsch, Allg. Gartenzeitung 3: 401. 1835

Kathleen Burt-Utley

John F. Utley

Herbs, terrestrial or among or on rocks, caulescent or without evident stems, occasionally stoloniferous. **Leaves** rosulate, straight to falcate; blade narrowly triangular, margins spinose. **Inflorescences** lateral [terminal or lateral], many-flowered, simple or compound and paniculate; branches capitate to lax and elongate; floral bracts broad [broad to narrow], conspicuous. **Flowers** dimorphic, functionally unisexual, staminate and pistillate on different plants; sepals distinct, symmetric; staminate petals distinct or appearing briefly connate basally from adnation of filaments to adjacent petals; stamens included or exserted; filaments adnate [free or adnate] to petals; pistillate petals distinct, triangular or ovate; ovary largely superior to largely inferior. **Capsules** ovoid, dehiscent. **Seeds** narrowly winged to almost wingless, plumose appendages absent.

Species ca. 50 (2 in the flora): North America, Mexico, Central America.

The Texas species are unusual in their scaly capsules and flowers with scaly sepals, petals, and ovaries.

SELECTED REFERENCE Burt-Utley, K. and J. F. Utley. 1987. Contributions toward a revision of *Hechtia* (Bromeliaceae). Brittonia 39: 37–43.

1. Staminate flowers with sepals 2.5–4 mm, petals 3.5–4.5 mm; pistillate flowers with sepals
   3–4 mm, petals 4.5–5 mm; capsules 6–8.5 mm. . . . . 1. *Hechtia glomerata*
1. Staminate flowers with sepals 4.5–5 mm, petals 7.5–9 mm; pistillate flowers with sepals
   4.5–7 mm, petals 7.5–9 mm; capsules 8.5–12.5 mm. 2. *Hechtia texensis*

1. Hechtia glomerata Zuccarini, Abh. Math-Phys. Cl. Königl. Bayer. Akad. Wiss. 3: 240, plate 6. 1840
  • Guapilla ⬚F⬚

Plants 0.7–2.8 m in flower. **Leaves** pungent; blade 21–98 × 2–6 cm. **Inflorescences** 2–3-pinnately compound, scaly; proximal primary bracts often inconspicuous, shorter than lateral branches; lateral branches laxly or densely flowered, appearing capitate or cylindric, 0.5–40 cm. **Staminate flowers** sessile, scaly; floral bracts 2.5–3.5(–5.5) mm, exceeded by sepals; sepals ovate, 2.5–4 mm; petals 3.5–4.5 mm. **Pistillate flowers** sessile, scaly; floral bracts 3.5–4.5(–6) mm, exceeded by sepals; sepals ovate, 3–4 mm; petals ovate, 4.5–5 mm; ovary almost wholly superior. **Capsules** 6–8.5 mm, scaly, glabrescent with age.

Flowering spring–summer. In gravel and on sandstone formations; 0–150 m; Tex.; Mexico; Central America.

2. Hechtia texensis S. Watson, Proc. Amer. Acad. Arts 20: 374. 1885 • False-agave

Hechtia scariosa L. B. Smith

Plants 0.7–1.3 m in flower. **Leaves** pungent; blade to 44 × 1.5–4.5 cm. **Inflorescences:** staminate and pistillate inflorescences 2–3-pinnately compound, scaly; proximal primary bracts inconspicuous, shorter than lateral branches; lateral branches laxly to densely flowered, occasionally appearing capitate,

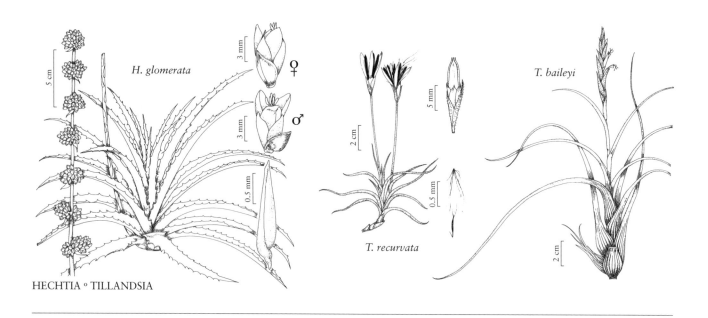

H. glomerata

T. baileyi

T. recurvata

HECHTIA ° TILLANDSIA

1–21.5 cm. **Staminate flowers** nearly sessile, scaly; floral bracts 4.5–7 mm, exceeded by sepals; sepals ovate to ovate-elliptic, 4.5–5 mm; petals 7.5–9 mm. **Pistillate flowers** subsessile, scaly; floral bracts 4–6 mm, exceeded by sepals; sepals ovate to broadly ovate, 4.5–7 mm; petals ovate, 7.5–10 mm; ovary almost wholly superior. **Capsules** 8.5–12.5 mm, scaly, glabrescent with age.

Flowering spring–early summer. Dry limestone slopes and bedrock, gravelly soils; 600–1150 m; Tex.; Mexico.

Hechtia texensis was considered to be endemic to Texas and known only from the type collection, while all other collections from trans-Pecos Texas, except those of *H. glomerata*, were recognized as *H. scariosa* (D. S. Correll and M. C. Johnston 1970; L. B. Smith and R. J. Downs 1974). *Hechtia texensis* was distinguished from *H. scariosa* by sepal shape and size, using pistillate sepals from *H. texensis* and staminate sepals from *H. scariosa*. When flowers of the same sex from each taxon were compared, it was impossible to distinguish *H. texensis* from *H. scariosa* using sepal shape or other floral characters (K. Burt-Utley and J. F. Utley 1987).

2. TILLANDSIA Linnaeus, Sp. Pl. 1: 286. 1753; Gen. Pl. ed. 5, 138. 1754 • [For the Swedish botanist E. Tillands, 1640–1693]

Harry E. Luther

Gregory K. Brown

Herbs, usually epiphytic, stemless to long caulescent. **Leaves** mostly many-ranked, rosulate, or occasionally 2-ranked and/or laxly arranged; blade linear to triangular or ligulate, margins entire, trichomes usually conspicuous. **Inflorescences** central, 1–many-flowered, 2-ranked; floral bracts mostly broad, conspicuous, rachis covered or exposed at anthesis. **Flowers** bisexual; sepals distinct or adaxial pair connate, usually symmetric; petals distinct; stamens included or exserted, filaments free; ovary superior. **Capsules** cylindric, dehiscent. **Seeds** with basal, white, plumose appendage.

Species ca. 550 (13 species and 2 described natural hybrids in the flora): widespread in the Neotropics.

SELECTED REFERENCES Gardner, C. S. 1982. Systematic Study of *Tillandsia* Subgenus *Tillandsia*. Ph.D. dissertation. Texas A&M University. Gardner, C. S. 1984. New species and nomenclatural changes in Mexican *Tillandsia*—I. Selbyana 7: 361–379. Luther, H. E. 1985. Notes on hybrid tillandsias in Florida. Phytologia 57: 175–176.

1. Leaves 2-ranked; inflorescence 1–2(–3)-flowered.
    2. Stems elongate, plants forming pendent festoons; inflorescence reduced to single, apparently sessile flower, scape concealed within leaf sheath; petals yellow-green. . . . . . . 1. *Tillandsia usneoides*
    2. Stems short, plants densely clustered to form spheric masses; inflorescence of (1–)2(–3) flowers, scape conspicuous; petals violet. . . . . 2. *Tillandsia recurvata*
1. Leaves many-ranked; inflorescence more than 3-flowered.
    3. Floral bracts spreading and/or small, exposing most of rachis at anthesis.
        4. Leaves 10–20, spirally twisted, banded silver; floral bracts 2.3–3.1 cm; inflorescences with 2–6 flowers per branch; corolla pink to dark rose. . . . . 14. *Tillandsia flexuosa*
        4. Leaves more than 20, not twisted or banded (rarely variegated with linear cream stripes); floral bracts 1.2–2 cm; inflorescence with 6–11 flowers per branch; corolla white. . . . . 15. *Tillandsia utriculata*
    3. Floral bracts imbricate, broad, covering all or most of rachis, rachis not visible at anthesis.
        5. Leaf sheath conspicuously inflated; plants pseudobulbous.
            6. Leaf blade channeled to involute; scapes 8–35 cm; floral bracts glabrous to inconspicuously scaly near apex only.
                7. Leaf blade and scape bracts twisted, reflexed. . . . . 6. *Tillandsia balbisiana*
                7. Leaf blade and scape bracts erect to spreading. . . . . 7. *Tillandsia* ×*smalliana*
            6. Leaf blade involute; scapes 1–15 cm; floral bracts densely scaly.
                8. Scape 7–15 cm; plants from extreme s Texas. . . . . 3. *Tillandsia baileyi*
                8. Scape 6 cm or less; plants from Florida.
                    9. Inflorescence linear to narrowly elliptic; base of floral bracts visible at anthesis. . . . . 4. *Tillandsia paucifolia*
                    9. Inflorescence broadly elliptic; base of floral bracts not visible at anthesis . . . . 5. *Tillandsia pruinosa*
        5. Leaf sheath flat or only slightly inflated; plants not pseudobulbous.
            10. Leaf blade obviously narrowly triangular, tapering evenly from base to apex.
                11. Leaves soft, brittle; inflorescence simple or laxly 2–3-branched, never palmate; floral bracts 0.6–0.9 cm wide. 12. *Tillandsia variabilis*
                11. Leaves stiff, leathery; inflorescence densely palmate to laxly 2(–3)-pinnate with 3–15 branches; floral bracts 1.2–2 cm wide. . . . . 13. *Tillandsia fasciculata*
            10. Leaf blade linear-triangular to filiform.
                12. Leaf sheath narrowly elliptic, 1.2–2 cm wide, slightly inflated. . . . . 8. *Tillandsia simulata*
                12. Leaf sheath broadly elliptic to triangular, 0.8–2.5 cm wide, flat.
                    13. Leaves finely appressed-scaly, appearing green to reddish green; sheath of scape bracts, especially distal ones, narrowing abruptly into blade; floral bracts green or tinged red, uniformly scaly; corolla lavender. . . 9. *Tillandsia setacea*
                    13. Leaves densely and coarsely appressed-scaly, appearing gray; sheath of scape bracts narrowing gradually into blade; floral bracts uniformly red to rose, scaly distally, becoming sparsely scaly toward base; corolla violet.
                        14. Leaves 15–30, not over 0.5 cm wide at mid-length; inflorescence simple or with 1–5 lateral branches. . . . . 10. *Tillandsia bartramii*
                        14. Leaves 20–50, 0.5–1 cm wide at mid-length; inflorescence 2-pinnate with 2–10 lateral branches.. . . . . 11. *Tillandsia* ×*floridana*

1. Tillandsia usneoides (Linnaeus) Linnaeus, Sp. Pl. ed. 2, 1: 411. 1762 • Spanish-moss, long-moss, black-moss, mousse espagnole, mousse

Renealmia usneoides Linnaeus, Sp. Pl. 1: 287. 1753; *Dendropogon usneoides* (Linnaeus) Rafinesque

Plants pendent in long festoons, flowering to 300 cm. **Stems** elongate. **Leaves** 4–8, 2-ranked, often twisted or contorted, gray to silver-gray, 1.5–3 × 0.1–0.2 cm, densely grayish-scaly; sheath pale, narrowly elliptic, not inflated, not forming pseudobulb, 0.2–0.4 cm wide; blade filiform, succulent, margins involute to nearly tubular, apex acute. **Inflorescences:** scape concealed within leaf sheath, appearing scapeless, pendent with shoot, ± 1 mm diam. **Floral bracts** enveloping flower, erect, green, broad (covering all or most of rachis, rachis not visible at anthesis), ovate, not keeled, 0.4–0.5 cm, thin-leathery, apex acute, surfaces densely grayish-scaly, venation even to slight. **Flowers** solitary, inconspicuous, apparently sessile; sepals free, ovate, not keeled, 0.6–0.7 cm, thin, veined, apex acute, surfaces glabrous; corolla spreading, petals spreading, yellow-green, elliptic, to 1 cm; stamens included; stigma included, simple-erect. **Fruits** to 2.5 cm.

Flowering summer. Epiphytic, occasionally on fences, telephone lines; 0–300 m; Ala., Ark., Fla., Ga., La., Miss., N.C., S.C., Tex., Va.; Mexico; West Indies; Central America; South America.

2. Tillandsia recurvata (Linnaeus) Linnaeus, Sp. Pl. ed. 2, 1: 410. 1762 • Ball-moss F

Renealmia recurvata Linnaeus, Sp. Pl. 1: 287. 1753; *Diaphoranthema recurvata* (Linnaeus) Beer

Plants in dense spheric clusters, flowering to 15 cm diam. **Stems** short. **Leaves** 4–10, 2-ranked, recurving, gray, 6–12 × 0.2–0.3 cm, densely pruinose-scaly; sheath pale, elliptic, not inflated, not forming pseudobulb, 4–8 mm wide; blade subulate, terete distally, succulent, margins involute to nearly tubular, apex acute to attenuate. **Inflorescences:** scape conspicuous, erect, 2–5 cm, ± 1 mm diam.; bracts 1–2, widely spaced, erect, inconspicuous, nearly foliaceous; sheath of bracts narrowing gradually into blade; spikes ascending, subpalmate, elliptic, compressed, 8–15 × 4–6 mm, apex acute; lateral branches absent. **Floral bracts** laxly imbricate, erect, green, tinged purple, broad (covering all or most of rachis, rachis not visible at anthesis), narrowly elliptic, not keeled, 0.8–1 cm, thin-leathery, apex acute, surfaces densely grayish-scaly, venation

even to slight. **Flowers** usually 2, conspicuous; sepals free, lanceolate, not keeled, 6–8 mm, thin, veined, apex acute, surfaces glabrous; corolla tubular; petals spreading toward apex, violet, elliptic, 0.7–1 cm; stamens included; stigma included, simple-erect. **Fruits** to 3 cm.

Flowering summer. Epiphytic to occasionally among or on rocks (Arizona, Texas), usually in bright, exposed habitats; 0–1500 m; Ariz., Fla., Ga., La., Tex.; Mexico; West Indies; Central America; South America.

3. Tillandsia baileyi Rose ex Small, Fl. S.E. U.S., 246, 1328. 1903 F

Plants usually several individuals in cluster, flowering to 20–40 cm. **Stems** short. **Leaves** 6–14 in small rosette, many-ranked, slightly contorted, gray-green to silver, 5–40 × 0.3–0.7 cm, densely appressed-grayish-scaly; sheath pale to nearly chestnut brown within, ovate, conspicuously inflated, passing gradually into blade, forming small pseudobulb, 1.5–2.5 cm wide; blade linear, semisucculent, margins involute, apex acute. **Inflorescences:** scape conspicuous, erect or ascending, 7–15 cm, 2–3 mm diam.; bracts densely imbricate, erect to spreading, like leaves but smaller; sheath of bracts narrowing gradually into blade; spikes usually ascending, pinnate, linear, compressed, 3–7 × 0.6–1 cm, apex acute to obtuse; branches rarely 1–2. **Floral bracts** imbricate, erect, pink to dark rose, broad (covering all or most of rachis, rachis not visible at anthesis), elliptic, not keeled, 1.5–2.3 cm, thin-leathery, base not visible at anthesis, apex broadly acute, surfaces densely grayish-scaly. **Flowers** 5–15, conspicuous; sepals with adaxial pair short-connate, lanceolate, keeled, 1.3–1.6 cm, papery, prominently veined, apex acute, surfaces scaly; corolla tubular; petals erect, purple, ligulate, to 3 cm; stamens exserted; stigma exserted, conduplicate-spiral. **Fruits** 2.5–4 cm.

Flowering spring. Epiphytic in dry thickets, woods; 5–100 m; Tex.; Mexico.

4. Tillandsia paucifolia Baker, Gard. Chron., ser. 2, 10: 748. 1878

Plants single or clustering, flowering to 15 cm. **Stems** short. **Leaves** 5–10, many-ranked, recurved to twisted, silvery gray, to 12 × 0.3–0.6 cm, grayish-scaly; sheath nearly chestnut brown within, ovate, conspicuously inflated, forming small pseudobulb, 1.5–3.5 cm wide; blade narrowly triangular, semisucculent, margins involute, apex obtuse to nearly acute. **Inflorescences:** scape short and conspicous, erect or ascending, 1–6 cm, 2–3 mm diam.; bracts densely imbricate, erect to spreading, like leaves but gradually smaller; sheath of bracts narrowing gradually into blade; single spikes, or with 2–4 lateral spikes, usually ascending, pinnate, linear to narrowly elliptic, compressed, 2–7 × 0.6–1 cm, apex acute to obtuse. **Floral bracts** imbricate, erect, pale pink, broad (covering all or most of rachis, rachis not visible at anthesis), elliptic, not keeled, 2–3 cm, thin-leathery, base visible at anthesis, apex broadly acute to obtuse, surfaces densely grayish-scaly. **Flowers** 2–15, conspicuous; sepals with adaxial pair short connate, lanceolate, keeled, to 2 cm, thin-leathery to papery, veined, apex acute, surfaces sparsely scaly; corolla tubular; petals erect, lavender-blue, ligulate, to 4 cm; stamens exserted; stigma exserted, conduplicate-spiral. **Fruits** to 4 cm.

Flowering spring–summer. Epiphytic in bright exposed habitats; 0–30 m; Fla.; Mexico; West Indies; Central America; South America.

The leaf blade of *Tillandsia paucifolia* is often shorter than the leaf sheath. The name *T. circinnata* has been misapplied to this species.

5. Tillandsia pruinosa Swartz, Fl. Ind. Occid. 1: 594. 1797

Plants usually single, rarely clustering, flowering to 10 cm. **Stems** short. **Leaves** 5–10, many-ranked, contorted or secund-spreading, gray-green to silver, 6–10 × 0.3–0.5 cm, densely pruinose-scaly, scales coarse; sheath dark chestnut brown within, ovate to elliptic, conspicuously inflated, forming small pseudobulb, 1.5–3 cm wide; blade linear-subulate, semisucculent, margins involute, apex attenuate. **Inflorescences:** scape inconspicuous, usually ascending, 1–3 cm, 2–3 mm diam.; bracts densely imbricate, erect to spreading, like leaves but gradually smaller; sheath of bracts narrowing abruptly into blade; single spikes, usually ascending, pinnate, broadly elliptic, compressed, 2–3 × 1.2–2.4 cm, apex acute to obtuse. **Floral bracts** imbricate, erect, pink, broad (covering all or most of rachis, rachis not visible at anthesis), ovate, keeled, 2–2.2 cm, thin-leathery, base not visible at anthesis, apex broadly acute, surfaces densely scaly, venation even to slight. **Flowers** 3–12, conspicuous; sepals free, elliptic, adaxial pair keeled, 1.2–1.8 cm, thin-leathery, veined, apex obtuse, surfaces glabrous; corolla tubular; petals erect, blue-violet, ligulate, to 3 cm; stamens exserted; stigma exserted, conduplicate-spiral. **Fruits** to 3.5 cm.

Flowering spring–summer. Epiphytic in shady, humid hammocks; 0–30 m; Fla.; West Indies; Central America; South America.

In the flora area, *Tillandsia pruinosa* is infrequent but locally abundant.

6. Tillandsia balbisiana Schultes f. in J. J. Roemer and J. A. Schultes, Syst. Veg. 7(2): 1212. 1830 [F]

Plants single or clustering, flowering to 75 cm. **Stems** short. **Leaves** 15–30, many-ranked, reflexed, twisted or contorted, gray, often flushed red, to 65 × 0.6–1.4 cm, appressed-grayish-scaly; sheath conspicuously rust-colored toward base, ovate to elliptic, conspicuously inflated, forming small pseudobulb, 2–4 cm wide; blade linear-triangular, leathery, channeled to involute, apex attenuate. **Inflorescences:** scape conspicuous, erect, 8–30 cm, 2–4 mm diam.; bracts densely imbricate, spreading, recurved and twisted like leaves; sheath of bracts narrowing gradually into blade; spikes erect, 2-pinnate, linear, compressed, 2–10 × 1 cm, apex acute; lateral branches 2–10 (rarely simple). **Floral bracts** imbricate, erect, green to red, broad (covering all or most of rachis, rachis not visible at anthesis), elliptic, keeled, 1.5–2 cm, leathery, base not visible at anthesis, apex acute, surfaces glabrous to inconspicuously scaly near apex only, venation even to slight. **Flowers** 5–30, conspicuous; sepals with adaxial pair connate, lanceolate, keeled, 1.5–2 cm, leathery, apex acute, surfaces glabrous; corolla tubular; petals erect, violet, ligulate, to 3.5 cm; stamens exserted; stigma exserted, conduplicate-spiral. **Fruits** to 4 cm.

Flowering spring–summer. Epiphytic on a variety of hosts in open woods, cypress swamps, coastal forest; 0–30 m; Fla.; Mexico; West Indies; Central America; South America.

7. Tillandsia ×smalliana H. Luther, Phytologia 57: 176. 1985

Plants single or clustering, flowering to 50 cm. **Stems** short. **Leaves** 20–40, many-ranked, erect to spreading, gray, 30–45 × 1–2.2 cm, appressed-grayish-scaly; sheath dark rust-colored, broadly elliptic, conspicuously inflated, forming small pseudobulb, 3–4 cm wide; blade narrowly triangular, leathery, channeled to involute, apex attenuate. **Inflorescences:** scape conspicuous, erect, 15–35 cm, 3–6 mm diam.; bracts densely imbricate, erect to spreading, like leaves but gradually smaller; sheath

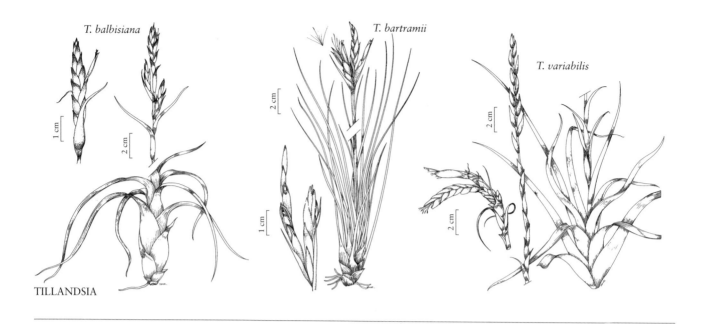

TILLANDSIA

of bracts narrowing gradually into blade; spikes erect, 2-pinnate, narrowly elliptic, compressed, 2–6 × 1–1.5 cm, apex acute; lateral branches 3–13. **Floral bracts** imbricate, erect, red, broad (covering all or most of rachis, rachis not visible at anthesis), elliptic, keeled, 2–2.5 cm, leathery, base not visible at anthesis, apex acute, surfaces glabrous, venation even to slight. **Flowers** 5–40, conspicuous; sepals with adaxial pair connate, lanceolate, keeled, 2–2.4 cm, thin-leathery, veined, apex acute, surfaces slightly scaly; corolla tubular, petals erect, violet, ligulate, to 5 cm; stamens exserted; stigma exserted, conduplicate-spiral. **Fruits** to 3.8 cm.

Flowering winter–summer. Epiphytic, usually on *Taxodium*, in swamps and well-lit hammocks; 0–30 m; Fla.; Central America.

Tillandsia ×*smalliana* has usually been misdetermined as *T. polystachia* (Linnaeus) Linnaeus, a common Caribbean species not known to occur in Florida.

The probable parentage of *Tillandsia* ×*smalliana* is *T. balbisiana* Schultes f. × *T. fasciculata* Swartz.

8. **Tillandsia simulata** Small, Man. S.E. Fl., 270, 1503. 1933 [E]

Plants single or clustering, flowering to 40 cm. **Stems** short. **Leaves** 15–30, many-ranked, erect to spreading, gray, 20–40 × 0.1–0.4 cm, densely appressed-grayish-scaly; sheath chestnut brown, narrowly elliptic, slightly inflated, not forming pseudobulb, 1.2–2 cm wide; blade very narrowly linear-triangular, leathery, margins involute, apex

attenuate. **Inflorescences:** scape conspicuous, erect, 10–15 cm, 4–6 mm diam.; bracts densely imbricate, erect, like leaves but gradually smaller; sheath of bracts narrowing abruptly into blade; spikes erect to spreading, palmate, linear, compressed, 2–5 × 1 cm, apex acute; lateral branches 1–5. **Floral bracts** imbricate, erect, rose, broad (covering all or most of rachis, rachis not visible at anthesis), elliptic, keeled, 1.4–1.8 cm, thin-leathery, base not visible at anthesis, apex acute, surfaces appressed-grayish-scaly, venation slight apically. **Flowers** 5–30, conspicuous; sepals with adaxial pair connate, elliptic, keeled, to 1.6 cm, thin-leathery, veined, apex obtuse, surfaces slightly scaly; corolla tubular, petals erect, violet, ligulate, 3–4.5 cm; stamens exserted; stigma exserted, conduplicate-spiral. **Fruits** to 3 cm.

Flowering spring. Epiphytic on a variety of hosts in swamps and moist hammocks, usually in strong light; 0–30 m; Fla.

Tillandsia simulata is a distinct species, but preserved material often is misdetermined as *Tillandsia bartramii* Elliott. L. B. Smith and R. J. Downs (1977) treated *T. simulata* as a synonym under *T. bartramii*.

9. **Tillandsia setacea** Swartz, Fl. Ind. Occid. 1: 593. 1797

Plants densely clustering, flowering to 30 cm. **Stems** short. **Leaves** 15–30, many-ranked, finely appressed, erect, green to reddish green, 20–30 × 0.1–0.4 cm, uniformly scaly throughout; sheath rust-colored, broadly triangular, flat, not forming pseudobulb, 0.8–1.8 cm wide; blade very narrowly linear-triangular, ribbed, leathery, margins involute, apex filiform-attenuate. **Inflorescences:** scape conspicuous, erect, 8–15 cm, 2–4 mm diam.; bracts densely imbricate, erect; sheath of bracts, especially upper ones, narrowing abruptly into blade; spikes erect, palmate (rarely simple), linear, compressed, 1–4 × 0.5–0.6 cm, apex acute; branches 1–5. **Floral bracts** imbricate, erect, green or tinged red, broad (covering all or most of rachis, rachis not visible at anthesis), ovate, keeled only toward apex, 0.8–1.2 cm, thin-leathery, base not visible at anthesis, apex attenuate to acute, surfaces appressed-pale scaly, venation slight. **Flowers** 3–15, conspicuous; sepals with adaxial pair connate, elliptic, keeled, 0.8–1 cm, thin-leathery, veined, apex acute, surfaces slightly scaly; corolla tubular, petals erect, lavender, ligulate, 1.8–2.2 cm; stamens exserted; stigma exserted, conduplicate-spiral. **Fruits** 2–3 cm.

Flowering spring–fall. Epiphytic on a variety of hosts in swamps and humid forests; 0–60 m; Fla., Ga.; Mexico; West Indies; Central America.

It is probable that Mexican and Central American materials represent one or more additional species.

10. Tillandsia bartramii Elliott, Sketch Bot. S. Carolina 1: 379. 1817 F

Tillandsia myriophylla Small

Plants densely clustering, flowering to 20–40 cm. **Stems** short. **Leaves** 15–30, many-ranked, erect to spreading, gray, 15–40 × 0.2–0.5 cm, coarsely appressed-scaly; sheath rust-colored toward base, broadly triangular, flat, not forming pseudobulb, 1–2 cm wide; blade linear-subulate, leathery, margins involute, apex attenuate. **Inflorescences:** scape conspicuous, erect, 8–15 cm, 2–4 mm diam.; bracts densely imbricate, erect; sheath of bracts narrowing gradually into blade; spikes erect, simple or palmate, linear, compressed, 2–4 × 1 cm, apex acute; simple or lateral branches 1–5. **Floral bracts** imbricate, erect, uniformly red to rose, broad (covering all or most of rachis, rachis not visible at anthesis), elliptic, keeled apically, 1.4–1.7 cm, thin-leathery, base not visible at anthesis, apex acute, surfaces scaly distally, becoming sparsely scaly toward base, venation slight. **Flowers** 5–20, conspicuous; sepals with adaxial pair connate, elliptic, keeled, to 1.5 cm, thin-leathery, veined, apex obtuse, surfaces scaly; corolla tubular, petals erect, violet, ligulate, 3–4.5 cm; stamens exserted; stigma exserted, conduplicate-spiral. **Fruits** 2.5–3 cm.

Flowering spring–summer. Epiphytic on a variety of hosts in swamps and hammocks; 0–60 m; Fla., Ga.; Mexico (Tamaulipas).

11. Tillandsia ×floridana (L. B. Smith) H. Luther, Phytologia 57: 175. 1985 E

Tillandsia fasciculata Swartz var. *floridana* L. B. Smith, Phytologia 15: 197. 1967

Plants usually clustering, flowering to 6 dm. **Stems** short. **Leaves** 20–50, many-ranked, erect to slightly spreading, gray, 20–50 × 0.5–1 cm, coarsely appressed-scaly; sheath dark chestnut brown toward base, triangular, flat, not forming pseudobulb, to 2.5 cm wide; blade narrowly linear-triangular, channeled, leathery, margins involute, apex attenuate. **Inflorescences:** scape conspicuous, erect, 15–30 cm, 3–6 mm diam.; bracts densely imbricate, erect, like leaves but gradually smaller; sheath of bracts narrowing gradually into blade; spikes erect, 2-pinnate, narrowly elliptic, compressed, 2–7 × 1.5 cm, apex acute; lateral branches 2–10. **Floral bracts** imbricate, erect, uniformly red to rose, broad (covering all or most of rachis, rachis not visible at anthesis), elliptic, keeled apically, 2–2.5 cm, thin, base not visible at anthesis, apex acute, surfaces scaly distally, becoming sparsely scaly toward base. **Flowers** 10–40, conspicuous; sepals with adaxial pair connate, lanceolate, keeled, 1.7–2.2 cm, thin-leathery, veined, apex acute, surfaces slightly scaly; corolla tubular, petals erect, violet, ligulate, to 4.5 cm; stamens exserted; stigma exserted, conduplicate-spiral. **Fruits** to 4 cm.

Flowering spring–summer. Epiphytic in swamps and river forests; 0–30 m; Fla.

This species is often confused with *Tillandsia bartramii* Elliott and *T. simulata* Small, which share its distribution and narrow leaves. Its probable parentage is *T. bartramii* Elliott × *T. fasciculata* Swartz.

12. Tillandsia variabilis Schlechtendal, Linnaea 18: 418. 1844 F

Tillandsia houzeavii Chapman; *T. valenzuelana* A. Richard

Plants usually single, rarely clustering, flowering to 40 cm. **Stems** short. **Leaves** 15–20, many-ranked, spreading, gray-green or flushed rose, 12–30 × 1–2 cm, finely appressed-scaly; sheath pale to nearly chestnut brown, ovate, flat, not forming pseudobulb, 2–4 cm wide; blade narrowly triangular, tapering evenly from base to apex, nearly plane to channeled, soft, brittle, margins involute, apex attentuate. **Inflorescences:** scape conspicuous, erect or ascending, 3–10 cm, 2–5 mm diam.; bracts densely imbricate, erect, blade often hanging, like leaves but gradually smaller; sheath of bracts narrowing abruptly into blade; spikes erect or ascending, never palmate, linear, compressed, 5–20 × 0.8–1.2 cm, apex acute; simple or laxly 2–3 lateral branches. **Floral bracts** laxly imbricate, erect, green, red, or purple, broad (covering all or most of rachis, rachis not visible at anthesis), elliptic, keeled toward apex, 1.8–2 × 0.6–0.9 cm, leathery, base visible in fruit, apex acute, surfaces glabrous. **Flowers** 5–30, conspicuous; sepals with adaxial pair connate, oblong, keeled, 1.5–1.8 cm, thin-leathery, slightly veined, apex obtuse, surfaces glabrous; corolla tubular, petals erect, lavender-blue, ligulate, to 3 cm; stamens exserted; stigma exserted, conduplicate-spiral. **Fruits** to 3 cm.

Flowering spring–fall. Epiphytic in moist, shaded habitats; 0–30 m; Fla.; Mexico; West Indies; Central America; South America.

13. Tillandsia fasciculata Swartz, Prodr., 56. 1788

Cardinal airplant

Plants clustering, flowering to 65 cm. **Stems** short. **Leaves** 20–50, many-ranked, erect to spreading, gray to gray-green, 25–50 × 1–2.5 cm, grayish-scaly; sheath dark rust colored toward base, broadly elliptic, flat, not forming pseudobulb, 3–4 cm wide; blade narrowly triangular, tapering evenly from base to apex, stiff, leathery, channeled to involute, apex attenuate. **Inflorescences:** scape conspicuous, erect or ascending, 10–35 cm, 4–8 mm diam.; bracts densely imbricate, erect to spreading, like leaves but gradually smaller; sheath of bracts narrowing gradually into blade; spikes erect to spreading, densely palmate to laxly 2(–3)-pinnate, narrowly elliptic, compressed, 5–20 × 1.5–2.5 cm, apex acute; lateral branches 3–15. **Floral bracts** imbricate, erect, red, red-yellow-green, or green, broad (covering all or most of rachis, rachis not visible at anthesis), elliptic, keeled, 2–4.8 × 1.2–2 cm, thin-leathery, base not visible at anthesis, apex acute, surfaces glabrous or slightly scaly toward apex, venation even to slight. **Flowers** 10–50, conspicuous; sepals with adaxial pair connate, lanceolate, to ½ keeled, to 4.2 cm, leathery, slightly veined, apex acute, surfaces glabrous to slightly scaly; corolla tubular, petals erect, violet (white), ligulate, 5–6 cm; stamens exserted; stigma exserted, conduplicate-spiral. **Fruits** to 4 cm.

Varieties 7 (3 in the flora): North America; Mexico; West Indies; Central America; South America.

SELECTED REFERENCE Luther, H. E. 1993. A new record for *Tillandsia* (Bromeliaceae) in Florida. Rhodora 95: 342–347.

1. Spikes of inflorescence rarely less than 10 cm; floral bracts to 3–4.8 cm. . . . . 13a. *Tillandsia fasciculata* var. *fasciculata*
1. Spikes of inflorescence, or at least their fertile portions, rarely over 10 cm; floral bracts 2–3 cm.
2. Inflorescence laxly palmate, spikes stipitate with elongate, slender, sterile bracteate bases . . . . 13b. *Tillandsia fasciculata* var. *clavispica*
2. Inflorescence densely palmate, spikes short-stipitate to subsessile, without elongate, slender, sterile bracteate bases. . . . . 13c. *Tillandsia fasciculata* Swartz var. *densispica*

13a. Tillandsia fasciculata Swartz var. fasciculata

Inflorescences palmate, slightly stipitate, with short sterile bracteate base, spikes rarely less than 10 cm. **Floral bracts** 3 cm–4.8 cm.

Flowering spring–summer. Epiphytic on a variety of hosts, hammocks; 0–30 m; Fla.; Mexico; West Indies; Central America; South America.

This variety is very difficult to find in Florida and possibly has been eradicated.

13b. Tillandsia fasciculata Swartz var. clavispica Mez in A. L. P. de Candolle and C. de Candolle, Monogr. Phan. 9: 683. 1896

Tillandsia bracteata Chapman

Inflorescences: spikes laxly palmate, each stipitate with elongate, slender, sterile bracteate base, fertile part rarely over 10 cm. **Floral bracts** to 3 cm.

Flowering spring–summer. Epiphytic on a variety of hosts, usually in bright light; 0–30 m; Fla.; Mexico; West Indies.

This variety appears to hybridize with *Tillandsia fasciculata* var. *densispica*.

13c. Tillandsia fasciculata Swartz var. densispica Mez in A. L. P. de Candolle and C. de Candolle, Monogr. Phan. 9: 683. 1896

Tillandsia hystricina Small; *T. wilsonii* S. Watson

Inflorescences: spikes densely palmate, each short-stipitate to subsessile, without elongate, slender, sterile bracteate base, fertile part rarely over 10 cm. **Floral bracts** to 3 cm.

Flowering spring–summer. Ep-

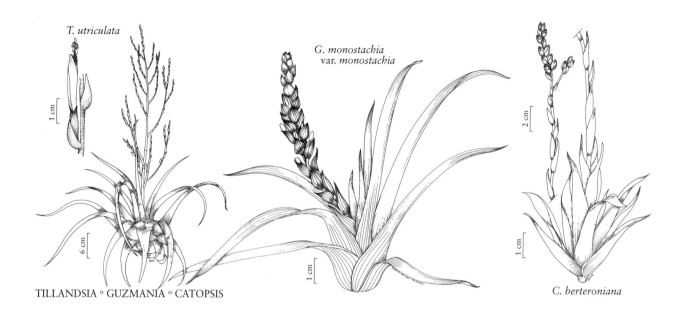

T. utriculata

G. monostachia
var. monostachia

TILLANDSIA ° GUZMANIA ° CATOPSIS

C. berteroniana

iphytic on a variety of hosts, usually in bright light; 0–30 m; Fla.; Mexico; West Indies; Central America.

This is the most common variety in central and southern Florida. It is quite variable in size and coloration.

Tillandsia fasciculata var. *floridana* L. B. Smith is apparently of hybrid origin and is treated above as *T.* ×*floridana*.

14. Tillandsia flexuosa Swartz, Prodr., 56. 1788

Tillandsia aloifolia Hooker

Plants single or in small clusters, flowering to 1 m. **Stems** short. **Leaves** 10–20, many-ranked, spreading to recurved, spirally twisted, gray or reddish, banded silver, to 40 × 1–2.5 cm, appressed-grayish-scaly; sheath dark chestnut brown, broadly elliptic to ovate, not inflated, not forming pseudobulb, 2–3 cm wide; blade narrowly triangular, leathery, channeled to involute, apex attenuate. **Inflorescences:** scape conspicuous, erect, 15–40 cm, 3–5 mm diam.; bracts laxly imbricate, erect to spreading, distal widely spaced, like leaves but gradually smaller; sheath of bracts narrowing abruptly into blade; spikes very laxly 2–6-flowered, erect to spreading, simple or 2-pinnate, linear in outline, 12–40 × 3–4.5 cm, apex acute; lateral branches 2–8. **Floral bracts** widely spaced, spreading with flowers, gray or reddish, exposing most of rachis at anthesis, elliptic, not keeled, 2.3–3.1 cm, thin-leathery, base not visible at anthesis, apex acute, sur-

faces glabrous to slightly scaly. **Flowers** 5–40, conspicuous; sepals free, elliptic, not keeled, 2–3 cm, leathery to thin-leathery, veined, apex obtuse, surfaces glabrous; corolla tubular, petals erect, apex spreading, pink to dark rose, ligulate, to 4 cm; stamens exserted; stigma exserted, conduplicate-spiral. **Fruits** to 7.5 cm.

Flowering summer. Epiphytic in exposed habitats often near the coast; 0–30 m; Fla.; West Indies; Central America; South America.

15. Tillandsia utriculata Linnaeus, Sp. Pl. 1: 286. 1753

• Giant wild-pine F

Plants single, flowering to 2 m. **Stems** short. **Leaves** 20–75, many-ranked, spreading and recurved, not twisted, gray-green (rarely variegated with linear cream stripes), to 1 m × 1.5–3.5 cm, finely appressed-scaly; sheath pale or slightly rust colored, ovate, not inflated, not forming pseudobulb, 6–15 cm wide; blade linear-triangular, leathery, channeled to involute, apex attenuate. **Inflorescences:** scape conspicuous, erect, 20–50 cm, 6–12 mm diam.; bracts densely imbricate proximally, often lax distally, erect to spreading, like leaves but gradually smaller; sheath of bracts narrowing gradually into blade; spikes very laxly 6–11-flowered, erect to spreading, 2–3-pinnate, linear, 15–40 × 10–15 cm, apex acute; branches 5–40 (rarely simple). **Floral bracts** widely spaced, erect, green or tinged purple, exposing most of rachis at anthesis, ovate, not keeled, 1.2–2 cm, leathery, venation slight, base visible at anthesis, apex acute, surfaces glabrous. **Flowers** 10–200, conspicuous; sepals free, elliptic, not keeled, 1.4–2 cm, thin-leathery, veined, apex obtuse, surfaces glabrous; corolla tubular, somewhat bilaterally symmetric, petals erect, slightly twisted, white, ligulate, to 4 cm; stamens exserted;

stigma exserted, conduplicate-spiral. **Fruits** to 4 cm. $n = 25$ (Puerto Rico, West Indies).

Flowering summer. Epiphytic on a variety of hosts, often in bright, exposed habitats, usually abundant; 0–30 m; Fla.; Mexico; West Indies; Central America; South America.

3. GUZMANIA Ruiz & Pavón, Fl. Peruv. 3: 37. 1802 • [For A. Guzman, an 18th-century Spanish naturalist]

Harry E. Luther

Gregory K. Brown

Herbs, usually epiphytic, stemless to rarely caulescent. **Leaves** many-ranked, usually ligulate, margins entire. **Inflorescences** 5–many-flowered, many-ranked, mostly 2-pinnate to less commonly single spike, flowers laxly to densely arranged; floral bracts broad, conspicuous, mostly obscuring rachis. **Flowers** bisexual; sepals distinct to connate over ½ length, usually symmetric; petals with claws adherent to subconnate petal, forming short tube, blade distinct; stamens usually included, adherent to adnate with petal claws; ovary superior. **Capsules** cylindric, dehiscent. **Seeds** with basal, usually tan-brown plumose appendage.

Species ca. 160 (1 in the flora): widespread in the moist Neotropics.

SELECTED REFERENCE Bennet, B. C. 1992. The Florida bromeliads: *Guzmania monostachia.* J. Bromeliad Soc. 42: 266–270.

1. Guzmania monostachia (Linnaeus) Rusby ex Mez in A. L. P. de Candolle and C. de Candolle, Monogr. Phan. 9: 905. 1896 (as monostachya)

Renealmia monostachia Linnaeus, Sp. Pl. 1: 287. 1753

Plants flowering to 4 dm high. **Leaves** many, rosulate, spreading, 30–40 cm; sheath pale or darkened toward base, ovate; blade ligulate, acute. **Inflorescences:** scape central, erect, shorter than leaves; bracts imbricate, erect; inflorescences densely many-flowered, erect, simple, cylindric, to 15 cm; floral bracts imbricate, most proximal bracts (fertile) pale green, striped dark green or reddish, distal bracts (mostly sterile) rose, salmon or nearly white [red or orange in tropical America], ovate, to 2.5 cm, acute. **Flowers:** sepals basally connate, green, symmetric, 1.4–1.6 cm; petals very slightly spreading, adherent for most of length, white, without petal appendages, 2.2–2.5 cm, apex obtuse; stamens included, filaments subadnate to petals; stigma included, simple-erect. **Fruits** 2–3 cm.

Varieties 3 (2 in the flora): Mexico; West Indies; Central America; South America.

Florida plants of this taxon appear to be self fertilizing.

1. Leaf blade of one color. . . . .
. . . . 1a. *Guzmania monostachia* var. *monostachia*
1. Leaf blade longitudinally green-and-white striped
. . . . 1b. *Guzmania monostachia* var. *variegata*

1a. Guzmania monostachia (Linnaeus) Rusby ex Mez var. **monostachia** [F]

Leaf blade of one color, bright green.

Flowering spring–summer. Epiphytic in shady, moist habitats; 0–30 m; Fla.; West Indies; Central America; South America.

This variety is seldom found but locally abundant.

1b. Guzmania monostachia (Linnaeus) Rusby ex Mez var. **variegata** M. B. Foster, Bull. Bromeliad Soc. 3: 30. 1953

Leaf blade longitudinally green-and-white striped.

Flowering spring–summer. Epiphytic in shady, moist habitats; 0–30 m; Fla.; Central America.

This variety is horticulturally valuable and threatened with overcollection. This species is uncommon in the flora.

4. CATOPSIS Grisebach, Fl. Brit. W. I., 599. 1864

Harry E. Luther

Gregory K. Brown

Herbs, epiphytic, without evident stems. **Leaves** many-ranked, rosulate; blade linear to triangular, base usually with chalklike bloom, margins entire, appressed-scaly. **Inflorescence** central, 5–many-flowered, always many-ranked, simple or compound, flowers laxly arranged; floral bracts small, inconspicuous, not obscuring rachis. **Flowers** bisexual or functionally unisexual; sepals distinct, strongly asymmetric; petals distinct; stamens included, usually in 2 unequal sets, filaments free; ovary superior. **Capsules** ovoid to ellipsoid, dehiscent. **Seeds** with light tan to brown, apical, plumose appendage.

Species 18 (3 in the flora): Fla.; West Indies; Mexico; Central America; South America.

1. Inflorescence 3–10-flowered, pendent; petals bright yellow; flowers nocturnal. . . . . 1. *Catopsis nutans*
1. Inflorescence 15–50-flowered, erect to pendulous; petals white; flowers diurnal.
    2. Leaves yellow-green, covered with conspicuous, white, chalky powder; floral bracts 6–8 mm. . . . . 2. *Catopsis berteroniana*
    2. Leaves bright green, not covered with conspicuous, white, chalky powder; floral bracts 4–5 mm. . . . . 3. *Catopsis floribunda*

1. Catopsis nutans (Swartz) Grisebach, Fl. Brit. W. I., 599. 1864

Tillandsia nutans Swartz, Prodr., 56. 1788

Plants at flowering 0.1–0.3 m. **Leaves** spreading, bright green, 8–15 cm, somewhat chalky, especially toward base; sheath pale, elliptic; blade subtriangular to narrow triangular, apex acuminate. **Inflorescences:** scapes decurved; bracts subfoliaceous, laxly imbricate; inflorescences single spikes (in flora), spikes 3–10-flowered, pendent; floral bracts green, elliptic, 1.4–1.5 cm. **Flowers** nocturnal; sepals green, broadly elliptic, 1–1.2 cm, thin-leathery, apex obtuse, surfaces glabrous; petals widely spreading, bright yellow, ligulate, 2–2.4 cm; style included, stigma erect, simple. **Fruits** ovoid, 1.2–2 cm.

Flowering fall–winter. Epiphytic in shady, humid hammocks; 0–30 m; Fla.; Mexico; West Indies; Central America; South America.

Catopsis nutans is rare in the flora.

2. Catopsis berteroniana (Schultes f.) Mez in A. L. P. de Candolle and C. de Candolle, Monogr. Phan. 9: 621. 1896 [F]

Tillandsia berteroniana Schultes f. in J. J. Roemer and J. A. Schultes, Syst. Veg. 7(2): 1221. 1830

Plants at flowering 0.4–1.3 m. **Leaves** erect, yellow-green, to 45 cm, covered with conspicuous, white, chalky powder, especially toward base; sheath pale, elliptic; blade triangular, apex acute. **Inflorescences:** scapes erect; bracts subfoliaceous, laxly imbricate; inflorescences 2-pinnate (rarely simple), with 2–8 lateral branches; spikes 15–50-flowered, erect, stipitate; floral bracts yellow-green, broadly ovate, 6–8 mm. **Flowers** diurnal; sepals yellow-green, obovate, 1–1.2 cm, leathery, apex obtuse, surfaces glabrous; petals erect to very slightly spreading, white, elliptic, 1–1.2 cm; style included, stigma erect, simple. **Fruits** ellipsoid, 1.2–1.4 cm.

Flowering fall–winter. Epiphytic on a variety of hosts, usually in strong light; 0–30 m; Fla.; Mexico; West Indies; Central America; South America.

3. Catopsis floribunda L. B. Smith, Contr. Gray Herb. 117: 5. 1937

Plants at flowering to 0.6 m. **Leaves** erect to spreading, bright green, 20–40 cm, not covered with conspicuous, white, chalky powder; sheath pale or slightly darkened toward base, elliptic; blade narrowly triangular, apex attenuate. **Inflorescences:** scapes erect or leaning; bracts foliaceous, imbricate, spreading; inflorescences usually 2-pinnate, with 5–15 lateral branches; spikes 15–50-flowered, erect to pendulous, stipitate; floral bracts green, ovate, 4–5 mm. **Flowers** diurnal; sepals yellow-green, broadly elliptic, 4–6 mm, leathery, apex obtuse, surfaces glabrous; petals erect to very slightly spreading, white, elliptic, 6–8 mm; style included, stigma erect, simple. **Fruits** ellipsoid, 8–10 mm.

Flowering fall–winter. Epiphytic on a variety of hosts in humid, shady situations; 0–30 m; Fla.; Mexico; West Indies; Central America; South America.

215. HELICONIACEAE (A. Richard) Nakai
• Heliconia, False Bird-of-paradise, or Lobster-claw Family

W. John Kress

Alan T. Whittemore

Herbs, perennial, from rhizomes. **True aerial stems** absent or weak and concealed in pseudostem. **Leaves** basal or basal and cauline, in 2 ranks, differentiated into basal sheath, petiole, and blade; sheaths overlapping, forming unbranched pseudostem, open, ligule absent; summit of petiole not differentiated; blade with lateral veins parallel, diverging from prominent midrib. **Inflorescences** 1 per aerial shoot, projecting from tip of pseudostem, pedunculate racemes of several- to many-flowered monochasial cymes (cincinni); bracts of main axis enclosing cincinni. **Flowers** bisexual, bilaterally symmetric; sepals and petals scarcely differentiated, sepals 3, petals 3, 2 sepals and 3 petals fused, remaining sepal distinct; fertile stamens 5, not petal-like; anthers 2-locular; 1 rudimentary staminode opposite free sepal; ovary inferior, 3-carpellate, 3-locular, all locules fertile; placentation basal; ovules 1 per locule; style standing away from stamens and staminode, filiform, stigma capitate or 3-lobed. **Fruits** 1–3-seeded drupes, sepals not persistent in fruit. **Seeds:** aril absent, endosperm copious, perisperm copious, embryo straight. $x = 12$.

Genera 1, species 225 (1 species in the flora): introduced; North America, Mexico, West Indies, Central America, South America, and Oceania.

The bird-of-paradise, *Strelitzia reginae* Aiton, of the closely related family Strelitziaceae, is an acaulescent rhizomatous herb often cultivated in the western and southern United States. It is not known to reproduce outside of cultivation, but it can persist for very long periods on abandoned ground in southern California. Its leaves are in two ranks; the sheaths do not overlap and do not form a pseudostem. The petioles are long, and the blades are glabrous but often strongly glaucous, and have parallel lateral veins diverging from a prominent midrib. The inflorescence is a single cincinnus, enclosed within a large bract, with the mature flowers projecting laterally. The perianth is large and showy, with three orange sepals and three dark blue petals; one petal is short, the other two are long and enfold the style and the five fertile stamens, forming a prominent arrow-shaped structure. The fruit is a many-seeded capsule.

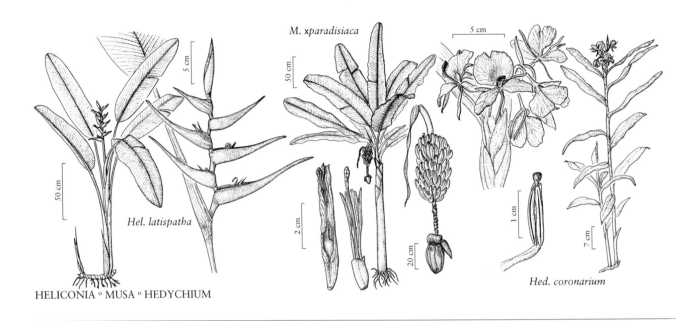

M. xparadisiaca

Hel. latispatha

Hed. coronarium

HELICONIA ° MUSA ° HEDYCHIUM

1. HELICONIA Linnaeus, Mant. Pl. 2: 147, 211. 1771, conserved name • False bird-of-paradise, héliconia [after Mount Helicon in southern Greece, regarded as the home of the Muses]   ⊡

Plants: pseudostems erect, in groups of [1–]5–50. **Leaves:** petiole long [short or nearly absent], base of blade unequal on either side of midrib. **Inflorescences** terminal, erect [pendent], raceme of cincinni; cincinni spiral [2-ranked]; cincinnal bract ± enclosing each cincinnus, brightly colored, leaflike. **Flowers** each subtended by membranous floral bract. **Fruits** blue [rarely red or orange] at maturity. **Seeds** surrounded by stony, roughened endocarp (pyrenes). $x = 12$.

Species 225 (1 in the flora): introduced, North America (Florida); Mexico, West Indies, Central America, South America, Pacific Islands (Oceania).

Several species of *Heliconia* are important ornamental and landscape plants in warmer areas of North America (e.g., Florida, Texas, and California). In addition to *H. latispatha*, three other species, *H. psittacorum* Linnaeus f., *H. metallica* Planchon & Linden ex Hooker, and *H. schiedeana* Klotzsch, may persist after cultivation and might be expected to disperse occasionally into disturbed areas. The first two species are much smaller than *H. latispatha* and all have flowers that are fully exposed at anthesis. *Heliconia metallica* has deep rose red flowers, whereas *H. psittacorum* has yellow or orange flowers. *Heliconia schiedeana* has red, strongly reflexed bracts and yellow flowers. All of the New World species of *Heliconia* are hummingbird pollinated.

SELECTED REFERENCES   Andersson, L. 1985. Revision of *Heliconia* subgen. *Stenochlamys* (Musaceae-Heliconioideae). Opera Bot. 82: 5–123.   Andersson, L. 1992. Revision of *Heliconia* subgen. *Taeniostrobus* and subgen. *Heliconia* (Musaceae-Heliconioideae). Opera Bot. 111: 1–98.   Berry, F. and W. J. Kress. 1991. *Heliconia*: An Identification Guide. Washington.   Daniels, G. S. and F. G. Stiles. 1979. The *Heliconia* taxa of Costa Rica. Keys and descriptions. Brenesia 15(suppl.): 1–150.   Kress, W. J. 1984. Systematics of Central American *Heliconia* (Heliconiaceae) with pendent inflorescences. J. Arnold Arbor. 65: 429–532.

1. **Heliconia latispatha** Bentham, Bot. Voy. Sulphur 6: 170. 1846 F I

Plants 2–4 m, pseudostem well-developed. **Leaves** 3–5 per shoot, green, longest blade 75–150 × 18–33 cm. **Inflorescences** erect, to 45 cm; cincinnal bracts spirally arranged, 8–13 per inflorescence, middle bract with outer surface orange and/or red and glabrous, 14–19.5 cm, 5–7 cm wide at base. **Flowers** 10–15 per cincinnus; perianth green, yellow, or orange with dark green sepal margins, sometimes sparsely puberulent along fused sepal margins, 3.5–4.6 cm, 7–8 mm wide at gibbous base, nearly straight. **Drupes** blue, 10–13 × 8–9 mm. $2n = 24$.

Flowering early fall (Sep). Disturbed areas; 0–4 m; introduced; Fla.; West Indies; native, Mexico, Central America, and South America.

Heliconia latispatha often persists in old gardens but only rarely spreads outside them.

216. MUSACEAE Jussieu

• Banana Family

Alan T. Whittemore

Treelike herbs, perennial, from rhizome [corm]. **True aerial stems** absent. **Leaves** basal, in several ranks, differentiated into basal sheath, petiole, and blade; sheaths overlapping, forming unbranched pseudostem, open, ligule absent; summit of petiole not differentiated; blade with lateral veins parallel, diverging from prominent midrib. **Inflorescences** 1 per aerial shoot, projecting from tip of pseudostem, pedunculate racemes of 12–20-flowered monochasial cymes (cincinni); bracts of main axis enclosing cincinni. **Flowers** unisexual (proximal flowers pistillate, distal flowers staminate), bilaterally symmetric; sepals and petals differentiated, sepals 3, petals 3, 3 sepals and 2 petals fused, remaining petal distinct; fertile stamens 5(–6), not petal-like; anthers 2-locular; occasionally 1 rudimentary staminode; ovary inferior, 3-carpellate, 3-locular, all locules fertile; placentation axile; ovules many per locule; style terminal, filiform; stigma 3-lobed. **Fruits** berries; sepals not persistent in fruit. **Seeds:** aril absent; endosperm copious; perisperm copious; embryo straight. $x = 9, 10, 11$.

Genera 3, species ca. 40 (1 genus, 1 species, and 1 stable hybrid in the flora): introduced; tropical parts of Africa, Asia, Australia, and Oceania; often persisting around gardens and plantations throughout the wet tropics.

1. **MUSA** Linnaeus, Sp. Pl. 2: 1043. 1753; Gen. Pl. ed. 5, 466. 1754 • Banana, bananier [Arabic *mouz*] ☐

Underground stems (corms) rhizomatous, short, pseudostems clustered, [0.5–]3–10 m. **Leaf blades** unlobed (older leaves often split to midrib), oblong or oblong-elliptic, [0.6–]2–3 × 0.3–0.6 m. **Inflorescences** pendent [erect]; pistillate flowers crowded, numerous; bracts of staminate flowers imbricate, forming budlike mass at apex of inflorescence. **Berries** cylindric, usually ± curved, weakly angled in cross section, [10–]20–35 cm, soft, fleshy. $x = 10, 11$.

Species ca. 30 (1 species and 1 stable hybrid in the flora): introduced; Asia (India to Japan and Indonesia), Australia (Queensland), Pacific Islands (Oceania); often persisting around gardens and plantations in North America, Mexico, West Indies, Central America, South America, Africa, Pacific Islands (Oceania).

Species of *Musa* are very important economically throughout the wet tropics. The fruits of several species are edible; they may be sweet (bananas) or starchy (plantains), and may be eaten raw or cooked. Some species are important fiber sources, especially *M. textilis* Née (abacá or Manila-hemp), and others are grown as ornamentals in subtropical and tropical regions. In addition, the plants have many minor uses in the tropics: banana leaves are used for wrapping and various other purposes, and the corms, the interior of the pseudostems, and the buds of staminate flowers are eaten as vegetables. Bananas used in North America are almost always sweet-fruited cultivars, imported from Central America to be eaten raw or used in cooking.

Prior to 1948, the taxonomy of cultivated bananas was not understood. Since then, it has become clear that most of the cultivated bananas are parthenocarpic diploids, triploids, and tetraploids ($2n$ = 22, 33, 44) derived either from *Musa acuminata* Colla, *M. balbisiana* Colla, or hybrids between them (*M.* ×*paradisiaca* Linnaeus). The most common crop bananas in North and Central America are triploid races of *M. acuminata* (genotype AAA) and triploid *M.* ×*paradisiaca* with two sets of chromosomes from *M. acuminata* and one from *M. balbisiana* (genotype AAB). Those two types are very similar morphologically; distinguishing them reliably requires numerical scoring of a large number of characters from the pseudostem, petiole, peduncle, bracts of staminate flowers, and staminate and pistillate flowers (N. W. Simmonds and K. Shepherd 1955), many of which are very difficult to score on herbarium material. The ranges given below, based on herbarium specimens, are tentative, and need to be checked in the field.

SELECTED REFERENCES Simmonds, N. W. and K. Shepherd. 1955. The taxonomy and origins of the cultivated bananas. J. Linn. Soc., Bot. 55: 302–312.   Stover, R. H. and N. W. Simmonds. 1987. Bananas, ed. 3. London.

1. Bracts of staminate flowers lanceolate or narrowly ovate, apex acute, abaxial surface yellow, red, or dull purple, adaxial surface yellow proximally, often yellow or dull purple distally; perianth of staminate flower white or cream. . . . . .1. *Musa acuminata*
1. Bracts of staminate flowers lanceolate to ovate, apex acute or broadly acute, abaxial surface purple, adaxial surface reddish purple or crimson; perianth of staminate flower white, cream, or pink. . . . . .2. *Musa* ×*paradisiaca*

1. **Musa acuminata** Colla, Mem. Reale Accad. Sci. Torino 25: 66. 1820 [I]

Musa cavendishii Lambert

Pseudostems heavily blotched with brown or black. **Petioles:** margins of adaxial groove erect, winged proximally. **Inflorescences:** pedicels short; bracts of staminate flowers lanceolate or narrowly ovate, apex acute, abaxial surface yellow, red, or dull purple, adaxial surface yellow proximally, often yellow or dull purple distally. **Staminate flowers** white or cream. **Pistillate flowers:** stigmas deep yellow or orange; each locule with 2 regular rows of ovules.

Flowering all year. Abandoned gardens and dis- turbed sites; 0–10 m; introduced; Fla.; Mexico; West Indies; Central America; South America; native, s Asia; Africa; Pacific Islands (Oceania).

2. **Musa** ×**paradisiaca** Linnaeus, Sp. Pl. 2: 1043. 1753 (as species) [F] [I]

Musa sapientum Linnaeus

Pseudostems moderately blotched. **Petioles:** margins of adaxial groove erect to incurved, ± winged proximally. **Inflorescences:** pedicels rather long; bracts of staminate flowers lanceolate to ovate, apex acute to broadly acute, abaxial surface purple, adaxial surface reddish purple or crimson. **Stam-**

inate flowers white, cream, or pink. **Pistillate flowers:** stigmas usually yellow; each locule with usually $2 \pm$ regular rows of ovules.

Flowering all year. Abandoned gardens and disturbed sites; 0–20 m; introduced; Fla.; Mexico; West Indies; Central America; South America; Africa; Asia; Pacific Islands (Oceania).

Plants of *Musa* ×*paradisiaca* combine the characters of the two parents, *M. acuminata* and *M. balbisiana*, in various ways. *Musa acuminata* is described above. In *M. balbisiana*, the blotching on the pseudostem is pale or absent, the margins of the adaxial groove of the petiole are incurved and not winged; the peduncle is glabrous, the pedicels are long, the bracts of staminate flowers are ovate and obtuse, widest at least 0.3 above the base, with the outer surface brownish purple and the inner surface uniformly bright crimson; the apex remains plane when bract spreads, and the bract scars are not very prominent. The flowers are often tinged with pink, the free tepal of the staminate flower is always plane, and the ovules are in 4 irregular rows per locule. The description of *M.* ×*paradisiaca* is based on plants with the AAB genome, by far the commonest hybrid bananas in cultivation. For data on characteristics of various other strains of banana, see N. W. Simmonds and K. Shepherd (1955).

217. ZINGIBERACEAE Lindley

• Ginger Family

Alan T. Whittemore

Herbs, perennial [annual], from tuberous rhizome [tuberous roots]. **Aerial stems** present or absent, unbranched. **Leaves** basal or basal and cauline, in 2 ranks, differentiated into basal sheath, petiole, and blade, or petiole absent; sheaths often overlapping to form pseudostem, open, ligule present, summit of petiole not differentiated; blade with lateral veins parallel, diverging from prominent midrib. **Inflorescences** 1 per aerial shoot, terminating short stem without laminate leaves or projecting from tip [side] of pseudostem, pedunculate racemes of flowers or of 2–7-flowered monochasial cymes (cincinni); bracts of main axis enclosing or subtending flowers or cincinni. **Flowers** bisexual, bilaterally symmetric; sepals and petals differentiated, sepals 3, connate, petals 3, connate, but sepals free from petals; fertile stamens 1, not petal-like; anthers 2-locular; 2 connate staminodes forming petal-like structure (lip) opposite fertile stamen, sometimes also with 2 petal-like or rudimentary distinct staminodes; ovary inferior, 3-carpellate, [1-]3-locular, all locules fertile; placentation axile, parietal, or ± basal; ovules many per locule; style held between pollen-sacs of anther, filiform; stigma funnelform, sometimes slightly 2-lobed; style base free from stamen and staminodes. **Fruits** capsules or berries; sepals often persistent in fruit. **Seeds:** aril present; endosperm copious; perisperm copious; embryo straight.

Genera ca. 50, species ca. 1300 (4 genera, 4 species in the flora): introduced; North America, Mexico, West Indies, Central America, South America, Asia, Africa, Australia.

Zingiberaceae are tropical, the most diverse family in the Old World. Species of Zingiberaceae are often spicy to the taste, and many of them have been used as medicines and flavorings. The most important commercial spices in the family are ginger, turmeric, and cardamom.

Species of the large tropical American genus *Costus* Linnaeus, sometimes included in Zingiberaceae but now usually placed in a separate family Costaceae, are cultivated as ornamentals in peninsular Florida and the Gulf Coast. They are known to propagate spontaneously within gardens in Florida and might be expected to escape in disturbed habitats. Species of *Costus* are perennial herbs with erect stems ca. 1 m tall. The leaves are inserted in many ranks and arranged in a very distinctive open spiral; they consist of a basal sheath, a very short petiole, and a blade; the sheaths do not overlap to form a pseudostem; and the blades have parallel lateral veins diverging from a prominent midrib. The inflorescence is a terminal spike with a large bract subtending each flower; the flowers are bisexual and bilaterally symmetric, with three connate sepals and three connate petals. There is a single fertile stamen and opposite it a petal-like lip representing four fused staminodes; the fruit is a 3-locular capsule. Costaceae differ from all related families in having a well-developed aerial stem with the leaf sheaths not overlapping to form a pseudostem and in the very distinctive leaf arrangement.

1. Inflorescences lax; bracts on main axis of inflorescence minute, scalelike; cincinni stalked, each cincinnus enclosed in large, conspicuous bracteole. . . . . 4. *Alpinia*, p. 308
1. Inflorescences dense; bracts on main axis of inflorescence lanceolate to reniform, 2–9 cm; cincinni sessile, each cincinnus enclosed in bract, bracteoles small, inconspicuous, hidden by bracts.
   2. Inflorescences projecting from tip of pseudostem; filament linear; anther long-exserted . . . . 1. *Hedychium*, p. 306
   2. Inflorescences terminating short stem with only scale leaves; filament rectangular or nearly absent; anther enclosed in corolla.
      3. Inflorescences cylindric; bracts spreading; proximal bracts ovate, apex obtuse; distal bracts larger, narrowly ovate, apex rounded; lateral staminodes petal-like. . . . . 2. *Curcuma*, p. 307
      3. Inflorescences conelike; bracts imbricate; proximal bracts reniform or very broadly ovate, apex broadly rounded; distal bracts smaller but otherwise similar to proximal bracts; lateral staminodes absent or reduced to small teeth adnate with lip. . . . . 3. *Zingiber*, p. 307

1. HEDYCHIUM J. König in A. J. Retzius, Observ. Bot. 3: 73. 1783 · [Greek *edys*, sweet, and *chion*, snow, for the fragrant white flowers] I

Pseudostems well developed, 1–3 m. **Inflorescences** projecting from tip of pseudostem, dense, conelike [lax]; bracts of main axis crowded [not crowded], [2.5–]4–6[–7] cm, ovate to lanceolate [or almost circular]; cincinni sessile, [1–]2–3[–6]-flowered, enclosed in bracts; bracteoles small, inconspicuous, hidden by bracts. **Flowers:** calyx cylindric, 3-toothed or-lobed, split down one side [not split]; corolla tube slender, lobes linear; filament linear, tubular-incurved, enclosing style; anther long-exserted, not spurred, terminal appendage none; lateral staminodes large, petal-like, lip oblong, plane, 2-lobed. **Fruits** capsule, globose. $x = 17$.

Species ca. 50 (1 in the flora): introduced; North America, Mexico, West Indies, Central America, South America, Australia; native, Asia and Madagascar.

1. Hedychium coronarium J. König in A. J. Retzius, Observ. Bot. 3: 73. 1783 · Butterfly-lily F I

Leaf blades oblanceolate or narrowly elliptic, 28–48 × 4–7 cm. **Inflorescences** erect, 15–19 × 12–17 cm; bracts of main axis green; proximal bracts ovate, concave, 4–5.5 × 2–4.5 cm, apex obtuse or apiculate; distal bracts lance-oblong or ovate, 4–6 × 1–2.5 cm, apex rounded or apiculate. **Flowers:** perianth and staminodes white.

Flowering summer–fall (Jun–Oct). Swamps, shores of lakes and streams; 0–50 m; introduced; Fla.; Mexico; West Indies; Central America; South America; native, Asia; Australia.

The flowers of *Hedychium coronarium* are very fragrant. Naturalized populations have been reported several times from southern Louisiana, but the specimens I have seen are sterile and cannot be identified with certainty. There is also an old report from near Brunswick, Georgia (J. K. Small, unpublished ms.), but I have seen no specimens from the state.

2. CURCUMA Linnaeus, Sp. Pl. 1: 2. 1753; Gen. Pl. ed. 5, 3. 1754 • Hidden-lily [the name in some East Indian language] [I]

Pseudostems absent. **Inflorescences** from rhizome branch without basal leaves [projecting among basal leaves], dense, cylindric; bracts of main axis crowded, [2–]4–9 cm, ovate to narrowly ovate or rectangular; cincinni sessile, 2–7-flowered, enclosed in bracts; bracteoles small, inconspicuous, hidden by bracts. **Flowers:** calyx short-cylindric, minutely 2–3-toothed, split down one side; corolla tube funnelform, lobes ovate or oblong; filament rectangular, plane; anther enclosed in corolla, base spurred, terminal appendage sometimes present, lingulate; lateral staminodes large, petal-like, lip concave, rounded or 2-fid. **Fruits** capsule, ellipsoid. $x = 11, 21$.

Species ca. 50 (1 in the flora): introduced; North America; native, s Asia, Australia.

1. Curcuma zedoaria (Christmann) Roscoe, Trans. Linn. Soc. London 8: 354. 1807 • Zedoary [F] [I]

Amomum zedoaria Christmann in G. F. Christmann and G. W. F. Panzer, Vollst. Pflanzensyst. 5: 12. 1779; *Curcuma pallida* Loureiro

Leaf blades narrowly ovate or elliptic, 45–67 × 15–22 cm. **Inflorescences** erect, 11–23 × 5–10 cm; bracts of main axis whitish proximally, green (proximal bracts) or pink (distal bracts) distally; proximal bracts ovate to rectangular, deeply saccate, 4–4.5 × 4 cm, apex obtuse or truncate-apiculate; distal bracts narrowly ovate, 8–9 × 4–4.5 cm, apex rounded. **Flowers:** perianth white or spotted with purple; staminodes pale yellow with yellow streak down center of lip.

Flowering spring (May). Disturbed lakeshore; 50 m; introduced; Fla.; native, Asia (ne India).

Curcuma zedoaria is grown commercially in Asia for its starchy rhizome, which is used as a condiment or tonic.

3. ZINGIBER Miller, Gard. Dict. Abr. ed. 4, vol. 3. 1754, orthography conserved • Ginger [the classical name, from Sanskrit *crngavera*] [I]

Pseudostems well developed, 1–2 m. **Inflorescences** terminating short stem with only scale leaves [projecting from side or tip of pseudostem], dense, conelike; bracts of main axis crowded, 1–3[–5], reniform or very broadly ovate [to lance-elliptical or lanceolate]; cincinni sessile, 1-flowered, enclosed in bracts; bracteoles small, inconspicuous, hidden by bracts. **Flowers:** calyx cylindric, shortly 3-lobed, split down one side; corolla tube cylindric, dilated distally, lobes lanceolate; filament short or nearly absent; anther enclosed within upper petal, not spurred, terminal appendage long; lateral staminodes absent or reduced to small teeth connate with lip, lip plane, entire, notched, or 3-lobed. **Fruits** capsule, ellipsoid. $x = 11$.

Species ca. 100 (1 in the flora): introduced; North America, Mexico, Africa, n Australia; native, s Asia.

The ginger of commerce, *Zingiber officinale* Roscoe, is native to southeast Asia; it is commonly cultivated throughout the tropics, and most of the commercial supply now comes from Jamaica. Ginger seldom flowers or fruits in cultivation although plants are known to spread vegetatively in the vicinity of abandoned gardens in some tropical areas.

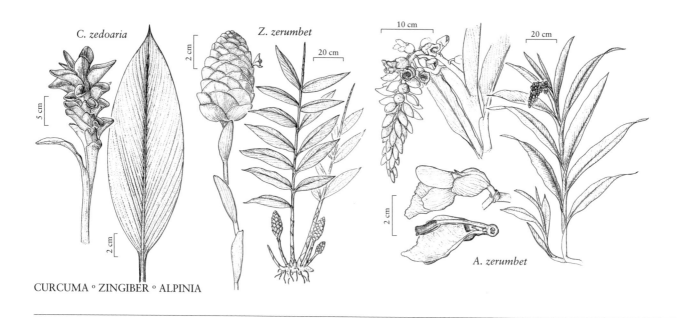

CURCUMA ○ ZINGIBER ○ ALPINIA

1. Zingiber zerumbet (Linnaeus) Smith, Exot. Bot. 2: 105. 1806 F I

Amomum zerumbet Linnaeus, Sp. Pl. 1: 1. 1753

Leaf blades oblanceolate or narrowly elliptic, 30–32 × 6–7 cm (smaller distally). **Inflorescences** erect, 7–11 × [3–]5–6 cm; bracts of main axis green when young, becoming red; proximal bracts reniform or very broadly ovate, concave, 2–3 × 3–4 cm, apex broadly rounded; distal bracts smaller but otherwise similar to proximal bracts, ca. 1 × 2 cm. **Flowers:** perianth and staminodes pale yellow.

Flowering fall (Oct). Disturbed areas; 50 m; introduced; Fla.; native, Asia (India and the Malay Peninsula).

In Asia, the rhizomes of *Zingiber zerumbet* are used as a spice in much the same way as those of *Z. officinale.* In North America, escaped populations are known only from two sites in Gainesville, Florida.

4. ALPINIA Roxburgh, Asiat. Res. 11: 350. 1810, conserved name • Ginger-lily [for Italian botanist Prosper Alpinus (1553–1617)] I

Pseudostems well developed, 1–3 m. **Inflorescences** projecting from tip of pseudostem, lax, paniculate; bracts of main axis remote [imbricate or not], minute [to 7 cm], scalelike [ovate to lance-oblong or lanceolate]; cincinni stalked, 1–3-flowered; bracteoles large, conspicuous [small or absent], enclosing cincinni. **Flowers:** calyx subcampanulate, shallowly 3-toothed, split down one side [not split]; corolla tube cylindric, lobes oblanceolate to elliptical; filament linear, plane; anther enclosed in corolla, not spurred, terminal appendage none; lateral staminodes absent or very small and connate with lip, lip ovate, tubular-incurved, notched. **Fruits** mostly indehiscent, globose. $x = 11, 12$.

Species ca. 230 (1 in the flora): introduced; North America, Mexico, West Indies, Central America, South America; native, Asia, Oceania.

Several species of *Alpinia* are grown as ornamentals in warm climates. Only *A. zerumbet* is known to spread outside cultivation, but at least three other species, *A. calcarata* Roscoe, *A.*

nigra (Gaertner) B. L. Burtt, and *A. officinarum* Hance, may persist for years in abandoned gardens in coastal Florida. All three of these species may be distinguished from *A. zerumbet* by having erect inflorescences, among other characters.

1. **Alpinia zerumbet** (Persoon) B. L. Burtt & R. M. Smith, Notes Roy. Bot. Gard. Edinburgh 31: 204. 1972 • Shellflower F I

Costus zerumbet Persoon, Syn. Pl. 1: 3. 1805; *Languas speciosum* (J. C. Wendland) Small; *Zerumbet speciosum* J. C. Wendland

Leaf blade lanceolate to narrowly elliptic, (20–)33–67 × (3–)7.5–11 cm. **Inflorescences** drooping, 15–30 × 6–10 cm; stalks of cincinni 0.3–3 cm, bracteoles sheathing, white proximally, pink distally. **Flowers:** lip yellow with red penciling, perianth and staminodes otherwise white or pink.

Flowering spring–summer (May–Sep). Disturbed hammocks and thickets; 0–30 m; introduced; Fla.; Central America; South America; native, Asia.

Alpinia zerumbet is commonly cultivated, but rarely escapes. The illegitimate name *Alpinia speciosa* (J. C. Wendland) K. Schumann and Hollrung 1887 [not *A. speciosa* (Blume) D. Dietrich 1839] has often been used for this species.

218. CANNACEAE Jussieu

• Canna Family

W. John Kress

Linda M. Prince

Herbs, perennial, from rhizomes. **Aerial stems** present, unbranched. **Leaves** cauline, 2-ranked, differentiated into basal sheath, petiole, and blade; sheaths overlapping, supporting stem, open, ligule absent; summit of petiole not differentiated; blade with lateral veins parallel, diverging from prominent midrib. **Inflorescences** 1 per aerial shoot, terminal on leafy shoot, pedunculate racemes or panicles of flowers or of 2-flowered monochasial cymes (cincinni); bracts of main axis subtending flowers or cincinni. **Flowers** bisexual, asymmetric; sepals and petals differentiated, sepals 3, distinct, petals 3, connate at base; fertile stamens 1, petal-like, anther marginal, 1-locular; staminodes (1–)3–4, petal-like, showy, unequal, anterior staminode (labellum) often broader than posterior staminodes; ovary inferior, 3-carpellate, 3-locular, all locules fertile; placentation axile; ovules few to numerous per locule; style standing away from stamens and staminode, petal-like; stigmatic area a marginal callosity; style, stamen, and staminodes basally connate into tube. **Fruits** capsules; sepals persistent in fruit. **Seeds:** aril absent; endosperm scanty; perisperm copious; embryo straight. $x = 9$.

Genera 1, species 10 (1 genus, 3 species in the flora): tropical and subtropical, primarily the Americas.

The flowers of Cannaceae are showy and brightly colored with foliaceous or petal-like sepals and have a conspicuously warty or spiny-fimbriate ovary. The fruits are large, broadly ellipsoid, loculicidal, and warty or spiny-fimbriate. The seeds are very hard and ovoid to globose.

Cannaceae are most closely related to the Marantaceae, the prayer-plants, with which they share several unusual reproductive features, such as asymmetric flowers, a reduction in the number of pollen-bearing stamens to a single bisporangiate anther, and secondary pollen presentation (P. F. Yeo 1993). The pollen grains are large and spheroid, and, like those of most members of the order Zingiberales, have a much reduced exine and a much expanded intine layer.

Cannaceae are known from several dubious fossils of vegetative organs dating to the late Cretaceous. The seeds are exceptionally hard and durable: a seed reported to be 600 years old and preserved in a South American archeological tomb germinated and flowered (J. C. Lerman and E. M. Cigliano 1971); yet, no fossilized seeds have been discovered.

Cannaceae are native to the American tropics and subtropics, occurring from sea level to mountainous slopes below 3000 m. One species (*Canna indica*) has become naturalized throughout the Asian and African tropics and subtropics, and several species and hybrids are commonly cultivated in the gardens of Europe, North America, and tropical zones. Cannas are found scattered in transitional habitats, both natural and artificial. In the wild, they grow along the edges of marshes and forest margins, often in water to 10 cm deep. They also thrive in roadside ditches and refuse sites, given adequate moisture.

SELECTED REFERENCES Grootjen, C. J. and F. Bouman. 1988. Seed structure in Cannaceae: Taxonomic and ecological implications. Ann. Bot. (Oxford), n. s. 61: 363–371. Kress, W. J. 1990. The phylogeny and classification of the Zingiberales. Ann. Missouri Bot. Gard. 77: 698–721. Kress, W. J. and D. E. Stone. 1982. Nature of the sporoderm in monocotyledons, with special reference to the pollen grains of *Canna* and *Heliconia*. Grana 21: 129–148. Maas, P. J. M. 1985. 195. Cannaceae. In: A. R. A. Görts-van Rijn, ed. 1985+. Flora of the Guianas. Series A: Phanerogams. 12+ fasc. Königstein. Fasc. 1, pp. 69–73. Maas, P. J. M. and H. Maas. 1988. 223. Cannaceae. In: G. Harling et al., eds. 1973+. Flora of Ecuador. 60+ nos. Göteborg. No. 32, pp. 1–9. Rogers, G. K. 1984. The Zingiberales (Cannaceae, Marantaceae, and Zingiberaceae) in the southeastern United States. J. Arnold Arbor. 65: 5–55. Woodson, R. E. Jr. and R. W. Schery. 1945. Cannaceae. In: R. E. Woodson Jr. et al., eds. 1943–1981. Flora of Panama. 41 fasc. St. Louis. [Ann. Missouri Bot. Gard. 32: 74–80.]

1. CANNA Linnaeus, Sp. Pl. 1: 1. 1753; Gen. Pl. ed. 5, 1. 1754 • [Greek *kanna*, reedlike plant]

Herbs, rhizomatous, 1–2[–5] m, forming small to large monotypic stands. **Leaves** green [bronze or magenta in hybrids and cultivars], often glaucous [cottony]; blade narrowly ovate to narrowly elliptic, 20–70 cm × 15–30 cm, base gradually or abruptly tapered, apex acute to acuminate. **Inflorescences:** peduncles green [magenta], often glaucous; bracts green [magenta], often glaucous; primary bracts to 30 cm, secondary bracts to 20 cm; floral bracts 0.5–3 × 0.3–1.5 cm, papery. **Flowers** nearly sessile, subtended by pedicel bract; sepals usually green [magenta], often less than half size of petals; petals sharply reflexed or not, green or brightly colored, 4–15 cm, generally shorter than staminodes; staminodes pale yellow to deep crimson red; labellum 3–9 × 4–10 cm; ovary green [magenta]. **Capsules** brown, 1.5–6 × 2–4.5 cm, warty, becoming papery. **Seeds** 5–25[–75] per capsule, medium to dark brown or black, 4–10 × 4–8 mm.

Species 10 (3 in the flora): tropical and subtropical Americas with some species naturalized and many cultivated hybrids.

Until recently taxonomists recognized more than 50 species in *Canna*, but that number has now been reduced to ten (P. J. M. Maas 1985; P. J. M. Maas and H. Maas 1988) based mainly on new concepts of biogeographical history, the extent of hybridization during cultivation, and the plasticity of morphologic features, especially in the highly polymorphic species *C. indica*.

Little has been published regarding pollination of these plants. The two North American species with pale yellow flowers, *Canna glauca* and *C. flaccida*, flower at dusk and may be pollinated by hawkmoths. Several neotropical species with bright red or orange flowers are hummingbird-pollinated. Nectar, which accumulates at the base of the floral tube, is the apparent reward in all cases. Pollen is shed from the bisporangiate anther onto the adjacent style before the flower opens (secondary pollen presentation), which usually results in self-pollination; thus, greenhouse-grown plants readily set seed. The large seed size and lack of reward for potential animal dispersal agents suggests that seeds are dispersed by gravity and water. Seeds can germinate and produce reproductive shoots in a single growing season.

Both *Canna flaccida* and *C. glauca*, as well as several Central American species, are hosts to the larva of the skipper butterfly *Calpodes ethlius* (Cramer), which builds protective tents by folding or rolling the apices of the leaves.

The most common use of *Canna* by Europeans and North Americans is as ornamentals. Both the hybrids and some species are grown for their striking foliage and large, delicate flowers. Two hybrids commonly grown in both temperate and tropical zones are *C. ×generalis* L. H. Bailey and *C. ×orchioides* L. H. Bailey. The former is apparently a cross between *C. indica* and *C. glauca*, or *C. iridiflora* Ruiz & Pavón and the latter between *C. ×generalis* and *C. flaccida* (L. H. Bailey 1924; J. W. Donahue 1965). These hybrids are usually sterile, although they may persist in cultivation.

The seeds of *Canna indica* are used as beads (especially for rosaries) and in gourds to form rattles. The rhizomes are used to make a form of arrowroot. The vegetation and rhizomes have been used as medicinals for both humans and domesticated animals.

SELECTED REFERENCES   Donahue, J. W. 1965. History, breeding and cultivation of the canna. Amer. Hort. Mag. 44: 84–91.   Gade, D. W. 1966. Achira, the edible canna, its cultivation and use in the Peruvian Andes. Econ. Bot. 20: 407–415.   Segeren, W. and P. J. M. Maas. 1971. The genus *Canna* in northern South America. Acta Bot. Neerl. 20: 663–680.   Tomlinson, P. B. 1961. The anatomy of *Canna*. J. Linn. Soc., Bot. 56: 467–473.   Young, A. M. 1982. Notes on the interaction of the skipper butterfly *Calpodes ethlius* (Lepidoptera: Hesperiidae) with its larval host plant *Canna edulis* (Cannaceae) in Mazatlan, State of Sinaloa, Mexico. J. New York Entomol. Soc. 90(2): 99–114.

1. Plants not glaucous; flowers red to yellow-orange, but never pure yellow (except in some hybrid cultivars). . . . . 1. *Canna indica*
1. Plants glaucous; flowers pure yellow.
   2. Inflorescence simple with fewer than 5 flowers; sepals 2.5–3.5 cm; petals strongly reflexed; staminodes broadly ovate, labellum broadly obovate, 8–9 cm wide; seeds nearly globose, less than 7 mm. . . . . 2. *Canna flaccida*
   2. Inflorescence simple or occasionally branched with more than 10 flowers; sepals 1–2 cm; petals erect; staminodes narrowly elliptic to narrowly ovate, labellum linear, 0.7 cm wide; seeds ovoid, more than 7 mm. . . . . 3. *Canna glauca*

1. **Canna indica** Linnaeus, Sp. Pl. 1: 1. 1753 • Indian-shot, platanillo   I

Canna coccinea Miller; *C. discolor* Lindley; *C. edulis* Ker Gawler; *C. lutea* Miller; *C. warscewiczii* Dietrich

Rhizomes fleshy. **Leaves:** sheath glabrous; blade narrowly ovate to ovate, 20–60 × 10–30 cm, base obtuse to narrowly cuneate, apex shortly acuminate to acute, abaxially and adaxially glabrous. **Inflorescences** racemes, sometimes branched, bearing 1–2-flowered cincinni, 6–20 per inflorescence; primary bracts to 15 cm; secondary bracts to 9 cm; floral bracts persistent, broadly obovate to narrowly (ob)ovate(-triangular), 0.5–3 × 0.5–1.5 cm, apex entire, often glaucous; bracteoles (ovate-)triangular, 0.5–2 × 0.3–0.8 cm, apex entire. **Flowers** red to yellow-orange, never pure yellow (except in some hybrid cultivars), 4.5–7.5 cm; pedicels 0.2–1 cm, to 1.5 cm in fruit; sepals narrowly triangular, 0.9–1.7 × 0.2–0.5 cm; petals erect, 4–6.5 cm, tube 0.5– 1.5 cm, lobes lanceolate to narrowly oblong, 3.5–5 × 0.4–0.7 cm; staminodes 3–4, narrowly obovate to spatulate, 4.5–7.5 cm, free part 0.3–0.5 cm wide, apex rounded, acute, or cleft; labellum reflexed, narrowly oblong, approximately equal to other staminodes. **Capsules** ellipsoid to nearly globose, 1.5–3 × 1.5–2 cm. **Seeds** black, globose to nearly globose, 5–8 × 4–6.7 mm diam. $2n = 18$.

Flowering primarily spring–summer; fruiting summer–early fall. Often, if not always, in secondary growth and waste places; 0–100 m; apparently introduced; Fla., La., S.C., Tex.

Canna indica is probably native to Neotropics and is now common throughout tropics and subtropics.

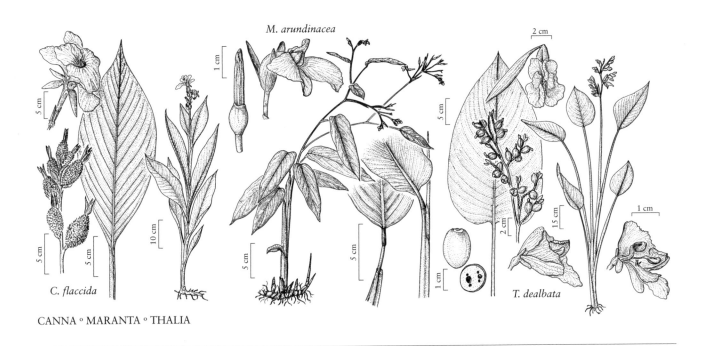

M. arundinacea

C. flaccida

T. dealbata

CANNA ° MARANTA ° THALIA

2. Canna flaccida Salisbury, Icon. Stirp. Rar., 3, plate 2. 1791 • Golden canna, bandana-of-the-everglades E F

Rhizomes fleshy. Leaves: sheath and blade glaucous; blade narrowly ovate to narrowly elliptic, 20–50 × 8–12 cm, base gradually narrowed into sheath, apex acute. Inflorescences racemes, simple, bearing 1-flowered cincinni, fewer than 5 flowers per inflorescence; primary bracts 10–30 cm; secondary bracts generally absent; floral bracts persistent, (broadly) ovate-triangular, 0.5–1 × 0.4–0.7 cm, glaucous, apex entire or irregularly lobed; bracteoles ovate-triangular, 0.3–2 cm × 4–8 mm, apex entire or irregularly lobed. Flowers pure yellow, 10–15 cm; pedicels short, to 0.5 cm; sepals narrowly oblong-elliptic, 2.5–3.5 × 0.4–0.6 cm; petals strongly reflexed, 10–12 cm, tube ca. 4 cm, lobes narrowly oblong-elliptic, 6–8 cm × 4–13 mm, base sharply reflexed; staminodes 3, broadly ovate, 9–11 × 8–9 cm; labellum not obviously reflexed, broadly ovate, ca. 10 × 8–9 cm. Capsules irregularly ellipsoid, 5–6 × 4–4.5 cm. Seeds brown, nearly globose, 6–6.5 × 5–6 mm. $2n = 18$.

Flowering primarily summer; fruiting shortly thereafter. Seasonally wet areas in open marshes, lake margins, and inundated pine flatwoods; 0–100 m; Ala., Fla., Ga., La., S.C.

Canna flaccida was reported from Texas (D. S. Correll and M. C. Johnston 1970), but no specimens have been seen to verify. Although no specimens have been seen from the area between the Louisiana site and the Alabama site, *Canna flaccida* is to be expected there.

3. Canna glauca Linnaeus, Sp. Pl. 1: 1. 1753 • Louisiana canna

Rhizomes far-creeping, 0.5–1.5 cm diam. Leaves: sheath and blade glaucous; blade narrowly ovate, 28–70 × 1.5–14 cm, base cuneate, apex very gradually narrowing to acute. Inflorescences racemes, simple or occasionally branched, bearing 2-flowered cincinni, more than 10 flowers per inflorescence; primary bracts 10–30 cm; secondary bracts 5–20 cm; floral bracts persistent, (broadly) ovate-triangular, 0.7–2.5 × 0.4–1 cm, apex entire or irregularly lobed, glaucous; bracteoles (broadly) ovate-triangular, 0.3–2 cm × 4–8 mm, apex entire or irregularly lobed. Flowers pure yellow, 7.5–10 cm; pedicels short, to 0.5 cm; sepals narrowly elliptic-triangular, 1.2–2 × 0.3–0.5 cm; petals erect, 5–9 cm, tube 1–2 cm, lobes narrowly ovate, 4–7 × 0.7–1.1 cm; staminodes 4, narrowly elliptic to narrowly ovate, 7.5–10 cm, free part 0.5–2.3 cm wide, apex sometimes slightly notched;

labellum strongly reflexed, linear, approximately equal to other staminodes. **Capsules** globose to ellipsoid, 2–5 × 2–4 cm. **Seeds** brown, ovoid, 7–10 × 6–8 mm. $2n = 18$.

Flowering summer; fruiting summer–fall (Jun–Sep).

Margins of marshes, swamps, ponds, and wet ditches; 0–100 m; Fla., La., S.C., Tex.; West Indies; Central America; South America.

Canna glauca is introduced in Florida and probably in South Carolina.

219. MARANTACEAE Petersen

• Arrowroot or Prayer-plant Family

Helen Kennedy

Herbs, perennial, from rhizomes. **Aerial stems** branched or unbranched. **Leaves** in a basal cluster, both basal and cauline [strictly cauline], alternate, differentiated into basal sheath, petiole (occasionally absent), and blade; sheaths often overlapping, supporting stem, open, ligule absent; summit of petiole distinctly differentiated as pulvinus (i.e., region of specialized cells controlling leaf movement, differing in color, often size, from petiole); blade with lateral veins parallel, diverging from prominent midrib. **Inflorescences** 1–several per aerial shoot, terminal or terminal and lateral on aerial stem [occasionally arising directly from rhizome], scapose or pedunculate [subsessile], compound, headlike or spikelike to diffuse or much branched, bracteate; each bract enclosing pair of flowers or cincinnus of paired flowers [solitary flowers (*Monotagma, Monophrynium*)]. **Flowers** bisexual, asymmetric, paired flowers mirror images of each other; sepals and petals differentiated, sepals 3, distinct [rarely connate basally (*Megaphyrnium*)], petals 3, corolla, staminodes, and style fused basally into tube; fertile stamens 1, occasionally petal-like; anthers 1-locular; staminodes [2–]3–4, outer staminodes [0–]1–2, petal-like, inner staminodes 2, one fleshy (callose staminode), one hooded (cucullate staminode), bearing 1(2 in *Thalia*) appendage(s), subterminal or medial, fingerlike or flaplike; ovary inferior, 3-carpellate, 3-locular, 1[–3] locule(s) fertile; placentation basal; ovules 1 per locule; terminal portion of style and stigma enclosed in cucullate staminode, explosively released, stout (appendaged in *Thalia*); stigma cup-shaped. **Fruits** fleshy [dry] capsules (rarely indehiscent) [berry or achene]; sepals often persistent in fruit. **Seeds:** aril present [rarely absent]; endosperm scanty or absent; perisperm copious; embryo curved.

Genera 31, species ca. 530 (2 genera, 3 species in the flora): lowland tropics of Asia and Africa, mainly (80%) in American tropics, occasionally subtropics, s United States to n Argentina.

Maranta arundinacea, Thalia dealbata, and *T. geniculata* are cultivated. Marantaceae are clearly monophyletic, uniquely defined by two vegetative traits: leaf venation of broadly sigmoidally-curved (S-shaped) lateral veins with numerous, regular, parallel cross veinlets (attached at right angles) between them and the pulvinus at the base of the leaf blade. The family can be recognized worldwide from a single leaf. Florally, Marantaceae are characterized by their highly modified staminodes and unusual pollination mechanism: explosive, secondary pollen presentation. During the bud stage, pollen is shed onto the back of the style, behind the stigma. At anthesis the style is under tension and is enfolded and held in place by the cucullate (hooded) staminode. Bees, probing for nectar in the flower, depress the appendage (trigger) of the cucullate staminode. This releases the style, which springs forward, moving in a single plane or helically, bringing the cup-shaped stigma in contact with pollen on the pollinator and, in the same motion, depositing fresh pollen on the same spot. Marantaceae are self-compatible but are mainly allogamous (out-crossing). In spite of the elaborate pollination system, some 8% of the species are known to be autogamous (self-pollinating), including all three species in the flora [very rarely cleistogamous]. Such self-pollen is deposited in the stigma during pollen transfer within the bud prior to anthesis, not during subsequent stylar movement.

Within the Zingiberales, Marantaceae are most closely related to Cannaceae, in which they were originally placed. The two families share similar leaf venation, presence of specialized cells at the base of the leaf (only a few cells abaxially in Cannaceae, not organized into a pulvinus), and secondary pollen presentation.

The leaves of the three North American species of Marantaceae are all homotropic (all leaves rolled in the same direction in the bud), though extraterritorial species may be antitropic (leaves on opposite sides of the stem rolled in opposite directions, half of the leaves with the left side of the blade larger than the right, opposite leaves with the right side of the blade larger). A prophyll is associated with each pair of flowers in the inflorescence. The North American species lack secondary bracts and bracteoles, although extraterritorial species may have a secondary bract and as many as four bracteoles, which may be modified as extrafloral nectaries. The perisperm of the seed has 1–2 canals.

The family is best known in horticulture for its elaborately patterned leaves, present in 20% of the species (H. Kennedy 1992). Such patterned-leaved species are popular as house plants, especially the ubiquitous *Maranta leuconeura*, in conservatories, and as ornamentals in tropical gardening. The most significant food plant is *Maranta arundinacea*, cultivated in tropical regions worldwide for arrowroot starch.

SELECTED REFERENCES   Andersson, L. 1976. The synflorescence of the Marantaceae. Organization and descriptive terminology. Bot. Not. 129: 39–48.   Andersson, L. 1981. The neotropical genera of Marantaceae. Circumscription and relationships. Nordic J. Bot. 1: 218–245.   Grootjen, C. J. 1983. Development of ovule and seed in Marantaceae. Acta Bot. Neerl. 32: 69–86.   Holttum, R. E. 1951. The Marantaceae of Malaya. Gard. Bull. Singapore 13: 254–296.   Kennedy, H. 1978. Systematics and pollination of the "closed-flowered" species of *Calathea* (Marantaceae). Univ. Calif. Publ. Bot. 71: 1–90, plates 1–20.   Kirchoff, B. K. 1983. Floral organogenesis in five genera of the Marantaceae and in *Canna* (Cannaceae). Amer. J. Bot. 70: 508–523.   Rogers, G. K. 1984. The Zingiberales (Cannaceae, Marantaceae, and Zingiberaceae) in the southeastern United States. J. Arnold Arbor. 65: 5–55.   Roscoe, W. [1824–]1828. Monandrian Plants of the Order Scitamineae. 15 parts. Liverpool.   Schumann, K. 1902. Marantaceae. In: H. G. A. Engler, ed. 1900–1953. Das Pflanzenreich. . . . 107 vols. Berlin. Vol. 11 [IV,48], pp. 1–184.   Tomlinson, P. B. 1961b. Morphological and anatomical characteristics of the Marantaceae. J. Linn. Soc., Bot. 58: 55–78.

1. Plants terrestrial; bracts persistent; flowers white; sepals more than 5 mm. . . . . 1. *Maranta*, p. 317
1. Plants aquatic; bracts deciduous; flowers purple; sepals less than or equal to 3 mm. . . . . 2. *Thalia*, p. 318

1. MARANTA Linnaeus, Sp. Pl. 1: 2. 1753; Gen. Pl. ed. 5, 2. 1754 • Arrowroot, prayer-plant [for Bartolomea Maranti, Venetian physician and botanist who lived during the mid 1500s] I

Plants terrestrial, prostrate, scandent, or upright, usually dying back to rhizome during dry season, 0.1–1.5(–1.8) m. **Rhizomes** occasionally swollen, storing starch. **Stems** branched or unbranched with basal and cauline leaves to highly branched above elongate, canelike stem (internode) with few or no basal leaves. **Leaves** homotropic [rarely antitropic]; sheath usually auriculate, not spongy; blade [patterned] plain green, ovate to elliptic. **Inflorescences** usually 2–several per shoot, spikelike, unbranched; bracts persistent, subtending 2–6 pedicellate flower pairs, herbaceous; prophylls keeled, membranous; secondary bracts absent; bracteoles usually absent. **Flowers** self-fertilizing [outcrossing], corolla white, staminodes white [purple]; sepals persistent in fruit, more than 5 mm, herbaceous; corolla tube [4–]12–14 mm, corolla lobes unequal; outer staminodes 2, petal-like; callose staminode apex usually petal-like; cucullate staminode with 1 appendage, medial [subterminal], flaplike [fingerlike]; stylar movement in single plane; style unappendaged. **Fruits** capsules, 1-seeded, obliquely ellipsoid, pericarp relatively thin, dehiscent. **Seeds** brown, ellipsoid, rugose; perisperm canal 1, distally branched; aril conspicuous, white.

Species 32 (1 in the flora): tropical and subtropical regions, s Mexico, West Indies, Central America, South America (to n Argentina).

SELECTED REFERENCES Andersson, L. 1986. Revision of *Maranta* subgen. *Maranta*. Nordic J. Bot. 6: 729–756. Anonymous. 1893. St. Vincent arrowroot. Bull. Misc. Inform. Kew 1893: 191–204. Hodge, W. H. and D. Taylor. 1957. The ethnobotany of the Island Caribs of Dominica. Webbia 12: 513–644. Morton, J. F. 1977. Wild Plants for Survival in South Florida, ed. 4. Miami. Purseglove, J. W. 1972. Tropical Crops. Monocotyledons. 2 vols. London. Pp. 336–342. Sturtevant, W. C. 1969. History and ethnography of some West Indian starches. In: P. J. Ucko and G. W. Dimbleby, eds. 1969. The Domestication and Exploitation of Plants and Animals. Chicago. Pp. 177–199.

1. Maranta arundinacea Linnaeus, Sp. Pl. 1: 2. 1753 • Arrowroot F I

Maranta sylvatica Roscoe ex Smith

Plants erect, 0.3–1.3 m. **Rhizomes** tuberous, thickened. **Stem** often branched above. **Leaves:** basal 4–8, cauline 1–8; sheath auriculate, 4.5–31 cm, margins and apex densely pilose to nearly glabrous; petiole often absent in cauline leaves, 3.5–20 cm; pulvinus 0.2–1.8 cm, adaxially tomentose; blade ovate, 3.5–35 × 3–11 cm, basal leaves largest, abaxially glabrous to minutely pilose, adaxially sparsely minutely pilose. **Inflorescences:** bracts 1–2(–3), 2.4–6 cm; common pedicel of flower pair 2.3–5.5 cm. **Flowers:** sepals 10–17 mm; corolla white, corolla tube curved, 12–14 mm; staminodes white; ovary tan, densely pubescent, rarely glabrous or nearly glabrous. **Fruits** green or tinged red-brown, 8 × 4–5 mm. $2n = 18, 48$.

Flowering summer–winter. Hammocks and moist soil; 0–10 m; introduced; Fla.; native, s Mexico, West Indies, Central America, n South America (w Ecuador).

Maranta arundinacea, cultivated for its starch-storing rhizome, is the mostly widely field-grown species of all the Marantaceae. In the early 1800s, it was commercially cultivated in Georgia and South Carolina (W. J. Titford 1812). This species is often referred to as St. Vincent arrowroot, after St. Vincent Island, West Indies, which is the major site of commercial production of the starch. The name arrowroot derives from its medicinal use as a poultice to remove arrow poison from wounds. The starch has also been used internally against ingested poison (R. Bentley and H. Trimen [1875–]1880, vol. 4, p. 265). Because the starch is so readily digested, it has been fed to infants, invalids, and those allergic to wheat. The cultivar 'Variegata,' with white and green variegated leaves, is often grown as an ornamental.

2. THALIA Linnaeus, Sp. Pl. 2: 1193. 1753; Gen. Pl. ed. 5, 3. 1754 • [For Johann Thal, German physician and botanist who lived during the mid 1500s]

Plants aquatic, upright, often dying back to rhizome in winter [dry season], 1–3.5 m. **Rhizome** not evidently enlarged or specialized for starch storage. **Stems** unbranched below inflorescence; leaves all basal or rarely with single cauline leaf borne above elongate (0.7–2.5 m) internode. **Leaves** homotropic; sheath not auriculate, spongy, containing prominent air spaces; blade plain green, ovate to elliptic. **Inflorescences** branched, branches short and upright to elongate and arching; rachis internodes conspicuously zigzagged; bracts deciduous, both bracts and pro-phylls falling with flower if fruit not set, leaving proximal portion of rachis bare, each bract subtending 1 flower pair, herbaceous to leathery; prophylls not evidently keeled, membranous; secondary bracts absent; bracteoles absent. **Flowers** self-fertilizing [outcrossing], pale to dark purple (corolla and staminodes); sepals persistent in fruit, 0.5–3 mm, membranous; corolla tube 1–6 mm, corolla lobes subequal to strongly unequal; outer staminode 1, petal-like, showy; callose staminode mainly fleshy, narrow apical rim petal-like; cucullate staminode with 2 appendages, subterminal, finger-like; stylar movement helical when tripped; styles with 1 ap-pendage, elongate, straplike. **Fruits** capsules, 1-seeded, nearly globose to ellipsoid, pericarp thin, indehiscent. **Seeds** dark brown, nearly globose or ellipsoid, smooth; perisperm canals 2, curved; aril reduced. *x* = 6.

Species 6 (2 in the flora): warm temperate to tropical regions, se North America, Mexico, West Indies, Central America, South America (to n Argentina), w Africa.

Thalia is not closely related to any of the other Neotropical genera (L. Andersson 1981), and it is the only genus of Marantaceae that is strictly aquatic. The cucullate staminode bearing two appendages, the helical stylar movement, and the appendaged style are unique within the family. Also unique is the prevalence of various air chambers in the leaf sheath and petiole, a specialization for the aquatic habitat, which gives the sheath its distinctive spongy texture. Dispersal by water is probably the most common means of dispersal: a gas-filled space between the seed and the fruit wall in *Thalia* makes the fruit buoyant (C. J. Grootjen 1983). Fruits of *T. geniculata* have also been reported to be eaten by ducks (W. L. McAtee 1915). Carpenter bees (*Xylocopa* spp.) are reported to pollinate *T. geniculata* in cultivation in Asia (R. Classen-Backhoff 1991).

The widely varying chromosome numbers reported for species of *Thalia* probably represent errors or misinterpretations. Chromosome counts in Marantaceae are difficult, and few taxa have been adequately studied.

SELECTED REFERENCES   Andersson, L. 1981b. Revision of the *Thalia geniculata* complex (Marantaceae). Nordic J. Bot. 1: 48–56.   Dressler, R. L. et al. 1987. Identification Manual for Wetland Plant Species of Florida. . . . Gainesville. P. 119.   Godfrey, R. K. and J. W. Wooten. 1979. Aquatic and Wetland Plants of Southeastern United States: Monocotyledons. Athens, Ga. Pp. 621–622.   Gris, A. 1859. Observations sur la fleur des Marantées. Ann. Sci. Nat., Bot., sér. 4, 12: 193–219, plates 11–14.   Koechlin, J. 1964. Marantacées. In: A. Aubréville and J.-F. Leroy. 1961+. Flore du Gabon. 35+ vols. Paris. Vol. 9, pp. 91–158.

1. Leaves glabrous adaxially; inflorescences lax, broadly spreading to pendent, paniclelike array; rachis internodes 5–20 mm; bracts not pruinose, green or streaked or tinged with purple. . . . . 1. *Thalia geniculata*
1. Leaves pilose at base adaxially; inflorescences erect, tightly clustered, compact array; rachis internodes 2–3 mm; bracts markedly pruinose, appearing whitish, red-brown to red-purple beneath waxy coating. . . . . 2. *Thalia dealbata*

1. Thalia geniculata Linnaeus, Sp. Pl. 2: 1193. 1753

• Fire-flag, arrowroot

Thalia divaricata Chapman; *T. trichocalyx* Gagnepain; *T. welwitschii* Ridley

Plants 1–3.5 m. **Leaves:** basal 2–6, cauline 0–1(–2); sheath green or occasionally red-purple, glabrous; petiole green or occasionally red-purple, glabrous; pulvinus caramel-colored, olive-green, or red-purple, 0.3–2.5 cm, glabrous; blade ovate to narrowly ovate, 19–60 × 4–26 cm, firm, stiff-papery, base rounded to subtruncate, apex acute to acuminate, occasionally obtuse with acuminate tip, abaxial surface green, faintly pruinose, glabrous, adaxial surface glabrous. **Inflorescences** lax, broadly spreading to pendent, paniclelike array, to ca. 0.6 × 1 m; scapes 0.8–2.5 m; rachis not pruinose, internodes 5–20 mm; bracts not pruinose, green or streaked or tinged with purple, narrowly ovate, 1.3–2.8 cm, herbaceous, sparsely to densely villous. **Flowers:** sepals 0.5–2 mm; outer staminode faint lavender to purple, 15–20 × 5–10 mm; callose staminode base yellow, apex purple, apical rim, reflexed, petal-like. **Fruits** ellipsoid, 9–12 × 6–7 mm. **Seeds** smooth dark brown to black, ellipsoid, 7–10 × 5–6 mm. 2*n* = 18 (Senegal) 2*n* = 26 (in cultivation).

Flowering summer–fall (late Jun–Dec); fruiting summer–winter (Aug–Jan). Lowlands in ponds, wet roadside ditches, swamps, marshes, cypress sloughs, margin of streams or lakes, full sun, often in regions with pronounced dry season; Fla., La.; Mexico; West Indies; Central America; South America (to Argentina and Paraguay); w Africa.

This species has the widest distribution known for any of the Marantaceae, occurring in both Africa and the Americas. Because of the marked lack of variation among the African populations, it is believed that its occurrence in west Africa was a historically recent, probably accidental, introduction (L. Andersson 1981b). The variation in pubescence and bract size within the American continent has been used as the basis for describing additional species or varieties (L. Andersson 1981b). Populations of *Thalia geniculata* with a striking red-purple coloration on the petiole, sheath, and pulvinus were described as *T. geniculata* f. *rheumoides* Shuey (A. G. Shuey 1975). Such homogeneous populations are to be expected in a mainly selfing species.

SELECTED REFERENCES Classen-Bockhoff, R. 1991. Untersuchungen zur Konstruktion des Bestäubungsapparates von *Thalia geniculata*, (Marantaceen). Bot. Acta 104: 183–193. Kirchoff, B. K. 1986. Inflorescence structure and development in the Zingiberales: *Thalia geniculata* (Marantaceae). Canad. J. Bot. 64: 859–864.

2. Thalia dealbata Fraser, Thalia dealbata [plate]. 1794

• Powdery thalia

Thalia barbata Small

Plants 0.7–2.5 m. **Leaves:** basal 2–5, cauline 0; sheath pruinose, glabrous; petiole pruinose, glabrous; pulvinus yellowish brown to red or purple-brown, 0.6–2.5 cm, glabrous; blade ovate, occasionally narrowly elliptic, 17–55 × 7–22 cm, firm, stiff-papery, base rounded, rarely obtuse, apex acuminate, abaxial surface pruinose, appearing whitish, glabrous, adaxial surface pilose to nearly villous in basal 0.5–1 cm of blade, including midrib, at apex, and along margin of broader side. **Inflorescences** erect, tightly clustered, compact array, 9–31 × 7–18 cm; scapes 0.5–1.9 m; rachis pruinose, internodes 2–3 mm; bracts markedly pruinose, appearing whitish, red-brown to red-purple beneath waxy coating, orbiculate, strongly cupped, 0.8–1.5 cm, stiff, leathery, glabrous. **Flowers:** sepals 1.5–2.5 mm; outer staminode dark purple, 12–15 × ca. 6 mm; callose staminode base white to pale purple, margins and apex dark purple, apical rim reduced, petal-like. **Fruits** nearly globose to broadly obovoid, 9–12 × 8–11 mm. **Seeds** dark brown to black, nearly globose to broadly ellipsoid, 8–10 × 7–9. 2*n* = 12 (in cultivation).

Flowering late spring–early fall (May–Sep); fruiting summer–fall (Jun–Oct). Coastal plain in swamps, streamsides, roadside ditches, and ponds; Ark., Ga., Ill., La., Miss., Mo., Okla., S.C., Tex.

Thalia dealbata is the only species of Marantaceae endemic to North America and the only one not found in the tropics. Its sister species, *T. multiflora* Horkel ex Körnicke, is similarly restricted, but to the southernmost part of the range, southern Brazil to Uruguay. It is possible that our species is the result of an early, chance long-distance dispersal event from a South American population. This might partially explain why the plants are autogamous in spite of the elaborate floral mechanism.

Thalia dealbata is grown as an ornamental in tropical and temperate gardens and is hardy as far north as Philadelphia, Pennsylvania, and Vancouver, British Columbia, when the rhizome is submersed during winter.

SELECTED REFERENCE Price, S. and G. Rogers. 1987. *Thalia dealbata* (Marantaceae), a Missouri surprise. Missouriensis 8: 73–78.

Literature Cited

Robert W. Kiger, Editor

This is a consolidated list of all works cited in volume 22, whether as selected references, in text, or in nomenclatural contexts. In citations of articles, both here and in the taxonomic treatments, and also in nomenclatural citations, the titles of serials are rendered in the abbreviated forms recommended in G. D. R. Bridson and E. R. Smith (1991). Cross references to the corresponding full serial titles are interpolated here alphabetically by abbreviated form. In nomenclatural citations (only), book titles are rendered in the abbreviated forms recommended in F. A. Stafleu and R. S. Cowan (1976–1988) and F. A. Stafleu and E. A. Mennega (1992+). Here, those abbreviated forms are indicated parenthetically following the full citations of the corresponding works, and cross references to the full citations are interpolated in the list alphabetically by abbreviated form. Two or more works published in the same year by the same author or group of coauthors will be distinguished uniquely and consistently throughout all volumes of Flora of North America by lower-case letters (b, c, d, . . .) suffixed to the date for the second and subsequent works in the set. The suffixes are assigned in order of editorial encounter and do not reflect chronological sequence of publication. The first work by any particular author or group from any given year carries the implicit date suffix "a"; thus, the sequence of explicit suffixes begins with "b". Some citations in this list have dates suffixed "b" but are not preceded by citations of "[a]" works for the same year, or have dates suffixed "c" but are not preceded by citations of "[a]" and/or "b" works for that year. In these cases, the missing "[a]" and "b" works are ones cited (and encountered first from) elsewhere in the *Flora* that are not pertinent here.

Abh. Auslandsk., Reihe C, Naturwiss. = Abhandlungen aus dem Gebiet der Auslandskunde. Reihe C: Naturwissenschaften.

Abh. Königl. Ges. Wiss. Göttingen = Abhandlungen der Königlichen Gesellschaft der Wissenschaften zu Göttingen.

Abh. Math.-Phys. Cl. Königl. Bayer. Akad. Wiss. = Abhandlungen der Mathematisch-physikalischen Classe der Königlich bayerischen Akademie der Wissenschaften.

Abh. Naturwiss. Vereine Bremen = Abhandlungen herausgegeben vom Naturwissenschaftlichen Vereine zu Bremen.

Abrams, L. and R. S. Ferris. 1923–1960. Illustrated Flora of the Pacific States: Washington, Oregon, and California. 4 vols. Stanford. (Ill. Fl. Pacific States)

Acta Amazon. = Acta Amazonica.

Acta Bot. Neerl. = Acta Botanica Neerlandica.

Actes Soc. Hist. Nat. Paris = Actes de la Société d'Histoire Naturelle de Paris.

Allen, G. E. 1976. Investigations and current status of insect enemies as biological control agents of aquatic weeds. In: C. K. Varshney and J. Rzóska, eds. 1976. Aquatic Weeds in South East Asia. The Hague. Pp. 299–306.

Allg. Gartenzeitung = Allgemeine Gartenzeitung.

Allioni, C. 1785. Flora Pedemontana sive Enumeratio Methodica Stirpium Indigenarum Pedemontii. 3 vols. Turin. (Fl. Pedem.)

Amer. Hort. Mag. = American Horticultural Magazine.

Amer. J. Bot. = American Journal of Botany.

Amer. J. Sci. Arts = American Journal of Science, and Arts.

Amer. Midl. Naturalist = American Midland Naturalist; Devoted to Natural History, Primarily That of the Prairie States.

Anales Inst. Biol. Univ. Nac. México = Anales del Instituto de Biológia de la Universidad Nacional de México.

Ancibor, E. 1979. Systematic anatomy of vegetative organs of the Hydrocharitaceae. Bot. J. Linn. Soc. 78: 237–266.

Anderson, E. 1952. Plants, Man and Life. Boston.

Anderson, E. 1954. A field survey of chromosome numbers in the species of *Tradescantia* closely related to *Tradescantia virginiana*. Ann. Missouri Bot. Gard. 41: 305–327.

Anderson, E. and K. Sax. 1936. A cytological monograph of the American species of *Tradescantia*. Bot. Gaz. 97: 433–476.

Anderson, E. and R. E. Woodson Jr. 1935. The species of *Tradescantia* indigenous to the United States. Contr. Arnold Arbor. 9: 1–132.

Anderson, L. C., C. D. Zeis, and S. F. Alam. 1974. Phytogeography and possible origins of *Butomus* in North America. Bull. Torrey Bot. Club 101: 292–296.

Anderson's Fl. Alaska—See: S. L. Welsh 1974

Andersson, L. 1976. The synflorescence of the Marantaceae. Organization and descriptive terminology. Bot. Not. 129: 39–48.

Andersson, L. 1981. The neotropical genera of Marantaceae. Circumscription and relationships. Nordic J. Bot. 1: 218–245.

Andersson, L. 1981b. Revision of the *Thalia geniculata* complex. Nordic J. Bot. 1: 48–56.

Andersson, L. 1985. Revision of *Heliconia* subgen. *Stenochlamys* (Musaceae-Heliconioideae). Opera Bot. 82: 5–123.

Andersson, L. 1986. Revision of *Maranta* subgen. *Maranta*. Nordic J. Bot. 6: 729–756.

Andersson, L. 1992. Revision of *Heliconia* subgen. *Taeniostrobus* and subgen. *Heliconia* (Musaceae-Heliconioideae). Opera Bot. 111: 1–98.

Anl. Wiss. Beobacht. Reisen—See: G. B. von Neumayer 1875

Ann. Bot. Fenn. = Annales Botanici Fennici.

Ann. Bot. (Genoa) = Annali di Botanica. (Genoa.)

Ann. Bot. (König & Sims) = Annals of Botany. [Edited by König & Sims.]

Ann. Bot. (Oxford) = Annals of Botany. (Oxford.)

Ann. Mag. Nat. Hist. = Annals and Magazine of Natural History, Including Zoology, Botany, and Geology.

Ann. Missouri Bot. Gard. = Annals of the Missouri Botanical Garden.

Ann. Mus. Natl. Hist. Nat. = Annales du Muséum National d'Histoire Naturelle. = ["National" dropped with vol. 5.]

Ann. Naturhist. Mus. Wien, B = Annalen des Naturhistorischen Museums in Wien. Serie B, Botanik und Zoologie.

Ann. Sci. Nat., Bot. = Annales des Sciences Naturelles. Botanique.

Annot. Zool. Bot. = Annotationes Zoologicae et Botanicae.

Annual Rev. Ecol. Syst. = Annual Review of Ecology and Systematics.

Anonymous. 1893. St. Vincent arrowroot. Bull. Misc. Inform. Kew 1893: 191–204.

Aquaphyte = Aquaphyte; Newsletter of the I P P C Aquatic Weed Program of the University of Florida.

Aquatic Bot. = Aquatic Botany; International Scientific Journal Dealing with Applied and Fundamental Research on Submerged, Floating and Emergent Plants in Marine and Freshwater Ecosystems.

Aquilo, Ser. Bot. = Aquilo. Ser. Botanica.

Arch. Bot. (Leipzig) = Archiv für die Botanik. (Leipzig.)

Arch. Hydrobiol. = Archiv für Hydrobiologie.

Arditti, J. and E. Rodriguez. 1982. *Dieffenbachia*: Uses, abuses and toxic constituents: A review. J. Ethnopharmacol. 5: 293–302.

Argue, C. L. 1974. Pollen studies in the Alismataceae (Alismaceae). Bot. Gaz. 135: 338–344.

Argue, C. L. 1976. Pollen studies in the Alismataceae with special reference to taxonomy. Pollen & Spores 18: 161–173.

Ark. Bot. = Arkiv för Botanik Utgivet av K. Svenska Vetenskapsakademien.

Aroideana = Aroideana; Journal of the International Aroid Society.

Arr. Brit. Pl. ed. 3—See: W. Withering 1796

Årsberätt. Bot. Arbeten Upptäckter = Årsberättelse om Botaniska Arbeten och Upptäckter.

Asiat. Res. = Asiatic Researches, or Transactions of the Society.

Aublet, J. B. 1775. Histoire des Plantes de la Guiane Françoise. . . . 4 vols. Paris. [Vols. 1 and 2: text, paged consecutively; vols. 3 and 4: plates.] (Hist. Pl. Guiane)

Aubréville, A. and J.-F. Leroy. 1961+. Flore du Gabon. 35+ vols. Paris.

Austin, D. F. 1978. Exotic plants and their effects in southeastern Florida. Environm. Conservation 5: 25–34.

Austin, D. F. 1978b. The coconut in Florida. Principes 22: 83–87.

Austin, D. F. 1985. *Commelina gigas* rediscovered and lost. Palmetto 5(4): 11.

Autik. Bot.—See: C. S. Rafinesque 1840

Backman, T. W. H. 1991. Genotypic and phenotypic variability of *Zostera marina* on the west coast of North America. Canad. J. Bot. 69: 1361–1371.

Bailey, L. H. 1924. Manual of Cultivated Plants. . . . New York and London.

Bailey, L. H. 1936. *Washingtonia*. Gentes Herb. 4: 51–82.

Balbis, G. B. [1804.] Miscellanea Botanica. . . . Turin. (Misc. Bot.)

Barabé, D. 1982. Vascularisation de la fleur de *Symplocarpus foetidus* (Araceae). Canad. J. Bot. 60: 1536–1544.

Barabé, D.and M. Labrecque. 1984. Vascularisation de la fleur de *Lysichitum camtschatcense* (Araceae). Canad. J. Bot. 62: 1971–1983.

Barnabas, A. D. 1994. Anatomical, histochemical and ultrastructural features of the seagrass *Phyllospadix scouleri* Hook. Aquatic Bot. 49: 167–182.

Barrett, O. W. and O. F. Cook. 1910. Promising root crops for the South. U.S.D.A. Bur. Pl. Industr. Bull. 164: 1–43.

Barrow, S. C. 1998. A monograph of *Phoenix* L. (Palmae: Coryphoideae). Kew Bull. 53: 513–575.

Barthlott, W. and D. Frölich. 1983. Mikromorphologie und Orientierungsmuster epicuticularer Wachs-Kristalloide: Ein neues systematisches Merkmal bei Monokotylen. Pl. Syst. Evol. 142: 171–185.

Barton, W. P. C. 1817–1819. Vegetable Materia Medica of the United States; or Medical Botany, Containing a Botanical, General, and Medical History, of Medicinal Plants Indigenous to the United States. 2 vols. in 4 parts each. Philadelphia. [Parts numbered consecutively throughout, volumes paged independently.] (Veg. Mater. Med. U.S.)

Bartonia = Bartonia; a Botanical Annual.

Bartram, W. 1791. Travels through North and South Carolina, Georgia, East and West Florida, the Cherokee Country, the Extensive Territories of the Muscogulges, or Creek Confederacy, and the Country of the Chactaws. . . . Philadelphia. (Travels Carolina)

Beal, E. O. 1960. *Sparganium* (Sparganiaceae) in the southeastern United States. Brittonia 12: 176–181.

Beal, E. O. 1960b. The Alismataceae of the Carolinas. J. Elisha Mitchell Sci. Soc. 76: 68–79.

Beal, E. O. 1977. A Manual of Marsh and Aquatic Vascular Plants of North Carolina, with Habitat Data. Raleigh. [North Carolina Agric. Exp. Sta. Techn. Bull. 247.]

Beal, E. O., S. S. Hooper, and K. Rataj. 1980. Misapplication of the name *Sagittaria longirostra* (Micheli) J. G. Smith to *S. australis* (J. G. Smith) Small. Kew Bull. 35: 369–371.

Beal, E. O., J. W. Wooten, and R. B. Kaul. 1982. Review of *Sagittaria engelmanniana* complex (Alismataceae) with environmental correlations. Syst. Bot. 7: 417–432.

Beccari, O. 1912[–1914]. Palme del Madagascar. . . . 5 fasc. Florence. [Fascicles paged consecutively.] (Palme Madagascar)

Beitr. Bot.—See: M. J. Schleiden 1844

Beitr. Kenntn. Najas—See: P. W. Magnus 1870

Beitr. Naturk.—See: J. F. Ehrhart 1787–1792

Bennet, B. C. 1992. The Florida bromeliads: *Guzmania monostachia*. J. Bromeliad Soc. 42: 266–270.

Bennett, B. C. and J. R. Hicklin. 1998. Uses of saw palmetto (*Serenoa repens*, Arecaceae) in Florida. Econ. Bot. 52: 381–393.

Bentham, G. 1839[–1857]. Plantas Hartwegianas Imprimis Mexicanas. . . . London. [Issued by gatherings with consecutive signatures and pagination.] (Pl. Hartw.)

Bentham, G. 1844[–1846]. The Botany of the Voyage of H.M.S. Sulphur, Under the Command of Captain Sir Edward Belcher . . . During the Years 1836–1842. 6 parts. London. [Parts paged consecutively.] (Bot. Voy. Sulphur)

Bentham, G. and J. D. Hooker. 1862–1883. Genera Plantarum ad Exemplaria Imprimis in Herbariis Kewensibus Servata Definita. 3 vols. London. (Gen. Pl.)

Bentley, R. and H. Trimen. [1875–]1880. Medicinal Plants . . . . 4 vols. in 42 parts. London.

Ber. Deutsch. Bot. Ges. = Berichte der Deutschen botanischen Gesellschaft.

Ber. Geobot. Inst. E. T. H. Stiftung Rübel. = Berichte des Geobotanischen Instituts der Eidg. techn. Hochschule Stiftung Rübel.

Berry, F. and W. J. Kress. 1991. *Heliconia*: An Identification Guide. Washington.

Bierzychudek, P. 1982. The demography of jack-in-the-pulpit, a forest perennial that changes sex. Ecol. Monogr. 52: 335–351.

Bigelow, J. 1824. Florula Bostoniensis. A Collection of Plants of Boston and Its Vicinity . . . , ed. 2. Boston. (Fl. Boston. ed. 2)

Bigley, R. E. and J. L. Barreca. 1982. Evidence for synonymizing *Zostera americana* den Hartog with *Zostera japonica* Aschers. & Graebn. Aquatic Bot. 14: 349–356.

Biochem. Syst. & Ecol. = Biochemical Systematics and Ecology.

Biodivers. & Conservation = Biodiversity and Conservation.

Biol. Ser. Bull. State Univ. Montana = Biological Series of the Bulletin of the State University of Montana.

Björkquist, I. 1968. Studies in *Alisma* L. II. Chromosome studies, crossing experiments and taxonomy. Opera Bot. 19: 1–138.

Blackburn, R. D. and L. W. Weldon. 1970. Control of *Hydrilla verticillata*. Hyacinth Control J. 8: 4–9.

Blackwell, W. H. and K. P. Blackwell. 1974. The taxonomy of *Peltandra* (Araceae). J. Elisha Mitchell Sci. Soc. 90: 137–140.

Blume, C. L. 1835[1836]–1848[1849]. Rumphia, sive Commentationes Botanicae Imprimis de Plantis Indiae Orientalis. . . . 4 vols. in 40 parts. Leiden etc. [Parts paged consecutively within volumes.] (Rumphia)

Blumea = Blumea; Tidjschrift voor die Systematiek en die Geografie der Planten (A Journal of Plant Taxonomy and Plant Geography).

Blytt, M. N. and A. G. Blytt. 1861–1877. Norges Flora. . . . 3 vols. + suppl. Oslo. (Norges Fl.)

Böcher, T. W. 1950. Contributions to the flora and plant geography of west Greenland II. Meddel. Grønland 147(7): 11–23.

Bogin, C. 1955. Revision of the genus *Sagittaria* (Alismataceae). Mem. New York Bot. Gard. 9: 179–233.

Boivin, B. 1967–1979. Flora of the prairie provinces. Part I (–IV). Phytologia 15–18, 22–23, 42–43: passim. [Reprinted as Provancheria 2–5 with auxiliary pagination.]

Bol. Ann. Cons. Ultramar. (Portugal) = Boletim e Annaes do Conselho Ultramarino. (Portugal.)

Bolick, M. 1981. *Tradescantia bracteata* Small and *T. occidentalis* (Britt.) Smyth. in the Great Plains. [Abstract.] In: Botanical Society of America. 1981. Abstracts of Papers To Be Presented at the Meetings of the Botanical Society of America and Certain Affiliated Groups at Indiana University, Bloomington, Indiana, 16–21 August 1981. Oxford, Ohio. P. 63.

Bolick, M. 1986. Commelinaceae. In: Great Plains Flora Association. 1986. Flora of the Great Plains. Lawrence, Kans. Pp. 1046–1049.

Bomhard, M. L. 1935. *Sabal louisiana*, the correct name for the polymorphic palmetto of Louisiana. J. Wash. Acad. Sci. 25: 35–44.

Borchsenius, F. 1997. Flowering biology of *Geonoma irena* and *G. cuneata* var. *soderoi* (Arecaceae). Pl. Syst. Evol. 208: 187–196.

Börner, C. 1912. Botanisch-systematische Notizen. Abh. Naturwiss. Vereine Bremen 21: 245–282.

Börner, C. [1912]b. Eine Flora für das deutsche Volk. . . . Leipzig. (Fl. Deut. Volk)

Bosse, J. F. W. 1840–1849[–1854]. Vollständiges Handbuch der Blumengärtnerei, oder genaue Beschreibung fast aller in Deutschland bekannt gewordenen Zierpflanzen . . . ., ed. 2. 5 vols. Hannover. [Vols. 1–3 same as ed. 1, vols. 4–5 new.] (Vollst. Handb. Bl.-Gärtn.)

Boston J. Nat. Hist. = Boston Journal of Natural History.

Bot. Acta = Botanica Acta; Berichte der Deutschen Botanischen Gesellschaft.

Bot. California—See: S. Watson 1876–1880

Bot. Centralbl. = Botanisches Centralblatt.

Bot. Gaz. = Botanical Gazette; Paper of Botanical Notes.

Bot. Helv. = Botanica Helvetica.

Bot. J. Linn. Soc. = Botanical Journal of the Linnean Society.

Bot. Jahrb. Syst. = Botanische Jahrbücher für Systematik, Pflanzengeschichte und Pflanzengeographie.

Bot. Mag. = Botanical Magazine; or, Flower-garden Displayed. . . . [Edited by Wm. Curtis.] [With vol. 15, 1801, title became Curtis's Botanical Magazine; or. . . .]

Bot. Mag. (Tokyo) = Botanical Magazine. [Shokubutsu-gaku Zasshi.] (Tokyo.)

Bot. Not. = Botaniska Notiser.

Bot. Rev. (Lancaster) = Botanical Review, Interpreting Botanical Progress.

Bot. Tidsskr. = Botanisk Tidsskrift.

Bot. Untersuch. (Berlin) = Botanische Untersuchungen. (Berlin.)

Bot. Voy. Sulphur—See: G. Bentham 1844[–1846]

Bot. Zeitung (Berlin) = Botanische Zeitung. (Berlin.)

Bot. Zhurn. (Moscow & Leningrad) = Botanicheskii Zhurnal. (Moscow and Leningrad.)

Botanical Society of America. 1981. Abstracts of Papers To Be Presented at the Meetings of the Botanical Society of America and Certain Affiliated Groups at Indiana University, Bloomington, Indiana, 16–21 August 1981. Oxford, Ohio. [Bot. Soc. Amer. Misc. Ser. 160.]

Botany (Fortieth Parallel)—See: S. Watson 1871

Bown, D. 1988. Aroids: Plants of the Arum Family. Portland.

Brashier, C. K. 1966. A revision of *Commelina* (Plum.) L. in the U.S.A. Bull. Torrey Bot. Club 93: 1–19.

Brayshaw, T. C. 1985. Pondweeds and Bur-reeds, and Their Relatives: Aquatic Families of Monocotyledons in British Columbia. Victoria. [Brit. Columbia Prov. Mus., Occas. Pap. 26.]

Bridson, G. D. R. and E. R. Smith. 1991. B-P-H/S. Botanico-Periodicum-Huntianum/Supplementum. Pittsburgh.

Briggs, B. G. and L. A. S. Johnson. 1968. The status and relationships of the Australian species of *Typha*. Contr. New South Wales Natl. Herb. 4: 57–78.

Britton, N. L. 1909. Scheuchzeriaceae. In: N. L. Britton et al., eds. 1905+. North American Flora. . . . 47+ vols. New York. Vol. 17, pp. 41–42.

Britton, N. L. et al., eds. 1905+. North American Flora. . . .

47+ vols. New York. [Vols. 1–34, 1905–1957; ser. 2, vols. 1–13, 1954+.] (N. Amer. Fl.)

Britton, N. L. and A. Brown. 1896–1898. An Illustrated Flora of the Northern United States, Canada and the British Possessions from Newfoundland to the Parallel of the Southern Boundary of Virginia, and from the Atlantic Ocean Westward to the 102d Meridian. . . . 3 vols. New York. (Ill. Fl. N. U.S.)

Britton, N. L. and A. Brown. 1913. An Illustrated Flora of the Northern United States, Canada and the British Possessions from Newfoundland to the Parallel of the Southern Boundary of Virginia, and from the Atlantic Ocean Westward to the 102d Meridian . . . , ed. 2. 3 vols. New York. (Ill. Fl. N. U.S. ed. 2)

Brittonia = Brittonia; a Journal of Systematic Botany. . . .

Brooks, R. E. 1989. A Revision of *Juncus* Subgenus *Poiophylli* (Juncaceae) in the Eastern United States. Ph.D. dissertation. University of Kansas.

Brown, G. K. and A. J. Gilmartin. 1989. Chromosome numbers in Bromeliaceae. Amer. J. Bot. 76: 657–665.

Brown, G. K. and A. J. Gilmartin. 1989b. Stigma types in Bromeliaceae—A systematic survey. Syst. Bot. 14: 110–132.

Brown, G. K. and R. G. Terry. 1992. Petal appendages in Bromeliaceae. Amer. J. Bot. 79: 1051–1071.

Brown, K. E. 1976. Ecological studies of the cabbage palm, *Sabal palmetto*. Principes 20: 3–10, 49–56, 98–115, 148–157.

Brown, M. L. and R. G. Brown. 1984. Herbaceous Plants of Maryland. College Park, Md.

Brown, R. 1810. Prodromus Florae Novae Hollandiae et Insulae van-Diemen. . . . London. (Prodr.)

Browne, E. T. Jr. and R. Athey. 1992. Vascular Plants of Kentucky: An Annotated Checklist. Lexington, Ky.

Bruggen, H. W. E. van. 1973. Revision of the genus *Aponogeton* (Aponogetonaceae): VI. The species of Africa. Bull. Jard. Bot. Natl. Belg. 43: 1–2, 193–233.

Bruggen, H. W. E. van. 1985. Monograph of the Genus *Aponogeton* (Aponogetonaceae). Stuttgart.

Brummitt, R. K. 1996. Report of the Committee for Spermatophyta: 44. Taxon 45: 671–681.

Brummitt, R. K. and C. E. Powell, eds. 1992. Authors of Plant Names. A List of Authors of Scientific Names of Plants, with Recommended Standard Forms of Their Names, Including Abbreviations. Kew.

Bruner, M. C. 1982. Water-lettuce, *Pistia stratiotes* L. Aquatics 4(3): 4–5.

Buchenau, F. 1868. Index Criticus Butomacearum, Alismacearum, Juncaginacearumque. . . . Bremen. [Reprinted/preprinted from Abh. Naturwiss. Vereine Bremen 1: 213–224. 1868; 2: 1–49. 1869.] (Index Crit. Butom. Alism. Juncag.)

Buchenau, F. 1890. Monographia Juncacearum. Bot. Jahrb. Syst. 12: 1–495, 622–623, plates 1–3.

Buchenau, F. 1906. Juncaceae. In: H. G. A. Engler, ed. 1900–1953. Das Pflanzenreich. . . . 107 vols. Berlin. Vol. 25[IV,26], pp. 1–284.

Bull. Acad. Roy. Sci. Bruxelles = Bulletins de l'Académie Royale des Sciences et Belles-lettres de Bruxelles.

Bull. Amer. Rock Gard. Soc. = Bulletin of the American Rock Garden Society.

Bull. Bromeliad Soc. = Bulletin of the Bromeliad Society.

Bull. Florida State Mus., Biol. Sci. = Bulletin of the Florida State Museum. Biological Sciences.

Bull. Heliconia Soc. Int. = Bulletin, Heliconia Society International.

Bull. Indo-Pacific Prehist. Assoc. = Bulletin of the Indo-Pacific Prehistory Association.

Bull. Inst. Franç. Études Andines = Bulletin, Institut Française d'Études Andines.

Bull. Jard. Bot. Natl. Belg. = Bulletin du Jardin Botanique National de Belgique.

Bull. Mar. Sci. = Bulletin of Marine Science.

Bull. Mar. Sci. Gulf Caribbean = Bulletin of Marine Science of the Gulf and Caribbean.

Bull. Misc. Inform. Kew = Bulletin of Miscellaneous Information, Royal Gardens, Kew.

Bull. Mus. Life Sci. Louisiana State Univ. = Bulletin of the Museum of the Life Sciences, Louisiana State University.

Bull. Natl. Mus. Canada = Bulletin of the National Museum of Canada.

Bull. Orto Bot. Regia Univ. Napoli = Bulletino dell'Orto Botanico della Regia Università di Napoli.

Bull. S. Calif. Acad. Sci. = Bulletin of the Southern California Academy of Sciences.

Bull. Sci. Soc. Philom. Paris = Bulletin des Sciences, par la Société Philomatique (de Paris).

Bull. Soc. Imp. Naturalistes Moscou = Bulletin de la Société Impériale des Naturalistes de Moscou.

Bull. Torrey Bot. Club = Bulletin of the Torrey Botanical Club.

Bull. U.S. Bur. Fish. = Bulletin of the United States Bureau of Fisheries.

Bull. U.S.D.A. = Bulletin of the U.S. Department of Agriculture.

Bullock, S. H. 1980. Dispersal of a desert palm by opportunistic frugivores. Principes 24: 29–32.

Burk, W. R. and R. L. Stuckey. 1994. Ronald L. Stuckey: His Role in the Ohio Academy of Science. Chapel Hill.

Burman, N. L. 1768. Flora Indica . . . Nec Non Prodromus Florae Capensis. Leiden and Amsterdam. (Fl. Indica)

Burt-Utley, K. and J. F. Utley. 1987. Contributions toward a revision of *Hechtia* (Bromeliaceae). Brittonia 39: 37–43.

Butts, E. H. 1959. *Livistona chinensis* naturalized in Florida. Principes 3: 133.

Byull. Moskovsk. Obshch. Isp. Prir., Otd. Biol. = Byulleten' Moskovskogo Obshchestva Ispytatelei Prirody. Otdel Biologicheskii.

Calam. Gram. Tripet. Linn.—See: J. F. Ehrhart 1785–1793

Calcutta J. Nat. Hist. = Calcutta Journal of Natural History, and Miscellany of the Arts and Sciences in India.

Callihan, R. H., S. L. Carson, and R. T. Dobbins. 1995. NAWEEDS, Computer-aided Weed Identification for North America. Illustrated User's Guide plus Computer Floppy Disk. Moscow, Idaho.

Camazine, S. and K. J. Niklas. 1984. Aerobiology of *Symplocarpus foetidus:* Interactions between spathe and spadix. Amer. J. Bot. 71: 843–850.

Campbell, D. H. 1897. A morphological study of *Najas* and *Zannichellia*. Proc. Calif. Acad. Sci., ser. 3, 1: 1–70.

Canad. Entomol. = Canadian Entomologist.

Canad. Field-Naturalist = Canadian Field-Naturalist.

Canad. J. Bot. = Canadian Journal of Botany.

Canad. J. Pl. Sci. = Canadian Journal of Plant Science.

Candolle, A. L. P. de and C. de Candolle, eds. 1878–1896. Monographiae Phanerogamarum. . . . 9 vols. Paris. (Monogr. Phan.)

Caribbean J. Sci. Math. = Caribbean Journal of Science and Mathematics.

Case, F. W. 1992. Plants for the bog garden. Bull. Amer. Rock Gard. Soc. 50: 129–144.

Castanea = Castanea; Journal of the Southern Appalachian Botanical Club.

Cat. Pl. Cub.—See: A. H. R. Grisebach 1866

Catalogue 1879—See: James Veitch and Sons 1879

Catling, P. M. and W. G. Dore. 1982. Status and identification of *Hydrocharis morsus-ranae* and *Limnobium spongia* (Hydrocharitaceae) in northeastern North America. Rhodora 84: 523–545.

Catling, P. M. and K. W. Spicer. 1987. The perennial *Juncus* of section *Poiophylli* in the Canadian prairie provinces. Canad. J. Bot. 65: 750–760.

Cavanilles, A. J. 1791–1801. Icones et Descriptiones Plantarum, Quae aut Sponte in Hispania Crescunt, aut in Hortis Hospitantur. 6 vols. Madrid. (Icon.)

Celarier, R. P. 1956. A new species of *Tradescantia* (Commelinaceae) from south Texas. Field & Lab. 24: 5–9.

Centr. Florida Palm Bull. = Central Florida Palm Bulletin.

Chapman, A. W. 1860. Flora of the Southern United States . . . , New York. (Fl. South. U.S.)

Chapman, A. W. 1897. Flora of the Southern United States . . . , ed. 3. Cambridge, Mass. (Fl. South. U.S. ed. 3)

Charlton, W. A. 1973. Studies in the Alismataceae. II. Inflorescences of Alismataceae. Canad. J. Bot. 51: 775–789.

Chase, M. W. et al. 1993. Phylogenetics of seed plants: An analysis of nucleotide sequences from the plastid gene *rbc*L. Ann. Missouri Bot. Gard. 80: 528–580.

Chase, S. S. 1947. Preliminary Studies in the Genus *Najas* in the United States. Ph.D. thesis. Cornell University.

Chase, S. S. 1947b. Polyploidy in an immersed aquatic angiosperm. [Abstract.] Amer. J. Bot. 34: 581–582.

Choris, L. [1821–]1822[–1823]. Voyage Pittoresque Autour du Monde. . . . 7 parts in 22 fasc. Paris. [Parts paged independently.] (Voy. Pittor.)

Christmann, G. F. and G. W. F. Panzer. 1777–1788. Des Ritters Carl von Linné . . . vollständiges Pflanzensystem. . . . 14 vols. in 15. Nuremberg. (Vollst. Pflanzensyst.)

Clancy, K. E. and M. J. Sullivan. 1988. Some observations on seed germination, the seedling, and polyembryony in the needle palm, *Rhapidophyllum hystrix*. Principes 32: 18–25.

Class-book Bot. ed. s.n.(k)—See: A. Wood 1868

Classen-Bockhof, R. 1991. Untersuchungen der Konstruktion des Bestäuberungsapparates von *Thalia geniculata* (Marantaceen). Bot. Acta 104: 183–193.

Clay, K. 1993. Size-dependent gender change in green dragon (*Arisaema dracontium*; Araceae). Amer. J. Bot. 80: 769–777.

Cock, A. W. A. M. de. 1980. Flowering, pollination and fruiting in *Zostera marina* L. Aquatic Bot. 9: 201–220.

Cody, W. J. 1975. *Schuechzeria palustris* L. (Scheuchzeriaceae) in northwestern North America. Canad. Field-Naturalist 89: 69–71.

Commelin. Ind.—See: J. K. Hasskarl 1870

Compt. Rend. Sommaire Séances Soc. Biogéogr. = Compte Rendu Sommaire des Séances de la Société de Biogéographie.

Consp. Fl. Groenland.—See: J. M. C. Lange 1880[–1894]

Contr. Arnold Arbor. = Contributions from the Arnold Arboretum of Harvard University.

Contr. Gray Herb. = Contributions from the Gray Herbarium of Harvard University. = [Some numbers reprinted from (or in?) other periodicals, e.g. Rhodora.]

Contr. Inst. Bot. Univ. Montréal = Contributions de l'Institut Botanique de l'Université de Montréal.

Contr. Mar. Sci = Contributions in Marine Science.

Contr. New South Wales Natl. Herb. = Contributions from the New South Wales National Herbarium.

Contr. Univ. Michigan Herb. = Contributions from the University of Michigan Herbarium.

Contr. U.S. Natl. Herb. = Contributions from the United States National Herbarium.

Cook, C. D. K. 1982. Pollinating mechanisms in the Hydrocharitaceae. In: J.-J. Symoens et al., eds. 1982. Studies on Aquatic Vascular Plants. Brussels. Pp. 1–15.

Cook, C. D. K. 1985. *Sparganium*: Some old names and their types. Bot. Jahrb. Syst. 107: 269–276.

Cook, C. D. K. and R. Lüönd. 1982. A revision of the genus *Hydrocharis* (Hydrocharitaceae). Aquatic Bot. 14: 177–204.

Cook, C. D. K. and R. Lüönd. 1982b. A revision of the genus *Hydrilla* (Hydrocharitaceae). Aquatic Bot. 13: 485–504.

Cook, C. D. K. and R. Lüönd. 1983. A revision of the genus *Blyxa* (Hydrocharitaceae). Aquatic Bot. 15: 1–52.

Cook, C. D. K. and M. S. Nicholls. 1986. A monographic study of the genus *Sparganium* (Sparganiaceae). Part. 1. Subgenus *Xanthosparganium* Holmberg. Bot. Helv. 96: 213–267.

Cook, C. D. K. and M. S. Nicholls. 1987. A monographic study of the genus *Sparganium* (Sparganiaceae). Part. 2. Subgenus *Sparganium*. Bot. Helv. 97: 1–44.

Cook, C. D. K., J. J. Symoens, and K. Urmi-König. 1984. A revision of the genus *Ottelia* (Hydrocharitaceae). 1. Generic considerations. Aquatic Bot. 18: 263–274.

Cook, C. D. K. and K. Urmi-König. 1983. A revision of the genus *Limnobium* including *Hydromystria* (Hydrocharitaceae). Aquatic Bot. 17: 1–27.

Cook, C. D. K. and K. Urmi-König. 1984. A revision of the genus *Ottelia* (Hydrocharitaceae). 2. The species of Eurasia, Australasia and America. Aquatic Bot. 20: 131–177.

Cook, C. D. K. and K. Urmi-König. 1984b. A revision of the genus *Egeria* (Hydrocharitaceae). Aquatic Bot. 19: 73–96.

Cook, C. D. K. and K. Urmi-König. 1985. A revision of the genus *Elodea* (Hydrocharitaceae). Aquatic Bot. 21: 111–156.

Cook, O. F. 1936. Royal palms in upper Florida. Science, ser. 2, 84: 60–61.

Cooper, J. G. 1861. On the forests and trees of Florida and the Mexican boundary. Rep. (Annual) Board Regents Smithsonian Inst. 1860: 439–442.

Cornett, J. W. 1986. *Washingtonia robusta* naturalized in southeastern California. Bull. S. Calif. Acad. Sci. 85: 56–57.

Cornett, J. W. 1987. The occurrence of the desert fan palm, *Washingtonia filifera,* in southern Nevada. Desert Pl. 8: 169–171.

Cornett, J. W. 1988. Naturalized populations of the desert fan palm, *Washingtonia filifera,* in Death Valley National Monument. In: C. A. Hall Jr. and V. Doyle-Jones, eds. 1988. Plant Biology of Eastern California. Los Angeles. Pp. 167–174.

Correll, D. S. 1968. Some additions to the flora of Texas—IV. Madroño 19: 187–192.

Correll, D. S. and H. B. Correll. 1972. Aquatic and Wetland Plants of Southwestern United States. Washington. [Reissued 1975 in 2 vols., Stanford.]

Correll, D. S. and M. C. Johnston. 1970. Manual of the Vascular Plants of Texas. Renner, Tex.

Cox, P. A. 1988. Hydrophilous pollination. Annual Rev. Ecol. Syst. 19: 261–280.

Cox, P. A., T. Elmqvist, and P. B. Tomlinson. 1990. Submarine pollination and reproductive morphology in *Syringodium filiforme* (Cymodoceaceae). Biotropica 22: 259–265.

Cox, P. A. and R. B. Knox. 1989. Two-dimensional pollination in hydrophilous plants: Convergent evolution in the genera *Halodule* (Cymodoceaceae), *Halophila* (Hydrocharitaceae), *Ruppia* (Ruppiaceae), and *Lepilaena* (Zannichelliaceae). Amer. J. Bot. 76: 164–175.

Craighead, F. C. Sr. and D. B. Ward. N.d. Endangered: Buccaneer palm. In: D. B. Ward, ed. N.d. Rare and Endangered Biota of Florida. Vol. 5. Plants. Gainesville. Pp. 54–55.

Croat, T. B. 1994. The use of the New World Araceae as drug plants. J. Jap. Bot. 69: 185–203.

Cronquist, A. 1981. An Integrated System of Classification of Flowering Plants. New York.

Cronquist, A., A. H. Holmgren, N. H. Holmgren, J. L. Reveal, P. K. Holmgren, and R. C. Barneby. 1972+. Intermountain Flora. Vascular Plants of the Intermountain West, U.S.A. 5+ vols. New York and London. [Vol. 1, 1972; vol. 3, part B, 1989; vol. 4, 1984; vol. 5, 1994; vol. 6, 1977.]

Crow, G. E. and C. B. Hellquist. 1981. Aquatic vascular plants of New England. Pt. 2. Typhaceae and Sparganiaceae. New Hampshire Agric. Exp. Sta. Bull. 517: 1–21.

Cunningham, S. A. 1995. Ecological constraints on fruit initiation by *Calyptrogyne ghiesbreghtiana* (Arecaceae): Floral herbivory, pollen availability, and visitation by pollinating bats. Amer. J. Bot. 82: 1527–1536.

Cycl.—See: A. Rees [1802–]1819–1820

Dahlgren, R. M. T. and H. T. Clifford. 1982. The Monocotyledons: A Comparative Study. New York.

Dahlgren, R. M. T., H. T. Clifford, and P. F. Yeo. 1985. The Families of the Monocotyledons. Structure, Evolution, and Taxonomy. Berlin etc.

Daniels, G. S. and F. G. Stiles. 1979. The *Heliconia* taxa of

Costa Rica. Keys and descriptions. Brenesia 15(suppl.): 1–150.

Darwiniana = Darwiniana; Carpeta del "Darwinion."

Daubs, E. H. 1965. A monograph of Lemnaceae. Illinois Biol. Monogr. 34.

Davis, J. I. 1995. A phylogenetic structure for the monocotyledons, as inferred from chloroplast DNA restriction site variation, and a comparison of measures of clade support. Syst. Bot. 20: 503–527.

De Junco—See: F. W. G. Rostkovius [1801]

Delile, A. 1812[1813–1829]. Description de l'Égypte, ou Recueil des Observations et des Recherches Qui Ont Été Faites en Égypte Pendant l'Expédition de l'Armée Française.... Histoire Naturelle. Tom. Second. Paris. (Descr. Egypte, Hist. Nat.)

Descr. Egypte, Hist. Nat.—See: A. Delile 1812[1813–1829]

Descr. Pl. Nouv.—See: É. P. Ventenat [1800–1803]

Desert Pl. = Desert Plants.

Deutschl. Fl.—See: G. F. Hoffmann [1791–1804]

Deutschl. Fl. ed. 3—See: J. C. Röhling et al. 1823–1839

Díaz-Miranda, D., D. Philcox, and P. Denny. 1981. Taxonomic clarification of *Limnobium* Rich. and *Hydromystria* G. W. F. Meyer (Hydrocharitaceae). Bot. J. Linn. Soc. 83: 311–323.

Dike, D. H. 1969. Contributions to the Biology of *Ottelia alismoides* (Hydrocharitaceae). M.S. thesis. University of Southwestern Louisiana.

Donahue, J. W. 1965. History, breeding and cultivation of the canna. Amer. Hort. Mag. 44: 84–91.

Dorn, R. D. 1977. Manual of the Vascular Plants of Wyoming. 2 vols. New York.

Doty, M. S. and B. C. Stone. 1967. Typification for the generic name *Halophila* Thouars. Taxon 16: 414–418.

Dransfield, J. and H. J. Beentje. 1995. Palms of Madagascar. Kew. (Palms Madagascar)

Dransfield, J. and N. W. Uhl. 1986. An outline of a classification of palms. Principes 30: 3–11.

Dray, F. A. Jr. and T. D. Center. 1989. Seed production by *Pistia stratiotes* L. (water lettuce) in the United States. Aquatic Bot. 33: 155–160.

Dressler, R. L. et al. 1987. Identification Manual for Wetland Plant Species of Florida.... Gainesville.

Dudley, M. G. 1937. Morphological and cytological studies of *Calla palustris*. Bot. Gaz. 98: 556–571.

Dugle, J. R. and T. P. Copps. 1972. Pollen characteristics of Manitoba cattails. Canad. Field-Naturalist 86: 33–40.

Duncan, W. H. and J. T. Kartesz. 1981. Vascular Flora of Georgia: An Annotated Checklist. Athens, Ga.

Dunn, C. P. and R. R. Sharitz. 1990. The history of *Murdannia keisak* (Commelinaceae) in the southeastern United States. Castanea 55: 122–129.

Duvall, M. R. et al. 1993b. Phylogenetic hypotheses for the monocotyledons constructed from *rbc*L data. Ann. Missouri Bot. Gard. 80: 607–619.

Duvall, M. R., G. H. Learn. Jr., L. E. Eguiarte, and M. T. Clegg. 1993. Phylogenetic analysis of *rbc*L sequences identifies *Acorus calamus* as the primal extant monocotyledon. Proc. Natl. Acad. Sci. U.S.A. 90: 4641–4644.

Eastwood, A. 1900. Some plants of Mendocino County new to the flora of California. Zoë 5: 58–60.

Ebinger, J. E. 1962. The varieties of *Luzula acuminata*. Rhodora 64: 74–83.

Ebinger, J. E. 1964. Taxonomy of the subgenus *Pterodes*, genus *Luzula*. Mem. New York Bot. Gard. 10(5): 279–304.

Eckardt, T. 1964. Pandanales. In: H. Melchior, ed. 1964. A. Engler's Syllabus der Pflanzenfamilien..., ed. 12. 2 vols. Berlin. Vol. 2, pp. 598–602.

Ecol. Monogr. = Ecological Monographs.

Ecol. Res. (Tokyo) = Ecological Research. (Tokyo.)

Ecology = Ecology, a Quarterly Journal Devoted to All Phases of Ecological Biology.

Econ. Bot. = Economic Botany; Devoted to Applied Botany and Plant Utilization.

Edwards's Bot. Reg. = Edwards's Botanical Register....

Ehrhart, J. F. 1785–1793. Calamariae, Gramina et Tripetaloideae Linnaei. [Exsiccatae.] 14 decades. Hannover. (Calam. Gram. Tripet. Linn.)

Ehrhart, J. F. 1787–1792. Beiträge zur Naturkunde.... 7 vols. Hannover and Osnabrück. (Beitr. Naturk.)

Eiseman, N. J. and C. McMillan. 1980. A new species of seagrass, *Halophila johnsonii*, from the Atlantic coast of Florida. Aquatic Bot. 9: 15–19.

Elliott, S. [1816–]1821–1824. A Sketch of the Botany of South-Carolina and Georgia. 2 vols. in 13 parts. Charleston. (Sketch Bot. S. Carolina)

Emory, W. H. 1857–1859. Report on the United States and Mexican Boundary Survey, Made under the Direction of the Secretary of the Interior. 2 vols. in parts. Washington. (Rep. U.S. Mex. Bound.)

Encycl.—See: J. Lamarck et al. 1783–1817

Endlicher, S. L. 1836–1840[–1850]. Genera Plantarum Secundum Ordines Naturales Disposita. 18 parts + 5 suppl. in 6 parts. Vienna. [Paged consecutively through suppl. 1(2); suppls. 2–5 paged independently.] (Gen. Pl.)

Engelmann, G. 1866–1868. Revision of the North American species of the genus *Juncus*, with a description of new or imperfectly known species. Trans. Acad. Sci. St. Louis 2(2, 3): 424–498.

Engler, H. G. A. 1886. Ueber die Familie der Typhaceen. Bot. Centralbl. 25: 127.

Engler, H. G. A., ed. 1900–1953. Das Pflanzenreich.... 107 vols. Berlin. [Sequence of volume (Heft) numbers (order of publication) is independent of the sequence of series and family (Roman and Arabic) numbers (taxonomic order).] (Pflanzenr.)

Engler, H. G. A., H. Harms, J. Mattfeld, H. Melchior, and E. Werdermann, eds. 1924+. Die natürlichen Pflanzenfamilien..., ed. 2. 26+ vols. Leipzig and Berlin. (Nat. Pflanzenfam. ed. 2)

Entomophaga = Entomophaga; Mémoire Hors Série.

Enum. Pl.—See: C. S. Kunth 1833–1850; M. Vahl 1804–1805

Enum. Stirp. Vindob.—See: N. J. Jacquin 1762

Enum. Syst. Pl.—See: N. J. Jacquin 1760.

Environm. Conservation = Environmental Conservation.

Ertter, B. 1986. The *Juncus triformis* complex. Mem. New York Bot. Gard. 39: 1–90.

Ervik, F. and J. P. Feil. 1997. Reproductive biology of the monoecious understory palm *Prestoea schultzeana* in Amazonian Ecuador. Biotropica 29: 309–317.

Erythea = Erythea; a Journal of Botany, West American and General.

Essai Monogr. Jonc.—See: J. J. C. de Laharpe [1825]

Evolution = Evolution, International Journal of Organic Evolution.

Exot. Bot.—See: J. E. Smith 1804–1805[–1808]

Faden, R. B. 1978. Review of the lectotypification of *Aneilema* R. Br. (Commelinaceae). Taxon 27: 289–298.

Faden, R. B. 1989. *Commelina caroliniana* (Commelinaceae): A misunderstood species in the United States is an old introduction from Asia. Taxon 38: 43–53.

Faden, R. B. 1993. *Tradescantia crassifolia* (Commelinaceae), an overlooked species in the southwestern United States. Ann. Missouri Bot. Gard. 80: 219–222.

Faden, R. B. 1993b. The misconstrued and rare species of *Commelina* (Commelinaceae) in the eastern United States. Ann. Missouri Bot. Gard. 80: 208–218.

Faden, R. B. 1998. Commelinaceae. In: K. Kubitzki et al., eds. 1990+. The Families and Genera of Vascular Plants. 4+ vols. Berlin etc. Vol. 4, pp. 109–128.

Faden, R. B. and D. R. Hunt. 1991. The classification of the Commelinaceae. Taxon 40: 19–31.

Fairchild Trop. Gard. Bull. = Fairchild Tropical Garden Bulletin.

Fassett, N. C. 1930. Preliminary reports of the flora of Wisconsin. Pandanales. Trans. Wisconsin Acad. Sci. 25: 183–187.

Fassett, N. C. 1955. *Echinodorus* in the American tropics. Rhodora 57: 133–156, 174–188, 202–212.

Fassett, N. C. and B. M. Calhoun. 1952. Introgression between *Typha latifolia* and *T. angustifolia*. Evolution 6: 367–379.

Felger, R. and M. B. Moser. 1973. Eelgrass (*Zostera marina* L.) in the Gulf of California: Discovery of its nutritional value by the Seri Indians. Science, ser. 2, 181: 355–356.

Fernald, M. L. 1918. The diagnostic character of *Vallisneria americana*. Rhodora 20: 108–110.

Fernald, M. L. 1922b. Notes on *Sparganium*. Rhodora 24: 26–34.

Fernald, M. L. 1923. The American variety of *Scheuchzeria palustris*. Rhodora 25: 177–179.

Fernald, M. L. 1932. The linear-leaved North American species of *Potamogeton* section *Axillares*. Mem. Amer. Acad. Arts, n. s. 17: 1–183.

Fernald, M. L. 1945b. Botanical specialties of Virginia: Some varieties and forms of *Juncus canadensis*. Rhodora 47: 127–131.

Fernald, M. L. 1946. The North American representatives of *Alisma plantago-aquatica*. Rhodora 48: 86–88.

Fernald, M. L. 1950. Gray's Manual of Botany, ed. 8. New York.

Fernald, M. L., A. C. Kinsey, and R. C. Rollins. 1958. Edible Wild Plants of Eastern North America. New York.

Fernald, M. L. and K. M. Wiegand. 1910. The North American variation of *Juncus effusus*. Rhodora 12: 81–93.

Fernald, M. L. and K. M. Wiegand. 1914. The genus *Ruppia* in eastern North America. Rhodora 16: 119–127.

Field & Lab. = Field & Laboratory.

Fieldiana, Bot. = Fieldiana: Botany.

Finlayson, C. M., J. Roberts, A. J. Chick, and P. J. M. Sale. 1983. The biology of Australian weeds. II. *Typha domingensis* Pers. and *Typha orientalis* Presl. J. Austral. Inst. Agric. Sci. 1983: 3–10.

Fisher, J. B., J. G. Chong, and A. N. Rao. 1989. Non-axillary branching in the palms *Eugeissona* and *Oncosperma* (Arecaceae). Bot. J. Linn. Soc. 99: 347–363.

Fl. Alaska Yukon—See: E. Hultén 1941–1950

Fl. Amer. Sept.—See: F. Pursh 1814

Fl. Bor.-Amer.—See: W. J. Hooker [1829–]1833–1840; A. Michaux 1803

Fl. Boston. ed. 2—See: J. Bigelow 1824

Fl. Bras.—See: C. F. P. von Martius et al. 1840–1906

Fl. Brit.—See: J. E. Smith 1800–1804

Fl. Brit. W. I.—See: A. H. R. Grisebach [1859–]1864

Fl. Calif.—See: W. L. Jepson 1909–1943

Fl. Carol.—See: T. Walter 1788

Fl. Cochinch.—See: J. de Loureiro 1790

Fl. Deut. Volk—See: C. Börner [1912]b

Fl. Flumin.—See: J. M. Vellozo 1825–1827[1829–1831]

Fl. Franç. ed. 3—See: J. Lamarck and A. P. de Candolle 1805[–1815]

Fl. Ind. ed. 1832—See: W. Roxburgh 1832

Fl. Ind. Occid.—See: O. P. Swartz 1797–1806

Fl. Indica—See: N. L. Burman 1768

Fl. Lapp.—See: G. Wahlenberg 1812

Fl. New York—See: J. Torrey 1843

Fl. Pedem.—See: C. Allioni 1785

Fl. Peruv.—See: H. Ruiz López and J. A. Pavón 1798–1802

Fl. Ross.—See: C. F. von Ledebour [1841]1842–1853

Fl. S.E. U.S.—See: J. K. Small 1903

Fl. Samojed. Cisural.—See: F. Ruprecht 1845b

Fl. Schles. ed. 3—See: F. Wimmer 1857

Fl. Sedin.—See: F. W. G. Rostkovius and W. L. E. Schmidt 1824

Fl. South. U.S.—See: A. W. Chapman 1860

Fl. South. U.S. ed. 3—See: A. W. Chapman 1897

Fl. Spa—See: A. L. S. Lejeune 1811–1813

Fl. Tellur.—See: C. S. Rafinesque 1836[–1838]

Flexner, S. B. and L. C. Hauck, eds. 1987. The Random House Dictionary of the English Language, ed. 2 unabridged. New York.

Flora = Flora; oder (allgemeine) botanische Zeitung. [Vols. 1–16, 1818–33, include "Beilage" and "Ergänzungsblätter"; vols. 17–25, 1834–42, include "Beiblatt" and "Intelligenzblatt."]

Florida Field Naturalist = Florida Field Naturalist: A Semiannual Journal of the Florida Ornithological Society.

Florida Sci. = Florida Scientist.

Folia Geobot. Phytotax. = Folia Geobotanica et Phytotaxonomica.

Fragm. Bot.—See: N. J. Jacquin [1800–]1809

Fraser, J. 1794. *Thalia ?dealbata*. Discover'd Growing in a Lake in North America, in the Year 1790. [Plate only.] [London.] [Artist: James Sowerby.] (Thalia dealbata [plate])

Gade, D. W. 1966. Achira, the edible canna, its cultivation and use in the Peruvian Andes. Econ. Bot. 20: 407–415.

Galeano-Garcés, G. 1986. Primer registro de dos géneros de palmas para la flora Colombiana. Mutisia 66: 1–4.

Gamerro, J. C. 1968. Observaciones sobre la biología floral y morfología de la Potamogetonácea *Ruppia cirrhosa* (Petagna) Grande (= *R. spiralis* L. ex Dum.). Darwiniana 14: 575–608.

Gard. Bull. Singapore = Gardens' Bulletin. Singapore.

Gard. Chron. = Gardener's Chronicle.

Gard. Dict. Abr. ed. 4—See: P. Miller 1754

Gardner, C. S. 1982. Systematic Study of *Tillandsia* Subgenus *Tillandsia*. Ph.D. dissertation. Texas A&M University.

Gardner, C. S. 1984. New species and nomenclatural changes in Mexican *Tillandsia*—I. Selbyana 7: 361–379.

Gart.-Zeitung (Berlin) = Garten-Zeitung. (Berlin.)

Gen. N. Amer. Pl.—See: T. Nuttall 1818

Gen. Nov. Madagasc.—See: L. M. A. du P. Thouars 1806

Gen. Pl.—See: G. Bentham and J. D. Hooker 1862–1883; S. L. Endlicher 1836–1840[–1850]

Gen. Pl. ed. 5—See: C. Linnaeus 1754

Gentes Herb. = Gentes Herbarum; Occasional Papers on the Kinds of Plants.

Ges. Naturf. Freunde Berlin Mag. Neuesten Entdeck. Gesammten Naturk. = Der Gesellschaft naturforschender Freunde zu Berlin Magazin für die neuesten Entdeckungen in der gesammten Naturkunde.

Giles, N. H. Jr. 1942. Autopolyploidy and geographical distribution in *Cuthbertia graminea* Small. Amer. J. Bot. 29: 637–645.

Gilmore, M. R. 1931. Dispersal by Indians a factor in the extension of discontinuous distribution of certain species of native plants. Pap. Michigan Acad. Sci. 13: 89–94.

Giorn. Bot. Ital. = Giornale Botanico Italiano.

Gleason, H. A. and A. Cronquist. 1991. Manual of Vascular Plants of Northeastern United States and Adjacent Canada, ed. 2. Bronx.

Godfrey, R. K. and P. Adams. 1964. The identity of *Sagittaria isoetiformis* (Alismataceae). Sida 1: 269–273.

Godfrey, R. K. and J. W. Wooten. 1979. Aquatic and Wetland Plants of Southeastern United States: Monocotyledons. Athens, Ga.

Goldman, D. H. 1999. Distribution update: *Sabal minor* in Mexico. Palms 43: 40–44.

Görts-van Rijn, A. R. A., ed. 1985+. Flora of the Guianas. Series A: Phanerogams. 12+ fasc. Königstein.

Grace, J. B. 1988. The effect of nutrient additions on mixtures of *Typha latifolia* L. and *Typha domingensis* Pers. along a water-depth gradient. Aquatic Bot. 31: 83–92.

Grace, J. B. and J. S. Harrison. 1986. The biology of Canadian weeds. 73. *Typha latifolia* L., *Typha angustifolia* L. and *Typha* ×*glauca* Godr. Canad. J. Pl. Sci. 66: 361–379.

Grace, J. B. and R. G. Wetzel. 1982. Niche differentiation between two rhizomatous plant species: *Typha latifolia* and *Typha angustifolia*. Canad. J. Bot. 60: 46–57.

Grana = Grana; an International Journal of Palynology Including World Pollen and Spore Flora.

Graves, A. H. 1908. The morphology of *Ruppia maritima*. Trans. Connecticut Acad. Arts 14: 59–170.

Gray, A. 1848. A Manual of the Botany of the Northern United States.... Boston, Cambridge, and London. (Manual)

Gray, A. 1856. A Manual of the Botany of the Northern United States..., ed. 2. New York. (Manual ed. 2)

Gray, A. 1867. A Manual of the Botany of the Northern United States..., ed. 5. New York and Chicago. [Pteridophytes by D. C. Eaton.] (Manual ed. 5)

Gray, A., S. Watson, and J. M. Coulter. 1890. A Manual of the Botany of the Northern United States..., ed. 6. New York and Chicago. (Manual ed. 6)

Grayum, M. H. 1987. A summary of the evidence and arguments supporting the removal of *Acorus* from the Araceae. Taxon 36: 723–729.

Grayum, M. H. 1990. Evolution and phylogeny of the Araceae. Ann. Missouri Bot. Gard. 77: 628–697.

Grear, J. W. Jr. 1966. Cytogeography of *Orontium aquaticum* (Araceae). Rhodora 68: 25–34.

Great Plains Flora Association. 1986. Flora of the Great Plains. Lawrence, Kans.

Greenwell, A. B. 1947. Taro—with special reference to its culture and uses in Hawaii. Econ. Bot. 1: 276–289.

Grey, W. F. and M. D. Moffler. 1978. Flowering of the seagrass *Thalassia testudinum* (Hydrocharitaceae) in the Tampa Bay, Florida area. Aquatic Bot. 5: 251–259.

Gris, A. 1859. Observations sur la fleur des Marantées. Ann. Sci. Nat., Bot., sér. 4, 12: 193–219, plates 11–14.

Grisebach, A. H. R. 1843–1845[–1846]. Spicilegium Florae Rumelicae et Bithynicae.... 2 vols. in 6 parts. Braunschweig. [Vols. paged independently but parts numbered consecutively.] (Spic. Fl. Rumel.)

Grisebach, A. H. R. [1859–]1864. Flora of the British West Indian Islands. 7 parts. London. [Parts paged consecutively.] (Fl. Brit. W. I.)

Grisebach, A. H. R. 1866. Catalogus Plantarum Cubensium Exhibens Collectionem Wrightianam Aliasque Minores ex Insula Cuba Missas. Leipzig. (Cat. Pl. Cub.)

Grootjen, C. J. 1983. Development of ovule and seed in Marantaceae. Acta Bot. Neerl. 205: 1–25.

Grootjen, C. J. and F. Bouman. 1988. Seed structure in Cannaceae: Taxonomic and ecological implications. Ann. Bot. (Oxford), n. s. 61: 363–371.

Hagström, J. O. 1916. Critical researches on the potamogetons. Kongl. Svenska Vetenskapsakad. Handl., n. s. 55(5): 1–281.

Hall, C. A. Jr. and V. Doyle-Jones, eds. 1988. Plant Biology of Eastern California. Los Angeles.

Hämet-Ahti, L. 1956. *Luzula piperi* (Cov.) M. E. Jones. An overlooked woodrush in western North America and eastern Asia. Aquilo, Ser. Bot. 3: 11–21.

Hämet-Ahti, L. 1971. A synopsis of the species of *Luzula*, subgenus *Anthelaea* Grisebach (Juncaceae) indigenous in North America. Ann. Bot. Fenn. 8: 368–381.

Hämet-Ahti, L. 1980. The *Juncus effusus* aggregate in eastern North America. Ann. Bot. Fenn. 17: 183–191.

Hämet-Ahti, L. 1986. North American races of *Juncus alpinoarticulatus*. Ann. Bot. Fenn. 23: 277–281.

Handb. Skand. Fl. ed. 12—See: C. J. Hartman et al. [1889]

Handel-Mazzetti, H. 1929–1937. Symbolae Sinicae. Botanische Ergebnisse der Expedition der Akademie der Eissenschaften in Wien nach Südwest-China. 1914/1918. 7 parts. Vienna. (Symb. Sin.)

Harley, K. L., R. C. Kassulke, D. P. Sands, and M. D. Day. 1990. Biological control of water lettuce, *Pistia strati-*

otes (Araceae) by *Neohydronomus affinis* (Coleoptera: Curculionidae). Entomophaga 35: 363–374.

Harling, G., B. Sparre, and L. Andersson, eds. 1973+. Flora of Ecuador. 60+ nos. Göteborg. [Nos. 1–4 published as Opera Bot., B, 1–4.]

Harlow, R. F. 1961. Fall and winter foods of Florida white-tailed deer. Quart. J. Florida Acad. Sci. 24: 19–38.

Harms, V. L. 1973. Taxonomic studies of North American *Sparganium*. I. *S. hyperboreum* and *S. minimum*. Canad. J. Bot. 51: 1629–1641.

Harries, H. C. 1978. The evolution, dissemination, and classification of *Cocos nucifera* L. Bot. Rev. (Lancaster) 44: 265–320.

Harries, H. C. 1992. Biogeography of the coconut *Cocos nucifera* L. Principes 36: 155–162.

Harrington, H. D. 1954. Manual of the Plants of Colorado. Denver. (Man. Pl. Colorado)

Harris, P. J. and R. D. Hartley. 1980. Phenolic constituents of the cell walls of monocotyledons. Biochem. Syst. & Ecol. 8: 153–160.

Harris, S. W. and W. H. Marshall. 1963. Ecology of water-level manipulations on a northern marsh. Ecology 44: 331–343.

Harrison, P. G. 1976. *Zostera japonica* (Aschers. & Graebn.) in British Columbia, Canada. Syesis 9: 359–360.

Harrison, P. G. 1979. Reproductive strategies in intertidal populations of two co-occurring seagrasses (*Zostera* spp.). Canad. J. Bot. 57: 2635–2638.

Hartman, C. J. et al. [1889.] Handbok i Skandinaviens Flora, Innefattande Sveriges, Norges, Finlands och Danmarks Ormbunkar och Fanerogamer. 1 part (only). Stockholm. [Unfinished work.] (Handb. Skand. Fl. ed. 12)

Hartog, C. den. 1959. A key to the species of *Halophila* (Hydrocharitaceae), with descriptions of the American species. Acta Bot. Neerl. 8: 484–489.

Hartog, C. den. 1964. An approach to the taxonomy of the sea-grass genus *Halodule* Endl. (Potamogetonaceae). Blumea 12: 289–312.

Hartog, C. den. 1970. The Sea-grasses of the World. Amsterdam.

Harvard Pap. Bot. = Harvard Papers in Botany.

Hasskarl, J. K. 1870. Commelinaceae Indicae, Imprimis Archipelagi Indici. . . . Vienna. (Commelin. Ind.)

Hatch, S. L., K. N. Gandhi, and L. E. Brown. 1990. Checklist of the Vascular Plants of Texas. College Station, Tex.

Hawkes, A. D. 1950. Notes on the palms 2. Saw palmetto *Serenoa repens* Small. Natl. Hort. Mag. 29: 93–95.

Hayden, A. 1939. Notes on *Typha angustifolia* in Iowa. Iowa State Coll. J. Sci. 13: 341–351.

Haynes, R. R. 1975. A revision of North American *Potamogeton* subsection *Pusilli* (Potamogetonaceae). Rhodora 76: 564–649.

Haynes, R. R. 1977. The Najadaceae in the southeastern United States. J. Arnold Arbor. 58: 161–170.

Haynes, R. R. 1978. The Potamogetonaceae in the southeastern United States. J. Arnold Arbor. 59: 170–191.

Haynes, R. R. 1979. Revision of North and Central American *Najas* (Najadaceae). Sida 8: 34–56.

Haynes, R. R. 1985. A revision of the clasping-leaved *Potamogeton* (Potamogetonaceae). Sida 11: 173–188.

Haynes, R. R. and L. B. Holm-Nielsen. 1985. A generic treatment of Alismatidae in the Neotropics. Acta Amazon. 15(suppl.): 153–193.

Haynes, R. R. and L. B. Holm-Nielsen. 1986. Notes on *Echinodorus* (Alismataceae). Brittonia 38: 325–332.

Haynes, R. R. and L. B. Holm-Nielsen. 1987. The Zannichelliaceae in the southeastern United States. J. Arnold Arbor. 68: 259–268.

Haynes, R. R. and L. B. Holm-Nielsen. 1992. The Limnocharitaceae. In: Organization for Flora Neotropica. 1968+. Flora Neotropica. 75+ nos. New York. No. 56.

Haynes, R. R. and L. B. Holm-Nielsen. 1994. The Alismataceae. In: Organization for Flora Neotropica. 1968+. Flora Neotropica. 75+ nos. New York. No. 64.

Haynes, R. R. and W. A. Wentz. 1974. Notes on the genus *Najas* (Najadaceae). Sida 5: 259–264.

Hébert, L. P. 1973. Contribution à l'Étude des Sparganiaceae Rudolphi en Amerique du Nord. M.S. thesis. Université de Montréal.

Hegelmaier, C. F. 1868. Die Lemnaceen. Eine monographische Untersuchung. . . . Leipzig. (Lemnac.)

Hegi, G. et al. 1964+. Illustrierte Flora von Mitteleuropa, ed. 3. 5+ vols. in 8+. Berlin and Hamburg.

Hellquist, C. B. 1984. Observations of *Potamogeton hillii* Morong in North America. Rhodora 86: 101–111.

Hellquist, C. B. and G. E. Crow. 1986. *Potamogeton* ×*haynesii* (Potamogetonaceae), a new species from northeastern North America. Brittonia 38: 415–419.

Hellquist, C. B., C. T. Philbrick, and R. L. Hilton. 1988. The taxonomic status of *Potamogeton lateralis* Morong (Potamogetonaceae). Rhodora 90: 15–20.

Henderson, A. 1986. A review of pollination studies in the Palmae. Bot. Rev. (Lancaster) 52: 221–259.

Henderson, A., G. Galeano, and R. G. Bernal. 1995. Field Guide to the Palms of the Americas. Princeton.

Hendricks, A. J. 1957. A revision of the genus *Alisma* (Dill.) L. Amer. Midl. Naturalist 58: 470–493.

Henkel, A. 1907. American root drugs. U.S.D.A. Bur. Pl. Industr. Bull. 107: 1–80.

Herb. Brit.—See: J. Hill 1769–1770

Hermann, F. J. 1975. Manual of the Rushes (*Juncus* spp.) of the Rocky Mountains and Colorado Basin. Fort Collins, Colo. [U.S.D.A. Forest Serv., Gen. Techn. Rep. RM-18.]

Hickman, J. C., ed. 1993. The Jepson Manual. Higher Plants of California. Berkeley, Los Angeles, and London.

Hill, J. 1769–1770. Herbarium Britannicum. . . . 2 vols. London. [Volumes paged consecutively.] (Herb. Brit.)

Hilmon, J. B. 1968. Autecology of Saw Palmetto (*Serenoa repens* (Bartr.) Small). Ph.D. dissertation. Duke University.

Hist. Nat. Palm.—See: C. F. P. von Martius et al. [1823–1853]

Hist. Pl. Dauphiné—See: D. Villars 1786–1789

Hist. Pl. Guiane—See: J. B. Aublet 1775

Hist. Stirp. Fl. Petrop.—See: F. Ruprecht 1845c

Hitchcock, C. L., A. Cronquist, M. Ownbey, and J. W. Thompson. 1955–1969. Vascular Plants of the Pacific Northwest. 5 vols. Seattle. (Vasc. Pl. Pacif. N.W.)

Hodge, W. H. and D. Taylor. 1957. The ethnobotany of the Island Caribs of Dominica. Webbia 12: 513–644.

Hoffmann, G. F. [1791–1804.] Deutschland Flora oder botanisches Taschenbuch. . . . 4 vols. Erlangen. (Deutschl. Fl.)

Holmes, W. C. 1978. Range extension for *Ottelia alismoides* (L.) Pers. (Hydrocharitaceae). Castanea 43: 193–194.

Holttum, R. E. 1951. The Marantaceae of Malaya. Gard. Bull. Singapore 13: 254–296.

Holub, J. 1997. *Stuckenia* Börner 1912—the correct name for *Coleogeton* (Potamogetonaceae). Preslia 69: 361–366.

Hooker, W. J. [1829–]1833–1840. Flora Boreali-Americana; or, the Botany of the Northern Parts of British America . . . . 2 vols. in 12 parts. London, Paris, and Strasbourg. (Fl. Bor.-Amer.)

Hort. Bot. Vindob.—See: N. J. Jacquin 1770–1776

Hort. Rev. = Horticultural Reviews.

Hotchkiss, N. and H. L. Dozier. 1949. Taxonomy and distribution of North American cattails. Amer. Midl. Naturalist 41: 237–254.

Hultén, E. 1941–1950. Flora of Alaska and Yukon. 10 vols. Lund and Leipzig. [Vols. paged consecutively and designated as simultaneous numbers of Lunds Univ. Årsskr. (= Acta Univ. Lund.) and Kungl. Fysiogr. Sällsk. Handl.] (Fl. Alaska Yukon)

Hultén, E. and H. St. John. 1931. The American species of *Lysichitum*. Svensk Bot. Tidskr. 25: 453–464.

Humboldt, A. von and A. J. Bonpland. [1805–]1808–1809 [–1817]. Plantae Aequinoctiales. . . . 2 vols. in 17 parts. Paris and Tübingen. [Parts numbered consecutively.] (Pl. Aequinoct.)

Humboldt, A. von, A. J. Bonpland, and C. S. Kunth. 1815[1816]–1825. Nova Genera et Species Plantarum Quas in Peregrinatione Orbis Novi Collegerunt, Descripserunt. . . . 7 vols. in 36 parts. Paris. (Nov. Gen. Sp.)

Hunt, D. R. 1975. The reunion of *Setcreasea* and *Separotheca* with *Tradescantia*. American Commelinaceae: I. Kew Bull. 30: 443–458.

Hunt, D. R. 1980. Sections and series in *Tradescantia*. American Commelinaceae: IX. Kew Bull. 35: 437–442.

Hunt, D. R. 1986. *Campelia, Rhoeo* and *Zebrina* united with *Tradescantia*. American Commelinaceae: XIII. Kew Bull. 41: 401–405.

Hunt, D. R. 1986b. Amplification of *Callisia* Loefl. American Commelinaceae: XV. Kew Bull. 41: 407–412.

Hunt, D. R. 1986c. A revision of *Gibasis* Rafin. American Commelinaceae: XII. Kew Bull. 41: 107–129.

Hunziker, A. T. 1981. *Hydromystria laevigata* (Hydrocharitaceae) en el centro de Argentina. Lorentzia 4: 5–8.

Hunziker, A. T. 1982. Observaciones biológicas y taxonómicas sobre *Hydromystria laevigata* (Hydrocharitaceae). Taxon 31: 472–477.

Huttleston, D. G. 1949. The three subspecies of *Arisaema triphyllum*. Bull. Torrey Bot. Club 76: 407–413.

Huttleston, D. G. 1953. A Taxonomic Study of the Temperate North American Araceae. Ph.D. dissertation. Cornell University.

Huttleston, D. G. 1981. The four subspecies of *Arisaema triphyllum*. Bull. Torrey Bot. Club 108: 479–481.

Hyacinth Control J. = Hyacinth Control Journal.

Icon.—See: A. J. Cavanilles 1791–1801

Icon. Fl. Germ. Helv.—See: H. G. L. Reichenbach et al. 1835–1914

Icon. Pl. Rar.—See: J. H. F. Link and C. F. Otto 1828[–1831]

Icon. Stirp. Rar.—See: R. A. Salisbury 1791

Ill. Bot. Himal. Mts.—See: J. F. Royle [1833–]1839[–1840]

Ill. Dict. Gard.—See: G. Nicholson 1884–1901

Ill. Fl. N. U.S.—See: N. L. Britton and A. Brown 1896–1898

Ill. Fl. N. U.S. ed. 2—See: N. L. Britton and A. Brown 1913

Ill. Fl. Pacific States—See: L. Abrams and R. S. Ferris 1923–1960

Ill. Hort. = Illustration Horticole.

Illinois Biol. Monogr. = Illinois Biological Monographs.

Index Crit. Butom. Alism. Juncag.—See: F. Buchenau 1868

Industr. Alimenticia = Industria Alimenticia.

Inst. Bot.—See: V. Petagna 1785–1787

Int. J. Pl. Sci. = International Journal of Plant Sciences.

Iowa State Coll. J. Sci. = Iowa State College Journal of Science.

Iter Hispan.—See: P. Loefling 1758

Iverson, J. B. 1979. Behaviour and ecology of the rock iguana *Cyclura carinata*. Bull. Florida State Mus., Biol. Sci. 24: 175–358.

J. Arnold Arbor. = Journal of the Arnold Arboretum.

J. Asiat. Soc. = Journal of the Asiatic Society.

J. Austral. Inst. Agric. Sci. = Journal of the Australian Institute of Agricultural Science.

J. Bombay Nat. Hist. Soc. = Journal of the Bombay Natural History Society.

J. Bot. = Journal of Botany, British and Foreign.

J. Bot. (Desvaux) = Journal de Botanique. [Edited by Desvaux.]

J. Bromeliad Soc. = Journal of the Bromeliad Society.

J. Elisha Mitchell Sci. Soc. = Journal of the Elisha Mitchell Scientific Society.

J. Ethnopharmacol. = Journal of Ethnopharmacology; Interdisciplinary Journal Devoted to Bioscientific Research on Indigenous Drugs.

J. Fish Biol. = Journal of Fish Biology.

J. Jap. Bot. = Journal of Japanese Botany.

J. Linn. Soc., Bot. = Journal of the Linnean Society. Botany.

J. Louisiana Soc. Hort. Res. = Journal of the Louisiana Society for Horticultural Research.

J. New York Bot. Gard. = Journal of the New York Botanical Garden.

J. New York Entomol. Soc. = Journal of the New York Entomological Society.

J. Phys. Chim. Hist. Nat. Arts = Journal de Physique, de Chimie, d'Histoire Naturelle et des Arts.

J. Wash. Acad. Sci. = Journal of the Washington Academy of Sciences.

J. Wildlife Managem. = Journal of Wildlife Management.

Jacobs, S. W. L. and M. A. Brock. 1982. A revision of the genus *Ruppia* (Potamogetonaceae) in Australia. Aquatic Bot. 14: 325–337.

Jacquin, N. J. 1760. Enumeratio Systematica Plantarum, Quas in Insulis Carabaeis Vicinaque Americes Continente Detexit Novas. . . . Leiden. (Enum. Syst. Pl.)

Jacquin, N. J. 1762. Enumeratio Stirpium Plerarumque, Quae

Sponte Crescunt in Agro Vindobonensi, Montibusque Confinibus. Vienna. (Enum. Stirp. Vindob.)

Jacquin, N. J. 1763. Selectarum Stirpium Americanarum Historia. . . . Vienna. (Select. Stirp. Amer. Hist.)

Jacquin, N. J. 1770–1776. Hortus Botanicus Vindobonensis . . . . 3 vols. Vienna. (Hort. Bot. Vindob.)

Jacquin, N. J. [1800–]1809. Fragmenta Botanica, Figuris Coloratis Illustrata. . . . 6 fasc. Vienna. (Fragm. Bot.)

James Veitch and Sons. 1879. Catalogue 1879. Chelsea. (Catalogue 1879)

Jarolimova, V. and J. Kirschner. 1995. Tetraploids in *Luzula multiflora* (Juncaceae) in Ireland. Folia Geobot. Phytotax. 30: 389–396.

Jepson, W. L. 1909–1943. A Flora of California. . . . 3 vols. in 12 parts. San Francisco etc. [Pagination consecutive within each vol.; vol. 1 page sequence independent of part number sequence (chronological); part 8 of vol. 1 (pp. 1–32, 579–index) never published.] (Fl. Calif.)

Jetton Castro, K. 1978. The Biology of *Commelinantia anomala*. M.S. thesis. Angelo State University.

Johnson, C. D., S. Zona, and J. A. Nilsson. 1995. Bruchid beetles and palm seeds: Recorded relationships. Principes 39: 25–35.

Johnson, E. A. and S. L. Williams. 1982. Sexual reproduction in seagrasses: Reports for five Caribbean species with details for *Halodule wrightii* Aschers. and *Syringodium filiforme* Kutz. Caribbean J. Sci. Math. 18: 61–70.

Judd, W. W. 1961. Insects and other invertebrates associated with flowering skunk cabbage, *Symplocarpus foetidus* (L.) Nutt., at Fanshawe Lake, Ontario. Canad. Entomol. 93: 241–249.

Kalm, P. 1770–1771. Travels into North America. . . . 3 vols. Warrington and London.

Kartesz, J. T. and K. N. Gandhi. 1992. Nomenclatural notes for the North American flora, X. Phytologia 72: 80–92.

Kaul, R. B. 1974. Ontogeny of foliar diaphragms in *Typha latifolia*. Amer. J. Bot. 61: 318–323.

Kennedy, H. 1978. Systematics and pollination of the closed-flowered species of *Calathea* (Marantaceae). Univ. Calif. Publ. Bot. 71: 1–90, plates 1–20.

Kennedy, H. 1992. Calatheas—their distribution, diversity, and potential for horticulture. Bull. Heliconia Soc. Int. 6(2): 3–4.

Kenton, A. 1982. A Robertsonian relationship in the chromosomes of two species of *Hydrocleys* (Butomaceae sens. lat.). Kew Bull. 36: 487–492.

Kew Bull. = Kew Bulletin.

Kinoshita, E. 1986. Size-sex relationship and sexual dimorphism in Japanese *Arisaema* (Araceae). Ecol. Res. (Tokyo) 1: 157–171.

Kirchoff, B. K. 1983. Floral organogenesis in five genera of the Marantaceae and in *Canna* (Cannaceae). Amer. J. Bot. 70: 508–523.

Kirchoff, B. K. 1986. Inflorescence structure and development in the Zingiberales: *Thalia geniculata* (Marantaceae). Canad. J. Bot. 64: 859–864.

Kirschner, J. 1990. *Luzula multiflora* and allied species (Juncaceae): A nomenclatural study. Taxon 39: 106–114.

Kirschner, J. 1993. Taxonomic survey of *Luzula* sect. *Luzula*

(Juncaceae) in Europe. Folia Geobot. Phytotax. 28: 141–182.

Klekowski, E. J. Jr. and E. O. Beal. 1965. A study of variation in the *Potamogeton capillaceus–diversifolius* complex (Potamogetonaceae). Brittonia 17: 175–181.

Klimstra, W. D. and A. L. Dooley. 1990. Foods of the Key deer. Florida Sci. 53: 264–273.

Klotz, L. H. 1992. On the biology of *Orontium aquaticum* L. (Araceae), golden club or floating arum. Aroideana 15: 25–33.

Knutson, R. M. 1972. Temperature measurements of the spadix of *Symplocarpus foetidus* (L.) Nutt. Amer. Midl. Naturalist 88: 251–254.

Koechlin, J. 1964. Marantacées. In: A. Aubréville and J.-F. Leroy. 1961+. Flore du Gabon. 35+ vols. Paris. Vol. 9, pp. 91–158.

Kongl. Svenska Vetenskapsakad. Handl. = Kongl[iga]. Svenska Vetenskapsakademiens Handlingar.

Kongl. Vetensk. Acad. Handl. = Kongl[iga]. Vetenskaps Academiens Handlingar.

Kortright, R. M. 1998. The Aquatic Plant Genus *Sagittaria* (Alismataceae): Phylogeny Based on Morphology. M.S. thesis. University of Alabama.

Kral, R. 1966. Eriocaulaceae of continental North America north of Mexico. Sida 2: 285–332.

Kral, R. 1966b. The genus *Xyris* (Xyridaceae) in the southeastern United States and Canada. Sida 2: 177–260.

Kral, R. 1983d. The Xyridaceae in the southeastern United States. J. Arnold Arbor. 64: 421–429.

Kral, R. 1989. The genera of Eriocaulaceae in the southeastern United States. J. Arnold Arbor. 70: 131–142.

Kress, W. J. 1984. Systematics of Central American *Heliconia* (Heliconiaceae) with pendent inflorescences. J. Arnold Arbor. 65: 429–532.

Kress, W. J. 1990. The phylogeny and classification of the Zingiberales. Ann. Missouri Bot. Gard. 77: 698–721.

Kress, W. J. and D. E. Stone. 1982. Nature of the sporoderm in monocotyledons, with special reference to the pollen grains of *Canna* and *Heliconia*. Grana 21: 129–148.

Kronfeld, E. M. 1889. Monographie der Gattung *Typha* Tourn. (Typhinae Agdh., Typhaceae Schur-Engl.). Verh. Zool.-Bot. Ges. Wien 39: 89–192.

Kubitzki, K., K. U. Kramer, P. S. Green, J. G. Rohwer, and V. Bittrich, eds. 1990+. The Families and Genera of Vascular Plants. 4+ vols. Berlin etc.

Kunth, C. S. 1833–1850. Enumeratio Plantarum. . . . 5 vols. in 6. Stuttgart and Tübingen. (Enum. Pl.)

Laharpe, J. J. C. de. [1825.] Essai d'une Monographie des Vraies Joncées, Comprenant les Genres *Juncus*, *Luzula* et *Abama*. . . . [Paris.] [Also published in Mém. Soc. Hist. Nat. Paris 3: 89–181. 1827.] (Essai Monogr. Jonc.)

Lakela, O. 1972. Field description of *Cuthbertia* (Commelinaceae) with description of a new form. Sida 5: 26–32.

Lakshmanan, C. 1951. A note on the occurrence of turions in *Hydrilla verticillata* Presl. J. Bombay Nat. Hist. Soc. 49: 802–804.

Lamarck, J. and A. P. de Candolle. 1805[–1815]. Flore Française, ou Descriptions Succinctes de Toutes les Plantes Qui Croissent Naturellement en France . . . , ed. 3. 5

tomes in 6 vols. Paris. [Tomes 1–4(2), 1805; tome 5, 1815.] (Fl. Franç. ed. 3)

Lamarck, J., J. Poiret, A. P. de Candolle, L. Desrousseaux, and M. Savigny. 1783–1817. Encyclopédie Méthodique. Botanique. . . . 13 vols. Paris and Liège. [Vols. 1–8, suppls. 1–5.] (Encycl.)

Lammers, T. G. and A. G. van der Valk. 1979. A checklist of the aquatic and wetland vascular plants of Iowa: II. Monocotyledons, plus a summary of the geographic and habitat distribution of all aquatic and wetland species in Iowa. Proc. Iowa Acad. Sci. 85: 121–163.

Lampe, K. F. and M. A. McCann. 1985. AMA Handbook of Poisonous and Injurious Plants. Chicago.

Landolt, E. 1986 The family of Lemnaceae—A monographic study, vol. 1. Veröff. Geobot. Inst. E. T. H. Stiftung Rübel Zürich 71.

Landolt, E. and R. Kandeler. 1987. The family of Lemnaceae—A monographic study, vol. 2. Veröff. Geobot. Inst. E. T. H. Stiftung Rübel Zürich 95.

Lange, J. M. C. 1880[–1894]. Conspectus Florae Groenlandicae. 3 parts. Copenhagen. [Parts paged consecutively.] (Consp. Fl. Groenland.)

Larson, G. E. 1993. Aquatic and Wetland Vascular Plants of the Northern Great Plains. Fort Collins, Colo. [U.S.D.A. Forest Serv., Gen. Techn. Rep. RM-238.]

Leafl. W. Bot. = Leaflets of Western Botany.

Ledebour, C. F. von. [1841]1842–1853. Flora Rossica sive Enumeratio Plantarum in Totius Imperii Rossici Provinciis Europaeis, Asiaticis, et Americanis Hucusque Observatarum. . . . 4 vols. Stuttgart. (Fl. Ross.)

Ledin, R. B., S. C. Kiem, and R. W. Read. 1959. *Pseudophoenix* in Florida. I. The native *Pseudophoenix sargentii*. Principes 3: 23–133.

Lehmann, N. L. and R. Sattler. 1992. Irregular floral development in *Calla palustris* (Araceae) and the concept of homeosis. Amer. J. Bot. 79: 1145–1157.

Lejeune, A. L. S. 1811–1813. Flore des Environs de Spa. . . . 2 vols. Liège. (Fl. Spa)

Lemnac.—See: C. F. Hegelmaier 1868

Lerman, J. C. and E. M. Cigliano. 1971. New carbon-14 evidence for six hundred years old *Canna compacta* seed. Nature 232: 568–570.

Les, D. H. 1983. Taxonomic implications of aneuploidy and polyploidy in *Potamogeton* (Potamogetonaceae). Rhodora 85: 301–323.

Les, D. H. and R. R. Haynes. 1995. Systematics of subclass Alismatidae: A synthesis of approaches. In: P. J. Rudall et al., eds. 1995. Monocotyledons: Systematics and Evolution. 2 vols. Kew. Vol. 2, pp. 353–377.

Les, D. H. and R. R. Haynes. 1996. *Coleogeton* (Potamogetonaceae), a new genus of pondweeds. Novon 6: 389–391.

Liliac.—See: P. J. Redouté 1802[–1815]

Lindquist, B. 1932. Taxonomical remarks on *Juncus alpinus* Villars and some related species. Bot. Not. 1932: 313–372.

Link, J. H. F. and C. F. Otto. 1828[–1831]. Icones Plantarum Rariorum Horti Regii Botanici Berolinensis. . . . 8 parts. Berlin. [Parts paged consecutively.] (Icon. Pl. Rar.)

Linnaea = Linnaea. Ein Journal für die Botanik in ihrem ganzen Umfange.

Linnaeus, C. 1753. Species Plantarum. . . . 2 vols. Stockholm. (Sp. Pl.)

Linnaeus, C. 1754. Genera Plantarum, ed. 5. Stockholm. (Gen. Pl. ed. 5)

Linnaeus, C. 1758[–1759]. Systema Naturae per Regna Tria Naturae . . . , ed. 10. 2 vols. Stockholm. (Syst. Nat. ed. 10)

Linnaeus, C. 1762–1763. Species Plantarum . . . , ed. 2. 2 vols. Stockholm. (Sp. Pl. ed. 2)

Linnaeus, C. 1767[–1771]. Mantissa Plantarum. 2 parts. Stockholm. [Mantissa [1] and Mantissa [2] Altera paged consecutively.] (Mant. Pl.)

Linnaeus, C. f. 1781[1782]. Supplementum Plantarum Systematis Vegetabilium Editionis Decimae Tertiae, Generum Plantarum Editionis Sextae, et Specierum Plantarum Editionis Secundae. Braunschweig. (Suppl. Pl.)

Lippincott, C. 1992. Restoring Sargent's cherry palm on the Florida Keys. Fairchild Trop. Gard. Bull. 47: 12–21.

Lippincott, C. 1995. Reintroduction of *Pseudophoenix sargentii* in the Florida Keys. Principes 39: 5–13.

Listabarth, C. 1992. A survey of pollination strategies in the Bactridinae (Palmae). Bull. Inst. Franç. Études Andines 21: 699–714.

Listabarth, C. 1993. Insect-induced wind pollination of the palm *Chamaedorea pinnatifrons* and pollination in the related *Wendlandiella* sp. Biodivers. & Conservation 2: 39–50.

Listabarth, C. 1993b. Pollination in *Geonoma macrostachys* and three congeners, *G. acaulis*, *G. gracilis*, and *G. interrupta*. Bot. Acta 106: 496–506.

Listabarth, C. 1994. Pollination and pollinator breeding in *Desmoncus*. Principes 38: 13–23.

Lockett, L. 1991. Native Texas palms north of the lower Rio Grande Valley: Recent discoveries. Principes 35: 64–71.

Loefling, P. 1758. Iter Hispanicum, Eller Resa til Spanska Ländern uti Europa och America, Förrättad Iffrån År 1751 til År 1756 . . . Utgifven Efter Dess Frånfälle af Carl Linnaeus. Stockholm. (Iter Hispan.)

Looman, J. 1976. Biological flora of the Canadian prairie provinces IV. *Triglochin* L., the genus. Canad. J. Pl. Sci. 56: 725–732.

Loureiro, J. de. 1790. Flora Cochinchinensis. . . . 2 vols. Lisbon. [Vols. paged consecutively.] (Fl. Cochinch.)

Lourteig, A. 1952. Mayacaceae. Notul. Syst. (Paris) 14: 30–33.

Lourteig, A. 1960. Distribution géographique des Mayacacées. Compt. Rend. Sommaire Séances Soc. Biogéogr. 36[323]: 31–35.

Löve, Á. and D. Löve. 1958. Biosystematics of *Triglochin maritimum* Agg. Naturaliste Canad. 85: 156–165.

Löve, D. and H. Lieth. 1961. *Triglochin gaspense*, a new species of arrow grass. Canad. J. Bot. 39: 1261–1272.

Lowden, R. M. 1982. An approach to the taxonomy of *Vallisneria* L. (Hydrocharitaceae). Aquatic Bot. 13: 269–298.

Lowden, R. M. 1986. Taxonomy of the genus *Najas* L. (Najadaceae) in the Neotropics. Aquatic Bot. 24: 147–184.

Luther, H. E. 1985. Notes on hybrid tillandsias in Florida. Phytologia 57: 175–176.

Luther, H. E. 1993. A new record for *Tillandsia* (Bromeliaceae) in Florida. Rhodora 95: 342–347.

Maas, P. J. M. 1985. 195. Cannaceae. In: A. R. A. Görts-van Rijn, ed. 1985+. Flora of the Guianas. Series A: Phanerogams. 12+ fasc. Königstein. Fasc. 1, pp. 69–73.

Maas, P. J. M. and H. Maas. 1988. 223. Cannaceae. In: G. Harling et al., eds. 1973+. Flora of Ecuador. 60+ nos. Göteborg. No. 32, pp. 1–9.

MacRoberts, D. T. 1977. New combinations in *Tradescantia*. Phytologia 37: 451–452.

MacRoberts, D. T. 1978. Notes on *Tradescantia*: *T. diffusa* Bush and *T. pedicellata* Celarier. Phytologia 38: 227–228.

MacRoberts, D. T. 1979. *Tradescantia ohiensis* Rafinesque var. *paludosa* (Anderson & Woodson) MacRoberts, comb. nov. Phytologia 42: 380–382.

MacRoberts, D. T. 1980. Notes on *Tradescantia* (Commelinaceae) V. *Tradescantia* of Louisiana. Bull. Mus. Life Sci. Louisiana State Univ. 4: 1–15.

MacRoberts, D. T. 1980b. Notes on *Tradescantia* IV (Commelinaceae): The distinction between *T. virginiana* and *T. hirsutiflora*. Phytologia 46: 409–416.

Madroño = Madroño; Journal of the California Botanical Society [from vol. 3: a West American Journal of Botany].

Maehr, D. S. 1984. The black bear as a seed disperser in Florida. Florida Field Naturalist 12: 40–42.

Maehr, D. S. and J. R. Brady. 1984. Food habits of Florida black bears. J. Wildlife Managem. 48: 230–235.

Mag. Hort. Bot. = Magazine of Horticulture, Botany and All Useful Discoveries and Improvements in Rural Affairs.

Magnus, P. W. 1870. Beiträge zur Kenntniss der Gattung *Najas* L. Berlin. (Beitr. Kenntn. Najas)

Malme, G. O. K. 1925. Xyridaceae der Insel Cuba. Ark. Bot. 19(19): 1–6.

Man. Pl. Colorado—See: H. D. Harrington 1954

Man. S.E. Fl.—See: J. K. Small 1933

Mant.—See: J. A. Schultes and J. H. Schultes 1822–1827

Mant. Pl.—See: C. Linnaeus 1767[–1771]

Manual—See: A. Gray 1848

Manual ed. 2—See: A. Gray 1856

Manual ed. 5—See: A. Gray 1867

Manual ed. 6—See: A. Gray et al. 1890

Marie-Victorin, Frère 1943. Les Vallisnéries Américaines. Contr. Inst. Bot. Univ. Montréal 46: 1–38.

Martin, W. C. and C. R. Hutchins. 1980. A Flora of New Mexico. 2 vols. Vaduz.

Martius, C. F. P. von. 1824. Palmarum Familia Ejusque Genera Denuo Illustrata. Munich. (Palm. Fam.)

Martius, C. F. P. von, A. W. Eichler, and I. Urban, eds. 1840–1906. Flora Brasiliensis. 15 vols. in 40 parts, 1304 fasc. Munich, Vienna, and Leipzig. [Volumes and5 parts numbered in systematic sequence, fascicles numbered independently in chronological sequence.] (Fl. Bras.)

Martius, C. F. P. von, H. von Mohl, and F. J. A. N. Unger. [1823–1853.] Historia Naturalis Palmarum. 3 vols. in 10 parts. Leipzig. [Parts numbered in chronological order, independently of volume sequence.] (Hist. Nat. Palm.)

Mason, H. L. 1957. A Flora of the Marshes of California. Berkeley.

Matthews, P. 1991. A possible tropical wildtype taro: *Colocasia esculenta* var. *aquatilis*. Bull. Indo-Pacific Prehist. Assoc. 11: 69–81.

Matuda, E. 1954. Las Araceas Mexicanas. Anales Inst. Biol. Univ. Nac. México 32: 147–155.

Mayo, S. J., J. Bogner, and P. C. Boyce. 1997. The Genera of Araceae. 1 vol. + laser disc. [London.]

McAtee, W. L. 1915. Eleven important wild-duck foods. Bull. U.S.D.A. 205: 1–25.

McClenaghan, L. R. Jr. and A. C. Beauchamp. 1986 Low genetic differentiation among isolated populations of the California fan palm (*Washingtonia filifera*). Evolution 40: 315–322.

McClintock, E. 1993. Arecaceae [Palmae]. In: J. C. Hickman, ed. 1993. The Jepson Manual. Higher Plants of California. Berkeley, Los Angeles, and London. P. 1105.

McMillan, C. 1991. Isozyme patterning in marine spermatophytes. Opera Bot. Belg. 4: 193–200.

McMillan, C., Y. Lipkin, and L. H. Bragg. 1975. The possible origin of peculiar *Thalia testudinum* reported from Texas as *Posidonia oceanica*. Contr. Mar. Sci. 10: 101–106.

McNaughton, S. J. 1966. Ecotype function in the *Typha* community-type. Ecol. Monogr. 36: 297–325.

Med. Fl.—See: C. S. Rafinesque 1828[–1830]

Med. Repos. = Medical Repository.

Meddel. Grønland = Meddelelser om Grønland, af Kommissionen for Ledelsen af de Geologiske og Geografiske Undersølgeser i Grønland.

Melchior, H., ed. 1964. A. Engler's Syllabus der Pflanzenfamilien . . . , ed. 12. 2 vols. Berlin.

Melet. Bot.—See: H. W. Schott and S. L. Endlicher 1832

Mém. Acad. Imp. Sci. St.-Pétersbourg, Sér. 6, Sci. Math. = Mémoires de l'Académie Impériale des Sciences de St.-Pétersbourg. Sixième Série. Sciences Mathématiques, Physiques et Naturelles.

Mém. Acad. Roy. Sci. Belgique, Cl. Sci. (8°) = Mémoires de l'Académie Royale des Sciences, Lettres et Beaux-arts de Belgique. Classe des Sciences. [In octavo.]

Mém. Acad. Roy. Sci. Hist. (Berlin) = Mémoires de l'Académie Royale des Sciences et Belles-lettres Depuis l'Avénement de Fréderic Guillaume II [later: III] au Thrône. Avec l'Histoire.

Mem. Amer. Acad. Arts = Memoirs of the American Academy of Arts and Science.

Mém. Cl. Sci. Math. Inst. Natl. France = Mémoires de la Classe des Sciences Mathématiques et Physiques de l'Institut National de France.

Mém. Mus. Hist. Nat. = Mémoires du Muséum d'Histoire Naturelle.

Mem. New York Bot. Gard. = Memoirs of the New York Botanical Garden.

Mem. Reale Accad. Sci. Torino = Memorie della Reale Accademia delle Scienze di Torino

Mem. Torrey Bot. Club = Memoirs of the Torrey Botanical Club.

Merriam-Webster. 1988. Webster's New Geographical Dictionary. Springfield, Mass.

Metcalfe, C. R., ed. 1960+. Anatomy of the Monocotyledons. 8+ vols. Oxford.

Meyer, E. 1822. Synopsis Juncorum Rite Cognitorum. Ad Inaugurandam Ejusdem Plantarum Generis Monographium. . . . Göttingen. (Syn. Junc.)

Meyer, E. 1823. Synopsis Luzularum Rite Cognitorum. Cum Additamentis Quibusdam ad Juncorum Synopsis Prius Editam. Göttingen. (Syn. Luzul.)

Meyer, G. F. W. 1818. Primitiae Florae Essequeboensis. . . . Göttingen. (Prim. Fl. Esseq.)

Michaux, A. 1803. Flora Boreali-Americana. . . . 2 vols. Paris and Strasbourg. (Fl. Bor.-Amer.)

Michigan Bot. = Michigan Botanist.

Miller, P. 1754. The Gardeners Dictionary. . . . Abridged . . . , ed. 4. 3 vols. London. (Gard. Dict. Abr. ed. 4)

Miller, V. J. 1983. Arizona's own palm: Washingtonia filifera. Desert Pl. 5: 99–104.

Misc. Bot.—See: G. B. Balbis [1804]

Mitra, E. 1966. Contributions to our knowledge of Indian freshwater plants 5. On the morphology, reproduction and autecology of Pistia stratiotes Linn. J. Asiat. Soc. 8: 115–135.

Moerman, D. E. 1986. Medicinal Plants of Native America. 2 vols. Ann Arbor. [Univ. Michigan, Mus. Anthropol., Techn. Rep. 19.]

Moffler, M. D., M. J. Durako, and W. F. Grey. 1981. Observations on the reproductive ecology of Thalassia testudinum (Hydrocharitaceae) in Tampa Bay, Florida. Aquatic Bot. 10: 183–187.

Moldenke, H. N. 1937. Eriocaulaceae. In: N. L. Britton et al., eds. 1905+. North American Flora. . . . 47+ vols. New York. Vol. 19, pp. 17–50.

Moldenke, H. N. 1967. Notes on new and noteworthy plants. XLVI. Phytologia 14: 325–326.

Monogr. Phan.—See: A. L. P. de Candolle and C. de Candolle 1878–1896

Moore, D. R. 1963. Distribution of the sea grass, Thalassia, in the United States. Bull. Mar. Sci. Gulf Caribbean 13: 329–342.

Moore, E. 1913. The potamogetons in relation to pond culture. Bull. U.S. Bur. Fish. 33: 251–291.

Moore, H. E. Jr. 1973. The major groups of palms and their distribution. Gentes Herb. 11: 27–141.

Morton, J. F. 1975. Cattails (Typha spp.)—weed problem or potential crop? Econ. Bot. 29: 7–29.

Morton, J. F. 1977. Wild Plants for Survival in South Florida, ed. 4. Miami.

Motley, T. J. 1994. The ethnobotany of sweet flag, Acorus calamus (Araceae). Econ. Bot. 48: 397–412.

Muenscher, W. C. 1944. Aquatic Plants of the United States. Ithaca, N.Y.

Müller-Doblies, D. 1970. Über die Verwandtschaft von Typha und Sparganium im Infloreszenz- und Blütenbau. Bot. Jahrb. Syst. 89: 451–562.

Müller-Doblies, U. 1969. Über die Blütenstände und Blüten sowie zur Embryologie von Sparganium. Bot. Jahrb. Syst. 89: 359–450.

Müller-Doblies, U. and D. Müller-Doblies. 1977. Typhaceae. In: G. Hegi et al. 1964+. Illustrierte Flora von Mitteleuropa, ed. 3. 5+ vols. in 8+. Berlin and Hamburg. Vol. 2, part 1(4), pp. 275–317.

Mulligan, G. A. and D. B. Munro. 1990. Poisonous Plants of Canada. Ottawa.

Murata, J. 1990. Present status of Arisaema systematics. Bot. Mag. 103: 371–382.

Murray, N. A. and D. M. Johnson. 1987. Taxonomic notes on Juncus pelocarpus s.l. Contr. Univ. Michigan Herb. 16: 179–187.

Mutisia = Mutisia; Acta Botanica Colombiana.

N. Amer. Fl.—See: N. L. Britton et al. 1905+

N. Amer. Sagittaria—See: J. G. Smith 1894

Nat. Pflanzenfam. ed. 2—See: H. G. A. Engler et al. 1924+

Natl. Hort. Mag. = National Horticultural Magazine.

Naturaliste Canad. = Naturaliste Canadien. Bulletin de Recherches, Observations et Découvertes se Rapportant à l'Histoire Naturelle du Canada.

Nature = Nature; a Weekly Illustrated Journal of Science.

Nauman, C. E. 1990. Intergeneric hybridization between Coccothrinax and Thrinax (Palmae: Coryphoideae). Principes 34: 191–198.

Nauman, C. E. and R. W. Sanders. 1991. An annotated key to the cultivated species of Coccothrinax. Principes 35: 27–46.

Neumayer, G. B. von. 1875. Anleitung zu wissenschaftlichen Beobachtungen auf Reisen. Berlin. (Anl. Wiss. Beobacht. Reisen)

New Fl.—See: C. S. Rafinesque 1836[–1838]b

New Hampshire Agric. Exp. Sta. Bull. = New Hampshire Agricultural Experiment Station. Bulletin.

Nicholson, G. 1884–1901. The Illustrated Dictionary of Gardening. . . . 4 vols. + suppl. London. (Ill. Dict. Gard.)

Nomencl. Bot. ed. 2—See: E. G. Steudel 1840–1841

Nomme, K. M. and P. G. Harrison. 1991. A multivariate comparison of the seagrasses Zostera marina and Zostera japonica in monospecific versus mixed populations. Canad. J. Bot. 69: 1984–1990.

Nordenskiöld, H. 1956. Cyto-taxonomical studies in the genus Luzula. II. Hybridization experiments in the campestris–multiflora complex. Hereditas (Lund) 42: 7–73.

Nordenskiöld, H. 1957. Hybridization experiments in the genus Luzula. III. The subg. Pterodes. Bot. Not. 110: 1–16.

Nordic J. Bot. = Nordic Journal of Botany.

Norges Fl.—See: M. N. Blytt and A. G. Blytt 1861–1877

Notes Roy. Bot. Gard. Edinburgh = Notes from the Royal Botanic Garden, Edinburgh.

Notizbl. Bot. Gart. Berlin-Dahlem = Notizblatt des Botanischen Gartens und Museums zu Berlin-Dahlem.

Notul. Syst. (Paris) = Notulae Systematicae, Herbier du Muséum de Paris. Phanérogamie.

Nov. Gen. Sp.—See: A. von Humboldt et al. 1815[1816]–1825

Novikov, V. S. 1990. Konspekt sistemy roda Luzula DC. (Juncaceae). (Survey of the system of the genus Luzula.)

Byull. Moskovsk. Obshch. Isp. Prir., Otd. Biol. 95(6): 63–70.

Novon = Novon; a Journal for Botanical Nomenclature.

Nuttall, T. 1818. The Genera of North American Plants, and Catalogue of the Species, to the Year 1817.... 2 vols. Philadelphia. (Gen. N. Amer. Pl.)

Observ. Bot.—See: A. J. Retzius [1779–]1791; H. F. Van Heurck 1870–1871; C. E. von Weigel 1772

Oesterr. Bot. Wochenbl. = Oesterreichisches botanisches Wochenblatt. Gemeinnütziges Organ für Botanik....

Oesterr. Bot. Z. = Oesterreichische botanische Zeitschrift. Gemeinütziges Organ für Botanik.

Ogden, E. C. 1943. The broad-leaved species of *Potamogeton* of North America north of Mexico. Rhodora 45: 57–105, 119–163, 171–214.

O'Hair, S. K. and M. P. Asokan. 1986. Edible aroids: Botany and horticulture. Hort. Rev. 8: 43–99.

Ohio J. Sci. = Ohio Journal of Science.

Opera Bot. = Opera Botanica a Societate Botanice Lundensi.

Opera Bot. Belg. = Opera Botanica Belgica.

Orbigny, A. D. d'. 1834[–1847]. Voyage dans l'Amerique Méridionale.... 8 vols. Paris and Strasbourg. (Voy. Amér. Mér.)

Organization for Flora Neotropica. 1968+. Flora Neotropica. 75+ nos. New York.

Orpurt, P. R. and L. L. Boral. 1964. The flowers and seeds of *Thalassia testudinum* König. Bull. Mar. Sci. Gulf Caribbean 14: 296–302.

Orzell, S. L. and E. L. Bridges. 1993. *Eriocaulon nigrobracteatum* (Eriocaulaceae), a new species from the Florida panhandle with a characterization of its poor fen habitat. Phytologia 74: 104–124.

Ostenfeld, C. H. 1905. Preliminary remarks on the distribution and the biology of the *Zostera* of the Danish seas. Bot. Tidsskr. 27: 123–125.

Packer, J. G. and G. S. Ringius. 1984. The distribution and status of *Acorus* (Araceae) in Canada. Canad. J. Bot. 62: 2248–2252.

Palm. Fam.—See: C. F. P. von Martius 1824

Palme Madagascar—See: O. Beccari 1912[–1914]

Palms Madagascar—See: J. Dransfield and H. J. Beentje 1995

Palms = Palms, newsletter of the Palm Society.

Palmy SSSR—See: S. G. Saakov 1954

Pap. Michigan Acad. Sci. = Papers of the Michigan Academy of Sciences, Arts and Letters.

Patterson, D. T. et al. 1989. Composite List of Weeds. Champaign.

Pellmyr, O. and J. M. Patt. 1986. Function of olfactory and visual stimuli in pollination of *Lysichiton americanum* (Araceae) by a staphylinid beetle. Madroño 33: 47–54.

Pennell, F. W. 1916. Notes on plants of the southern United States—I. Bull. Torrey Bot. Club 43: 96–111.

Pennell, F. W. 1937. *Commelina communis* in the eastern United States. Bartonia 19: 19–22.

Pennell, F. W. 1938. What is *Commelina communis*? Proc. Acad. Nat. Sci. Philadelphia 90: 31–39.

Perkins, K. D. and W. W. Payne. 1978. Guide to the Poisonous and Irritant Plants of Florida. Gainesville.

Persoon, C. H. 1805–1807. Synopsis Plantarum.... 2 vols. Paris and Tubingen. (Syn. Pl.)

Petagna, V. 1785–1787. Institutiones Botanicae.... 5 vols. Naples. (Inst. Bot.)

Petersen, G. 1989. Cytology and systematics of Araceae. Nordic J. Bot. 9: 119–166.

Peterson, B. 1991. A comparison of some central Florida acrocomias. Centr. Florida Palm Bull. 11(1): 11–12.

Peterson, B. 1991b. *Acrocomia* naturalized in central Florida. Principes 35: 110–111.

Pflanzenr.—See: H. G. A. Engler 1900–1953

Pharm. J. = Pharmaceutical Journal.

Phillips, R. C. 1964. Comprehensive Bibliography of *Zostera marina*. Washington. [U.S.D.I. Fish Wildlife Serv., Special Sci. Rep. Wildlife 79.]

Phillips, R. C. 1967. On species of the seagrass, *Halodule*, in Florida. Bull. Mar. Sci. 17: 672–676.

Phillips, R. C. 1979. Ecological notes on *Phyllospadix* (Potamogetonaceae) in the northeast Pacific. Aquatic Bot. 6: 159–170.

Phillips, R. C. and C. P. McRoy, eds. 1980. Handbook of Seagrass Biology: An Ecosystem Perspective. New York.

Phillips, R. C. and R. F. Shaw. 1976. *Zostera noltii* Hornem. in Washington, U.S.A. Syesis 9: 355–358.

Phytologia = Phytologia; Designed to Expedite Botanical Publication.

Pl. Aequinoct.—See: A. von Humboldt and A. J. Bonpland [1805–]1808–1809[–1817]

Pl. Hartw.—See: G. Bentham 1839[–1857]

Pl. Med. (Stuttgart) = Planta Medica. Zeitschrift für Arzneipflanzenanwendung und Arzneipflanzenforschung [subtitle varies]. [Supplement to: Hippokrates.]

Pl. Syst. Evol. = Plant Systematics and Evolution.

Pl. Syst. Evol., Suppl. = Plant Systematics and Evolution. Supplementum.

Plowman, T. 1969. Folk uses of New World aroids. Econ. Bot. 23: 97–122.

Pollen & Spores = Pollen et Spores.

Posluszny, U. and R. Sattler. 1976. Floral development of *Najas flexilis*. Canad. J. Bot. 54: 1140–1151.

Posluszny, U. and P. B. Tomlinson. 1977. Morphology and development of floral shoots and organs in certain Zannichelliaceae. Bot. J. Linn. Soc. 75: 21–46.

Précis Découv. Somiol.—See: C. S. Rafinesque 1814

Preslia = Preslia. Věstník (Časopis) Československé Botanické Společnosti.

Preston, C. D. 1995. Pondweeds of Great Britain and Ireland. London.

Price, S. and G. Rogers. 1987. *Thalia dealbata* (Marantaceae), a Missouri surprise. Missouriensis 8: 73–78.

Prim. Fl. Esseq.—See: G. F. W. Meyer 1818

Principes = Principes; Journal of the (International) Palm Society.

Proc. Acad. Nat. Sci. Philadelphia = Proceedings of the Academy of Natural Sciences of Philadelphia.

Proc. Amer. Acad. Arts = Proceedings of the American Academy of Arts and Sciences.

Proc. Biol. Soc. Wash. = Proceedings of the Biological Society of Washington.

Proc. Calif. Acad. Sci. = Proceedings of the California Academy of Science.

Proc. Davenport Acad. Nat. Sci. = Proceedings of the Davenport Academy of Natural Sciences.

Proc. Iowa Acad. Sci. = Proceedings of the Iowa Academy of Science.

Proc. Natl. Acad. Sci. U.S.A. = Proceedings of the National Academy of Sciences of the United States of America.

Prodr.—See: R. Brown 1810; O. P. Swartz 1788

Prov. Agric. Hort. Ill. = Provence Agricole et Horticole Illustrée; Organe de l'Agriculture et de l'Horticulture Méridionales.

Publ. Inst. Invest. Geogr. Fac. Filos. Letras Univ. Buenos Aires, A = Publicaciones del Instituto de Investigaciones Geográficas, Facultad de Filosofia y Letras, Universidad de Buenos Aires. Serie A, Memorias Originales y Documentos.

Purseglove, J. W. 1972. Tropical Crops. Monocotyledons. 2 vols. London.

Pursh, F. 1814. Flora Americae Septentrionalis; or, a Systematic Arrangement and Description of the Plants of North America. 2 vols. London. (Fl. Amer. Sept.)

Quart. J. Florida Acad. Sci. = Quarterly Journal of the Florida Academy of Sciences.

Rafinesque, C. S. 1814. Précis des Découvertes et Travaux Somiologiques.... Palermo. (Précis Découv. Somiol.)

Rafinesque, C. S. 1828[–1830]. Medical Flora; or, Manual of the Medical Botany of the United States of North America. 2 vols. Philadelphia. (Med. Fl.)

Rafinesque, C. S. 1836[–1838]. Flora Telluriana.... 4 vols. Philadelphia. (Fl. Tellur.)

Rafinesque, C. S. 1836[–1838]b. New Flora and Botany of North America.... 4 parts. Philadelphia. (New Fl.)

Rafinesque, C. S. 1840. Autikon Botanikon. 3 parts. Philadelphia. [Parts paged consecutively.] (Autik. Bot.)

Ramey, V. 1981. Typha—not just another weed. Aquaphyte 1: 1–2.

Ramp, P. F. 1989. Natural History of Sabal minor: Demography, Population Genetics and Reproductive Ecology. Ph.D. dissertation. Tulane University.

Ramp, P. F. and L. B. Thien. 1995. A taxonomic history and reexamination of Sabal minor in the Mississippi Valley. Principes 39: 77–83.

Rataj, K. 1972. Revision of the genus Sagittaria. Part II. (The species of West Indies, Central and South America). Annot. Zool. Bot. 78.

Rataj, K. 1975. Revizion [sic] of the Genus Echinodorus Rich. Prague.

Ray, T. S. 1987. Leaf types in the Araceae. Amer. J. Bot. 74: 1359–1372.

Ray, T. S. 1988. Survey of shoot organization in the Araceae. Amer. J. Bot. 75: 56–84.

Read, R. W. 1968. A study of Pseudophoenix (Palmae). Gentes Herb. 10: 169–213.

Read, R. W. 1975. The genus Thrinax (Palmae: Coryphoideae). Smithsonian Contr. Bot. 19: 1–98.

Redouté, P. J. 1802[–1815]. Les Liliacées.... 8 vols. in 80 parts. Paris. [Parts and plates numbered consecutively.] (Liliac.)

Rees, A. [1802–]1819–1820. The Cyclopaedia; or, Universal Dictionary of Arts, Sciences, and Literature.... 39 vols. in 79 parts. London. [Pages unnumbered.] (Cycl.)

Reese, G. 1962. Zur intragenerischen Taxonomie der Gattung Ruppia L. Z. Bot. 50: 237–264.

Reichenbach, H. G. L. et al. 1835–1914. Icones Florae Germanicae et Helveticae.... 25 vols. Leipzig and Gera. [Vol. 1 originally published under the title: Agrostographia Germanica....] (Icon. Fl. Germ. Helv.)

Rep. (Annual) Board Regents Smithsonian Inst. = Report (Annual) of the Board of Regents, Smithsonian Institution.

Rep. (Annual) Missouri Bot. Gard. = Report (Annual) of the Missouri Botanical Garden.

Rep. U.S. Mex. Bound.—See: W. H. Emory 1857–1859

Res. Stud. State Coll. Wash. = Research Studies of the State College of Washington.

Retzius, A. J. [1779–]1791. Observationes Botanicae.... 6 vols. Leipzig. (Observ. Bot.)

Reveal, J. L. 1977. Juncaginaceae. In: A. Cronquist et al. 1972+. Intermountain Flora. Vascular Plants of the Intermountain West, U.S.A. 5+ vols. New York and London. Vol. 6, pp. 18–22.

Reveal, J. L. 1977b. Potamogetonaceae. In: A. Cronquist et al. 1972+. Intermountain Flora. Vascular Plants of the Intermountain West, U.S.A. 5+ vols. New York and London. Vol. 6, pp. 24–42.

Reynoso, F. A. 1976. Importancia económica de la palma real Dominicana (Roystonea hispaniolana). Agroconocimiento 1: 8–9.

Reznicek, A. A. and R. S. W. Bobbette. 1976. The taxonomy of Potamogeton subsection Hybridi in North America. Rhodora 78: 650–673.

Rhodora = Rhodora; Journal of the New England Botanical Club.

Richardson, F. D. 1980. Ecology of Ruppia maritima L. in New Hampshire (U.S.A.) tidal marshes. Rhodora 82: 403–439.

Roberts, M. L., R. L. Stuckey, and R. S. Mitchell. 1981. Hydrocharis morsus-ranae (Hydrocharitaceae) new to the United States. Rhodora 83: 147–148.

Roemer, J. J., J. A. Schultes, and J. H. Schultes. 1817[–1830]. Caroli a Linné... Systema Vegetabilium... editione XV.... 7 vols. Stuttgart. (Syst. Veg.)

Rogers, G. K. 1983. The genera of Alismataceae in the southeastern United States. J. Arnold Arbor. 64: 387–424.

Rogers, G. K. 1984. The Zingiberales (Cannaceae, Marantaceae, and Zingiberaceae) in the southeastern United States. J. Arnold Arbor. 65: 5–55.

Röhling, J. C., F. C. Mertens, and W. D. J. Koch. 1823–1839. Deutschlands Flora, ed. 3. 5 vols. Frankfurt am Main. (Deutschl. Fl. ed. 3)

Rohweder, O. 1956. Commelinaceae in die Farinosae in der Vegetation von El Salvador. Abh. Auslandsk., Reihe C, Naturwiss. 18: 98–178.

Rohweder, O. 1962. Zur systematischen Stellung der Commelinaceen-Gattung Commelinantia Tharp. Ber. Deutsch. Bot. Ges. 75: 51–56.

Roscoe, W. [1824–]1828. Monandrian Plants of the Order Scitamineae. . . . 15 parts. Liverpool.

Röst, L. C. M. 1979. Biosystematic investigations with *Acorus* 4. Communication. A synthetic approach to the classification of the genus. Pl. Med. (Stuttgart) 37: 289–307.

Rostkovius, F. W. G. [1801]. Dissertatio Botanica Inauguralis de *Junco*. . . . Halle. (De Junco)

Rostkovius, F. W. G. and W. L. E. Schmidt. 1824. Flora Sedinensis Exhibens Plantas Phanerogamas Spontaneas nec non Plantas Praecipuas Agri Swinemundii. . . . Stettin. (Fl. Sedin.)

Rowlatt, U. and H. Morshead. 1992. Architecture of the leaf of the greater reed mace, *Typha latifolia* L. Bot. J. Linn. Soc. 110: 161–170.

Roxburgh, W. 1832. Flora Indica; or, Descriptions of Indian Plants. Serampore. (Fl. Ind. ed. 1832)

Royle, J. F. [1833–]1839[–1840]. Illustrations of the Botany and Other Branches of the Natural History of the Himalayan Mountains and of the Flora of Cashmere. . . . 1 vol. text and 1 vol. plates. London. (Ill. Bot. Himal. Mts.)

Rubtzoff, P. 1964. Notes on the genus *Alisma*. Leafl. W. Bot. 10: 90–95.

Rudall, P. J. et al., eds. 1995. Monocotyledons: Systematics and Evolution. 2 vols. Kew.

Ruebens, C. 1968. Industrialización del palmiche en Cuba. Industr. Alimenticia 1: 8–25.

Ruiz Lopez, H. and J. A. Pavon. 1798–1802. Flora Peruviana, et Chilensis, sive Descripciones, et Icones Plantarum Peruvianarum, et Chilensium. . . . 3 vols. [Madrid.] (Fl. Peruv.)

Rumphia—See: C. L. Blume 1835[1836]–1848[1849]

Ruprecht, F. 1845b. Flores Samejedorum Cisuralensium. St. Petersburg. [Alt. title: Beiträge zur Pflanzenkunde des Russischen Reiches. . . . Zweite Lieferung.] (Fl. Samojed. Cisural.)

Ruprecht, F. 1845c. In Historiam Stirpium Florae Petropolitanae Diatribae. St. Petersburg. [Alt. title: Beiträge zur Pflanzenkunde des Russischen Reiches. . . . Vierte Lieferung.] (Hist. Stirp. Fl. Petrop.)

Saakov, S. G. 1954. Palmy i Ikh Kultura v SSSR. Moscow. (Palmy SSSR)

Sachet, M.-H. and F. R. Fosberg. 1973. Remarks on *Halophila* (Hydrocharitaceae). Taxon 22: 439–443.

Salisbury, R. A. 1791. Icones Stirpium Rariorum Descriptionibus Illustratae. London. (Icon. Stirp. Rar.)

Sanders, L. L. and C. J. Burk. 1992. A naturally-occurring population of putative *Arisaema triphyllum* subsp. *stewardsonii* × *A. dracontium* hybrids in Massachusetts. Rhodora 94: 340–347.

Sastroutomo, S. S. 1981. Turion formation, dormancy and germination of curly pondweed, *Potamogeton crispus* L. Aquatic Bot. 10: 161–173.

Sattler, R. and V. Singh. 1973. Floral development of *Hydrocleis nymphoides*. Canad. J. Bot. 51: 2455–2458.

Scariot, A. O., E. Lleras, and J. D. Hay. 1991. Reproductive biology of the palm *Acrocomia aculeata* in central Brazil. Biotropica 23: 12–22.

Schleiden, M. J. 1844. Beiträge zur Botanik. Leipzig. (Beitr. Bot.)

Schott, H. W. and S. L. Endlicher. 1832. Meletemata Botanica. Vienna. (Melet. Bot.)

Schultes, J. A. and J. H. Schultes. 1822–1827. Mantissa. . . . Systematis Vegetabilium Caroli a Linné ex Editione Joan. Jac. Roemer . . . et Jos. Aug. Schultes. . . . 3 vols. Stuttgart. (Mant.)

Schumann, K. 1902. Marantaceae. In: H. G. A. Engler, ed. 1900–1953. Das Pflanzenreich. . . . 107 vols. Berlin. Vol. 11[IV,48], pp. 1–184.

Science = Science; an Illustrated Journal [later: a Weekly Journal Devoted to the Advancement of Science]. [American Association for the Advancement of Science.]

Scoggan, H. J. 1957. Flora of Manitoba. Bull. Natl. Mus. Canada 140: 1–619.

Scoggan, H. J. 1978–1979. The Flora of Canada. 4 parts. Ottawa. [Natl. Mus. Nat. Sci. Publ. Bot. 7.]

Scribailo, R. W. and P. B. Tomlinson. 1992. Shoot and floral development in *Calla palustris*. Int. J. Pl. Sci. 153: 1–13.

Segeren, W. and P. J. M. Maas. 1971. The genus *Canna* in northern South America. Acta Bot. Neerl. 20: 663–680.

Selbyana = Selbyana; Journal of the Marie Selby Botanical Gardens.

Select. Stirp. Amer. Hist.—See: N. J. Jacquin 1763

Shaffer-Fehre, M. 1991. The endotegmen tuberculae: An account of little-known structures from the seed coat of the Hydrocharitoideae (Hydrocharitaceae) and *Najas* (Najadaceae). Bot. J. Linn. Soc. 107: 169–188.

Shaffer-Fehre, M. 1991b. The position of *Najas* within the subclass Alismatidae (Monocotyledones) in the light of new evidence from seed coat structures in the Hydrocharitoideae (Hydrocharitales). Bot. J. Linn. Soc. 107: 189–209.

Shinners, L. H. 1956. *Elodea* correct without being conserved. Rhodora 58: 162.

Shinners, L. H. 1962. *Aneilema* (Commelinaceae) in the United States. Sida 1: 100–101.

Shireman, J. V. and M. J. Maceina. 1981. The utilization of grass carp, *Ctenopharyngodon idella* Val., for *Hydrilla* control in Lake Baldwin, Florida. J. Fish Biol. 19: 629–636.

Short, F. T. and M. L. Cambridge. 1984. Male flowers of *Halophila engelmanii*: Description and flowering ecology. Aquatic Bot. 18: 413–416.

Shuey, A. G. 1975. A red-petioled form of *Thalia geniculata* L. from central Florida. Rhodora 77: 210–212.

Shuey, A. G. and R. P. Wunderlin. 1977. The needle palm: *Rhapidophyllum hystrix*. Principes 21: 47–59.

Shull, J. M. 1925. *Spathyema foetida*. Bot. Gaz. 79: 45–59.

Sida = Sida; Contributions to Botany.

Simmonds, N. W. and K. Shepherd. 1955. The taxonomy and origins of the cultivated bananas. J. Linn. Soc., Bot. 55: 302–312.

Simpson, B. B., J. L. Neff, and G. Dieringer. 1986. Reproductive biology of *Tinantia anomala* (Commelinaceae). Bull. Torrey Bot. Club 113: 149–158.

Simpson, B. J. 1988. A Field Guide to Texas Trees. Austin.

Sinclair, C. 1967. Studies on the Erect Tradescantias. Ph.D. dissertation. University of Missouri.

Sitzungsber. Ges. Naturf. Freunde Berlin = Sitzungsberichte der Gesellschaft Naturforschender Freunde zu Berlin.

Sketch Bot. S. Carolina—See: S. Elliott [1816–]1821–1824

Small, J. A. 1959. Skunk cabbage, *Symplocarpus foetidus*. Bull. Torrey Bot. Club 86: 413–416.

Small, J. K. 1903. Flora of the Southeastern United States. . . . New York. (Fl. S.E. U.S.)

Small, J. K. 1929. Palmetto-with-a-stem—*Sabal deeringiana*. J. New York Bot. Gard. 30: 278–284.

Small, J. K. 1931. The fanleaf palm—*Washingtonia filifera*. J. New York Bot. Gard. 32: 33–43.

Small, J. K. 1933. Manual of the Southeastern Flora, Being Descriptions of the Seed Plants Growing Naturally in Florida, Alabama, Mississippi, Eastern Louisiana, Tennessee, North Carolina, South Carolina and Georgia. New York. (Man. S.E. Fl.)

Small, J. K. 1937. Facts and fancies about our royal palm. J. New York Bot. Gard. 38: 49–58.

Smith, D. 1972. Fruiting in the saw palmetto. Principes 16: 31–33.

Smith, J. E. 1800–1804. Flora Britannica. 3 vols. London. [Vols. paged consecutively. Vols. 1 and 2, 1800; vol. 3, 1804.] (Fl. Brit.)

Smith, J. E. 1804–1805[–1808]. Exotic Botany. . . . 2 vols. London. (Exot. Bot.)

Smith, J. G. 1894. North American Species of *Sagittaria* and *Lophotocarpus*. . . . St. Louis. [Preprinted from Rep. (Annual) Missouri Bot. Gard. 6: 27–64, plates 1–29. 1895.] (N. Amer. Sagittaria)

Smith, L. B. and R. J. Downs. 1974. Pitcairnioideae. In: Organization for Flora Neotropica. 1968+. Flora Neotropica. 75+ nos. New York. No. 14(1).

Smith, L. B. and R. J. Downs. 1977. Tillandsioideae. In: Organization for Flora Neotropica. 1968+. Flora Neotropica. 75+ nos. New York. No. 14(2).

Smith, L. B. and R. J. Downs. 1979. Bromelioideae. In: Organization for Flora Neotropica. 1968+. Flora Neotropica. 75+ nos. New York. No. 14(3).

Smith, S. G. 1967. Experimental and natural hybrids in North American *Typha* (Typhaceae). Amer. Midl. Naturalist 78: 257–287.

Smith, S. G. 1987. *Typha*: Its taxonomy and the ecological significance of hybrids. Arch. Hydrobiol. 27: 129–138.

Smithsonian Contr. Bot. = Smithsonian Contributions to Botany.

Snogerup, S. 1993. A revision of *Juncus* subgen. *Juncus* (Juncaceae). Willdenowia 23: 23–73.

Snogerup, S. and B. Snogerup. 1996. *Juncus pervetus* Fernald, a misunderstood N. American species. Ann. Naturhist. Mus. Wien, B 98(suppl.): 423–426.

Soros-Pottruff, C. L. and U. Posluszny. 1994. Developmental morphology of reproductive structures of *Phyllospadix* (Zosteraceae). Int. J. Pl. Sci. 155: 405–420.

SouthW. Naturalist = Southwestern Naturalist.

Sp. Pl.—See: C. Linnaeus 1753; C. L. Willdenow et al. 1797–1830

Sp. Pl. ed. 2—See: C. Linnaeus 1762–1763

Spic. Fl. Rumel.—See: A. H. R. Grisebach 1843–1845[–1846]

Sprengel, K. [1824–]1825–1828. Caroli Linnaei . . . Systema Vegetabilium. Editio Decima Sexta. . . . 5 vols. Göttin-gen. [Vol. 4 in 2 parts, each paged independently; vol. 5 by A. Sprengel.] (Syst. Veg.)

St. John, H. 1916. A revision of the North American species of *Potamogeton* of the section *Coleophylli*. Rhodora 18: 121–138.

St. John, H. 1920. The genus *Elodea* in New England. Rhodora 22: 17–29.

St. John, H. 1961. Monograph of the genus *Egeria* Planchon. Darwiniana 12: 293–307.

St. John, H. 1962. Monograph of the genus *Elodea* (Hydrocharitaceae): Part I. The species found in the Great Plains, the Rocky Mountains, and the Pacific states and provinces of North America. Res. Stud. State Coll. Wash. 30(2): 19–44.

St. John, H. 1965. Monograph of the genus *Elodea*: Part 4 and summary: The species of eastern and central North America. Rhodora 67: 1–35, 155–180.

Stace, C. A. 1970. Anatomy and taxonomy in *Juncus* subgenus *Genuini*. Bot. J. Linn. Soc. 63(suppl. 1): 75–81.

Stafleu, F. A. and R. S. Cowan. 1976–1988. Taxonomic Literature: A Selective Guide to Botanical Publications and Collections with Dates, Commentaries and Types, ed. 2. 7 vols. Utrecht, Antwerp, The Hague, and Boston.

Stafleu, F. A. and E. A. Mennega. 1992+. Taxonomic Literature. A Selective Guide to Botanical Publications and Collections with Dates, Commentaries and Types. Supplement. 5+ vols. Königstein.

Steudel, E. G. 1840–1841. Nomenclator Botanicus Enumerans Ordine Alphabetico Nomina atque Synonyma tum Generica tum Specifica. . . . 2 vols. Stuttgart and Tübingen. (Nomencl. Bot. ed. 2)

Steudel, E. G. [1853–]1855. Synopsis Plantarum Glumacearum. . . . 2 vols. in 10 fasc. Stuttgart. [Fascicles numbered consecutively but vols. paged independently.] (Syn. Pl. Glumac.)

Stoddard, A. A. 1989. The phytogeography and paleoflo[r]istics of *Pistia stratiotes* L. Aquatics 11(3): 21–24.

Stover, R. H. and N. W. Simmonds. 1987. Bananas, ed. 3. London.

Stuckey, R. L. 1968. Distributional history of *Butomus umbellatus* (flowering-rush) in the western Lake Erie and Lake St. Clair region. Michigan Bot. 7: 134–142.

Stuckey, R. L. 1979. Distributional history of *Potamogeton crispus* L. (curly pondweed) in North America. Bartonia 46: 22–42.

Stuckey, R. L. 1985. Distributional history of *Najas marina* (spiny naiad) in North America. Bartonia 51: 2–16.

Stuckey, R. L. 1994. Map of the known distribution of *Butomus* (flowering-rush) in North America. In: W. R. Burk and R. L. Stuckey. 1994. Ronald L. Stuckey: His Role in the Ohio Academy of Science. Chapel Hill. P. 85.

Stuckey, R. L. and D. P. Salamon. 1987. *Typha angustifolia* in North America: A foreigner masquerading as a native. [Abstract.] Ohio J. Sci. 87: 4.

Stuckey, R. L., G. Schneider, and M. L. Roberts. 1990. *Butomus umbellatus* L.: Notes from the German literature and North American field studies. Ohio J. Sci. 90(2): 5–6.

Stuckey, R. L., J. R. Wehrmeister, and R. J. Bartolotta. 1978. Submersed aquatic vascular plants in ice-covered ponds of central Ohio. Rhodora 80: 203–208.

Sturtevant, W. C. 1969. History and ethnography of some West Indian starches. In: P. J. Ucko and G. W. Dimbleby, eds. 1969. The Domestication and Exploitation of Plants and Animals. Chicago. Pp. 177–199.

Summa Veg. Scand.—See: O. P. Swartz 1814

Suppl. Pl.—See: C. Linnaeus f. 1781[1782]

Svensk Bot. Tidskr. = Svensk Botanisk Tidskrift Utgifven af Svenska Botaniska Föreningen.

Swartz, O. P. 1788. Nova Genera & Species Plantarum seu Prodromus. . . . Stockholm, Uppsala, and Åbo. (Prodr.)

Swartz, O. P. 1797–1806. Flora Indiae Occidentalis. . . . 3 vols. Erlangen and London. (Fl. Ind. Occid.)

Swartz, O. P. 1814. Summa Vegetabilium Scandinaviae Systematice Coordinatorum. . . . Stockholm. (Summa Veg. Scand.)

Symb. Antill.—See: I. Urban 1898–1928

Symb. Sin.—See: H. Handel-Mazzetti 1929–1937

Symoens, J.-J., S. S. Hooper, and P. Compère, eds. 1982. Studies on Aquatic Vascular Plants. Brussels.

Syn. Junc.—See: E. Meyer 1822

Syn. Luzul.—See: E. Meyer 1823

Syn. Pl.—See: C. H. Persoon 1805–1807

Syn. Pl. Glumac.—See: E. G. Steudel [1853–]1855

Syst. Bot. = Systematic Botany; Quarterly Journal of the American Society of Plant Taxonomists.

Syst. Nat. ed. 10—See: C. Linnaeus 1758[–1759]

Syst. Veg.—See: J. J. Roemer et al. 1817[–1830]; K. Sprengel [1824–]1825–1828

Tarver, D. P. et al. 1978. Aquatic and Wetland Plants of Florida. Tallahassee.

Taxon = Taxon; Journal of the International Association for Plant Taxonomy.

Taylor, N. 1909. Zannichelliaceae. In: N. L. Britton et al., eds. 1905+. North American Flora. . . . 47+ vols. New York. Vol. 17, pp. 13–27.

Thalia dealbata [plate]—See: J. Fraser 1794

Tharp, B. C. 1922. Commelinantia, a new genus of the Commelinaceae. Bull. Torrey Bot. Club 49: 269–275.

Thieret, J. W. 1969b. Sagittaria guayanensis (Alismaceae) in Louisiana: New to the United States. Sida 3: 445.

Thieret, J. W. 1972. Aquatic and marsh plants of Louisiana: A checklist. J. Louisiana Soc. Hort. Res. 13: 1–45.

Thieret, J. W. 1975. The Mayacaceae in the southeastern United States. J. Arnold Arbor. 56: 248–255.

Thieret, J. W. 1982. The Sparganiaceae in the southeastern United States. J. Arnold Arbor. 63: 341–356.

Thieret, J. W. 1988. The Juncaginaceae in the southeastern United States. J. Arnold Arbor. 69: 1–23.

Thieret, J. W., R. R. Haynes, and D. H. Dike. 1969. Blyxa aubertii (Hydrocharitaceae) in Louisiana: New to North America. Sida 3: 343–344.

Thieret, J. W. and J. O. Luken. 1996. The Typhaceae in the southeastern United States. Harvard Pap. Bot. 8: 27–56.

Thompson, S. A. 1995. Systematics and Biology of the Araceae and Acoraceae of Temperate North America. Ph.D. dissertation. University of Illinois.

Thorne, R. F. 1992b. Classification and geography of the flowering plants. Bot. Rev. (Lancaster) 58: 225–348.

Thorne, R. F. 1993. Juncaginaceae. In: J. C. Hickman, ed. 1993. The Jepson Manual. Higher Plants of California. Berkeley, Los Angeles, and London. Pp. 1166–1168.

Thorne, R. F. 1993b. Potamogetonaceae. In: J. C. Hickman, ed. 1993. The Jepson Manual. Higher Plants of California. Berkeley, Los Angeles, and London. Pp. 1304–1310.

Thorne, R. F. 1993c. Hydrocharitaceae. In: J. C. Hickman, ed. 1993. The Jepson Manual. Higher Plants of California. Berkeley, Los Angeles, and London. Pp. 1150–1151.

Thouars, L. M. A. du P. 1806. Genera Nova Madagascariensia. . . . Paris. (Gen. Nov. Madagasc.)

Timme, S. L. and R. B. Faden. 1984. Tradescantia longipes Anderson & Woodson (Commelinaceae) in the southeastern United States. Castanea 490: 83–85.

Titford, W. J. 1812. Sketches toward a Hortus Botanicus Americanus. London.

Tomlinson, P. B. 1961. The anatomy of Canna. J. Linn. Soc., Bot. 56: 467–473.

Tomlinson, P. B. 1961b. Morphological and anatomical characteristics of the Marantaceae. J. Linn. Soc., Bot. 58: 55–78.

Tomlinson, P. B. 1969. Eriocaulaceae. In: C. R. Metcalfe, ed. 1960+. Anatomy of the Monocotyledons. 8+ vols. Oxford. Vol. 3, pp. 146–192.

Tomlinson, P. B. 1982. Helobiae (Alismatidae). In: C. R. Metcalfe, ed. 1960+. Anatomy of the Monocotyledons. 8+ vols. Oxford. Vol. 7.

Tomlinson, P. B. 1990. The Structural Biology of Palms. Oxford.

Tomlinson, P. B. and U. Posluszny. 1976. Generic limits in the Zannichelliaceae (sensu Dumortier). Taxon 25: 273–279.

Tomlinson, P. B. and U. Posluszny. 1978. Aspects of floral morphology and development in the seagrass Syringodium filiforme (Cymodoceaceae). Bot. Gaz. 139: 333–345.

Tompkins, T. M. and J. Taylor. 1983. Hybridization in Typha in Genessee County, Michigan. Michigan Bot. 22: 127–131.

Torrey, J. 1843. A Flora of the State of New York. . . . 2 vols. Albany. (Fl. New York)

Torreya = Torreya; a Monthly Journal of Botanical Notes and News.

Trans. Acad. Sci. St. Louis = Transactions of the Academy of Science of St. Louis.

Trans. Connecticut Acad. Arts = Transactions of the Connecticut Academy of Arts and Sciences.

Trans. Kansas Acad. Sci. = Transactions of the Kansas Academy of Science.

Trans. Linn. Soc. London = Transactions of the Linnean Society of London.

Trans. Sapporo Nat. Hist. Soc. = Transactions of the Sapporo Natural History Society. [Sapporo Hakubutsu Gakkai Kaiho.]

Trans. Wisconsin Acad. Sci. = Transactions of the Wisconsin Academy of Sciences, Arts and Letters.

Travels Carolina—See: W. Bartram 1791

Treiber, M. 1980. Biosystematics of the *Arisaema triphyllum* Complex. Ph.D. dissertation. University of North Carolina.

Triest, L. 1988. A revision of the genus *Najas* L. (Najadaceae) in the Old World. Mém. Acad. Roy. Sci. Belgique, Cl. Sci. (8°), n. s. 22: 1–172, 29 plates.

Triest, L., J. van Geyt, and V. Ranson. 1986. Isozyme polymorphism in several populations of *Najas marina* L. Aquatic Bot. 24: 373–384.

Trudy Imp. S.-Peterburgsk. Bot. Sada = Trudy Imperatorskago S.-Peterburgskago Botanicheskago Sada

Tucker, G. C. 1989. The genera of Commelinaceae in the southeastern United States. J. Arnold Arbor. 70: 97–130.

Turner, B. L. 1983. New taxa of *Tradescantia* from northcentral Mexico. Phytologia 52: 369–371.

Turner, C. E. 1980. *Ottelia alismoides* (L.) Pers. (Hydrocharitaceae)—U.S.A., California. Madroño 27: 177.

Tutin, T. G., V. H. Heywood, N. A. Burges, D. H. Valentine, S. M. Walters, D. A. Webb, et al., eds. 1964–1980. Flora Europaea. 5 vols. Cambridge.

U.S.D.A. Bur. Pl. Industr. Bull. = U S Department of Agriculture. Bureau of Plant Industry. Bulletin.

U.S.D.A. Farmers Bull. = U S Department of Agriculture. Farmers Bulletin.

Ucko, P. J. and G. W. Dimbleby, eds. 1969. The Domestication and Exploitation of Plants and Animals. Chicago.

Uemura, S., K. Ohkawara, G. Kudo, N. Wada, and S. Higashi. 1993. Heat-production and cross-pollination of the Asian skunk cabbage *Symplocarpus renifolius* (Araceae). Amer. J. Bot. 80: 635–640.

Uhl, N. W. et al. 1990. Phylogenetic analyses within Coryphoideae (Palmae). [Abstract.] Amer. J. Bot. 77(suppl.): 160–161.

Uhl, N. W. et al. 1995. Phylogenetic relationships among palms: Cladistic analysis of morphological and chloroplast DNA restriction site variation. In: P. J. Rudall et al., eds. 1995. Monocotyledons: Systematics and Evolution. 2 vols. Kew. Vol. 2, pp. 623–662.

Uhl, N. W. and J. Dransfield. 1987. Genera Palmarum. Lawrence, Kans.

Uhl, N. W. and J. Dransfield. 1999. Genera Palmarum after ten years. Mem. New York Bot. Gard. 83: 245–253.

Univ. Calif. Publ. Bot. = University of California Publications in Botany.

University of Chicago Press. 1993. Chicago Manual of Style, ed. 14. Chicago.

Urban, I., ed. 1898–1928. Symbolae Antillanae seu Fundamenta Florae Indiae Occidentalis. . . . 9 vols. Berlin etc. (Symb. Antill.)

Urbanska-Worytkiewicz, K. 1980. Cytological variation within the family of Lemnaceae. Veröff. Geobot. Inst. E. T. H. Stiftung Rübel Zürich 70: 30–101.

Vahl, M. 1804–1805. Enumeratio Plantarum. . . . 2 vols. Copenhagen. (Enum. Pl.)

Van Heurck, H. F. 1870–1871. Observationes Botanicae. . . . 2 fasc. Antwerp and Berlin. [Fascicles paged consecutively.] (Observ. Bot.)

Van Vierssen, W. 1982. The ecology of communities dominated by *Zannichellia* taxa in western Europe. I. Characterization and autecology of the *Zannichellia* taxa. Aquatic Bot. 12: 103–155.

Varshney, C. K. and J. Rzóska, eds. 1976. Aquatic Weeds in South East Asia. The Hague.

Vasc. Pl. Pacif. N.W.—See: C. L. Hitchcock et al. 1955–1969

Veg. Mater. Med. U.S.—See: W. P. C. Barton 1817–1819

Vellozo, J. M. 1825–1827[1829–1831]. Florae Fluminensis, seu Descriptionum Plantarum Praefectura Fluminensi Sponte Nascentium. . . . 1 vol. text + 11 vols. plates + indexes. Rio de Janeiro. (Fl. Flumin.)

Ventenat, É. P. [1800–1803.] Description des Plantes Nouvelles et Peu Connues Cultivés dans le Jardin de J. M. Cels. . . . 10 parts. Paris. [Plates numbered consecutively.] (Descr. Pl. Nouv.)

Verh. Naturf. Vereins Brünn. = Verhandlungen des Naturforschenden Vereins in Brünn.

Verh. Zool.-Bot. Ges. Wien = Verhandlungen der Zoologisch-botanischen Gesellschaft in Wien.

Verhoeven, J. T. A. 1979. The ecology of *Ruppia*-dominated communities in western Europe. I. Distribution of *Ruppia* representatives in relation to their autecology. Aquatic Bot. 6: 197–268.

Veröff. Geobot. Inst. E. T. H. Stiftung Rübel Zürich = Veröffentlichungen des Geobotanischen Institutes der Eidg. Techn. Hochschule, Stiftung Rübel, in Zürich.

Viinikka, Y. 1973. The occurrence of B chromosomes and their effect on meiosis in *Najas marina*. Hereditas (Lund) 75: 207–212.

Villars, D. 1786–1789. Histoire des Plantes de Dauphiné. 3 vols. Grenoble, Lyon, and Paris. (Hist. Pl. Dauphiné)

Vollst. Handb. Bl.-Gärtn.—See: J. F. W. Bosse 1840–1849 [–1854]

Vollst. Pflanzensyst.—See: G. F. Christmann and G. W. F. Panzer 1777–1788

Voss, E. G. 1958. Confusion in *Alisma*. Taxon 7: 130–133.

Voss, E. G. 1966. Nomenclatural notes on monocots. Rhodora 68: 435–463.

Voy. Amér. Mér.—See: A. D. d'Orbigny 1834[–1847]

Voy. Pittor.—See: L. Choris [1821–]1822[–1823]

Vuille, F.-L. 1987. Reproductive biology of the genus *Damasonium* (Alismataceae). Pl. Syst. Evol. 157: 63–71.

Wahlenberg, G. 1812. Flora Lapponica Exhibens Plantas Geographice et Botanice Consideratas. . . . Berlin. (Fl. Lapp.)

Walter, T. 1788. Flora Caroliniana, Secundum Systema Vegetabilium Perillustris Linnaei Digesta. . . . London. (Fl. Carol.)

Wang J.-K. and S. Higa, eds. 1983. Taro, a Review of *Colocasia esculenta* and Its Potentials. Honolulu.

Ward, D. B., ed. N.d. Rare and Endangered Biota of Florida. Vol. 5. Plants. Gainesville.

Watson, S., W. H. Brewer, and A. Gray. 1876–1880. Geological Survey of California. . . . Botany. . . . 2 vols. Cambridge, Mass. (Bot. California)

Watson, S., D. C. Eaton, et al. 1871. United States Geological Expolration [sic] of the Fortieth Parallel. Clarence King, Geologist-in-charge. [Vol. 5] Botany. By Sereno Watson. . . . Washington. [Botanical portion of larger work by C. King.] [Botany (Fortieth Parallel)]

Webbia = Webbia; Raccolta di Scritti Botanici.

Weigel, C. E. von. 1772. Observationes Botanicae. . . . Greifswald. (Observ. Bot.)

Weimarck, H. 1946. Studies in the Juncaceae. With special reference to the species in Ethiopia and the Cape. Svensk Bot. Tidskr. 40: 141–178.

Welsh, S. L. 1974. Anderson's Flora of Alaska and Adjacent Parts of Canada. Provo. (Anderson's Fl. Alaska)

Wentz, W. A. and R. L. Stuckey. 1971. The changing distribution of the genus *Najas* (Najadaceae) in Ohio. Ohio J. Sci. 71: 292–302.

Wiegand, K. M. 1900. *Juncus tenuis* Willd., and some of its North American allies. Bull. Torrey Bot. Club 27: 511–527.

Wilder, G. J. 1974. Symmetry and development of *Limnobium spongia* (Hydrocharitaceae). Amer. J. Bot. 61: 624–642.

Wilhelm, G. S. and R. H. Mohlenbrock. 1986. The rediscovery of *Potamogeton floridanus* Small (Potamogetonaceae). Sida 11: 340–346.

Willdenow, C. L., C. F. Schwägrichen, and J. H. F. Link. 1797–1830. Caroli a Linné Species Plantarum. . . . Editio Quarta. . . . 6 vols. Berlin. [Vols. 1–5(1), 1797–1810, by Willdenow; vol. 5(2), 1830, by Schwägrichen; vol. 6, 1824–1825, by Link.] (Sp. Pl.)

Williams, K. A. 1919. A botanical study of skunk cabbage, *Symplocarpus foetidus*. Torreya 19: 21–29.

Williams, S. L. 1995. Surfgrass (*Phyllospadix torreyi*) reproduction: Reproductive phenology, resource allocation, and staminate rarity. Ecology 76: 1953–1970.

Wilson, K. A. 1960. The genera of the Arales in the southeastern United States. J. Arnold Arbor. 41: 47–72.

Wimmer, F. 1857. Flora von Schlesien . . . , ed. 3. Breslau. (Fl. Schles. ed. 3)

Withering, W. 1796. An Arrangement of British Plants . . . , ed. 3. 4 vols. London. (Arr. Brit. Pl. ed. 3)

Wood, A. 1868. A Class-book of Botany. . . . New York and Troy, N.Y. [Class-book Bot. ed. s.n.(k)]

Woodson, R. E. Jr. and R. W. Schery. 1945. Cannaceae. In: R. E. Woodson Jr. et al., eds. 1943–1981. Flora of Panama. 41 fasc. St. Louis. [Ann. Missouri Bot. Gard. 32: 74–80.]

Woodson, R. E. Jr., R. W. Schery, et al., eds. 1943–1981. Flora of Panama. 41 fasc. St. Louis. [Fascicles published as individual issues of Ann. Missouri Bot. Gard. and aggregating 8 nominal parts + introduction and indexes.]

Wooten, J. W. 1973 Taxonomy of seven species of *Sagittaria* from eastern North America. Brittonia 24: 64–74.

Wunderlin, R. P. 1982. Guide to the Vascular Plants of Central Florida. Tampa.

Yeo, P. F. 1993. Secondary pollen presentation: Form, function and evolution. Pl. Syst. Evol., Suppl. 6: 204–208.

Yeo, R. R. 1964. Life history of common cattail. Weeds 12: 284–288.

Young, A. M. 1982. Notes on the interaction of the skipper butterfly *Calpodes ethlius* (Lepidoptera: Hesperiidae) with its larval host plant *Canna edulis* (Cannaceae) in Mazatlan, State of Sinaloa, Mexico. J. New York Entomol. Soc. 90(2): 99–114.

Young, R. A. 1936. The dasheen: A southern root crop for home use and market. U.S.D.A. Farmers Bull. 1396: 1–38.

Z. Bot. = Zeitschrift für Botanik.

Zanoni, T. A. 1991. The royal palm on the island of Hispaniola. Principes 35: 49–54.

Zoë = Zoë; a Biological Journal.

Zona, S. 1985. A new species of *Sabal* (Palmae) from Florida. Brittonia 37: 366–368.

Zona, S. 1987. Phenology and pollination biology of *Sabal etonia* (Palmae) in southeastern Florida. Principes 31: 177–182.

Zona, S. 1990. A monograph of *Sabal* (Arecaceae: Coryphoideae). Aliso 12: 583–666.

Zona, S. 1991. Notes on *Roystonea* in Cuba. Principes 35: 225–233.

Zona, S. 1996. *Roystonea* (Arecaceae: Arecoideae). In: Organization for Flora Neotropica. 1968+. Flora Neotropica. 75+ nos. New York. No. 71, pp. 1–36.

Zona, S. 1997. The genera of Palmae (Arecaceae) in the southeastern United States. Harvard Pap. Bot. 2: 71–107.

Zona, S. 1999. New perspectives on generic limits and relationships in the Ptychospermatinae (Palmae: Arecoideae). Mem. New York Bot. Gard. 83: 255–263.

Zona, S. and A. Henderson. 1989. A review of animal-mediated seed dispersal of palms. Selbyana 11: 6–21.

Zona, S. and W. S. Judd. 1986. *Sabal etonia* (Palmae): Systematics, distribution, ecology, and comparisons to other Florida scrub endemics. Sida 11: 417–427.

Zona, S. and R. Scogin. 1988. Flavonoid aglycones and C-glycosides of the palm genus *Washingtonia* (Arecaceae: Coryphoideae). SouthW. Naturalist 33: 498.

Index

Names in *italics* are synonyms, casually mentioned hybrids, or plants not established in the flora. Page numbers in **boldface** indicate the primary entry for a taxon. Page numbers in *italics* indicate an illustration. Roman type is used for all other entries, including author names, vernacular names, and accepted scientific names for plants treated as established members of the flora.